Thomas Schäpers
Semiconductor Spintronics

Also of Interest

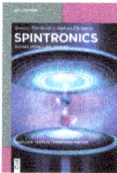

Spintronics
Tomasz Blachowicz, Andrea Ehrmann, 2019
ISBN 978-3-11-049062-6, e-ISBN 978-3-11-049063-3

Electrons in Solids
Mesoscopics, Photonics, Quantum Computing, Correlations, Topology
Hendrik Bluhm, Thomas Brückel, Markus Morgenstern,
Gero von Plessen, Christoph Stampfer, 2019
ISBN 978-3-11-043831-4, e-ISBN 978-3-11-043832-1

Plasma and Plasmonics
Kushsal Shah, 2018
ISBN 978-3-11-056994-0, e-ISBN 978-3-11-057003-8

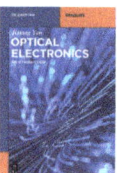

Optical Electronics
An Introduction
Jixiang Yan, 2019
ISBN 978-3-11-050049-3, e-ISBN 978-3-11-050060-8

Thomas Schäpers

Semiconductor Spintronics

2nd edition

DE GRUYTER

Physics and Astronomy Classification Scheme 2010
Primary: 85.75.-d, 85.35.-p, 73.63.-b, 73.43.-f, 75.20.-g; Secondary: 85.75.Hh, 85.35.Be, 85.35.Ds, 85.35.Gv, 73.63.Hs, 73.63.Kv, 73.63.Nm

Author
Thomas Schäpers
Forschungszentrum Jülich
Peter Grünberg Institut 9
52425 Jülich
Germany
th.schaepers@fz-juelich.de

ISBN 978-3-11-063887-5
e-ISBN (PDF) 978-3-11-063900-1
e-ISBN (EPUB) 978-3-11-063932-2

Library of Congress Control Number: 2021932015

Bibliographic information published by the Deutsche Nationalbibliothek
The Deutsche Nationalbibliothek lists this publication in the Deutsche Nationalbibliografie; detailed bibliographic data are available on the Internet at http://dnb.dnb.de.

© 2021 Walter de Gruyter GmbH, Berlin/Boston
Cover image: kindly provided by Sebastian Heedt
Typesetting: VTeX UAB, Lithuania
Printing and binding: CPI books GmbH, Leck

www.degruyter.com

To Gitta, Antonia, and Paul

Preface to the second edition

In recent years, it has become apparent that topological states in solid-state materials are becoming increasingly important for spinelectronic applications. Already in the first edition this was taken into account by discussing in detail the physical properties of topological insulators. In the new issue, this part has now been extended to include Landau quantization at high magnetic fields. The possibilities to include topological properties go even further. Combining low-dimensional materials with strong spin-orbit interaction with superconductors, so-called Majorana states can be generated under certain circumstances. These states emerge in pairs that are usually widely separated from each other. The exchange of these Majorana states follows special rules which differ from those of normal fermions or bosons. These are the so-called non-abelian anyons. This special property makes Majorana states very interesting for a new kind of quantum computing, the topological quantum computing. In order to take this recent development into account, the second edition now contains a new chapter which deals with the topic of Majorana states in detail.

Besides the electrical characterization of materials relevant for spin electronics, optical investigations play an equally important role. In order to better take this into account, the corresponding part has been significantly expanded in the new edition. The physical principles of the Faraday effect and the various spin dephasing processes are now explained in detail. In addition, resonant spin amplification is discussed, which is exploited in particular for long spin dephasing times.

Especially concerning the extension of the latter part, I am indebted to Bernd Beschoten, who provided me with appropriate material and was available for intensive discussions around the optical experiments. Furthermore, I thank Hans Lüth as a discussion partner around the topic of Majorana fermions. Last but not least, I would like to thank Dr. Vivien Schubert of de Gruyter for her very competent coordination of the second edition.

Jülich, March 2021   Thomas Schäpers

https://doi.org/10.1515/9783110639001-201

Preface to the first edition

In the past decades we observed a continuous increase of performance and complexity of electronic circuits. This progress is quantified by Moore's law [1], which predicts a doubling of the number of transistors in an integrated circuit every two years. Although it is not a law in a strict sense, it proved to be valid over more than four decades. The tremendous number of transistor in state-of-the-art integrated circuits comes along with a corresponding shrinkage of device dimensions, which nowadays have feature sizes in the order of a few tens of nanometers. It is foreseeable that this development cannot continue forever. In the *International Technology Roadmap for Semiconductors* [2] these future challenges are addressed and possible solutions are proposed. In this context, spin electronic devices, i. e. devices which make use of the electron spin, are mentioned as a possible alternative to solely charge-based devices.

The aim of this textbook is to introduce the various materials, mechanisms, and concepts of spintronic devices. We restrict ourself to semiconductor-based structures and leave out pure metal-based devices, which have an older history and are already used in various applications. The development of semiconductor spintronic devices is still in its infant stage. Different device concepts are proposed, but they have not yet found their way into the production of integrated circuits. Since the development of spintronics is still on a conceptual level, the emphasis of this textbook is on the underlying physical phenomena. The book should put students and researchers into a position to understand the basic concepts of spin electronic devices. For many of the discussed effects a profound knowledge of low-dimensional semiconductor structures is mandatory. Therefore, an introduction to these systems is provided which serves as a basis for the different phenomena discussed in the subsequent chapters. In addition to subjects which have in the meantime become well established, such as spin injection or spin manipulation, we also include recent developments. Especially for the new field of topological insulators it cannot even be foreseen what impact these materials will have on future spinelectronic devices. In any case, these materials, which distinguish themselves from other materials by their strict spin momentum locking, bring a new twist into the research on spintronic devices. Last but not least, in recent years a new computation paradigm has emerged which is based on quantum mechanical states. In the field of quantum computation, two-level systems, so-called quantum bits, are used as basic elements. Here, semiconductor spintronics is a very interesting candidate for realizing these systems, since the electron with its two spin states is a natural candidate for a quantum bit. Moreover, the mature semiconductor nanofabrication technology helps to scale up these systems.

The current textbook is based on lecture notes of the spintronics course for masters students held at RWTH Aachen University. It was a great pleasure for me to give these lectures together with Bernd Beschoten from RWTH Aachen. As one of the real experts in optical methods, he especially contributed to my understanding of the var-

https://doi.org/10.1515/9783110639001-202

ious optical phenomena in spintronics. I am very grateful for the many stimulating discussions with colleagues and students in Jülich and Aachen. In particular, I enjoyed the lively and inspiring discussions with Hans Lüth and Andreas Bringer about various aspects on quantum and spin transport in nanostructures. During these discussions I was reminded as to why I studied physics in the first place, and how much fun it is. Especially Andreas Bringer helped a great deal in figuring out the sometimes intricate theoretical concepts around spin-orbit coupling. I acknowledge the support and encouragement from colleagues in Jülich and Aachen: Gustav Bihlmayer, Stefan Blügel, Hendrik Bluhm, Nataliya Demarina, David DiVincenzo, Detlev Grützmacher, Gernot Güntherodt, Hilde Hardtdegen, Fabian Hassler, Mihail (Mike) Lepsa, Beata Kardynal, Gregor Mussler, Claus Schneider, and Christoph Stampfer. During various conferences, workshops, collaborations, and visits I enjoyed having discussions on various hot topics in spintronics and quantum transport, namely with Carlos Egues, Sigurður Erlingsson, Michele Governale, Ewelina Hankiewicz, Stefan Kettemann, Jia Grace Lu, Andrei Manolescu, Junsaku Nitta, Angela Rizzi, Björn Trauzettel, Paul Wenk, Roland Winkler, and Ulrich Zülicke. Many colleagues provided figures for this book: Irene Aguilera, Yulieth Arango, Bernd Beschoten, Gustav Bihlmayer, Hartmut Buhmann, Hilde Hardtdegen, Sebastian Heedt, Takaaki Koga, Jia Grace Lu, Martina Luysberg, Gregor Mussler, Junsaku Nitta, Lukasz Plucinski, and Lieven Vandersypen. Their experimental and theoretical results helped greatly to better illustrate the physical effects discussed in the book. The cover picture is based on a graphics by Sebastian Heedt. I thank my students for critical reading of the manuscript and many useful comments to improve and clarify explanations and derivation. I am also very grateful for the very professional support and patience of Silke Hutt and Dr. Konrad Kieling from De Gruyter. I was aware that transferring lecture notes to a textbook is quite an effort, but, one is still surprised how much work it is in the end. Finally I thank my wife Gitta, who accepted the extra time I needed to work on the manuscript, in addition to the time which is required in the normal daily life of a researcher.

Jülich, October 2015   Thomas Schäpers

Contents

1 Introduction

In spintronics the intrinsic magnetic property of the electron, its spin, is used for switching purpose in an electronic circuit instead of its charge. The operation principle of spintronic devices is based on completely different physical phenomena compared to their charge-based counterparts. In this textbook, we will have a closer look at the physical mechanisms of which spintronic devices make use. The roots of spintronic devices lie in the field of magneto-electronic devices. As a matter of fact, metallic magneto-electronic devices are already well established in information technology. The basic physical phenomena are giant magnetoresistance (GMR) or tunneling magnetoresistance (TMR). These structures contain ferromagnetic layers which are either separated by a metallic layer or by a tunneling barrier, respectively. By keeping the magnetization in one layer fixed and changing the magnetization in the other one with respect to the first, the resistance is changed. This mechanism can be used for switching or detection purposes. In the meantime both effects have been implemented in applications, e. g. GMR or TMR devices are used as read heads in state-of-the-art hard disc drives and were responsible for the huge increase of storage capacity in recent years. Because of its significance the Noble prize in physics was awarded to Albert Fert and Peter Grünberg in 2007: "For the discovery of magnetoresistance" [3, 4]. The GMR and TMR effect can also be employed in solid-state memories, i. e. magnetoresistive random access memories (MRAMs). The memory cells can be programmed by changing the magnetization of one magnetic layer with respect to the other by means of a magnetic field generated by two crossing current carrying lines. More recent switching schemes use spin-transfer torque. Instead of employing an external magnetic field, here the switching is achieved by using a spin-polarized current through the memory cell itself. The advantage of magnetic memories is that they are nonvolatile, i. e. the stored information remains even after the power supply is switched off. MRAM chips are already commercially available.

So far, magnetic device structures are mainly used for data storage, while the data processing itself is performed by semiconductor devices. Here, the work horse is the field-effect transistor. The vast majority of these transistors is made from silicon as a semiconductor material. During the last decades one could observe a continuous shrinkage of device dimensions following the so-called Moore's law [1]. This allowed the integration of more and more devices on a chip and the design of more complex circuits as well as denser solid-state memories. However, it is foreseeable that the miniaturization of Si-based circuits will reach its limits [2, 5]. Therefore, various alternative concepts are pursued, e. g. by extending the material base by using SiGe or by implementing novel device concepts like the tunnel field-effect transistor. Above that, more revolutionary concepts are also considered. One of them is semiconductor spintronics, where the electron spin is the entity, which is exploited for information processing. Using the electron spin in a device promises to perform digital switching sequences faster

https://doi.org/10.1515/9783110639001-001

with less power consumption [6, 7, 8, 9, 10, 11]. A prominent example of a spintronic device is the spin field-effect transistor (spin FET), proposed by Datta and Das [6]. Although the field of semiconductor spintronics has existed for some years already, these kind of devices have not yet found their way into electronic circuits. However, in the meantime a working spin FET as a demonstrator has been realized [12].

Let us have a look on different physical mechanisms employed in spintronic devices. Figure 1.1 (a) shows a schematic of a spin field-effect transistor.

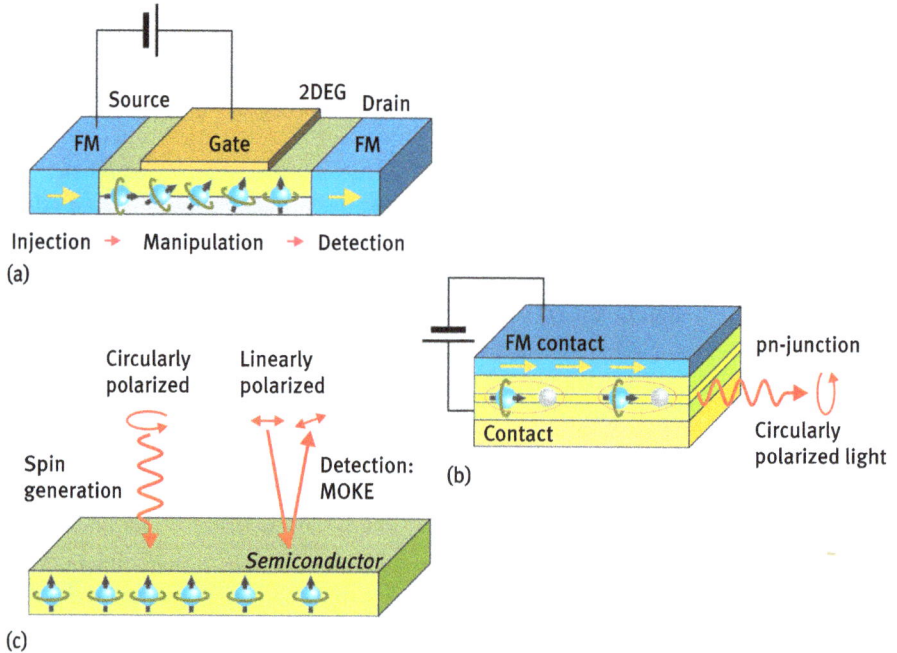

Figure 1.1: (a) Schematic illustration of a spin field-effect transistor. Spins are injected into the two-dimensional electron gas (2DEG) in the semiconductor. The spin orientation is manipulated by biasing the gate electrode. (b) Spin light emitting diode (spin LED). Spin-polarized electrons are supplied from the top ferromagnetic electrode. The unpolarized holes are provided by the bottom contact. Electron hole recombination results in the emission of circularly polarized light. (c) Spin-polarized electrons can by generated by circularly polarized light. The spin orientation can be detected by the magnetooptical Kerr effect (MOKE). Here, the polarization direction of a reflected linearly polarized beam of light is changed in the presence of spin-polarized carriers in the semiconductor.

Spin-polarized carriers are supplied from the ferromagnetic source contact by driving a current between source and drain. The spins are injected into the semiconductor material, keeping their initial spin polarization in the ferromagnet. As we will discuss in Chapter 6, this is a formidable task. Owing to the conductance mismatch and to imperfect interfaces, spin injection is often quite inefficient. In Chapter 4, we will introduce diluted magnetic semiconductors, which might be an interesting alternative to metallic ferromagnetic injectors, owing to their better conductance matching. Once the spin-polarized carriers are injected into the two-dimensional electron gas in the semiconductor, their spin orientation is manipulated by means of the Rashba effect [13]. Applying a voltage to the gate electrode results in a change of the electric field to which the electrons in the semiconductor are exposed. The electrons propagating in that electric field experience an effective magnetic field. The electron spin precesses about this effective magnetic field. The degree of spin precession depends on the strength of the electric field and thus on the gate voltage. The underlying physical mechanisms of a spin FET are introduced in Chapter 7. After passing the semiconductor section, the spin orientation is detected by the ferromagnetic drain contact. Ideally electrons can only enter where the spin orientation matches the magnetization in the drain contact. By biasing the gate the spin orientation of the carriers reaching the drain contact can be controlled. This is the switching scheme of the spin field-effect transistor. In an alternative concept electron interference can also be employed for switching purposes. This will be discussed in Chapter 8.

Owing to the inefficiency of purely electric spin injection and detection, often optical means are utilized as an alternative to measure spin-polarized carriers. A typical example is given in Figure 1.1 (b), where a spin light emitting diode (spin LED) is shown. Spin-polarized electrons are transferred from a ferromagnetic electrode into a quantum well layer. Here, the electrons recombine with unpolarized holes provided by the bottom contact. Owing to the spin polarization of the electrons, circularly polarized light is emitted. This is due to angular momentum conservation during the recombination process. Thus, the emission of circularly polarized light can be used to verify spin injection from a ferromagnetic electrode into a semiconductor. Moreover, the spin LED can also directly be used in applications, where it serves as an emitter for circularly polarized light.

One can even move one step further and perform all-optical experiments. As an inverse process, circularly polarized light can also be employed to generate spin-/polarized electrons in direct band gap semiconductors. Here, spin-polarized electron-hole pairs are excited by the irradiated light. Due to strong spin orbit coupling in the valence band, the spin polarization of the holes decays quickly, while the electron spin polarization remains. This is a very efficient method to generate spin-polarized electrons, compared to electrical spin injection. The presence of spin-polarized carriers in a semiconductor can also be detected by optical means, i. e. by the magnetooptical Kerr effect (MOKE). Here, a beam of linearly polarized light is reflected at the surface. In the presence of spin-polarized carriers, the polarization of the reflected beam is rotated. The rotation angle is proportional to the magnetization in the semiconductor, and thus it can be employed as a parameter to quantify the magnetization in a semiconductor. As a matter of fact, optical means are very powerful means to generate or detect spin-polarized carriers. However, conceptually one runs into problems when it comes to miniaturization. This can be achieved much more easily by purely electrical means. Indeed, in this textbook we will mostly focus on electrical phenomena, whereas optical effects will only be covered when they serve in experiments to generate or measure spin-polarized carriers.

Recently spintronics has made a large leap towards novel phenomena and materials, such as the quantum spin Hall effect or topological insulators [14, 15, 16]. Regarding the first, the strong spin orbit coupling in a HgTe/CdTe heterostructure results in an inversion between the conduction and valence bands. As a consequence, one-dimensional channels are formed at the edge of the sample, similarly to what is known for the quantum Hall effect. The crucial difference is that these edge channels are formed at zero magnetic field, whereas in the case of the quantum Hall effect a magnetic field is required. Furthermore, the transport on the edges is spin polarized. In this sense it is related to the spin Hall effect discussed later in Chapter 9, where spin-dependent scattering of propagating electrons also results in spin-polarized carriers at the edge of the sample.

Very similar effects to the quantum spin Hall effect are observed in three/dimensional topological insulators, where due to the very strong spin orbit coupling a conductive two-dimensional surface channel is formed. This surface channel is inherently stable, i. e. topologically protected. Quantum spin Hall systems and three-dimensional topological insulators are completely new states of matter. The related physical effects have not yet been fully explored. In any case, they promise a huge potential for future spintronic devices. The properties of these materials are discussed in Chapters 10 and 11.

Apart from more conventional circuits a new scheme for information processing has emerged which is directly based on quantum mechanics. Instead of 0 and 1 as two states, a bit (cf. Figure 1.2 (a)), i. e. the smallest unit of information in a conventional computer, in quantum computing the quantum bit (qubit), is constituted by a quantum mechanical two-level system. This is illustrated in Figure 1.2 (b) [17].

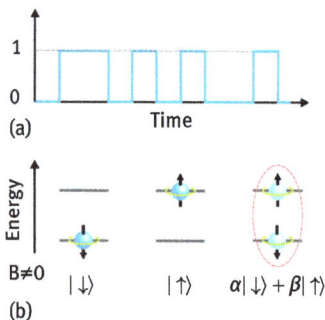

Figure 1.2: (a) Example of changing the bit value between the two values 0 and 1 in a conventional computer in the course of time. (b) Representation of a quantum bit by a spin-1/2 two-level system. The level splitting is due to the Zeeman effect in a magnetic field. Apart from the two basis states $|\downarrow\rangle$ and $|\uparrow\rangle$, superposition states $\alpha|\downarrow\rangle + \beta|\uparrow\rangle$ can also be realized.

The increase of computational power, e. g. for factorizing large numbers [18], originates from the fact that the corresponding algorithms make use of quantum mechanical superposition and entangled states. Spin-1/2 systems are ideally suited to realize a qubit, since the Zeeman split states naturally represent the required two-level system. In practice, this two-level system can be realized in a semiconductor quantum dot [19, 20]. Here, a single electron is electrostatically confined in the dot. Quantum computation algorithms are implemented by changing the spin orientation by means of an ac magnetic field or by coupling two adjacent quantum dots. The basic principles of quantum computation as well as quantum dot qubits are discussed in Chapter 12. Furthermore, in Chapter 13 an alternative concept, i. e. topological quantum computation, is discussed which is based on Majorana fermions.

For all these issues it is clear that a profound knowledge of semiconductor systems is required. In particular, many spintronic devices are based on low-dimensional systems such as two-dimensional electron gases, quantum wires, or quantum dots. Therefore, we will begin by introducing these systems in the next chapter. Furthermore, some basic knowledge of magnetism and magnetic electrodes is necessary, which is provided in Chapters 3 and 5.

2 Low-dimensional semiconductor structures

2.1 Overview

Semiconductors can be classified as materials which usually have a much lower conductivity than metals but in contrast to insulators do have a finite conductance at nonzero temperatures. In Figure 2.1 (a)–(c), a schematic of the electronic states, i.e. bands, is shown for all different kinds of materials. In metals the states are filled up to the Fermi energy E_F. The empty states above E_F allow electronic transport. In insulators the lower band is fully occupied. Since in a fully occupied band the electron occupation cannot be rearranged, i.e. by applying an electric field, no transport is possible here. The upper band is separated from the occupied lower band by a large energy band gap ($E_g > 4\,\text{eV}$). Thus, no thermal excitation is possible. In a semiconductor, the band gap is smaller ($0.15\,\text{eV} \leq E_g \leq 4\,\text{eV}$). In that case, electrons can be thermally excited into the upper band. These excited electrons can participate in the transport. Furthermore, the unoccupied states in the lower band, i.e. the holes, contribute to the transport as well.

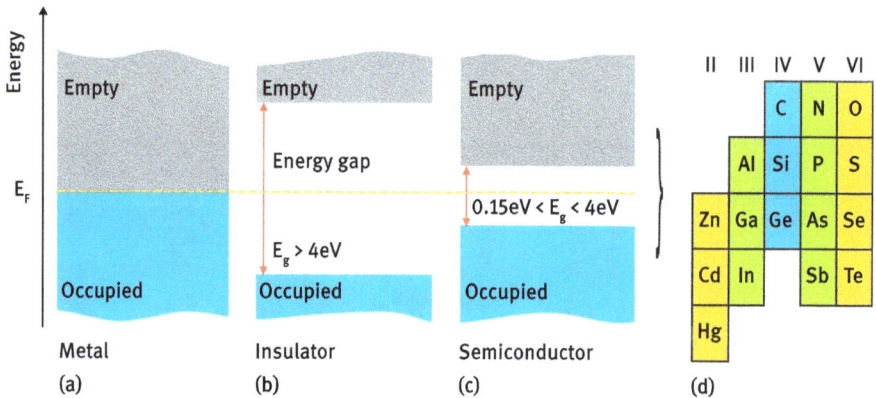

Figure 2.1: Comparison between different materials. (a) Metal: the band is partially filled up to the Fermi energy E_F, so that empty states are available for electron transport. (b) Insulator: no transport is possible, since the lower band is fully occupied and no empty states are available for transport. The upper band is empty. The band gap is so large that no thermal excitation is possible. (c) Semiconductor: the band gap is sufficiently small for the electrons to get thermally excited into the upper band, and hence electron transport is possible. (d) Chemical elements of the periodic table which constitute semiconductor materials.

What is so special about semiconductors? First, intrinsic, i.e. pure, semiconductors usually have low conductivity, which depends strongly on temperature. Second, in semiconductor crystals with a small amount of incorporated foreign atoms, so-called

https://doi.org/10.1515/9783110639001-002

dopants, the conductance can be increased by several orders of magnitude. Further-more, the type of carriers can be changed between electrons and holes. Third, in semi-conductors the field-effect can be employed. Here, the electron concentration can be changed by means of an electric field. These properties are unique for semiconductors and are not present in metals owing to their much higher electron concentration.

Depending on the application, different types of semiconductors are used in spin electronics:

- Doping of bulk semiconductors is used to create diluted magnetic semiconductors (DMS).
- Semiconductor layer systems consisting of layers with different materials, i. e. het-erostructures, are employed in spin light emitting diodes.
- Low-dimensional semiconductor structures, i. e. two-, one-, and zero-dimensio-nal structures find their applications in spin transistors and in devices for spin-based quantum computation.

In this chapter, we will restrict ourself to the semiconductor properties relevant to spintronic applications. More general information can be found in text books dedi-cated to semiconductor physics [21] or on semiconductor nanostructures [22, 23].

2.2 Bulk semiconductors

As illustrated in Figure 2.1 (d), chemical elements from group II to VI are the basis of a large variety of semiconductor materials which are employed in charge-based elec-tronics as well as in spin electronic:

- Elementary semiconductors, i. e. Si or Ge: especially Si is the most popular mate-rial for electronics in particular for integrated circuits.
- Compound semiconductors: here, one can distinguish between IV-IV (group IV and group IV) semiconductors, e. g. SiC; III-V semiconductors, e. g. AlP, AlAs, GaAs, InP, InAs, InSb, or GaN, and II-VI semiconductors, e. g. ZnO, CdS, CdSe, or CdTe. In particular, III-V compound semiconductors are interesting for optoelec-tronics and high speed electronics, due to their direct band gap and low effective electron mass.
- Alloys of the semiconductors mentioned above, e. g. Si_xGe_{1-x}, $Al_xGa_{1-x}As$, $Al_xGa_{1-x}Sb$, $In_xGa_{1-x}As$, or $Al_xGa_{1-x}N$.

2.2.1 Band structure

In the periodic potential of a crystal the discrete atomic levels are transfered into en-ergy bands. The energy-momentum dispersion $E_v(\vec{k})$ of a band structure describes how the energy E depends on the wave vector $\vec{k}$ in a certain band v. In most semiconductors, i. e. Si, the upper bands originate from s-type atomic orbitals. These bands are called conduction bands. Separated by the energy gap E_g the valence bands are found. These bands usually originate from p-type atomic orbitals. As we will see later, this has implications for the resulting angular momentum state. Typical band structures of Si and GaAs are shown in Figure 2.2. The band structure is plotted for different symmetry directions in the Brillouin zone. The Γ-point corresponds to $k = 0$. In most cases, electron and hole states close to the Γ-point are considered.

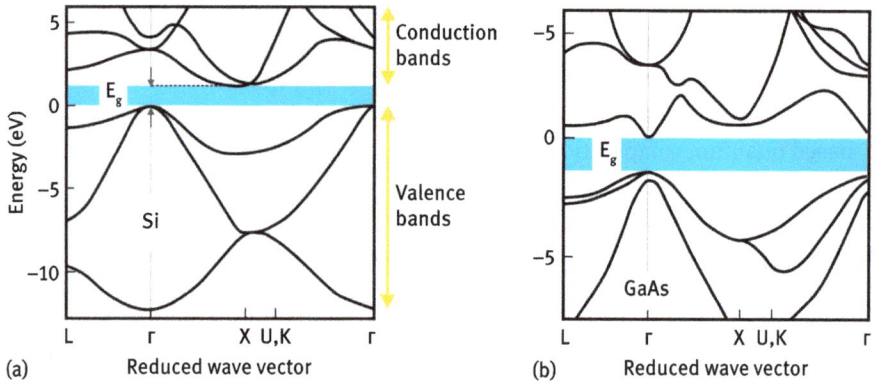

Figure 2.2: Band structure of (a) Si and (b) GaAs, with E_g the band gap.

The elementary semiconductors Si and Ge have an indirect band gap, which means that the maximum of the valence band at the Γ-point in the Brillouin zone does not match in k-space to the minimum of the conduction band. The conduction band minimum is found at a finite k-vector. As an example, the band structure of the indirect semiconductor Si is shown in Figure 2.2 (a). As a consequence no direct optical transitions of electrons from the conduction band down to the valence band are possible, since the photon cannot provide the momentum required. The semiconductors Si and Ge crystallize in the diamond lattice, which is depicted in Figure 2.3 (a).

Most III-V semiconductors have a direct band gap. Here, the maximum of the valence band is aligned in k-space to the minimum of the conduction band. A typical example is the band structure of GaAs shown in Figure 2.2 (b). In this case direct optical transitions are possible, which is the reason why they are commonly used in optoelectronics. The wave length of the photons emitted during transitions is determined by the band gap E_g. For example, in spintronics, direct band gap semiconductors are used to realize a spin light emitting diode. However, some III-V semiconductors have an indirect band gap, such as AlAs. Most III-V semiconductors crystallize in the zinc blende lattice, as shown in Figure 2.3 (b), e. g. GaAs or InP; however some are found in the wurtzite lattice, e. g. GaN.

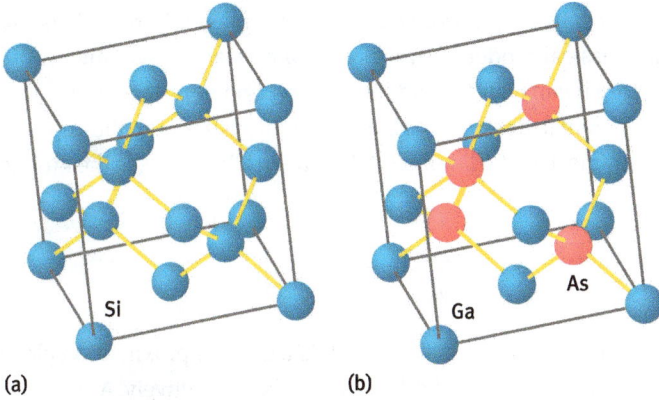

(a)   (b)

Figure 2.3: (a) Diamond lattice of Si and (b) zinc blende lattice of GaAs.

2.2.2 Effective mass

Due to the periodic potential landscape in a crystal the carriers in a semiconductor possess an effective mass m^* which differs from the free electron mass m_0. This can be understood as follows. The acceleration of a particle for the one-dimensional case, i. e. the change of the group velocity $v_g = (1/\hbar)(dE/dk)$ with time, can be written and rearranged as

$$\frac{dv_g}{dt} = \frac{1}{\hbar}\frac{d}{dt}\left(\frac{dE}{dk}\right) \tag{2.1}$$

$$= \frac{1}{\hbar}\frac{d^2E}{dkdt} \tag{2.2}$$

$$= \frac{1}{\hbar^2}\frac{d^2E}{dk^2}\frac{d(\hbar k)}{dt}. \tag{2.3}$$

In the last term we can identify $d(\hbar k)/dt$ as the force F, with k the wavenumber. With

$$F = \frac{d(\hbar k)}{dt} \tag{2.4}$$

we can write

$$F = m^*\frac{dv_g}{dt}, \tag{2.5}$$

where the effective mass m^* can be interpreted as

$$m^* = \left(\frac{1}{\hbar^2}\frac{d^2E}{dk^2}\right)^{-1}. \tag{2.6}$$

The effective mass is proportional to the inverse curvature of the band dispersion $E(k)$. Thus, m^* is a direct consequence of the band structure and the corresponding periodic potential of the crystal lattice. In simple words, the effective mass of a particle is the mass it carries in a semiclassical model of transport in a crystal. Depending on the details of the band structure it can also depend on the direction of propagation.

In many cases it is sufficient to describe the band structure by a parabolic approximation around extremal points, i. e. the Γ-point (cf. Figure 2.4). In that case the kinetic energy can be written as

$$E = \frac{\hbar^2 k^2}{2m^*}.$$ (2.7)

Here, one obtains a constant effective mass m^*, which is a good approximation close the maxima or minima of the valence and conduction bands, respectively. As one can see in Figure 2.4, the valence band is twofold degenerate at the Γ-point.

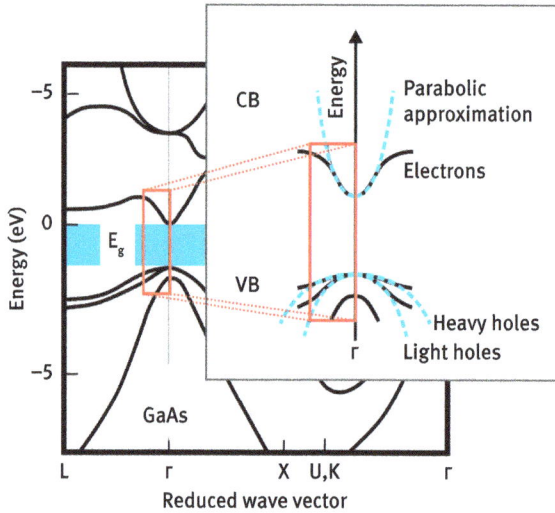

Figure 2.4: Effective mass in the parabolic approximation in the conduction band (CB) and valence band (VB). The valence band is degenerate at the Γ-point. The two bands emerging from there have a different curvature, and this results in a different effective mass for the holes, i. e. heavy and light holes.

Whereas, for finite k-vectors two valence bands with different energy-momentum dispersion emerge. The corresponding parabolic dispersions have different curvatures and thus results in different effective hole masses. Holes in the band with the smaller curvature have a larger effective mass (heavy holes), while the holes in the other band possess a lighter effective mass (light holes). In most cases the effective mass of the electrons is smaller than the effective mass of the holes in the valence band. Finally, one should keep in mind that at zero magnetic field and in the absence of spin-orbit coupling each state belonging to a given wave vector in a band is double degenerate owing to the two spin states.

2.2.3 Density of states

The density of states $D(E)$ is a measure for number of states for a given energy interval. It depends crucially on the dimensionality of the system. In order to illustrate the meaning of $D(E)$ we consider a one-dimensional wire of finite length L extended along the x-direction, with infinitely high walls at its boundary. In the transverse direction, the confining potential $V(y,z)$ should lead to energy eigenvalues E_{ij} due to quantization. Since we assume that the electron cannot leave the system along the x-direction, the wave function must vanish at the boundary. The longest electron wave which fits this condition has a wavelength of $\lambda_1 = 2L$, while for the next state the wavelength λ_2 corresponds to L. The associated wavevectors are calculated from

$$k_l = \frac{2\pi}{\lambda_l} = l\frac{\pi}{L}, \quad l = 1, 2, 3, \ldots \tag{2.8}$$

One finds, that the wave vectors are equally spaced so that the density of states in the k-space is simply given by

$$D(k_x) = g_s \frac{1}{L} \frac{dN_k}{dk_x} = g_s \frac{1}{\pi}. \tag{2.9}$$

Here, $D(k_x)$ was calculated per unit length scale. Furthermore, the spin degeneracy factor g_s was introduced, which is 2 for a spin-degenerated system with the same energy for spin up and down. From the expression of the kinetic energy given by equation (2.7) the density of states in energy space $D(E)$ can be deduced:

$$\begin{aligned}
D(E) &= \frac{1}{L}\frac{dN}{dE} \\
&= g_s \frac{1}{L}\frac{dN}{dk_x}\frac{dk_x}{dE} \\
&= \sum_{ij} g_s D(k_x)\sqrt{\frac{m^*}{2\hbar^2(E-E_{ij})}} \\
&= \sum_{ij} \frac{g_s}{\pi}\sqrt{\frac{m^*}{2\hbar^2}}\frac{1}{\sqrt{E-E_{ij}}}.
\end{aligned} \tag{2.10}$$

For a one-dimensional system the density of state has thus a $1/\sqrt{E}$ dependence, with singularities at $E = E_{ij}$, as shown in Figure 2.5 (c).

In a similar fashion the density of states can be calculated for a two- and three-dimensional system. As can be inferred from Table 2.1, for a two-dimensional structure

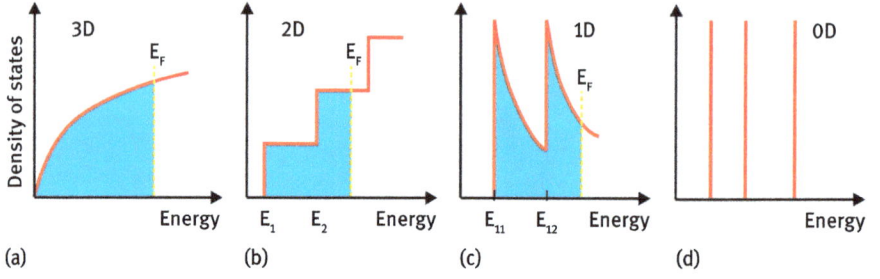

Figure 2.5: (a)–(d) Schematic illustration of the density of states as a function of energy for three-dimensional (3D) to zero-dimensional (0D) systems, respectively. At zero temperature the states are filled up to the Fermi energy.

Table 2.1: Density of states $D(E)$, Fermi energy E_F, and Fermi wavevector k_F for different dimensions, with n_{3D}, n_{2D} and n_{1D} the carrier concentrations in different dimensions and g_s the spin degeneracy factor.

	3D	2D	1D
$D(E)$	$g_s \frac{1}{2\pi^2} \frac{m^*}{\hbar^3} \sqrt{2m^* E}$	$g_s \frac{m^*}{2\pi\hbar^2}$	$\frac{g_s}{\pi} \sqrt{\frac{m^*}{2\hbar^2}} \frac{1}{\sqrt{E}}$
E_F	$\frac{\hbar^2}{m^*} (\frac{3\pi^2}{\sqrt{2}} \frac{1}{g_s})^{2/3} n_{3D}^{2/3}$	$\frac{2\pi\hbar^2}{m^* g_s} n_{2D}$	$\frac{\pi^2\hbar^2}{2g_s^2 m^*} n_{1D}^2$
k_F	$\sqrt[3]{\frac{6\pi^2 n_{3D}}{g_s}}$	$\sqrt{\frac{4\pi n_{2D}}{g_s}}$	$\frac{\pi n_{1D}}{g_s}$

the density of states is constant. In the case where more than one level E_j exists, $D(E)$ has a step-like form, as shown in Figure 2.5 (b). For a three-dimensional system the density of states follows a $\sqrt{E}$ dependence (cf. Figure 2.5 (a)). Only discrete levels are present in a zero-dimensional system, a so-called dot structure, so that the density of states can be described by a sequence of δ-functions, as depicted in Figure 2.5 (d).

Due to the Pauli principle a given energy level can only be occupied by a single electron. Thus at zero temperature ($T = 0$) the electrons are stacked in energy filling one after each other. The maximum energy of the electrons is defined as Fermi energy E_F. Its value is determined by the density of conduction electrons n. For a two-dimensional system with only one subband occupied from $E_1 = 0$ up to the Fermi energy E_F can be obtained by integrating the density of states $D(E)$

$$n_{2D} = \int_{E_1}^{E_F} D(E)\, dE, \tag{2.11}$$

which results in

$$E_F = \frac{2\pi\hbar^2}{m^* g_s} n_{2D}. \tag{2.12}$$

Similarly, the Fermi energy can be calculated for the three- and one-dimensional situation. The result is summarized in Table 2.1 and also shown in Figure 2.5. A crucial parameter which characterizes a nanoelectronic system is the Fermi wave number k_F belonging to an electron at E_F:

$$k_F = \frac{1}{\hbar}\sqrt{2m^*E_F}. \tag{2.13}$$

The expressions for the Fermi wave number in systems with different dimensions are given in Table 2.1. The Fermi wavelength λ_F can be obtained from k_F by

$$\lambda_F = \frac{2\pi}{k_F}. \tag{2.14}$$

As can be seen from the formulas given in Table 2.1, the Fermi wavelength increases with decreasing electron concentration. This has important consequences if metallic and semiconducting systems are compared. In the latter case the Fermi wavelength can be comparable to dimensions accessible by lithography because of the low electron concentration.

2.2.4 Intrinsic semiconductors

An intrinsic semiconductor is a pure semiconductor where the free carriers are exclusively provided by thermal excitation across the band gap between the valence and conduction band. At zero temperature no free carriers are present. As illustrated in Figure 2.6, at finite temperatures electrons are excited from the valence band into the conduction band, while leaving a hole in the valence band. Thus, in an intrinsic semiconductor the number of negatively charged electrons n and positively charged holes p is the same. Owing to the thermal generation of carriers the number of carriers and thus the conductance depends strongly on temperature.

The occupation of states in the conduction band and the number of unoccupied states in the valence band is governed by the Fermi distribution function

$$f(E,T) = \frac{1}{1 + \exp\frac{E-E_F}{k_B T}}, \tag{2.15}$$

with E the electron energy, E_F the Fermi energy, k_B the Boltzmann factor and, T the temperature (cf. Figure 2.6).

The concentration of the thermally excited electrons in the conduction band can be calculated by integrating over the density of states in the conduction band $D_c(E)$ and the Fermi distribution function $f(E,T)$:

$$n = \int_{E_c}^{\infty} D_c(E)f(E,T)\,dE. \tag{2.16}$$

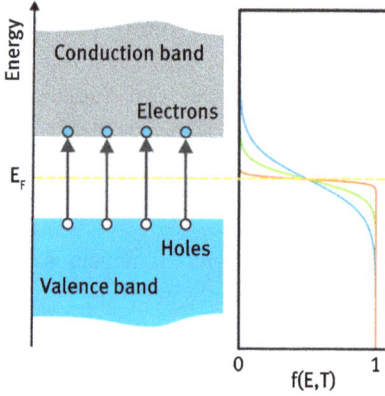

Figure 2.6: Thermal excitation of electrons from the valence band into the conduction band. In the valence band holes are left. On the right side the Fermi distribution function is schematically shown for different temperatures. The larger the temperature, the broader the distribution function around E_F is.

The integration is performed from the conduction band edge E_c to ∞. The latter simplification of the integral boundaries is acceptable, because the Fermi distribution function diminishes for larger energies. The hole concentration is determined correspondingly by integration over the density of states $D_v(E)$ in the valence band and the unoccupied states $1 - f(E, T)$:

$$p = \int_{-\infty}^{E_v} D_v(E)\left[1 - f(E, T)\right] dE. \tag{2.17}$$

The electron and hole concentration can be calculated by approximating the Fermi distribution function by a Boltzmann function, as illustrated in Figure 2.7. This is allowed if the Fermi energy E_F is sufficiently far apart from the conduction and valence band edges E_c and E_v, respectively. In this approximation one obtains the following expression for the electron and hole concentrations:

$$n = \left(\frac{2\pi m_n^* k_B T}{h^2}\right)^{3/2} \exp\left(-\frac{E_c - E_F}{k_B T}\right) \tag{2.18}$$

and

$$p = \left(\frac{2\pi m_h^* k_B T}{h^2}\right)^{3/2} \exp\left(\frac{E_v - E_F}{k_B T}\right), \tag{2.19}$$

respectively [24]. Here, m_n^* and m_h^* are the effective electron and hole masses, respectively.

In most cases the intrinsic carrier concentration in a semiconductor is too low for applications in electronic devices. For silicon the intrinsic electron concentration is

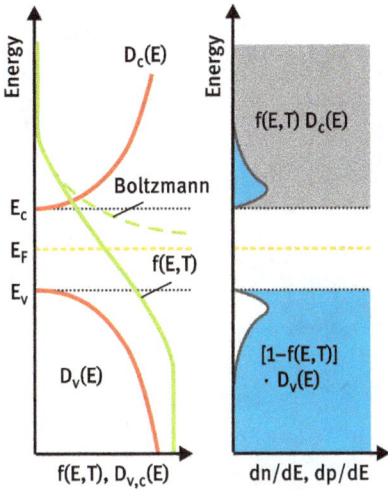

Figure 2.7: Hole and electron occupation in the valence and conduction band, respectively. The occupations are determined by the Fermi distribution function and the corresponding densities of states. The dashed line represents the approximation by the Boltzmann function.

$n = 1.5 \times 10^{10}$ cm^{-3} at 300 K, which is an extremely low value compared to electron concentrations found in metals.

2.3 Doped semiconductors

The carrier concentration in a semiconductor can be increased by several orders of magnitude compared to the intrinsic case if electrically active dopants are incorporated in the crystal. There are two different types: donors, which provide additional free electrons, and acceptors which provide holes.

In the case of a group IV semiconductor, i. e. Si or Ge, donor atoms are usually group V atoms, where four of the five outer electrons are used for the binding to the neighboring atoms in the diamond-type crystal, while the fifth electron is only weakly bound (cf. Figure 2.8 (a)). Possible donor atoms in Si are P or As. The energy level of this additional electron is usually located just a few meV below the conduction band edge, so that a thermal excitation into the conduction band is much more likely than in the case of an excitation across the band gap. This kind of semiconductor is called *n*-type.

In case of acceptors for Si a dopant atom with three electrons in its outer shell is used. Only three electrons are provided for the bonding in the crystal. As can be seen in Figure 2.8 (b), the forth electron can be gained by thermal excitation from the valence band, where a hole is left in the valence band. A semiconductor, where the current is carried by holes is called *p*-type.

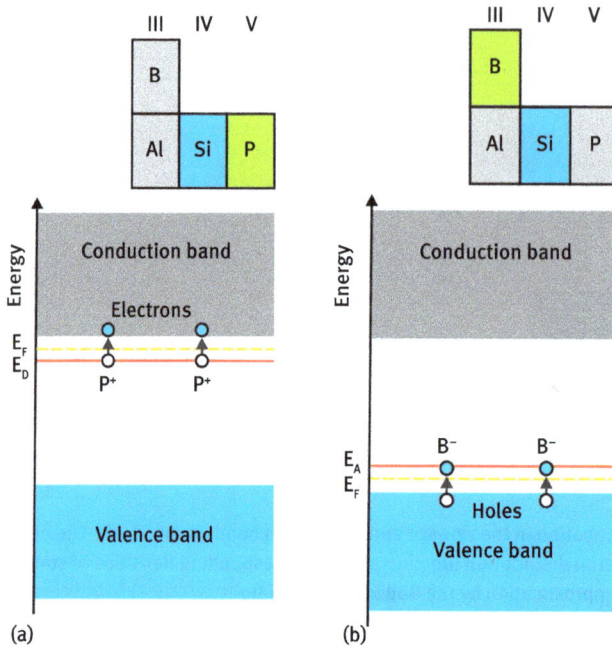

Figure 2.8: (a) Donors in Si: P as a group V element has five electrons in its outer shell, where only four are required for the bonding to the neighboring Si atoms. The fifth electron is only loosely bonded and can easily be excited into the conduction band. (b) Acceptors in Si: B as a group III element has three electrons in its outer shell. An additional electron is provided from the valence band by thermal excitation. A hole remains in the valence band.

The binding energy of an electron to the donor atom can be easily estimated by means of the hydrogen model, where the free electron mass m_0 is replaced by the effective electron mass m^* and ϵ_0 by $\epsilon_0\epsilon_r$, with ϵ_r the relative dielectric constant. The binding energy of a donor electron can thus be written as

$$E_\nu^D = -\frac{m^* e^4}{2(4\pi\epsilon_0\epsilon_r\hbar)^2}\frac{1}{\nu^2},\qquad(2.20)$$

with $\nu = 1, 2, 3, \dots$ Since m^* is usually smaller than m_0 and ϵ_r is larger than 1, the binding energy is much smaller compared to the one in a hydrogen atom. For Si one obtains a value of 30 meV for the lowest level. However, for some dopants the binding energy can be larger, i. e. in diluted magnetic semiconductors which are discussed later.

Similarly, the typical Bohr radius can be estimated by replacing m_0 by m^* and ϵ_0 by $\epsilon_0\epsilon_r$

$$r = \epsilon_0\epsilon_r\frac{\hbar^2}{\pi m^* e^2}.\qquad(2.21)$$

One finds that the valence electron is delocalized over about 1000 lattice atoms. For diluted magnetic semiconductors the overlap of these orbitals is important for the magnetic coupling.

The thermal excitation of carriers in an n-type semiconductor can be subdivided into three parts [24]. At low temperatures one finds an exponential dependence which depends on the donor-binding energy. With increasing temperature this range is followed by the so-called exhaustion range, where the electron concentration n corresponds to the doping concentration $n \approx N_D = $ const. At even higher temperatures intrinsic conduction takes over with the carrier concentration determined by the thermal excitation over the band gap E_g. Most doped semiconductors are in the intrinsic range at room temperature.

2.4 Transport

Depending on the dimensions of the sample, the number of scattering centers, or the temperature the carrier transport can be in a different regime. Especially at low temperatures the wave properties of the electrons are relevant, where interference effects lead to additional features in the transport characteristics. Here, we will first discuss the classical transport, where the contribution of the electron phase is neglected. Subsequently, the characteristic length scales and different transport regimes are introduced. Interference effects will not be addressed here, since a special chapter, Chapter 8, is devoted to these phenomena.

2.4.1 Classical diffusive transport

In the presence of an electric field $\vec{\mathcal{E}}$ the electrons acquire a drift velocity at time t:

$$\langle \vec{v}(t) \rangle = -e\vec{\mathcal{E}}t/m^*. \tag{2.22}$$

On average, after a time τ_e, the elastic scattering time, the electrons will be scattered so that the resulting average velocity, the drift velocity v_d, can be written as

$$\vec{v}_d \equiv \langle \vec{v}(\tau_e) \rangle = -\frac{e\tau_e}{m^*}\vec{\mathcal{E}}. \tag{2.23}$$

By using this expression the electron mobility μ_e of the electrons can be defined:

$$\mu_e = \frac{e\tau_e}{m^*}, \tag{2.24}$$

which quantifies how the average velocity of the electrons depends on the applied electric field $\vec{\mathcal{E}}$. The electrical current density $\vec{j}$ connected to the drift velocity is given by

$$\vec{j} = -en\vec{v}_d, \tag{2.25}$$

where n is the electron concentration. The conductivity σ is a measure of how large the current density is, if an electric field is applied:

$$\vec{j} = \sigma\vec{\mathcal{E}}. \tag{2.26}$$

From equations (2.23) and (2.24) we can infer that the conductivity is given by

$$\sigma = en\mu_e = \frac{e^2 n \tau_e}{m^*}. \tag{2.27}$$

In some situations it is more convenient to refer to the resistivity ϱ of a conductor. This is defined as the inverse of the conductivity $\varrho = 1/\sigma$.

In the above discussion it was assumed that all electrons take part in transport. From a quantum mechanical point of view this is not correct, since here the transport is carried only by electrons close to the Fermi energy, which can be transferred into vacant states close to E_F. Electrons in the center of the Fermi sphere will not find a vacant state they can occupy. However, even in a rigorous quantum mechanical treatment we would obtain the same result for the conductance. This can be understood by a simple argument [25]. By applying an electric field the Fermi sphere is slightly displaced. The amount of electrons taking part in the transport can be estimated by $n(v_d/v_F)$, with v_F the Fermi velocity. These electrons have a velocity of approximately v_F so that we obtain for the current density

$$j = -e\left(n\frac{v_d}{v_F}\right)v_F, \tag{2.28}$$

which is identical to equation (2.25).

2.4.2 Einstein relation

For an electron gas at zero temperature the sum of drift current $\vec{j}_{\mathcal{E}} = \sigma\vec{\mathcal{E}}$ and diffusion current $\vec{j}_D = eD\nabla n$ must vanish at thermodynamic equilibrium:

$$\vec{j}_{\mathcal{E}} + \vec{j}_D = \sigma\vec{\mathcal{E}} + eD\vec{\nabla}n = 0. \tag{2.29}$$

The quantity $\mathcal{D}$ is the diffusion constant. Thermodynamic equilibrium implies that the electrochemical potential μ is constant $\vec{\nabla}\mu = 0$. In general, the electrochemical potential is defined by the sum of the electrostatic potential energy $-eV$ and the Fermi energy, or chemical potential, E_F:

$$\mu = -eV + E_F, \tag{2.30}$$

where E_F is measured from the bottom of the conduction band. By using the definition of μ we can write the gradient of μ as

$$\vec{\nabla}\mu = e\vec{\mathcal{E}} + \frac{1}{D(E_F)}\vec{\nabla}n. \tag{2.31}$$

Here, we made use of $dE_F/dn = 1/D(E_F)$. Inserting the expression for $\vec{\nabla}n$ resulting from equation (2.29), we can write

$$\vec{\nabla}\mu = \left(e - \frac{1}{D(E_F)}\frac{\sigma}{\mathcal{D}e}\right)\vec{\mathcal{E}}. \tag{2.32}$$

As pointed out above, at thermal equilibrium the electrochemical potential is constant, $\nabla\mu = 0$. Since the electric field $\vec{\mathcal{E}}$ is not necessarily zero, it implies for the conductance that

$$\sigma = e^2 D(E_F)\mathcal{D}. \tag{2.33}$$

This is the Einstein relation, which connects the conductivity σ to the diffusion constant $\mathcal{D}$. Only if the conductivity and the diffusion constant are related by the Einstein relation will the current be zero in equilibrium, as required. Thus, for the determination of the conductance one first calculates the diffusion constant at the Fermi energy. Note that the Einstein relation in the form given above is only valid for a degenerate electron gas.

Let us consider a two-dimensional electron gas. If we take the expression for the conductance and make use of $D_{2D} = n_{2D}/E_F$ and of equation (2.27), we obtain the following relation for the diffusion constant:

$$\mathcal{D}_{2D} = \frac{1}{2}v_F^2\tau_e. \tag{2.34}$$

This can be generalized to the dimensionality $d = 1, 2,$ or 3:

$$\mathcal{D}_d = \frac{1}{d}v_F^2\tau_e. \tag{2.35}$$

2.4.3 Mobility

The temperature dependence of the conductance in an n-type doped semiconductor is determined by the temperature dependence of the electron concentration $n(T)$ and the electron mobility $\mu_e(T)$:

$$\sigma(T) = en(T)\mu_e(T). \tag{2.36}$$

The mobility is a function of the temperature-dependent elastic scattering time $\tau_e(T)$ and the effective electron mass m^*:

$$\mu_e(T) = \frac{e\tau_e(T)}{m^*}. \tag{2.37}$$

At higher temperatures phonon scattering limits the mobility [24]:

$$\mu_e^{ph} \propto T^{-3/2}, \tag{2.38}$$

while at lower temperatures ionized impurity scattering dominates:

$$\mu_e^{ii} \propto T^{+3/2}. \tag{2.39}$$

Figure 2.9 shows a typical dependence of the mobility on temperature of an n-type GaAs layer. At low temperatures the mobility is limited by ionized impurity scattering. Neutral impurities also contribute to scattering, but to a much smaller extent. At higher temperatures phonon scattering takes over. For GaAs, with acoustic and optical phonons, scattering by polar-optical phonons is the most prominent contribution.

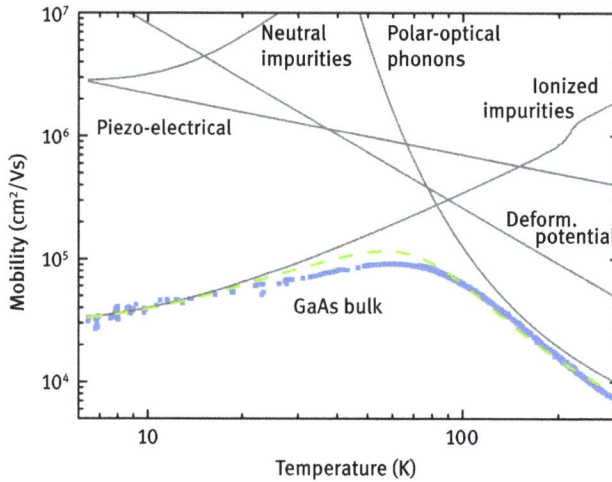

Figure 2.9: Measured mobility for a GaAs layer as a function of temperature. The solid lines correspond to the calculated contributions of the different scattering mechanisms. The dashed line represents the calculated total mobility.

2.4.4 Characteristic length scales

Nano- and spin-electronic systems can be classified by relating their sizes to specific characteristic length scales. These length scales determine in which fashion the carriers propagate through the conductor. Below we will introduce the elastic and inelastic mean free paths which result from scattering processes occurring in the sample. A length scale which gives information about the loss of the phase memory is the phase-coherence length.

2.4.4.1 Elastic mean free path

The elastic mean free path l_e is a measure of the distance between two elastic scattering events. These scattering events occur due to the fact that the conductor is not an ideal conductor but rather contains irregularities in the lattice, e. g. impurities or dislocations. The scattering events are considered to be elastic, which means that the electron does not change its energy. A typical example is the scattering of an electron at a charged impurity. Due to the large difference of the masses of the scattering partners, effectively no energy is transferred from the electron during the scattering event, whereas its momentum can change largely. The elastic mean free path is defined by assuming that the propagation of the electron in the initial direction is suppressed completely. In some cases many scattering events are necessary to fulfill this condition. The scattering mechanisms are sometimes classified as large angle and small angle scattering. The elastic mean free path can be calculated from the scattering time τ_e between successive scattering events:

$$l_e = \tau_e v_F, \tag{2.40}$$

where $v_F = \hbar k_F / m^*$ is the Fermi velocity. For semiconductors τ_e can be obtained from the electron mobility given by equation (2.24).

2.4.4.2 Inelastic mean free path

A further effect beside the above discussed static irregularities are the nonstationary scattering events. A typical example are lattice vibrations, i. e. in the quantum picture phonons. An electron propagating within a crystal will be scattered by these lattice vibrations. On the other hand, a moving electron can excite lattice vibrations and lose a certain amount of its energy. Since an energy transfer occurs, these scattering events are considered to be inelastic. Again we can define an inelastic scattering length l_{in} as a measure for the average distance between this type of scattering events. Beside electron-phonon scattering, electron-electron interaction is another process, where a considerable amount of energy is exchanged between both scattering partners.

2.4.4.3 Phase-coherence length

Another length scale which is of importance is the phase-coherence length l_φ. This parameter is a measure for the distance the electron propagates before its phase is randomized. By an elastic scattering event with a static scattering center, the phase of an electron is usually not randomized. This does not mean that the electron phase is not modified by the scattering event. The crucial point is that the phase is shifted by exactly the same amount if the electron would travel the same path a second time. This is in strong contrast to inelastic scatting events, e. g. electron-phonon scattering, where the scattering target changes with time. Here the phase shift an electron acquires is different each time, since the scattering mechanism is statistical in space and time. However, one must be careful to directly identify l_φ with l_{in} since they are not always identical, e. g. spin-flip scattering is considered to be phase-breaking, while it can be elastic at the same time. In the diffusive transport regime the phase-coherence length is given by

$$l_\varphi = \sqrt{\mathcal{D}\tau_\varphi},$$

(2.41)

with $\mathcal{D}$ being the diffusion constant and τ_φ the phase breaking time.

2.4.4.4 Transport regimes

By comparing the definitions given above with the dimension L of the sample and the Fermi wavelength λ_F, different transport regimes can be classified. In case that the elastic mean free path l_e is smaller than the dimensions of the sample, many scattering events occur. The carriers are traveling diffusively through the crystal (cf. Figure 2.10 (a)). If the phase coherence length l_φ is shorter than l_e and the size of the system, the transport is said to be classical. In contrast, if l_φ is larger than l_e, quantum mechanical effects due to the wave nature of the electrons can be expected (Table 2.2). This diffusive regime is thus called the quantum regime. In case that l_e is larger than the dimensions of the sample the electrons can transverse the system without any scattering. This regime is called ballistic (cf. Figure 2.10 (b)). Depending on the magnitude of the Fermi wavelength λ_F in comparison to the dimension of the sample the transport can either be regarded as classical ballistic or quantum ballistic (Table 2.2).

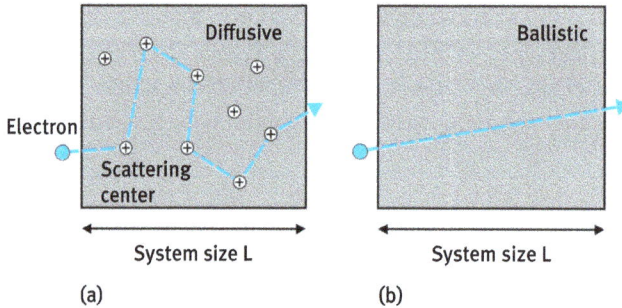

Figure 2.10: Illustration of the diffusive and ballistic transport regime.

Table 2.2: Comparison of the different transport regimes.

Diffusive	Classical	$\lambda_F, l_\varphi, l_e \ll L$
	Quantum	$\lambda_F, l_e \ll L, l_\varphi$
Ballistic	Classical	$\lambda_F \ll L < l_\varphi, l_e$
	Quantum	$\lambda_F, L < l_e < l_\varphi$

As introduced above, transport is considered to be diffusive if the elastic mean free path l_e is much smaller than the dimensions of the structure. The carriers are propagating randomly through the structure. In the classical diffusive transport regime, as discussed in the previous sections the phase-coherence length l_φ is also smaller than l_e, so that any interference effects can be neglected. In the quantum limit the phase coherence length exceeds the elastic mean free path. Here, even in large scale samples the electron interference effects can lead to additional contributions to the resistance, i. e. localization effects or conductance fluctuations. More details on these phenomena are given in Chapter 8.

2.5 Layer systems

By means of layer systems combining different semiconductor materials the possibilities for electronic devices and for studying interesting fundamental effects can be increased enormously. We first begin to describe the growth of these layer systems. Subsequently, the origin and the properties of two-dimensional electron gases are discussed.

2.5.1 Semiconductor heterostructures

Semiconductor heterostructures are objects where two different semiconductor materials are combined, usually in form of layer systems. These layer systems are fabricated by using epitaxy. In contrast to the ordinary deposition of materials, i. e. the deposition of amorphous layers, during epitaxial growth the crystal structure is maintained. This restricts the combination of semiconductor materials significantly, since the lattice constant of both materials has to match. In Figure 2.11, the band gap of different semiconductors is given as a function of the lattice constant. As one can see here, only for very few elementary or binary semiconductors do the lattice constants match. The most prominent example is GaAs/AlAs, where GaAs has the smaller band gap compared to AlAs. Matching of lattice constants can also be achieved by using ternary or quaternary alloys, e. g. $Ga_{0.47}In_{0.53}As/InP$.

Semiconductor layer systems can be fabricated by using molecular beam epitaxy (MBE) (cf. Figure 2.12). Here, a single-crystal wafer is transferred into a vacuum chamber. The substrate is heated to a certain temperature and exposed to molecular beams of those materials which constitute the deposited layer system, i. e. As and Ga for a GaAs layer. The flux of the molecular beam can be switched off by means of shutters located in front of the effusion cells. In order to obtain a homogeneous layer growth, the substrate is rotated. The growth chamber is internally cooled by liquid nitrogen cryo-shields to maintain a sufficiently good vacuum during growth. A specialty of MBE is that the layer growth can be monitored by reflection high energy electron diffraction

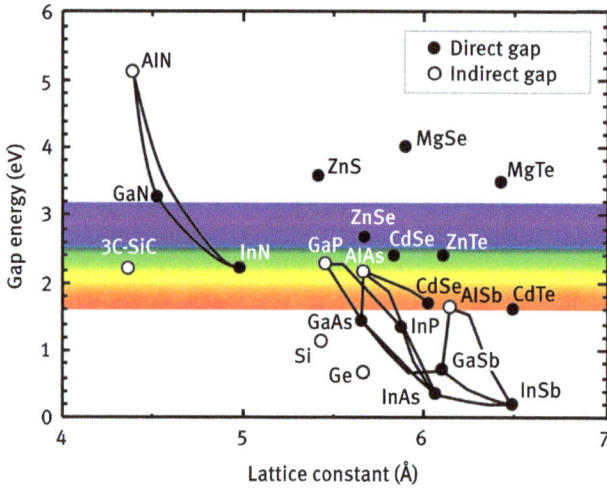

Figure 2.11: Energy band gap as a function of the lattice constant for various semiconductors. The colored stripe illustrates the color of the emitted photons, which are generated by direct band gap transitions.

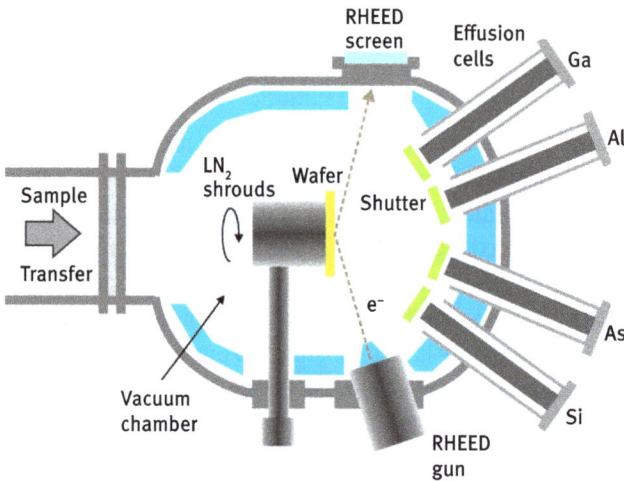

Figure 2.12: Schematic illustration of a molecular beam epitaxy chamber. The wafer is fixed on a rotating sample holder. The molecular beam of the effusion cells can be switched on and off by the shutters. The liquid nitrogene (LN$_2$) shrouds are needed to maintain a low pressure in the chamber. By the RHEED system, the growth can be monitored.

(RHEED). Here, the layer-by-layer growth is monitored by means of an electron beam reflected on the substrate.

Heterostructures containing phosphorus are mostly grown by means of metal-organic vapor phase epitaxy (MOVPE). A schematic of the MOVPE process is shown in

Figure 2.13. Here, the metal-organic precursors, e. g. trimethylindium (TMIn: $In(CH_3)_3$), trimethylgallium (TMGa: $Ga(CH_3)_3$), AsH_3, or PH_3, are added to the carrier gas, i. e. H_2 or N_2, which flows in a reactor over the surface of the wafer. By heating the substrate, the precursors are decomposed, and an epitaxial layer is grown.

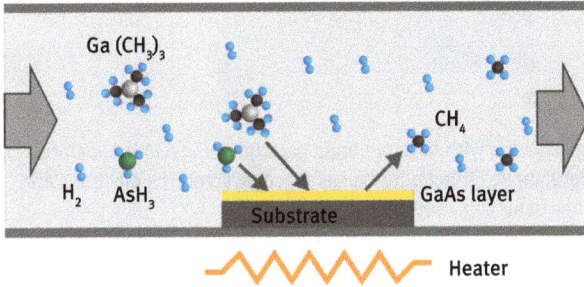

Figure 2.13: Schematic illustration of a metal-organic vapor phase epitaxy growth chamber. In this example trimethylgallium $Ga(CH_3)_3$ and AsH_3 is added to the H_2 carrier gas.

2.5.2 Two-dimensional electron gases

Many concepts of spintronic devices rely on two-dimensional electron gases (2DEG) in heterostructures [26]. Here, an n-doped semiconductor layer with a larger band gap, e. g. $Al_{0.3}Ga_{0.7}As$ (material A), is grown epitaxially on a semiconductor layer with a lower band gap, e. g. GaAs (material B) [27]. The initial situation before combining both materials is depicted in Figure 2.14 (a). After merging electrons are transferred from the n-doped layer into the other semiconductor with the lower band gap, owing to the adjustment of the Fermi level. The band bending due to the band offset of both materials takes care that these electrons are only found at the interface of both layers. The electrons are thus trapped in a potential well. This is situation is depicted in Figure 2.14 (b).

The electrons can only move freely along the interface; therefore these structures are called two-dimensional electron gases. In order to suppress impurity scattering, the impurities are separated by a spacer layer from the two-dimensional electron gas (modulation doping). If the doping layer is only a few monolayers thick it is called a δ-doping layer. A typical layer system containing a two-dimensional electron gas in an AlGaAs/GaAs heterostructure is depicted in Figure 2.15.

In Figure 2.16 mobility of a two-dimensional electron gas in an AlGaAs/GaAs heterostructure is compared to the one of a doped bulk GaAs semiconductor. In a bulk semiconductor the electrons are directly scattered at the positively charged donor atoms which provided the electrons. In case of a modulation-doped heterostructure the layer with the dopant atoms is spatially separated by the undoped spacer

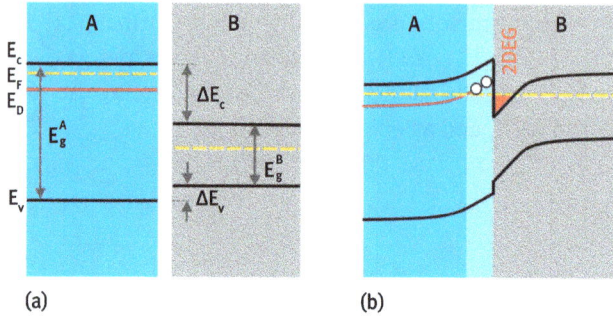

(a)   (b)

Figure 2.14: (a) Two semiconducting materials with different band gaps (material A and B) before they are combined in a heterostructure. (b) Resulting potential in a layer system. The electrons are accumulated at the interface in the low band gap material.

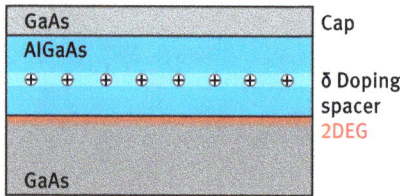

Figure 2.15: Layer system of a two-dimensional electron gas in an AlGaAs/GaAs heterostructure. The two-dimensional electron gas is separated by an undoped spacer layer from the δ-doping layer. The layer system is capped by a GaAs layer to prevent oxidation.

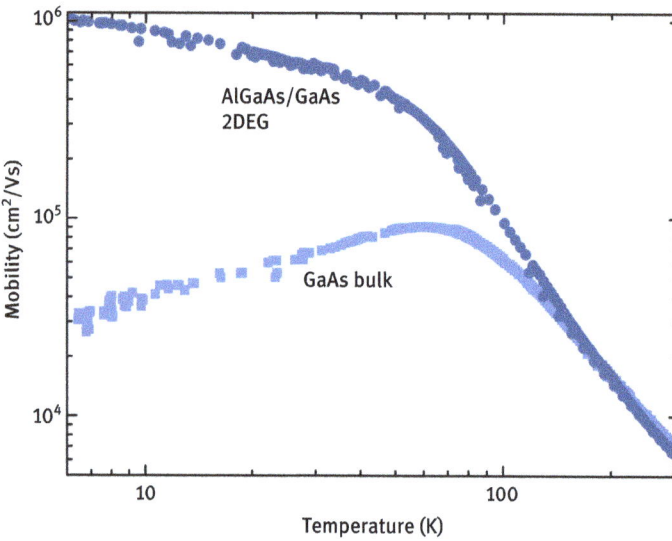

Figure 2.16: Mobility as a function of temperature for a two-dimensional electron gas (2DEG) in an AlGaAs/GaAs heterostructure in comparison to a bulk GaAs layer.

layer from the 2DEG. Consequently, at low temperatures, where the ionized impurity scattering dominates, the scattering is reduced considerably, i. e. the mobility is enhanced.

In $Al_{0.3}Ga_{0.7}As$/GaAs heterostructures grown by MBE the mobility in the 2DEG can be as high as 10^7 cm^2/Vs at temperatures $T < 4$ K [28]. Beside the AlGaAs/GaAs material system it is also possible to realize a 2DEG in an InGaAs/InP heterostructure. At an indium content of 53 % InGaAs is lattice matched to InP. A considerable improvement with regard to higher mobility and electron concentration is achieved if the 2DEG is realized in a 10 nm thick strained InGaAs layer with an indium content of 77 %. For this type of structures low temperature (6 K) mobilities up to 450 000 cm^2/Vs at an electron concentration of 6.0×10^{11} cm^{-2} were achieved [29, 30]. These In-containing heterostructures often serve as a basis for the spin field-effect transistor, because of the strong spin-orbit coupling. Further details are given in Chapter 7.

2.6 Quantum wires and nanowires

Quasi one-dimensional structures can be defined by lithographic means. Here, we will first explain the most common method to fabricate these structures from two-dimensional electron gases using electron beam lithography. Alternatively, one-dimensional structures, i. e. semiconductor nanowires, can be created directly by a bottom-up approach. In short ballistic one-dimensional structures quantized conductance can be observed. The mechanisms resulting in this very unique effect are discussed below.

2.6.1 Electron beam lithography

In order to fabricate one- or even zero-dimensional semiconductor structures, mostly electron beam lithography is used. This is due to the fact that for the observation of quantum effects the size of the sample must be in the order of a few tens of a nanometer. Structures with these dimensions usually are not realized by conventional optical lithography. Here, a resolution of about 1 μm can be obtained by standard optical lithography using ultra-violet (UV) light. However, with very advanced optical lithography methods for very large scale integration using short wavelength sources (deep UV, extreme UV), structures with sizes of about 100 nm can be achieved.

In the electron beam lithography process a focused beam of electrons is scanned over the surface of the sample, which was covered before by an electron beam sensitive resist (Poly(methyl)methacrylate, PMMA) (cf. Figure 2.17 (a)) [31]. The PMMA resist consists of long chains of molecules, which are cracked by the high energy electrons. Developing the sample in methylisobuthylketon (MIBK) after the electron beam ex-

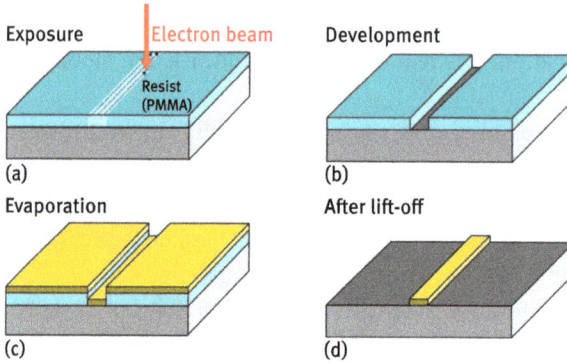

Figure 2.17: Fabrication of metallic electrodes by lift-off: (a) exposure of a PMMA layer by electron beam lithography; (b) resist after development; (c) deposition of a metal layer by evaporation; (d) remaining metal after rinsing in acetone.

posure results in openings in the resist at the positions where the electron beam was scanned (cf. Figure 2.17 (b)).

There are many ways to use the pattered resist layer for subsequent processing steps. One possibility is to use the resist as a mask, in order to deposit metallic gate structures on a semiconductor structure. By applying a negative voltage to these gates, the 2DEG can be depleted, so that the electron flow in the semiconductor is restricted only to those areas which are not covered by the gates. One of the common methods to define metal structures to the surface is called lift-off. Here, the sample is covered completely with a metal film deposited e. g. by electron beam evaporation (cf. Figure 2.17 (c)). In most cases a two-layer resist is used, where the upper layer has a higher molecular weight. During the development of the PMMA layer system this layer is solved more slowly in the developer, so that the opening of the upper layer is slightly smaller than the opening of the lower layer. This special shape ensures that the deposited metal film is easily separated at the edges of the resist pattern, when the sample is rinsed in acetone. Acetone is used to dissolve the PMMA resist, so that the metal layers covering the resist layer are removed. Only the metal directly deposited on the semiconductor surface is left after this processing step (Figure 2.17 (d)).

Beside the depletion of the two-dimensional electron gas by means of a gate electrode, the conductive area can also be pattered directly by various etching methods. Here, the PMMA layer itself or layers of other materials, e. g. metals or dielectrics, can be used as an etching mask. The etching process can either be performed by wet chemical etching or by dry etching methods, e. g. reactive ion etching or ion beam etching. In Figure 2.18 (a) a structure is shown in which the layer system is only shallowly etched. Here, the two-dimensional electron gas is depleted, due to the lifting of the band profile caused by the surface potential. An advantage of this method is that the conductive layer is not exposed to air. In some cases the heterostructure is

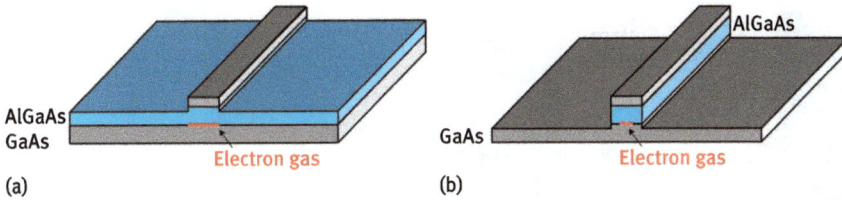

AlGaAs

AlGaAs
GaAs

Electron gas

(a)

GaAs

Electron gas

(b)

Figure 2.18: Wire structure in a two-dimensional electron gas defined by (a) shallow mesa etching and (b) deep mesa etching.

etched below the heterointerface (deep mesa etching), as illustrated in Figure 2.18 (b). Since the surface potential in some semiconductors, e. g. GaAs, leads to a depletion zone, the effective electronic width of the structure is smaller than the geometrical width.

2.6.2 Semiconductor nanowires

Nanoscaled wire structures can also be fabricated directly by self-organized growth without using elaborate lithographical means. One possible approach is to use the vapor-liquid-solid (VLS) growth mode to form a nanowire [32]. Here, a nanometer-size gold particle is employed as a seed particle for the growth. In many cases InAs is used as a semiconductor material. The underlying reason is that at the surface an electron accumulation layer is formed due to the Fermi level pinning within the conduction band (cf. Figure 2.19 (a)) [33]. This property is only found in few semiconductors, e. g. InAs, InN, or InSb. It ensures that even for low nanowire diameters the nanowires contain conductive electrons. For most other semiconductors a depletion layer is formed at the surface, e. g. for GaAs or Si, owing to the Fermi level pinning within the band gap (cf. Figure 2.19 (b)).

In addition to the VLS method nanowires can also be fabricated by selective-area metal-organic vapor phase epitaxy (SA-MOVPE) [34, 35]. Here, a prepatterned substrate is used, where the nanowires are grown only at those positions where holes are present in a dielectric mask layer. By this method the location of the nanowires can be determined, which is an important prerequisite for device processing (cf. Figure 2.20 (a)).

For transport measurements the as-grown nanowires are usually separated from the substrate and transferred to another substrate equipped with alignment markers. By using electron beam lithography these nanowires are contacted. A scanning electron micrograph of a typical contacted InAs semiconductor nanowire is shown in Figure 2.20 (b).

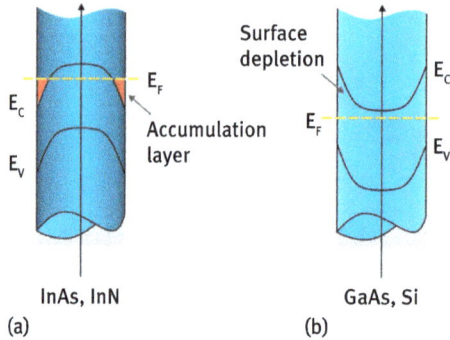

InAs, InN (a)

GaAs, Si (b)

Figure 2.19: (a) Schematic illustration of a semiconductor nanowire with a surface accumulation layer. The accumulation layer is formed because of the downwards bending of the conduction and valence bands due to the pinning of the Fermi level at the surface. Typical materials are InAs or InN. (b) Nanowire with a depletion layer at the surface, i. e. as found for GaAs or Si. Here, the bands are bent upwards at the surface.

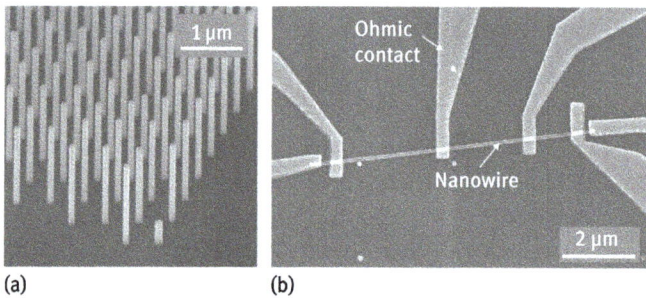

(a) (b)

Figure 2.20: (a) Scanning electron beam micrograph of a field of InAs nanowires grown by selective-area MOVPE (image provided by H. Hardtdegen, Forschungszentrum Jülich). (b) InAs nanowire with ohmic contacts.

2.6.3 Split-gate quantum point contacts

In one-dimensional systems quantized conductance can be observed. This effect can best be demonstrated by means of a split-gate quantum point contact. As shown in Figure 2.21, two opposite gate fingers, separated by a distance of a few hundreds of nanometers, are placed on a 2DEG [36, 37]. Split-gate electrodes are usually prepared by means of electron beam lithography. The distance of the split-gates is comparable with the Fermi wave length λ_F, since λ_F of a two-dimensional electron gas is typically a few tens of a nanometer. In Figure 2.22 an AlGaAs/GaAs sample with a number of split-gate point contacts is shown.

By applying a sufficiently large negative voltage to the gate fingers the underlying two-dimensional electron gas is depleted. Only a small opening between the gate fin-

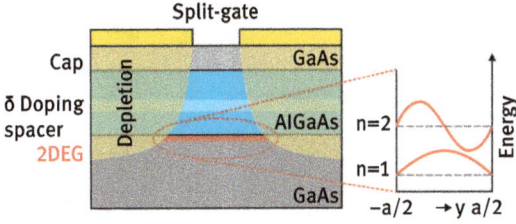

Figure 2.21: Cross section of a split-gate point contact. By applying a sufficiently large negative gate voltage the electron gas underneath the gate fingers is depleted. The right scheme illustrates the quantization in the 2DEG perpendicular to the channel.

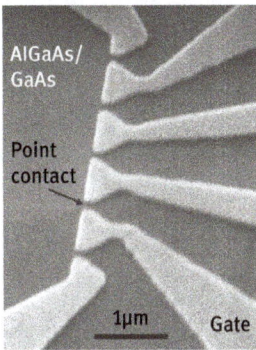

Figure 2.22: Scanning electron beam micrograph of the surface of an AlGaAs/GaAs heterostructure with five split-gate point contacts.

gers remains for the electrons to move from one side to the opposite side of the opening. By varying the gate voltage it is possible to control the effective opening width. A less negative bias voltage enlarges the depletion area and thus reduces the opening width. For a sufficiently large negative bias voltage and a sufficiently small electrode separation of the split-gate electrodes the opening can even be closed completely, i. e. pinched-off.

Owing to the depletion area underneath the split-gate electrodes, we can assume that the electrons in the 2DEG are confined in a potential well along the y-axis, while the opening is assumed to be along the x-axis. If the potential profile in the plane of the 2DEG is expressed by $V(x,y)$, the Hamilton operator has the following form:

$$\hat{H} = -\frac{\hbar^2}{2m^*}\left(\frac{\partial^2}{\partial x^2} + \frac{\partial^2}{\partial y^2}\right) + V(x,y). \tag{2.42}$$

For most applications it is sufficient to assume an approximated potential profile. For relatively low gate voltages, the potential profile in the center of the point contact can be approximated by a rectangular potential well:

$$V(y) = \begin{cases} V_0, & |y| \le a/2 \\ \infty, & |y| > a/2. \end{cases} \quad (2.43)$$

In order to determine the energy dispersion, we solve the Schrödinger equation

$$\hat{H}\psi(x,y) = E\psi(x,y), \quad (2.44)$$

with $\psi(x,y)$ the electron wavefunction. The electron propagation along the x-direction is not affected by the potential, i. e. we can assume plane wave solutions. Thus, we can express $\psi(x,y)$ as

$$\psi(x,y) = e^{ik_x x}\varphi(y), \quad (2.45)$$

with k_x the wave vector. Regarding $\varphi(y)$ we have to solve the Schrödinger equation for a particle-in-a-box, which gives for the boundary conditions $\varphi(-a/2) = \varphi(a/2) = 0$ the following odd and even solutions, respectively,

$$\varphi_n(y) = \begin{cases} \sqrt{\frac{2}{a}}\cos(k_n y), & n = 1,2,3,\dots, \\ \sqrt{\frac{2}{a}}\sin(k_n y), & n = 2,4,6,\dots, \end{cases} \quad (2.46)$$

with $k_n = n\pi/a$. The energy dispersion is given by the quantized particle-in-a-box energy eigenvalues and the parabolic energy-momentum dispersion for the electron propagation perpendicular to the point contact opening (x-direction)

$$E_n = \frac{n^2\pi^2\hbar^2}{2m^*a^2} + \frac{\hbar^2 k_x^2}{2m^*}, \quad n = 1,2,3,\dots. \quad (2.47)$$

In order to explain the quantized conductance we refer to Figure 2.23. Here, a one-dimensional channel is connected between two contacts with electrochemical potentials μ_1 and μ_2.

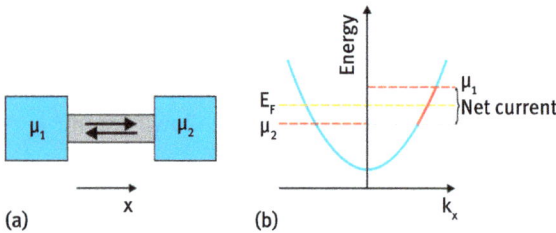

Figure 2.23: (a) One-dimensional channel between two reservoirs. (b) Energy-momentum dispersion of a one-dimensional channel with the electrochemical potentials μ_1 and μ_2 of the right and left reservoir, respectively.

As mentioned above, along the conductor the electrons can propagate freely, which is described by a parabolic energy dispersion (cf. Figure 2.23). On the right-hand side branch the electrons are moving in right direction, while on the left-hand side the electron propagate in left direction. The right-moving electrons are supplied by the contact up to the electrochemical potential μ_1, while the left-moving electrons are supplied up to μ_2. For the net current the motion electrons filled up to μ_2 are compensated by electrons moving in the opposite direction. However, in between μ_2 and μ_1 the right-moving electrons do not find their counterparts. Thus the current can be written as

$$I_{1D} = e \int_{\mu_2}^{\mu_1} \widetilde{D}_{1D} v(E)\, dE, \qquad (2.48)$$

with $\widetilde{D}_{1D}$ being the one-dimensional density of states for electrons flowing in only one direction and $v(E)$ their velocity. The density of states can be expressed as

$$\widetilde{D}_{1D}(E) = \frac{1}{2} D_{1D} = \frac{1}{2} \frac{2}{\pi} \sqrt{\frac{m^*}{2\hbar^2}} \frac{1}{\sqrt{E}} = \frac{2}{hv(E)}. \qquad (2.49)$$

Thus $\widetilde{D}_{1D}(E)$ is inversely proportional to $v(E)$. As a consequence, the integrand is energy independent. Therefore, the net current is directly proportional to the difference of the electrochemical potential:

$$I_{1D} = \frac{2e}{h}(\mu_1 - \mu_2). \qquad (2.50)$$

In order to calculate the conductance, the current has to be divided by the applied voltage $(\mu_1 - \mu_2)/e$:

$$G = \frac{I_{1D}}{(\mu_1 - \mu_2)/e} = \frac{2e^2}{h}. \qquad (2.51)$$

As one can see, the conductance is constant and depends only on fundamental constants. Here, we considered only a single one-dimensional channel. As can be seen in Figure 2.21, depending on the size of the opening more than one channel might be occupied below the Fermi energy. The number of occupied channels can be changed by varying the voltage applied to the split-gate electrodes. For N occupied channels the conductance is given:

$$G = N\frac{2e^2}{h}. \qquad (2.52)$$

The theoretical approach used here is based on the Landauer–Büttiker model [38, 39]. In our case we assumed ideal transmission of the carrier between the two reservoirs. In the more generalized picture, finite transmission and reflection coefficients are included.

In Figure 2.24 a measurement of a split-gate point contact is shown. The measurement was performed at low temperatures. As can be seen, for increasing negative gate voltages the conductance decreases in steps of $2e^2/h$.

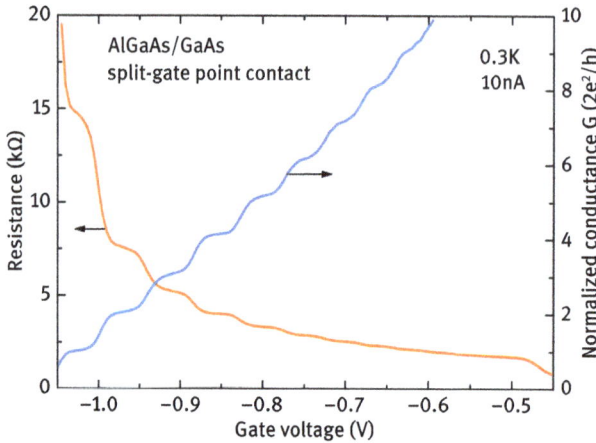

Figure 2.24: Measurement of the resistance R (left axis) of a split-gate point contact. The conductance G (right axis) is determined by inverting R.

2.7 Zero-dimensional structures: quantum dots

Two-dimensional electron gases are also often employed to define zero-dimensional structures, i. e. quantum dots. Here, a number of gates on the surface are used to define an electrically isolated island (cf. Figure 2.25). By means of electrostatically controllable openings the quantum dot can be filled or emptied, so that a current can flow through the dot. Since the electrons are confined in all directions only discrete electron states are formed, like in an atom (cf. Figure 2.5 (d)). What makes a quantum dot so interesting is that, owing to its small size, the capacitance C is very low. As a result, the corresponding charging energy

$$E_c = \frac{e^2}{2C} \tag{2.53}$$

is very high. Thus, the energy required to fill the quantum dot with an electron is relatively large. Typical values are in the order of a few 100 μV up to a few meV, so that only at temperatures below around 4 K or lower can signatures of this charging effects be observed in transport.

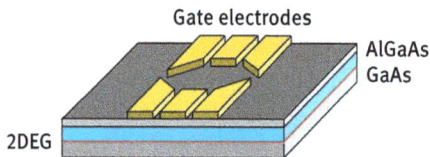

Figure 2.25: Schematic illustration of a quantum dot formed by gate electrodes on the surface of an AlGaAs/GaAs heterostructure.

In order to understand the transport properties of a quantum dot one has to look at the energy difference which is required to change the number of electrons in the dot from N to $N + 1$:

$$\mu_{\text{dot}}(N + 1) - \mu_{\text{dot}}(N) = \frac{e^2}{C} + \Delta E. \tag{2.54}$$

Here, in addition to the charging energy discussed above the energy ΔE is included, which is due to quantization effects. The latter contribution is particularly important in semiconductors because of the large Fermi wavelength. Quantization effects already occur at dimensions of a few tens of a nanometer. The electrical circuit is sketched in Figure 2.26. The dot island is separated by tunnel barriers from the source and drain contacts. These barriers are formed by biasing the left and right split-gates of the structure shown in Figure 2.25. The tunneling resistance between the dot island and the source and drain contacts has to be sufficiently large, $R > e^2/h$, in order to maintain a sufficient localization of carriers in the dot. The electronic levels in the dot can be controlled by a gate. This is usually done by one of the gates defining the effective size of the quantum dot, e. g. by changing the gate voltage on the gates in the middle of the structure shown in Figure 2.25. In order to observe charging effects, the total capacitance $C = \sum C_i$ has to be sufficiently small so that the charging energy given by equation (2.53) is larger than $k_B T$. Here, C_i are the capacitances between the dot and the source and drain contacts as well as between the dot and the gate electrode.

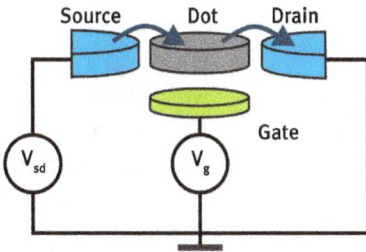

Figure 2.26: Schematic illustration of a gate-controlled quantum dot island connected to source and drain contacts.

2.7.1 Transport at small source-drain bias voltages

We first discuss the case where only a small constant voltage is applied between the source and drain contacts, compared to the charging energy. In Figure 2.27 (a) the electrochemical potentials of the source and drain contacts are located in between the dot levels for N and $N + 1$.

In order to obtain electron transport through the dot, the level at $\mu_{\text{dot}}(N+1)$ needs to be occupied. However, because of the low temperature, no thermal excitation to that

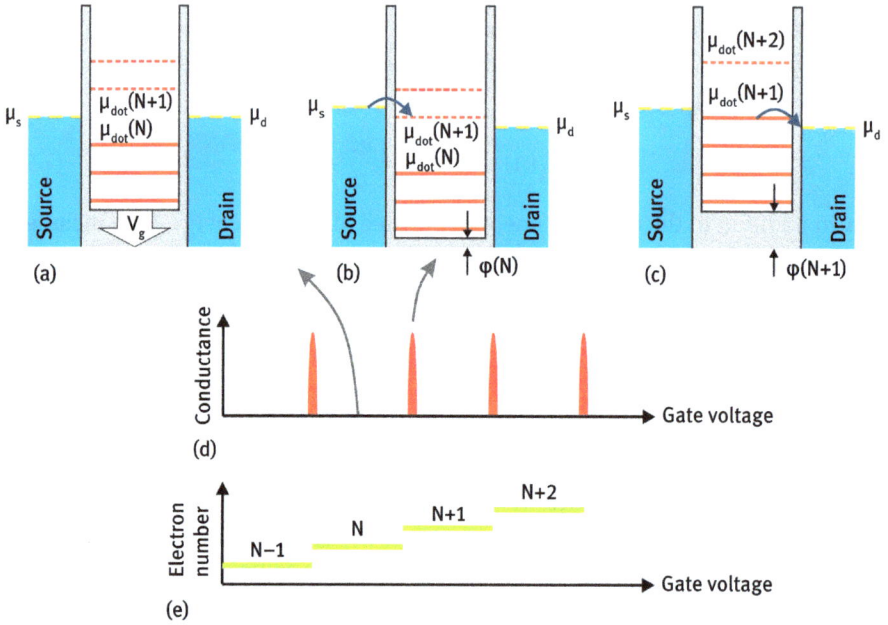

Figure 2.27: Schematic illustration of Coulomb blockade and single electron tunneling. (a) The electron transport is blocked. (b), (c) The gate voltage V_g was adjusted so that an unoccupied level is located between the electrochemical potentials of the source and drain contacts. Now an electron can enter the dot. The altered electrostatic energy, i. e. shifting of the energy levels, makes sure that no second additional electron can enter the dot. The corresponding conductance peaks, and occupation numbers are shown in panels (d), (e), respectively.

level is possible. As a consequence the transport is blocked. As shown in Figure 2.27 (d) the conductance through the dot is zero. This phenomena is called Coulomb blockade. The situation changes if the levels of the dot are pulled downwards by applying an appropriate gate voltage (cf. Figure 2.27 (b) and (c)). Now the level at $\mu_{dot}(N+1)$ is located in between the electrochemical potentials μ_s and μ_d of the source and drain contacts, so that an additional electron can tunnel into the dot. The additional electron in the dot changes its electrostatics, with the consequence that all levels in the dot are shifted upwards. The level shift blocks the occupation of the dot with $N + 2$ electrons. The $(N+1)$-th electron can leave the quantum dot by tunneling to the drain contact. Since only a single electron can be transported between the source and drain contact, this process is called single electron tunneling. If the conductance of the quantum dot is recorded as a function of gate voltage, zero conductance is measured in the Coulomb blockade regime, while peaks are observed if single electron tunneling is allowed as illustrated in Figure 2.27 (d). After crossing the single electron tunneling range the electron number in the quantum dot is changed by one (cf. Figure 2.27 (e)).

A semiconductor quantum dot realized in an InAs nanowire is shown in Figure 2.28. The quantum dot is defined by gate fingers crossing the nanowire. By apply-

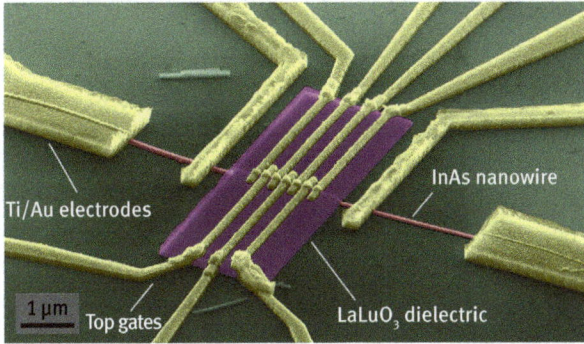

Figure 2.28: InAs nanowire with a set of gate fingers crossing the wire. The gate fingers are separated from the wire by an insulating dielectric layer ($LaLuO_3$). The source and drain contacts are placed on the terminals of the wire. Image provided by S. Heedt, Forschungszentrum Jülich.

ing a negative gate voltage to one of these gates a potential barrier is raised. Usually a quantum dot is formed by applying a negative gate voltage to a pair of gates. This leads to two potential barriers separating the dot from the leads. In between these gates another gate finger is employed to tune the potential within the quantum dot. This gate is often called a plunger gate. The corresponding measurements of the conductance as a function of gate voltage at a small drain-source bias is shown in Figure 2.29. As can be seen here, each time the Coulomb blockade is lifted, a peak in the current is found.

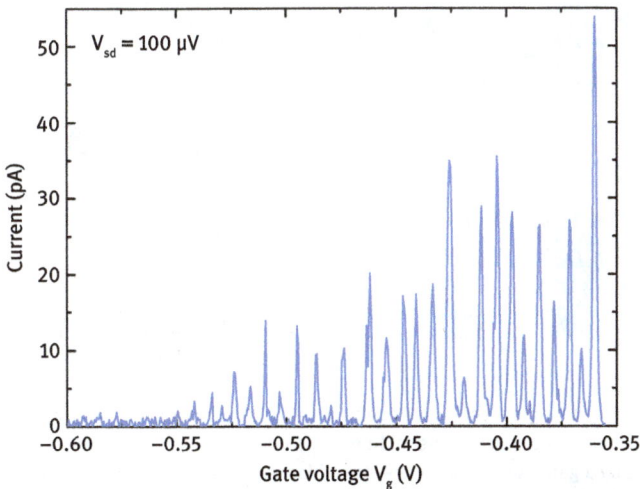

Figure 2.29: Current as a function of gate bias voltage V_g for a quantum dot structure based on an InAs nanowire. The source-drain bias V_{sd} was fixed at 100 μV. The bath temperature was 50 mK. Graph provided by S. Heedt, Forschungszentrum Jülich.

2.7.2 Transport as a function of source-drain bias voltage

After addressing the transport through a quantum dot as a function of gate voltage, now the case where the source-drain bias voltage V_{sd} is changed. In Figure 2.30 (a) the situation is shown where initially zero bias ($V_{sd} = 0$) is applied. Here, the electrochemical potentials of the source and drain contacts are located in the middle of the Coulomb gap of the quantum dot. In order to measure single electron tunneling, the source-drain bias V_{sd} has to be as large as the Coulomb gap. In Figure 2.30 (b) the situation is shown where the levels in the quantum dot are pulled downwards by applying a gate voltage. In this case a smaller source-drain bias voltage is required to initiate single electron tunneling. The required voltage corresponds to the energetic distance between the zero-bias electrochemical potentials and the lowest unoccupied level in the dot. In Figure 2.30 (c) the resulting conductance is schematically shown as a function of source-drain voltage and gate voltage. The largest source-drain bias until the quantum dot gets conductive is required in the symmetric case (Figure 2.30 (a)). From that point on by changing the gate voltage in both directions the required source-drain voltage gets linearly smaller. The white areas, corresponding to the Coulomb blockade regime are called Coulomb diamonds. Each Coulomb diamond is connected to a fixed number of electrons in the dot $\ldots N - 1, N, N + 1, \ldots$.

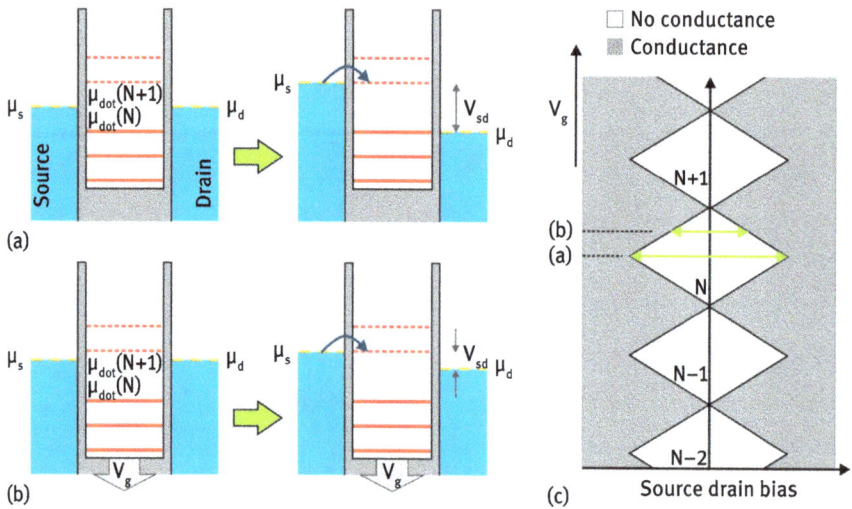

Figure 2.30: Schematic illustration of single electron tunneling under a source-drain bias V_{sd}. (a) The zero-bias electrochemical potentials are located in the middle of the Coulomb gap. (b) The levels in the dots are pulled downwards by a gate. A smaller source-drain bias is required for single electron tunneling, compared to the upper case. (c) Conductance as a function of source-drain bias voltage and gate voltage. In the white diamond-shaped areas the conductance is zero (Coulomb blockade). Each Coulomb diamond corresponds to a fixed number $\ldots N - 1, N, N + 1, \ldots$ of electrons in the dot.

In Figure 2.31 a typical measurement of Coulomb diamonds is shown. Here, the quantum dot was defined by placing gate fingers across an InAs semiconductor nanowire.

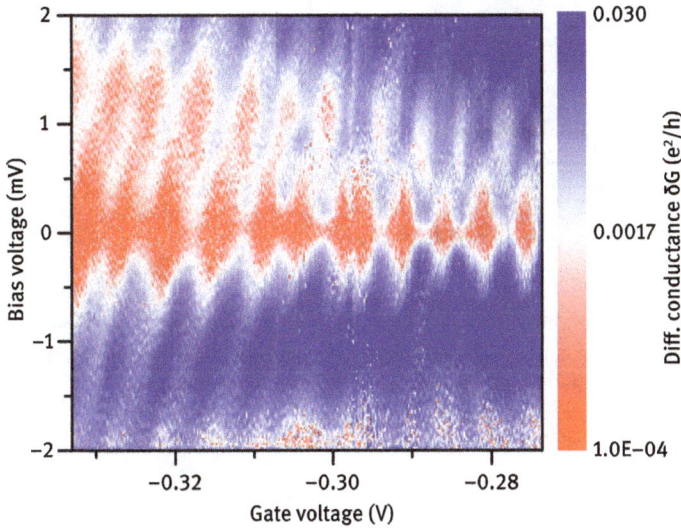

Figure 2.31: Differential conductance of a quantum dot formed in an InAs nanowire as a function of gate voltage and source-drain bias at 50 mK. An image of the corresponding sample can be found in Figure 2.28. Graph provided by S. Heedt, Forschungszentrum Jülich.

2.8 Transport in a quantizing magnetic field

Magneto-transport is a very powerful tool to gain information on the spin properties of semiconductors, e. g. by measuring oscillations in the longitudinal resistance information on the strength of spin-orbit coupling can be obtained. Furthermore, the concepts discussed here are important for the quantum spin Hall effect discussed in Chapter 10. In Figure 2.32 a typical Hall bar sample used for magneto-transport measurements is shown. The structure is based on a two-dimensional electron gas in a semiconductor heterostructure. The bias current I flows between the left and right contacts. The longitudinal resistance is obtained by measuring the voltage V_{xx} along the current path, while the Hall voltage V_H is measured across. The magnetic field is applied perpendicularly to the plane of the two-dimensional electron gas. A large magnetic field leads to an additional quantization of the electronic levels, the so-called Landau quantization. Below we will explain the origin of this quantization and the consequences for the electronic transport.

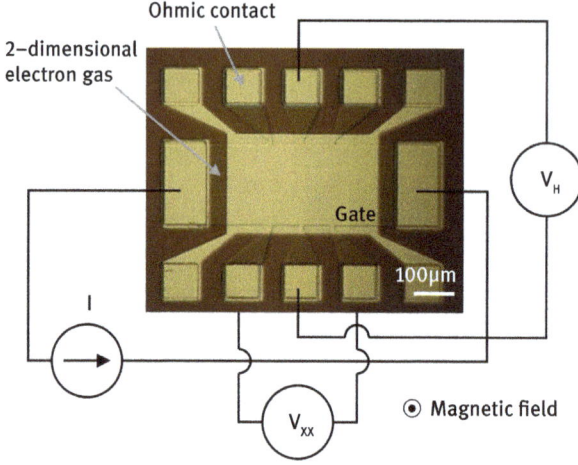

Figure 2.32: Hall bar mesa structure with voltage probes to measure the Hall voltage V_H and the voltage V_{xx} for the longitudinal resistance, respectively. The magnetic field is oriented perpendicularly to the surface plane. By means of a gate electrode the electron concentration in the two-dimensional electron gas can be adjusted.

2.8.1 Landau quantization

We assume a magnetic field perpendicular to the plane of the two-dimensional electron gas $\vec{B} = B\vec{e}_z$. In the presence of a magnetic field we have to replace the momentum operator $\hat{\vec{p}} = -i\hbar\vec{\nabla}$ by $\vec{\pi} = -i\hbar\vec{\nabla} + e\vec{A}$ in the Hamiltonian, with the vector potential $\vec{A}$ defined by $\vec{B} = \vec{\nabla} \times \vec{A}$. We assumed an electron with charge $-e$ ($e = +1.602 \times 10^{-19}$ C). With the vector potential in a particular gauge given by $\vec{A} = (0, Bx, 0)$ the Hamiltonian can be written as

$$\hat{H} = \frac{1}{2m^*}[\hat{p}_x^2 + (\hat{p}_y + eBx)^2], \tag{2.55}$$

with the kinetic momenta $\hat{p}_x = -i\hbar\partial/\partial x$ and $(\hat{p}_y + eA_y) = (-i\hbar\partial/\partial y + eBx)$. The solution of the Schrödinger equation $\psi_{nk}(x, y)$ can be separated in $\psi_{nk}(x, y) = e^{ik_y y}\varphi_{nk}(x)$, which leads to the eigenvalue problem for the function $\varphi_{nk}(x)$:

$$\frac{1}{2m^*}\left[-\hbar^2\frac{\partial^2}{\partial x^2} + \left(-i\hbar\frac{\partial}{\partial y} + eBx\right)^2\right]e^{ik_y y}\varphi_{nk}(x) = Ee^{ik_y y}\varphi_{nk}(x),$$

$$\left[-\frac{\hbar^2}{2m^*}\frac{\partial^2}{\partial x^2} + \frac{1}{2m^*}(\hbar k_y + eBx)^2\right]e^{ik_y y}\varphi_{nk}(x) = Ee^{ik_y y}\varphi_{nk}(x),$$

$$\left[-\frac{\hbar^2}{2m^*}\frac{\partial^2}{\partial x^2} + \frac{1}{2}m^*\left(\frac{eB}{m^*}\right)^2\left(\frac{\hbar k_y}{eB} + x\right)^2\right]\varphi_{nk}(x) = E\varphi_{nk}(x). \tag{2.56}$$

With

$$\omega_c = \frac{eB}{m^*} \tag{2.57}$$

as the cyclotron frequency, the center coordinate given by $x_0 = -k_y l_m^2$ and the magnetic length defined by

$$l_m = \sqrt{\frac{\hbar}{eB}}, \tag{2.58}$$

we arrive at the Schrödinger equation of a harmonic oscillator:

$$\left[-\frac{\hbar^2}{2m^*}\frac{\partial^2}{\partial x^2} + \frac{m^*}{2}\omega_c^2(x - x_0)^2\right]\varphi_{nk}(x) = E\varphi_{nk}(x). \tag{2.59}$$

The magnetic confinement potential is given by

$$V(x) = \frac{m^*}{2}\omega_c^2(x - x_0)^2. \tag{2.60}$$

The eigenfunctions and energy eigenvalues of equation (2.59) can most conveniently be gained by making use of so-called ladder operators [40], which are defined as

$$\hat{a}^\dagger = \frac{l_m}{\sqrt{2}}\left(\frac{x - x_0}{l_m^2} - \frac{\partial}{\partial x}\right), \tag{2.61}$$

$$\hat{a} = \frac{l_m}{\sqrt{2}}\left(\frac{x - x_0}{l_m^2} + \frac{\partial}{\partial x}\right). \tag{2.62}$$

As we will see below, using these operators the energy of the system can be lowered or increased by an energy quantum. Later on we will employ the more generalized form, the creation and annihilation operators, to create and annihilate a particle, respectively. Making use of the ladder operators, the Hamiltonian operator can be written very compactly

$$\hat{H} = \left(\hat{a}^\dagger\hat{a} + \frac{1}{2}\right)\hbar\omega_c. \tag{2.63}$$

If we express the discrete eigenstates in the Dirac notation by $|n\rangle$, with $n = 0, 1, 2, \ldots$, the Schrödinger equation reads as

$$\hat{H}|n\rangle = E_n|n\rangle, \tag{2.64}$$

with E_n the energy eigenvalues of the state $|n\rangle$. The operators $\hat{a}$ and $\hat{a}^\dagger$ obey the commutation relation

$$[\hat{a}, \hat{a}^\dagger] = \hat{a}\hat{a}^\dagger - \hat{a}^\dagger\hat{a} = 1, \tag{2.65}$$

which is directly obtained from the commutation relation of the spatial and momentum operators. In order to find the relation between adjacent eigenstates we apply $a^\dagger$ to the Schrödinger equation (2.64)

$$\hat{a}^\dagger\hat{H}|n\rangle = \hbar\omega_c\hat{a}^\dagger\left(\hat{a}^\dagger\hat{a} + \frac{1}{2}\right)|n\rangle = E_n\hat{a}^\dagger|n\rangle. \tag{2.66}$$

By employing the commutator relation (2.65) the relation $\hat{a}^\dagger\hat{a}^\dagger\hat{a} = \hat{a}^\dagger\hat{a}(\hat{a}^\dagger - 1)$ holds. This brings us to

$$\hbar\omega_c\left(\hat{a}^\dagger\hat{a}\hat{a}^\dagger - \frac{1}{2}\hat{a}^\dagger\right)|n\rangle = E_n\hat{a}^\dagger|n\rangle, \tag{2.67}$$

which can be written as

$$\hbar\omega_c\left(\hat{a}^\dagger\hat{a} - \frac{1}{2}\right)\hat{a}^\dagger|n\rangle = E_n\hat{a}^\dagger|n\rangle. \tag{2.68}$$

By using the definition of $\hat{H}$ given by equation (2.63), one ends up with a similar Schrödinger equation as given by equation (2.64)

$$\hat{H}\hat{a}^\dagger|n\rangle = (E_n + \hbar\omega_c)\hat{a}^\dagger|n\rangle, \tag{2.69}$$

with the only difference that the energy eigenvalue is increased by $\hbar\omega_c$ for the eigenstate $\hat{a}^\dagger|n\rangle$ compared to the energy E_n for $|n\rangle$. Thus, application of $\hat{a}^\dagger$ transforms the state $|n\rangle$ to the next higher state $|n + 1\rangle$. In complete analogy it can be shown that applying the operator $\hat{a}$ to $|n\rangle$ results in a decrease of the energy eigenvalue by $\hbar\omega_c$ and consequently to a transformation to the next lower state $|n - 1\rangle$. With the proper normalization one can write [40]

$$\hat{a}^\dagger|n\rangle = \sqrt{n + 1}|n + 1\rangle, \tag{2.70}$$
$$\hat{a}|n\rangle = \sqrt{n}|n - 1\rangle. \tag{2.71}$$

Once knowing the ground state $|0\rangle$ any higher state can be generated by means of

$$|n\rangle = \frac{1}{\sqrt{n!}}(\hat{a}^\dagger)^n|0\rangle. \tag{2.72}$$

Using the relations of the operators $\hat{a}$ and $\hat{a}^\dagger$ given by (2.71) and (2.70), respectively, one gets

$$\hat{a}\hat{a}^\dagger|n\rangle = n|n\rangle. \tag{2.73}$$

Inserting this into the Hamiltonian expressed by equation (2.63) directly results in the energy eigenvalues [40]

$$E_n = \hbar\omega_c\left(n + \frac{1}{2}\right). \tag{2.74}$$

The ground state $|0\rangle$ can be calculated by making use of $\hat{a}|0\rangle = 0$, i. e.

$$\frac{l_m}{\sqrt{2}}\left(\frac{x - x_0}{l_m^2} + \frac{\partial}{\partial x}\right)\varphi_{0k}(x) = 0 \tag{2.75}$$

which results in

$$\varphi_{0k}(x) = \frac{\pi^{1/4}}{\sqrt{b}} \exp\left(-\frac{(x-x_0)^2}{2b^2}\right),$$ (2.76)

with $b = \sqrt{\hbar/m^*\omega_c}$ the characteristic length of the harmonic oscillator. According to equation (2.72), applying $\hat{a}^\dagger$ multiple times can then be used to obtain the following n^{th} eigenfunction

$$\varphi_{nk}(x) = \frac{1}{\sqrt{b}} \frac{\pi^{-1/4}}{\sqrt{2^n n!}} H_n\left(\frac{x-x_0}{b}\right) \exp\left(-\frac{(x-x_0)^2}{2b^2}\right) \quad n = 0,1,2,\ldots,$$ (2.77)

with $H_n(x)$ being the Hermite polynomials of integer order n.

2.8.2 Shubnikov–de Haas oscillations

In Figure 2.33 the energy eigenvalues, also called Landau levels, for a two-dimensional electron gas are shown as a function of the magnetic field. The level splitting at larger magnetic fields originates from the Zeeman effect. The effect will be discussed in detail in the next chapter. Each time a Landau level crosses the Fermi level a maximum is observed in the magnetoresistance $R_{xx} = V_{xx}/I$. The oscillations which are periodic in $1/B$ are called Shubnikov–de Haas oscillations.

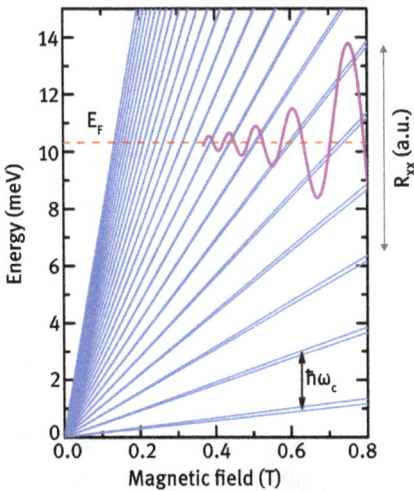

Figure 2.33: Landau level fan, i. e. Landau levels vs. magnetic field, of a two-dimensional electron gas. The small splitting of the Landau levels is due to the Zeeman effect. Each time the Landau levels cross the Fermi energy a maximum in the oscillating magnetoresistance R_{xx} is observed.

For a two-dimensional system with a parabolic energy-momentum dispersion the energy eigenvalues are given by equation (2.74). The energy difference between adjacent levels is constant and given by

$$\Delta E = E_{n+1} - E_n = \hbar \omega_c. \tag{2.78}$$

A maximum of the magnetoresistance $R(B)$ is observed whenever a level E_n crosses the Fermi level E_F, i. e.

$$E_n \doteq E_F \Rightarrow E_F = \frac{e\hbar B}{m^*}\left(n + \frac{1}{2}\right). \tag{2.79}$$

Here we made use of the definition of ω_c. We can rewrite the expression given above as

$$n + \frac{1}{2} = \frac{m^* E_F}{e\hbar} \frac{1}{B}, \tag{2.80}$$

which results in a frequency for the oscillations of $R(B)$ plotted vs. $1/B$ of

$$f_{1/B} = \frac{m^* E_F}{e\hbar}. \tag{2.81}$$

Since equation (2.80) corresponds to maxima in the resistance we can write for the Shubnikov–de Haas oscillations in the longitudinal magnetoresistance:

$$\Delta R_{xx} \sim \cos\left[2\pi\left(f_{1/B}\frac{1}{B} - \frac{1}{2}\right)\right] \tag{2.82}$$

$$\sim \cos\left(2\pi f_{1/B}\frac{1}{B} - \pi\right).$$

The last expression directly shows that the oscillations of R_{xx} contain a phase factor of π, which is different for a system with linear energy-momentum dispersion, as discussed later on in Section 11.6.4.

Using the relation between the Fermi energy and carrier concentration given by equation (2.12) one finds that the frequency $f_{1/B}$ is directly proportional to the carrier concentration

$$f_{1/B} = \frac{h}{2e}\frac{2}{g_s}n_{2D}. \tag{2.83}$$

Thus, from the frequency of the Shubnikov–de Haas oscillations plotted versus $1/B$, the carrier concentration can be extracted. We will see later, that in the presence of spin-orbit coupling the energy spectrum shown in Figure 2.33 is modified. This also implies that the oscillation pattern of R_{xx} changes.

2.8.3 Magnetic edge states

In Section 2.8.1 we assumed an infinite two-dimensional electron gas exposed to a perpendicular magnetic field. Of course if a real sample is considered, the two-dimensional electron gas will be confined. Usually a boundary is described by a potential barrier. Let us consider the sample boundaries in the x-direction, by assuming a potential $V(x)$, which increases at the boundaries. Inserting this potential into the Hamiltonian for a free electron in a magnetic field, equation (2.55) gives

$$\hat{H} = \frac{1}{2m^*}[\hat{p}_x^2 + (\hat{p}_y + eBx)^2] + V(x). \tag{2.84}$$

Close to the boundary the total confinement potential is not only given by the parabolic magnetic potential but also by the boundary potential (cf. Figure 2.34 (a)). Generally, the confinement is stronger close to the boundaries leading to an increase of the energy levels. The energy of the Landau levels follows the potential profile, as shown schematically in Figure 2.34 (b).

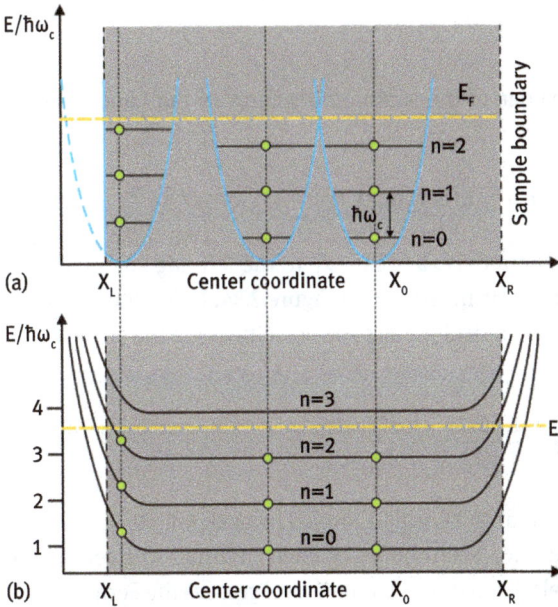

Figure 2.34: (a) Confinement of carriers due to the magnetic field (parabolic contribution) and due to the edge potential. Owing to the combination of both contributions at the edge, the confinement is stronger, compared to the center part of the sample. This leads to an increase of the level separation. (b) Schematic illustration of the energy spectrum of a two-dimensional electron gas in a magnetic field as a function of the harmonic-oscillator wave function with the center coordinate at x_0 with infinite potential walls at x_L and x_R.

In the classical image the electrons move in skipping orbits along the upper and lower edge of the sample, owing to the deflection at the edge, as depicted in Figure 2.35 (a), whereas, at the center the electrons classically move in circles with the cyclotron radius [41].

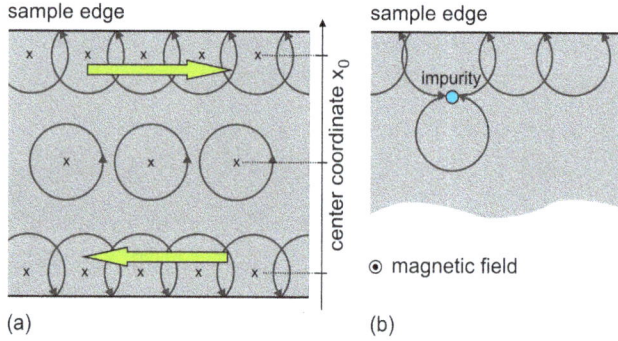

(a)   (b)

Figure 2.35: (a) Classical skipping orbits of electrons along the upper and lower interface of the sample. At the center of the sample the electrons move in closed circles. The center coordinates of the orbits are indicated. (b) Suppression of backscattering over distances being large compared to the cyclotron radius.

For a hard wall potential at the edge of the sample the energy of the Landau levels depends on the center coordinate x_0:

$$E_{nk} = E(n, x_0(k_y)). \tag{2.85}$$

With decreasing distance to the boundary, i. e. $x_0 - x_L$ to the left edge or $x_R - x_0$ to the right edge, the energy increases, as illustrated in Figure 2.34. Using the above expression for E_{nk}, in a quantum mechanical picture the velocity along the edge can be expressed as

$$v_{nk} = \frac{1}{\hbar} \frac{dE_{nk}}{dk_y} = \frac{1}{\hbar} \frac{dE_{nk}}{dx_0} \frac{dx_0}{dk_y}, \tag{2.86}$$

which is proportional to the slope of the Landau level. The assignment of the edge channels is illustrated in Figure 2.36. The number of edge channels is given by the number of occupied Landau levels. Each time a Landau level crosses the Fermi energy E_F an edge channel is formed.

By changing the magnetic field strength the Landau level separation given by $\hbar\omega_c$ is changed. This alters the number of occupied Landau levels and correspondingly the number of edge channels. This is shown in Figure 2.37 (a), where three edge channels are present. With increasing magnetic field the Landau level separation increases, so that now only two Landau levels are occupied below the Fermi energy. As shown in

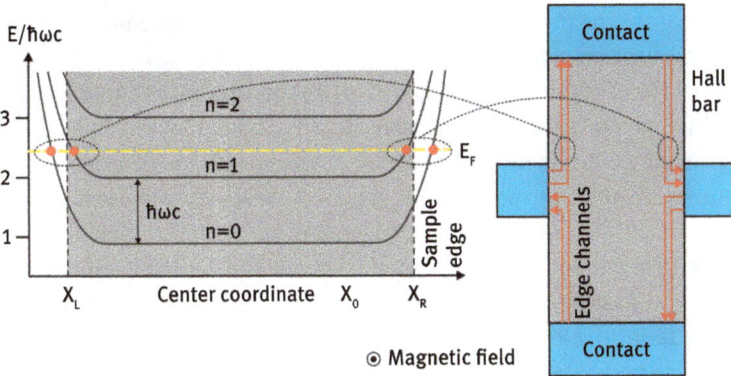

Figure 2.36: Assignment of edge channels: Each time a Landau level is raised at the boundary, it crosses the Fermi level, and an edge channel is formed. For the case shown here, two Landau levels are occupied in the center of the sample, and thus two edge channels are formed. The edge channels follow the sample boundary. Similar to the classical case, the direction of propagation is determined by the magnetic field orientation. Note that the center of the cyclotron motion can be outside the physical boundary of the sample.

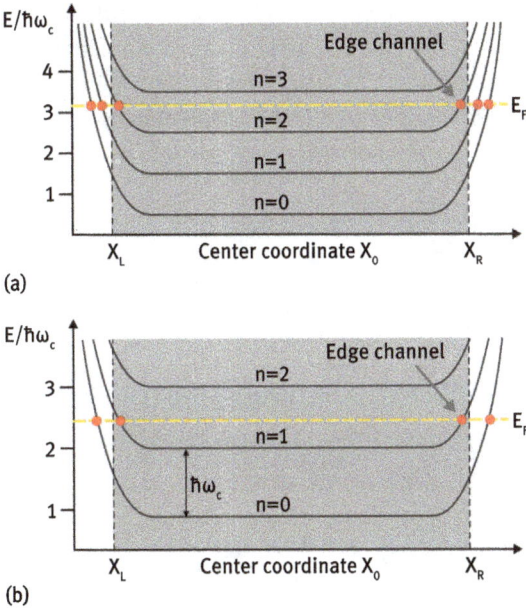

(a)

(b)

Figure 2.37: Edge channels at the boundary. (a) Three edge channels are present at each side of the sample. (b) After increasing the magnetic field, two edge channels are left.

Figure 2.37 (b), two edge channels are formed. Generally, by increasing the magnetic field the energetic separation of the Landau levels increases and thus the number of edge channels decreases. The total number of edge channels N is determined by the Fermi energy, i. e. the electron concentration.

Only electrons at the edge contribute to a current. At opposite edges of the sample the velocities are in opposite directions. Electrons in bulk Landau levels have no net velocity, since the energy does not dependent on x_0, and thus dE_{nk}/dx_0 is zero. Consequently, we can regard the transport at high magnetic fields as being exclusively carried by channels at the edge of the sample. As long as one of the Landau levels in the middle of the sample does not cross the Fermi level, no transfer of electrons from one edge to the other is possible, since no states are available for the scattering process. Thus, back-scattering of electrons moving in one direction in an edge channel to an edge channel on the opposite side with electrons moving in the opposite direction is suppressed. The suppression of back-scattering can also be understood in a classical picture, as illustrated in Figure 2.35 (b). After scattering at an impurity, the electron will resume its initial direction of propagation due to the bending of the trajectories by the Lorentz force. Thus an electron in an edge channel on one side of the sample cannot be scattered into an edge channel on the opposite side, due to this mechanism, as long as the cyclotron radius is small compared to the mean distance of the scattering centers.

The edge channels are of one-dimensional nature, and thereby the density of states along a Landau level E_n corresponds to the momentum density of states per unit length for a one-dimensional conductor $D(k) = 1/\pi$. Here we took spin degeneracy into account but regarded motion in only one direction. The density of states $D_n(E)$ of the n-th edge channel is related to the velocity by

$$D_n(E) = D(k)\left(\frac{dk}{dE}\right)_n = \frac{1}{\pi\hbar v_{nk}}. \tag{2.87}$$

We first consider a situation with only a single edge channel, as depicted in Figure 2.38.

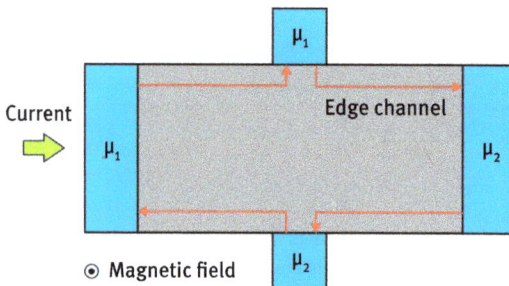

Figure 2.38: Hall bar with a current I_{1D} flowing from the left-hand to the right-hand contact. The Hall voltage is measured between the upper and lower contact. It is assumed that only a single edge channel is present.

The net current resulting from the current difference in the upper and lower edge channels is given by

$$I_{1D} = e \int_{\mu_2}^{\mu_1} D_n(E) v_{nk}(E)\, dE. \tag{2.88}$$

Here it was assumed that the upper edge channel is occupied up to the electrochemical potential μ_1, while the lower edge channel is occupied up to μ_2. Since the density of states is inversely proportional to the velocity, the current can be written as

$$I_{1D} = \frac{2e}{h}(\mu_1 - \mu_2). \tag{2.89}$$

As shown in Figure 2.38, the upper voltage probe acquires the electrochemical potential μ_1, owing to the supply of carriers from the edge channel occupied up to μ_1. Similarly, the lower contact acquires the electrochemical potential μ_2. Thus, the measured voltage, i. e. the Hall voltage V_H, is given by

$$V_H = (\mu_1 - \mu_2)/e. \tag{2.90}$$

Finally, the Hall resistance R_H is given by

$$R_H = \frac{V_H}{I_{1D}} = \frac{(\mu_1 - \mu_2)/e}{I_{1D}} = \frac{h}{2e^2}, \tag{2.91}$$

where the expression for the current in the edge channel was inserted. As can be inferred from equation (2.91), the Hall resistance only depends on basic constants. For more than one edge channel involved in the transport the net current increases with the number N of edge channels. Instead of equation (2.91) one obtains

$$R_H = \frac{(\mu_1 - \mu_2)/e}{I_{1D}} = \frac{h}{2e^2}\frac{1}{N}. \tag{2.92}$$

As we know from the previous discussion, the number of edge channels decreases with increasing magnetic field because of the increasing energetic distance between the Landau levels. This results in a stepwise increase of the Hall resistance with increasing magnetic field. The phenomenon is called the quantum Hall effect and was discovered by von Klitzing [42]. The stepwise increase of R_H can be seen in Figure 2.39, where a typical quantum Hall effect measurement on a two-dimensional electron gas in an InGaAs/InP heterostructure is shown.

Clear plateaus are observed in the Hall resistance R_H corresponding to the values given in equation (2.92). The appearance of the plateaus in the quantum Hall effect is due to the aforementioned suppression of backscattering of electrons propagating in the edge channel.

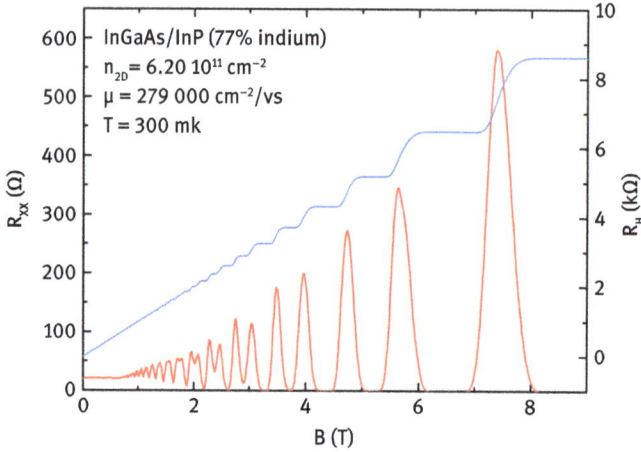

Figure 2.39: Quantum Hall effect measurements and Shubnikov–de Haas oscillations of a two-dimensional electron gas in an InGaAs/InP heterostructure.

As can be seen in Figure 2.39, the longitudinal resistance R_{xx} shows $1/B$-periodic oscillations, which can be assigned to the Shubnikov–de Haas oscillations discussed in Sect. 2.8.2. From the oscillation period the carrier concentration can be extracted by using equation (2.83). At larger magnetic fields R_{xx} is zero in between two peaks. This can be explained by the suppression of backscattering. As long as the Fermi level is in between two Landau levels, backscattering from one edge to the other is suppressed. Since no electrons get lost from or added to an edge channel, its electrochemical potential is constant along the edge. Thus, when the voltage drop is measured between two contacts along one edge, no voltage difference is found. We can assign a so-called filling factor to the magnetic field position where the Shubnikov–de Haas oscillations have a minimum or where the signal is zero at large magnetic field. Here, the Fermi level is in between two Landau levels, i. e. the Landau levels below the Fermi level are completely filled. The filling factor corresponds to the number of fully occupied Landau levels.

2.9 Summary

- In semiconductors the transport is carried by electrons and holes. The carrier concentration and thus the conductance can be enhanced by doping.
- At the boundary of a semiconductor heterostructure a two-dimensional electron gas can be formed. By modulation doping the mobility and elastic mean free path is increased.
- In ballistic one-dimensional structures the conductance is quantized in steps of $2e^2/h$.

- In quantum dot structures the electrons are confined in all three directions. The transport is governed by single electron tunneling and Coulomb blockade.
- In high-mobility two-dimensional electron gases the quantum Hall effect can be observed. The transport can be explained within the edge channel model within the framework of the Landauer–Büttiker model.

Exercises

Problem 2.1. The Hamiltonian of a quantum point contact with a rectangular opening is given by equation (2.42). Here a one-dimensional quantum well with infinitely high barriers at $y = 0$ and a was assumed, with a potential $V(y) = 0$ within the quantum well. Solve the Schrödinger equation under these boundary conditions. Assume a width of the well $a = 10$ nm and an effective electron mass of $m^* = 0.063m_0$ for GaAs. Calculate the binding energy of the first three levels in eV.

Problem 2.2. Derive an expression for the density of states for a two-dimensional system at a certain energy E.

Problem 2.3. The electron density of copper is approximately $n_{3D} = 8.4 \times 10^{22}$ cm^{-3}, and the specific resistance is $\rho = 1.7 \times 10^{-6}$ Ω cm. Calculate the elastic scattering time τ_e and the electron mobility μ_e. We assume that the electron mass corresponds to the free electron mass m_0.

Problem 2.4. Let us assume a point contact in a two-dimensional electron system, where the constriction possesses a potential step of V_0 referred to the reservoirs on both sides. The two-dimensional reservoirs are filled up to the Fermi energy E_F. Calculate an expression for the geometrical acceptance angle for electrons with wave vector $|k| = k_F$ passing the point contact opening.

Problem 2.5. Show explicitly that by applying the ladder operator $\hat{a}$ to a harmonic oscillator state $|n\rangle$ the energy E_n is lowered by the energy quantum $\hbar\omega_c$.

3 Magnetism in solids

3.1 Definitions and basics

Magnetic materials, in particular metallic materials, are very important for state of the art information technology. So far, these materials are mostly employed in data storage, while data processing is mainly performed by semiconductor materials. Typical applications for magnetic materials are hard disc drives for mass storage or magnetic random access memories (MRAM) as nonvolatile memories. In semiconductor spintronic devices magnetic materials are mostly used as spin injectors and detectors. Below we will introduce the basic physics of magnetic materials starting with para- and diamagnetism and finally discuss the ferromagnetic phase. The general properties of these magnetic states are also relevant for describing the characteristics of diluted magnetic semiconductors. These materials are introduced in the next chapter.

3.1.1 Definitions

Let us begin with defining some quantities describing magnetic phenomena. We first consider a situation where no additional material is present. In this case we can relate the magnetic field, the outer induction $\vec{B}_0$, to the magnetic field strength $\vec{H}$ by

$$\vec{B}_0 = \mu_0 \vec{H}, \tag{3.1}$$

with μ_0 the induction constant given by

$$\mu_0 = 4\pi \times 10^{-7} \frac{\text{Vs}}{\text{Am}}. \tag{3.2}$$

The magnetic field strength is usually produced by a magnetic coil. If n is the number of windings per meter and the current through the coil is I, then H is given by $H = nI$. The SI unit for the magnetic field is Tesla, $[B] = \text{T} = \text{Vs/m}^2$, while the unit for the magnetic field strength is $[H] = \text{A/m}$.

3.1.2 Magnetization

The magnetic state of matter is described by the magnetization

$$\vec{M} = \vec{\mu}\frac{N}{V}, \tag{3.3}$$

with $\vec{\mu}$ the magnetic moment, N the number of identical magnetic moments and V the volume. As shown in Figure 3.1, classically, for a current loop the magnetic moment is defined as

$$\vec{\mu} = I \cdot \vec{F}, \tag{3.4}$$

where I is the current in the loop and $\vec{F}$ the area surrounded by the loop.

https://doi.org/10.1515/9783110639001-003

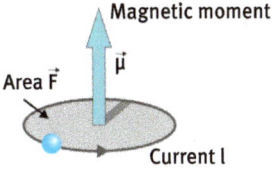

Figure 3.1: Magnetic moment $\vec{\mu}$ of a current loop. The current encircles the area $\vec{F}$.

The magnetic moment $\vec{\mu}$ is a vector pointing along the normal vector of the plane $\vec{F}$. For an electron moving around a circle with radius r the magnetic moment can be written as

$$\vec{\mu} = -\frac{1}{2}e\vec{r} \times \vec{v}, \tag{3.5}$$

with $\vec{v}$ being the electron velocity. Here we assumed $-e$ with ($e > 0$) for the electron charge, so that the electron velocity is opposite to the current direction. The corresponding microscopic mechanism is the closed-loop charge current in an atom due to the angular momentum of the electrons in the atomic shell. Thus, as we will see below, $\vec{\mu}$ will depend on the angular momentum quantum number. The energy of an object with a magnetic moment in a magnetic field is given by

$$U = -\vec{\mu} \cdot \vec{B}. \tag{3.6}$$

In solid state materials the magnetization often depends linearly on the field B_0:

$$\mu_0 \vec{M} = \chi \vec{B}_0, \tag{3.7}$$

where χ is the susceptibility. Both $\vec{H}$ and $\vec{M}$ contribute to the total magnetic field $\vec{B}$:

$$\vec{B} = \mu_0(\vec{H} + \vec{M}) = \vec{B}_0 + \chi \vec{B}_0 = (1 + \chi)\vec{B}_0 = \mu_0(1 + \chi)\vec{H}. \tag{3.8}$$

3.1.3 Magnetic moments of electrons in atomic orbitals

We first discuss the paramagnetism due to the orbital angular momentum. The orbital angular momentum operator is defined as

$$\hat{\vec{L}} = -i\hbar(\hat{\vec{r}} \times \hat{\vec{\nabla}}) = \hat{\vec{r}} \times \hat{\vec{p}}, \tag{3.9}$$

with $\hat{\vec{r}}$ being the position operator and $\hat{\vec{p}} = -i\hbar\vec{\nabla}$ the momentum operator. The orbital angular momentum is quantized and characterized by its quantum number l, where $l = 0, 1, 2, \ldots$ For a given l the z-component of the angular momentum is also quantized, with values m ranging from

$$m = -l, -l + 1, \ldots, l - 1, l. \tag{3.10}$$

Applying the operators $\hat{L}^2$ and $\hat{L}_z$ to a state $|\psi_{l,m}\rangle$ with the quantum numbers l, m the following eigenvalue equations hold:

$$\hat{L}^2|\psi_{l,m}\rangle = l(l+1)\hbar^2|\psi_{l,m}\rangle, \tag{3.11}$$

$$\hat{L}_z|\psi_{l,m}\rangle = m\hbar|\psi_{l,m}\rangle. \tag{3.12}$$

In complete analogy to equation (3.5) the magnetic moment of a state with angular momentum $\hat{\vec{L}}$ can be expressed as

$$\hat{\vec{\mu}} = -\frac{e}{2m_0}\hat{\vec{r}} \times \hat{\vec{p}} = -\frac{\mu_B}{\hbar}\hat{\vec{L}}. \tag{3.13}$$

The magnetic moment and the orbital angular momentum are connected by the Bohr magneton:

$$\mu_B = \frac{e\hbar}{2m_0} = 9.2741 \times 10^{-24}\,\frac{J}{T}. \tag{3.14}$$

Usually the orbitals of an atom are occupied by many electrons, with a total orbital angular momentum $\hat{\vec{L}}$ given by the sum of all individual orbital angular momenta:

$$\hat{\vec{L}} = \sum_i \hat{\vec{r}}_i \times \hat{\vec{p}}_i = \sum_i \hat{\vec{L}}_i. \tag{3.15}$$

In this case the total magnetic moment of an atom is given by

$$\hat{\vec{\mu}} = -\frac{e}{2m_0}\sum_i \hat{\vec{r}}_i \times \hat{\vec{p}}_i = -\frac{\mu_B}{\hbar}\hat{\vec{L}}. \tag{3.16}$$

3.1.4 The electron spin

Beside the orbital angular momentum, the electron also has an intrinsic angular momentum, the electron spin. The corresponding quantum number is half integer, and thus $s = 1/2$. The state of the spin can be described by a vector, the so-called spinor. The two states can be represented by the following vectors, corresponding to spin up and down along the z-direction:

$$|\uparrow\rangle = \begin{pmatrix} 1 \\ 0 \end{pmatrix}, \quad |\downarrow\rangle = \begin{pmatrix} 0 \\ 1 \end{pmatrix}. \tag{3.17}$$

These spinors are the eigenvectors of the spin operator of the z-direction:

$$\hat{S}_z = \frac{\hbar}{2}\hat{\sigma}_z, \tag{3.18}$$

with σ_z being one of the Pauli spin matrices given by

$$\sigma_x = \begin{pmatrix} 0 & 1 \\ 1 & 0 \end{pmatrix}, \quad \sigma_y = \begin{pmatrix} 0 & -i \\ i & 0 \end{pmatrix}, \quad \sigma_z = \begin{pmatrix} 1 & 0 \\ 0 & -1 \end{pmatrix}. \tag{3.19}$$

The eigenstates of the spin operators of the x- and y-directions are given by superpositions of the spinors given in equation (3.17), e. g. the eigenstates for the x-direction are given by

$$\frac{1}{\sqrt{2}}(|\uparrow\rangle + |\downarrow\rangle) = \frac{1}{\sqrt{2}}\begin{pmatrix} 1 \\ 1 \end{pmatrix}, \quad \frac{1}{\sqrt{2}}(|\uparrow\rangle - |\downarrow\rangle) = \frac{1}{\sqrt{2}}\begin{pmatrix} 1 \\ -1 \end{pmatrix}. \tag{3.20}$$

This can be easily confirmed by multiplying these states with the spin operator $\hat{S}_x = (\hbar/2)\hat{\sigma}_x$. For the sake of completeness, the 2×2 unit matrix is defined as

$$\sigma_0 = \begin{pmatrix} 1 & 0 \\ 0 & 1 \end{pmatrix}. \tag{3.21}$$

Generally a spin state can be described by a superposition of the basis vectors given in equation (3.17):

$$|s\rangle = c_0|\uparrow\rangle + c_1|\downarrow\rangle = \begin{pmatrix} c_0 \\ c_1 \end{pmatrix}, \tag{3.22}$$

with $c_0^2 + c_1^2 = 1$. Besides the magnetic moment due to the orbital angular momentum the electron spin also results in a magnetic moment

$$\hat{\mu}_s = -g_0 \frac{\mu_B}{\hbar} \hat{\vec{S}}, \tag{3.23}$$

with g_0 being the Landé g-factor

$$g_0 = 2.0023, \tag{3.24}$$

in case of a free electron. For a spin-1/2 particle and this value of the g-factor the magnetic moment corresponds to the Bohr magneton $\mu_s \approx \mu_B$. In a solid state material the g-factor can largely deviate from the value of free electrons. The two spin states are illustrated in Figure 3.2 (a).

Based on the expression for the magnetic energy given by equation (3.6) the Zeeman Hamiltonian for an electron spin in an external magnetic field can be written as

$$\hat{H}_Z = \frac{1}{2}\mu_B g_0 \hat{\vec{\sigma}} \cdot \vec{B}, \tag{3.25}$$

where $\hat{\vec{\sigma}} = (\hat{\sigma}_x, \hat{\sigma}_y, \hat{\sigma}_z)$ contains the Pauli spin matrices. If we assume that the magnetic field of strength B_0 is applied along the z-direction, the time-independent Schrödinger equation is given by

$$\frac{1}{2}\mu_B g_0 \begin{pmatrix} B_0 & 0 \\ 0 & -B_0 \end{pmatrix} \begin{pmatrix} c_0 \\ c_1 \end{pmatrix} = E \begin{pmatrix} c_0 \\ c_1 \end{pmatrix}, \tag{3.26}$$

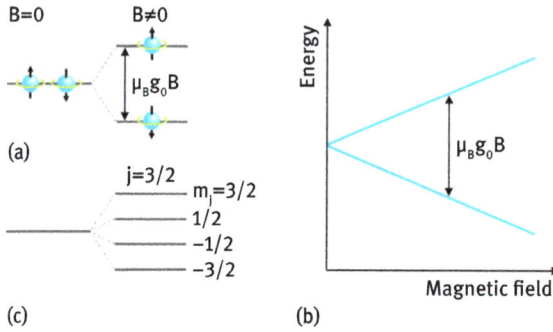

Figure 3.2: (a) Splitting in a magnetic field for electrons with opposite spins. (b) Linearly increasing spin splitting due to the Zeeman effect. (c) Spin splitting in a magnetic field for a total angular momentum $j = 3/2$.

assuming a general spin state given by equation (3.22). Since both components of the spinor are decoupled, the eigenstates are simply $|\uparrow\rangle$ and $|\downarrow\rangle$ with the energy eigenvalues

$$E_Z = \pm\frac{1}{2}\mu_B g_0 B_0. \tag{3.27}$$

The linear energy splitting of $s\mu_B g_0 B_0$ with increasing magnetic field for electrons with opposite spin orientations is shown in Figure 3.2 (b).

We can also generalize the Zeeman Hamiltonian for a magnetic field $\vec{B}$ pointing in an arbitrary direction. Here, it is most convenient to use polar coordinates, with the polar and azimuthal angles, θ and ϕ, respectively, as given in Figure 3.3.

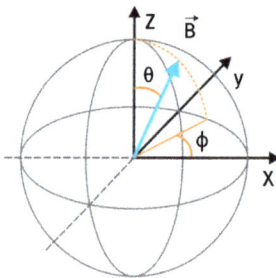

Figure 3.3: External magnetic field $\vec{B}$ in polar coordinates.

$$\vec{B} = B_0 \begin{pmatrix} \sin\theta\cos\phi \\ \sin\theta\sin\phi \\ \cos\theta \end{pmatrix}. \tag{3.28}$$

In this case the Zeeman Hamiltonian can be expressed as

$$\hat{H}_Z = \frac{1}{2}\mu_B g_0 B_0 \begin{pmatrix} \cos\theta & e^{-i\phi}\sin\theta \\ e^{i\phi}\sin\theta & -\cos\theta \end{pmatrix}.$$ (3.29)

It can be straightforwardly shown that the eigenfunctions of the corresponding Schrödinger equation are

$$|\psi_+\rangle = \begin{pmatrix} \cos(\theta/2) \\ e^{i\phi}\sin(\theta/2) \end{pmatrix}$$ (3.30)

and

$$|\psi_-\rangle = \begin{pmatrix} -\sin(\theta/2) \\ e^{i\phi}\cos(\theta/2) \end{pmatrix}.$$ (3.31)

We will use this representation further below when we discuss spin-orbit coupling related subjects.

3.2 Classification

There are many different phenomena related to magnetism. Common to all of them is their pure quantum mechanical nature. The different phenomena are summarized in the scheme shown in Figure 3.4. First, one can distinguish between magnetism originating from bound electrons in a shell and from free conductive electrons. For both the response to an outer field can be either paramagnetic, with $\chi > 0$, or diamagnetic, $\chi < 0$. Both phenomena can exist at the same time. In the paramagnetic case, already existing magnetic moments are aligned in the field, while in the diamagnetic case, counter-flowing currents are generated which shield the field. For sufficiently strong electron-electron interaction, phase transitions can occur. Typical examples are the ferromagnetic or the antiferromagnetic state.

3.3 Paramagnetism

In paramagnetic materials, given magnetic moments are aligned along an outer magnetic field. As mentioned above paramagnetism can be due to existing localized magnetic moments in atoms or due to the alignment of free carriers in a conductor.

3.3.1 Paramagnetism of localized moments

Paramagnetism can originate from the orbital angular momentum $\vec{L}$ of the electrons, described by the orbital angular momentum quantum number l, or from the electron spin $\vec{S}$.

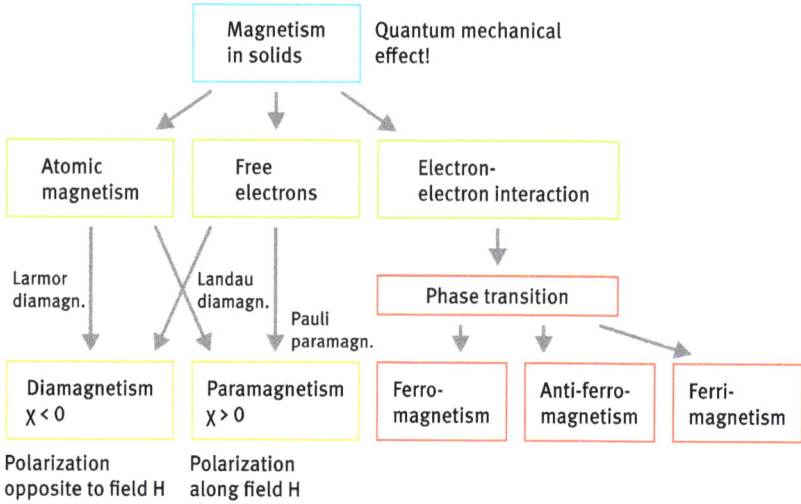

Magnetism in solids — Quantum mechanical effect!

Atomic magnetism | Free electrons | Electron-electron interaction

Larmor diamagn. | Landau diamagn. | Phase transition
Pauli paramagn.

Diamagnetism $\chi < 0$ | Paramagnetism $\chi > 0$ | Ferro-magnetism | Anti-ferro-magnetism | Ferri-magnetism

Polarization opposite to field H | Polarization along field H

Figure 3.4: Overview of magnetic properties and phenomena in solid state materials.

Let us begin with a two level system, i. e. a spin-1/2 system. The occupation of these two levels in thermal equilibrium can be described by the Boltzmann statistics:

$$\frac{n_1}{n} = \frac{\exp\left(\mu B_0/k_B T\right)}{\exp\left(\mu B_0/k_B T\right) + \exp\left(-\mu B_0/k_B T\right)}, \tag{3.32}$$

$$\frac{n_2}{n} = \frac{\exp\left(-\mu B_0/k_B T\right)}{\exp\left(\mu B_0/k_B T\right) + \exp\left(-\mu B_0/k_B T\right)}, \tag{3.33}$$

with n_1, n_2 being the concentrations of occupied states in each level, n the total concentration and μ the magnetic moment. The net magnetization is determined by the difference of both contributions:

$$M = (n_1 - n_2)\mu = n\mu \tanh\left(\frac{\mu B_0}{k_B T}\right), \tag{3.34}$$

with $M_0 = n\mu$ being the saturation magnetization. If the magnetic field is small and/or the temperature is high, so that

$$\frac{\mu B_0}{k_B T} \ll 1 \tag{3.35}$$

holds, one obtains

$$M \cong \frac{n\mu^2 B_0}{k_B T}. \tag{3.36}$$

For the susceptibility defined by $\chi = \mu_0 M/B_0$ one arrives at the so-called Curie law:

$$\chi = \frac{C}{T}, \tag{3.37}$$

where χ is inversely proportional to the temperature, with C the Curie constant. In Figure 3.5 (a) and (b) the dependence of the normalized magnetization is shown as a function of temperature and magnetic field, respectively. At very low temperatures, no thermal excitation occurs, and therefore only the lower state is occupied. As a result, the magnetization is saturated (cf. Figure 3.5 (a)). Furthermore, as can be seen in Figure 3.5 (b), for large magnetic fields the magnetization saturates as well. Here, the reason is that at large fields the level separation is becoming too large, so that no electrons are excited to the upper level.

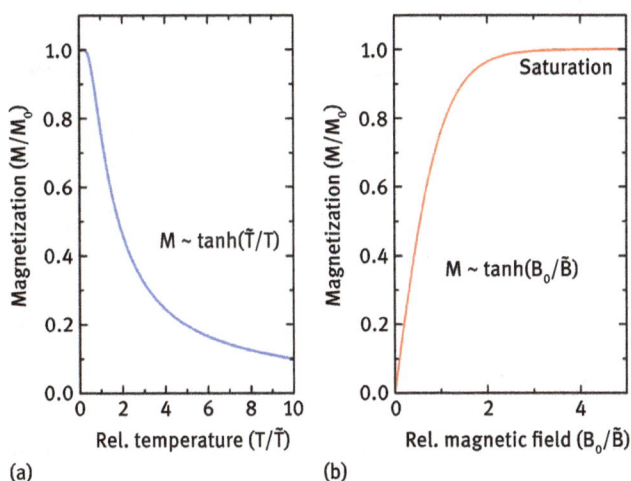

(a) (b)

Figure 3.5: Magnetization normalized to the saturation value M_0 as a function of relative temperature $\tilde{T} = \mu B_0 / k_B$ and magnetic field $\tilde{B} = k_B T / \mu$.

Let us discuss the situation where we have to account for both contributions, i. e. the orbital angular momentum $\hat{L}$ and the spin $\hat{S}$, to the magnetization. The total angular momentum $\hat{J}$ of an electron is given by

$$\hat{J} = \hat{L} + \hat{S}. \tag{3.38}$$

The total angular moment is described by the quantum number $j = l \pm s$ for $l = 1, 2, 3, \dots$ and $j = 1/2$ for $l = 0$, with $s = 1/2$. The magnetic quantum numbers m_j are in the range of $-j$ to j in steps of 1. As an example for a total angular momentum $\hat{J}$ with a quantum number $j = 3/2$ for $l = 1$ and $s = 1/2$ one finds a splitting into four energy levels according to $m_j = +3/2, +1/2, -1/2$, and $-3/2$ (cf. Figure 3.2 (c)).

In case of many electrons in a free atom the total orbital momentum $\vec{L}$, the total spin $\vec{S}$, and the total angular momentum $\vec{J}$ are constituted by the contribution of all electrons:

$$\hat{L} = \sum_i \hat{L}_i, \tag{3.39}$$

$$\hat{\vec{S}} = \sum_i \hat{\vec{S}}_i, \tag{3.40}$$

$$\hat{\vec{J}} = \sum_i \hat{\vec{J}}_i. \tag{3.41}$$

The corresponding magnetic moment $\hat{\vec{\mu}}_j$ can be written as

$$\hat{\vec{\mu}}_j = g_j \mu_B \hat{\vec{J}} \tag{3.42}$$

with the quantization in z-direction

$$(\hat{\vec{\mu}}_j)_z = g_j j \mu_B m_j \tag{3.43}$$

and $m_j = -j, -(j-1), \ldots, j-1, j$. Thus the degeneracy is $(2j+1)$. The corresponding Landé factor is given by

$$g_j = 1 + \frac{j(j+1) + s(s+1) - l(l+1)}{2j(j+1)}. \tag{3.44}$$

For the absolute value of the magnetic moment one obtains

$$|\vec{\mu}_j| = g_j \frac{|e|}{2m_0} \sqrt{j(j+1)} \hbar = g_j \mu_B \sqrt{j(j+1)}. \tag{3.45}$$

The dependence of the magnetization on the temperature and on the magnetic field is described by

$$M = n g_j \mu_B j B_j \left(\frac{g_j j \mu_B B_0}{k_B T} \right), \tag{3.46}$$

with the Brillouin function defined by

$$B_j(x) = \frac{2j+1}{2j} \coth\left(\frac{2j+1}{2j} x \right) - \frac{1}{2j} \coth\left(\frac{1}{2j} x \right). \tag{3.47}$$

In Figure 3.6 the dependence of the normalized magnetization M/M_0 on $x = g_j j \mu_B B_0 / (k_B T)$ is shown. Owing to the larger number of states for larger j the increase of the magnetization with increasing x is weaker.

From the expression of the magnetization equation (3.46) the susceptibility of materials with magnetic atoms can be calculated:

$$\chi = \frac{\partial M}{\partial H} = \mu_0 \frac{\partial M}{\partial B_0}. \tag{3.48}$$

For small fields and/or high temperatures, i. e.

$$g_j \mu_B B_0 \ll k_B T, \tag{3.49}$$

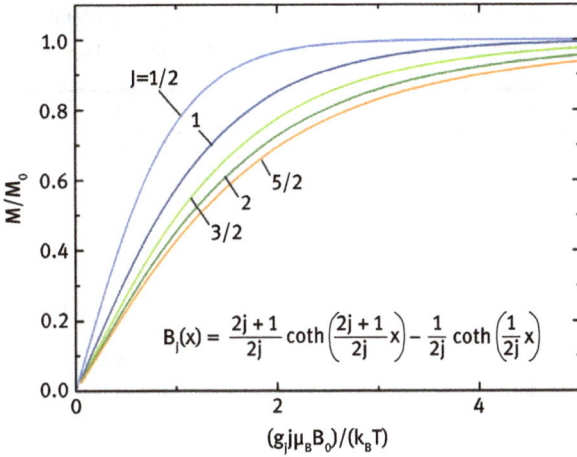

Figure 3.6: Normalized magnetization M/M_0 as a function of $g_j j \mu_B B_0/(k_B T)$.

the Brillouin function can be approximated by

$$B_j(x) \approx \frac{j+1}{3j}x + \mathcal{O}(x^3),\tag{3.50}$$

where $x = g_j j \mu_B B_0/(k_B T)$. Once again we obtain the Curie law:

$$\chi = \mu_0 n \frac{(g_j \mu_B)^2}{3} \frac{j(j+1)}{k_B T}.\tag{3.51}$$

Thus, the susceptibility is again inversely proportional to the temperature.

3.3.2 Hund's rule

By means of the so-called Hund's rules the electron occupation of the different orbitals is determined:
- The total spin s has the maximum value allowed by the Pauli principle.
- The total orbital angular momentum l has a maximum value, which is consistent with the value of s defined above.
- The total angular momentum j is given by $|l - s|$ for a shell which is less than half-filled and $l + s$ for a shell which is more than half filled. In case of a half-filled shell one obtains $l = 0$ and $j = s$ according the to two rules given above.

The orbital filling of $3d$-transition metals is shown in Table 3.1. As one can see there, for Mn the shell is half-filled, resulting in the maximum magnetic moment μ_s. Please note that in a crystal the orbital angular momentum usually does not contribute to the magnetic moment. Transition metals with a large magnetic moment, e. g. Cr, Mn, or Fe, are important as dopants in diluted magnetic semiconductors (see Chapter 4).

Table 3.1: Occupation of the 3d-orbitals of 3d-transition metals, the total spin s, the total orbital angular momentum l, the total angular momentum j, and the magnetic moment of the spin μ_s.

		2	1	0	−1	−2	s	l	j	$\mu_s(\mu_B)$
Sc	$4s^2 3d^1$	↓					1/2	2	3/2	1
Ti	$4s^2 3d^2$	↓	↓				1	3	2	2
V	$4s^2 3d^3$	↓	↓	↓			3/2	3	3/2	3
Cr	$4s^1 3d^5$	↓	↓	↓	↓	↓	5/2	0	5/2	5
Mn	$4s^2 3d^5$	↓	↓	↓	↓	↓	5/2	0	5/2	5
Fe	$4s^2 3d^6$	↓↑	↓	↓	↓	↓	2	2	4	4
Co	$4s^2 3d^7$	↓↑	↓↑	↓	↓	↓	3/2	3	9/2	3
Ni	$4s^2 3d^8$	↓↑	↓↑	↓↑	↓	↓	1	3	4	2
Cu	$4s^1 3d^{10}$	↓↑	↓↑	↓↑	↓↑	↓↑	0	0	0	0

3.3.3 Pauli paramagnetism

In addition to bound electrons, free electrons in a conductor are also a source of paramagnetism, which is called Pauli paramagnetism. In case that no magnetic field is applied, the number of free electrons with opposite spins is identical. If a magnetic field is applied, the energy is shifted depending on the spin orientation according to

$$U = -\mu B = \pm \frac{1}{2}\mu_B g_0 B. \tag{3.52}$$

This situation is shown in Figure 3.7, where the density of states $D(E)$ for electrons of the spin orientation parallel and antiparallel to B is shifted by the energy difference

$$\Delta E = g_0 \mu_B B. \tag{3.53}$$

For each spin orientation the states are filled up to the Fermi energy. Due to the energy difference ΔE the number of electrons of both spin orientations differs by the amount

$$n = \frac{1}{2}D(E_F)\Delta E. \tag{3.54}$$

At low temperatures, so that $k_B T \ll E_F$ holds, the magnetization is given by the product of n and the magnetic moment of the electrons μ:

$$M = n \cdot \mu \tag{3.55}$$

$$= \frac{1}{2}D(E_F)\Delta E \frac{1}{2}g_0 \mu_B \tag{3.56}$$

$$= \frac{1}{2}D(E_F)g_0 \mu_B B_0 \frac{1}{2}g_0 \mu_B. \tag{3.57}$$

The corresponding susceptibility can be expressed by

$$\chi = \mu_0 \frac{M}{B_0} = \mu_0 \frac{g_0^2}{4}\mu_B^2 D(E_F) \propto \mu_0 \mu_B^2 D(E_F). \tag{3.58}$$

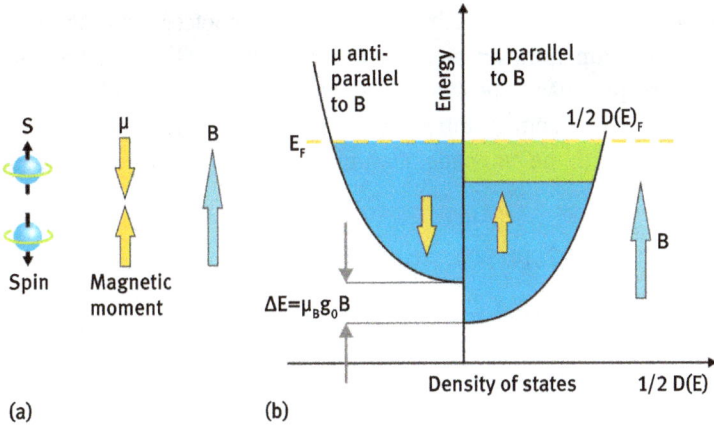

Figure 3.7: (a) Orientation of the spin and corresponding magnetic moments in a magnetic field. (b) Shifting of the density of states for free electrons in magnetic field.

In addition to the paramagnetic contribution, free electrons also have a diamagnetic contribution, which will not be discussed here. In total one obtains for the susceptibility of free electrons

$$\chi \propto \mu_0 \mu_B^2 D(E_F)\left[1 - \frac{1}{3}\left(\frac{m_0}{m^*}\right)^2\right]. \qquad (3.59)$$

Typical values of the susceptibility of conductors with free electrons are relatively low, e. g. for Na the susceptibility is $\chi_{Na} \approx 10^{-6}$ (S. I.). Thus, the response to an outer magnetic field is very small.

3.4 Collective magnetism

From the previous sections we learned that often the para- and diamagnetic contributions are very small. A much stronger response is found if electron-electron interaction comes into play, i. e. in the ferro- or antiferromagnetic state. First, we introduce the quantum mechanical origin of these phenomena, i. e. the exchange interaction. The ferromagnetic state in 3d-transition metals will be discussed later on in detail.

3.4.1 Exchange interaction

From the Pauli principle it follows that the total wave function must be antisymmetric with regard to the exchange of particles. For the sake of simplicity we shall start with a two electron system, or more precisely with the electron wave function of a H_2

molecule. In order to fulfil the Pauli principle, there are two choices. First, the spatial part of the total wave function can be antisymmetric: Ψ_a. In that case the spin contribution of the two spins has to be symmetric. The options are threefold, i. e. a spin-triplet state with the total spin quantum number $s = 1$ fulfils this requirement. Here, the following states with the magnetic quantum numbers $m_s = 1$, 0, and -1 are possible:

$$|T_+\rangle = |\uparrow\uparrow\rangle, \tag{3.60}$$

$$|T_0\rangle = \frac{1}{\sqrt{2}}(|\uparrow\downarrow\rangle + |\downarrow\uparrow\rangle), \tag{3.61}$$

$$|T_-\rangle = |\downarrow\downarrow\rangle. \tag{3.62}$$

This situation is illustrated in Figure 3.8 (a).

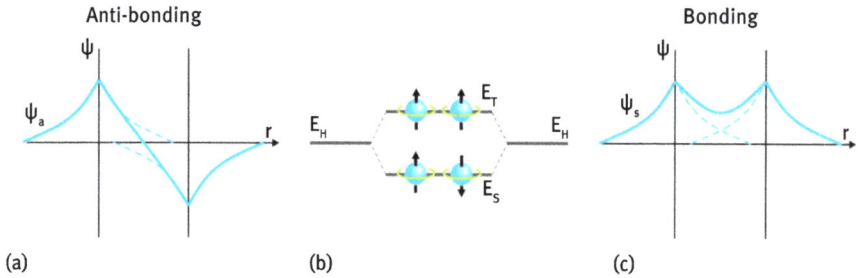

Figure 3.8: Bonding in a H_2 molecule. (a) Spin-triplet antibonding state with an antisymmetric spatial wave function Ψ_a. (b) Energy splitting between the singlet and triplet state. The initial energy of the hydrogen atom is E_H. (c) Spin-singlet bonding state with a symmetric spatial wave function Ψ_s.

The energy of that state is given by

$$E_T = \langle \Psi_a | \hat{H} | \Psi_a \rangle. \tag{3.63}$$

Secondly, the spatial part of the total wave function can be symmetric (cf. Figure 3.8 (c)). Now the spin contribution needs to be antisymmetric, i. e. in a spin singlet state:

$$|S\rangle = \frac{1}{\sqrt{2}}(|\uparrow\downarrow\rangle - |\downarrow\uparrow\rangle), \tag{3.64}$$

with the total spin of $s = 0$. The energy related to that state is given by

$$E_S = \langle \Psi_s | \hat{H} | \Psi_s \rangle. \tag{3.65}$$

In the center between the two hydrogen cores, the symmetric orbital state Ψ_s has a finite electron probability which is not the case for the antisymmetric orbital state Ψ_a.

For the previous case this leads to a better screening of the positively charged hydrogen cores and consequently to a lower energy. In total, energy is gained compared to the two separated H atoms, so that the H_2 molecule is formed. This is the reason why this state is called the bonding state. In contrast, the energy of the state Ψ_a is higher than the initial energy, so that this state is called the antibonding state. The important fact in our context is that the energetic state is directly connected to the spin configuration, i. e. singlet or triplet state. Interestingly, by its roots the energetic difference of the exchange interaction is due to the Coulomb interaction, as outlined above.

The next step is to map the Coulomb interaction to a spin system and extend it to a larger number of particles. This is done by using the parameterized Heisenberg Hamiltonian:

$$\hat{H}_{ex} = -\sum_{i,j} J_{ij} \vec{S}_i \cdot \vec{S}_j. \tag{3.66}$$

Here, $\vec{S}_i$ and $\vec{S}_j$ are two neighboring spins on the lattice site i and j, respectively. The coupling parameter is given by the energy difference between the spin singlet and triplet configuration

$$J_{ij} = E_S - E_T. \tag{3.67}$$

Depending on the sign of J_{ij} two cases can be distinguished:

$$J_{ij} \begin{cases} > 0 & \text{ferromagnetic coupling,} \\ < 0 & \text{anti ferromagnetic coupling} \end{cases} \tag{3.68}$$

In the ferromagnetic case the energy is lowered if the spins are aligned parallel, while in the antiferromagnetic case the energy is lowered for an antiparallel configuration of neighboring spins. A typical example of a Heisenberg system is a linear chain of magnetic $3d$-atoms coupled by exchange interaction between neighboring magnetic moments (cf. Figure 3.9).

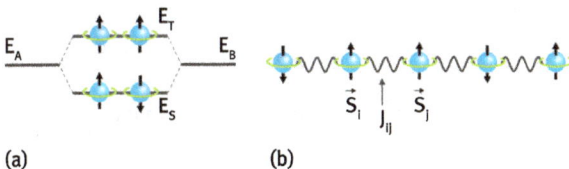

(a) (b)

Figure 3.9: (a) Energy splitting between a singlet and triplet spin state. (b) Linear chain of $3d$-atoms.

The sign and strength of the exchange parameter J_{ij} is determined by the details of the coupling. One can distinguish between the following coupling mechanisms which are illustrated in Figure 3.10:

(a) Direct exchange   Fe, Co, Ni

(b) Super exchange   Eu[4f] – O[2p] – Eu[4f]

(c) Indirect exchange   Gd[4f] – e⁻ – Gd[4f]

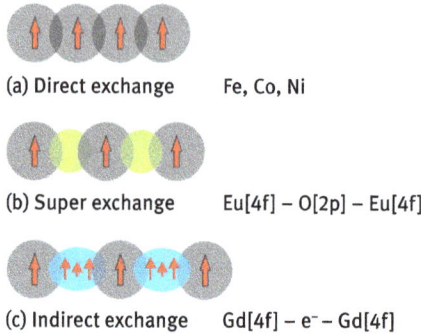

Figure 3.10: Different forms of exchange interaction between localized magnetic moments.

- Direct exchange: this coupling is maintained by the direct overlap of wavefunctions which possess a certain spin. Typical examples are $3d$-transition metals like Fe, Co, or Ni.
- Super exchange: here, localized magnetic moments are coupled by intermediate non-magnetic ions like in EuO.
- Indirect exchange or Ruderman–Kittel–Kasuya–Yoshida (RKKY) interaction [43, 44, 45]: In this case the magnetic moments are coupled by the spin of polarized free electrons as e. g. in Gd.

Depending on the strength and sign of the magnetic coupling different collective magnetic phenomena can be observed (cf. Figure 3.11):

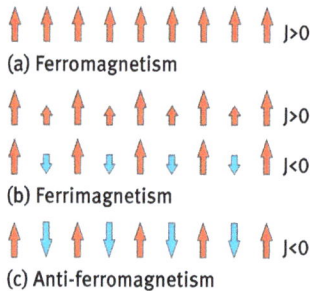

(a) Ferromagnetism

(b) Ferrimagnetism

(c) Anti-ferromagnetism

Figure 3.11: Collective magnetic phenomena: ferromagnetism, ferrimagnetism, and antiferromagnetism.

- Ferromagnetism, with a coupling parameter $J > 0$.
- Ferrimagnetism, where the magnetic moment on neighboring sites have a different magnitude. Here, a net magnetic moment can be observed. The coupling parameter between the neighboring sites can be either positive $J > 0$ or negative $J < 0$.

– Antiferromagnetic coupling with $J < 0$. The neighboring magnetic moments have the same magnitude but opposite orientation.

3.4.2 Stoner model

In a ferromagnetic state at temperatures below the ferromagnetic transition temperature, the Curie temperature T_C, all spins are aligned parallel to each other. Regarding the available states this means that only one kind of states, e. g. the spin-up states are occupied, while the spin-down states are empty. As illustrated in Figure 3.12, for a given number of electrons, states with a higher kinetic energy have to be occupied for the spin polarized case $(T < T_C)$ compared to the spin degenerated case $(T > T_C)$.

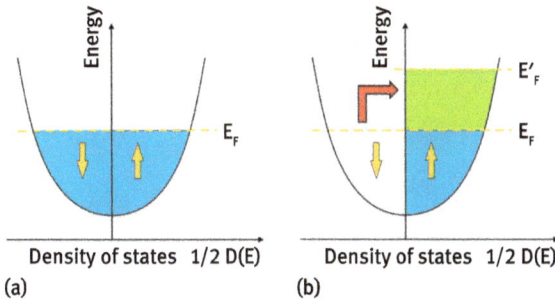

Figure 3.12: (a) Occupation of states with both spin directions at $T > T_C$, with being $D(E)$ the density of states. (b) Corresponding occupation at $T < T_C$ where states of only one spin orientation are occupied.

Thus, only in the case where the decrease of potential energy due to exchange interaction is higher than the increase of kinetic energy a ferromagnetic state is formed where all spins are polarized in one direction.

The increase of kinetic energy, which has to be overcompensated, can be quantified for a three-dimensional free electron system. As illustrated in Figure 3.13 (a), if each k-state is occupied with both spin orientations, the Fermi wavenumber is given by

$$k_F^{\uparrow\downarrow} = \left(3\pi^2\frac{N}{V}\right)^{1/3},$$ (3.69)

while for an occupation with only one spin orientation, as shown in Figure 3.13 (b), one obtains for the wavenumber

$$k_F^{\uparrow} = \left(3\pi^2\frac{2N}{V}\right)^{1/3}.$$ (3.70)

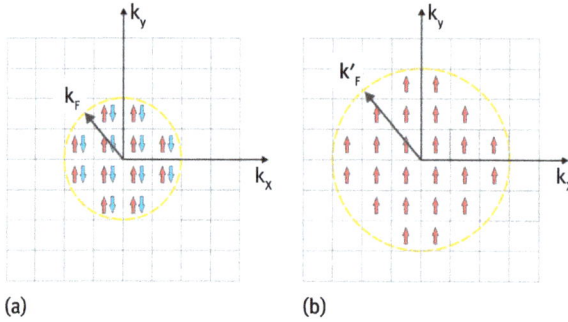

Figure 3.13: Occupation in k-space with electrons with both spin orientations (a) and with up-spin electrons only (b).

Thus, the difference in the Fermi wavenumbers is given by

$$k_F^\uparrow = 2^{1/3} k_F^{\uparrow\downarrow} = 1.26 k_F^{\uparrow\downarrow}. \tag{3.71}$$

The according difference in the kinetic energy

$$E_{\text{kin}} = \frac{\hbar^2 k_F^2}{2m_0} \tag{3.72}$$

is given by

$$E_{\text{kin}}^\uparrow = 2^{2/3} E_{\text{kin}}^{\uparrow\downarrow} = 1.59 E_{\text{kin}}^{\uparrow\downarrow}. \tag{3.73}$$

Thus for free spin-polarized carriers the kinetic energy at the Fermi level is larger by a factor of 1.59 compared to a system where each k-state is occupied with both spin orientations.

In most metals the decrease in potential energy due to exchange interaction is smaller than the increase of kinetic energy, so that no ferromagnetic state is formed. However, for some $3d$-transition metals the increase of kinetic energy due to spin polarization is sufficiently small that it can be overcompensated by the lowering due to exchange interaction. The underlying reason is the band structure. In $3d$-transition metals the $4s$-electrons are delocalized while the $3d$-electrons are close to the core and more localized. This is illustrated in Figure 3.14. The more delocalized the electrons are in the $4s$-orbitals, the broader the corresponding band is, owing to the larger wave function overlap. For localized $3d$-electrons the overlap of neighboring $3d$-orbitals is small. As a consequence the width of the corresponding band is small. This also means that the density of states is large. For some $3d$-transition metals this band is crossed by the Fermi energy E_F. In this case the increase of kinetic energy after spin polarization is relatively low, since due to the very large density of states there are many states available close to the Fermi energy.

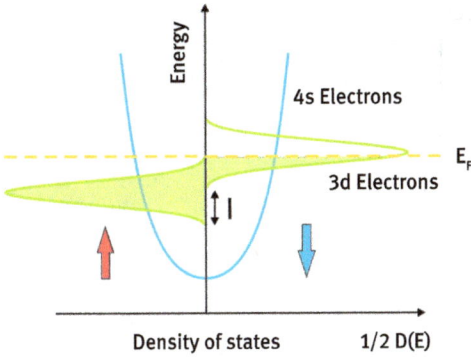

Figure 3.14: Schematic illustration of the density of states of ferromagnetic $3d$-transition metals. Due to exchange interaction the density of states for the d-bands are split by I. The spin-up d-bands are completely filled, while the spin-down band is only partly filled.

In most ferromagnetic materials not all carriers are aligned along only one direction. As illustrated in Figure 3.14, the bands corresponding to opposite spin directions are rather shifted by a certain energy. Due to the energy shift and the filling up to the same Fermi energy for both spin orientations, there is an imbalance of the concentration of spin-up and spin-down electrons, i. e. $n_\uparrow$ and $n_\downarrow$, respectively. For the corresponding energies we can write

$$E_\uparrow(\vec{k}) = E(\vec{k}) - \frac{In_\uparrow}{n}, \tag{3.74}$$

$$E_\downarrow(\vec{k}) = E(\vec{k}) + \frac{In_\downarrow}{n}, \tag{3.75}$$

with n being the total concentration. The parameter I is the Stoner parameter, which quantifies the energy lowering due to the electron correlation, i. e. the electron exchange. In order to observe ferromagnetism, the Stoner criterion has to be fulfilled:

$$I \cdot \widetilde{D}(E_F) > 1, \tag{3.76}$$

with $\widetilde{D}(E_F)$ being the density of states per atom and spin orientation at the Fermi energy. In Figure 3.15 the Stoner parameter I, the density of states per atom $\widetilde{D}(E_F)$ and the product of both is plotted for different $3d$-transition metals [46]. One finds that only for Fe, Co, and Ni the Stoner criterion is fulfilled. Indeed, all three metals are ferromagnets.

As an example the density of states of Fe is shown in Figure 3.16. The density of states originating from the $3d$-states can be identified as being relatively narrow, compared to the density of states originating from the $4s$-states being spread to a larger range. Owing to the ferromagnetic coupling the densities of states for both spin orientations are shifted. The larger fraction is called minority spins, while the smaller

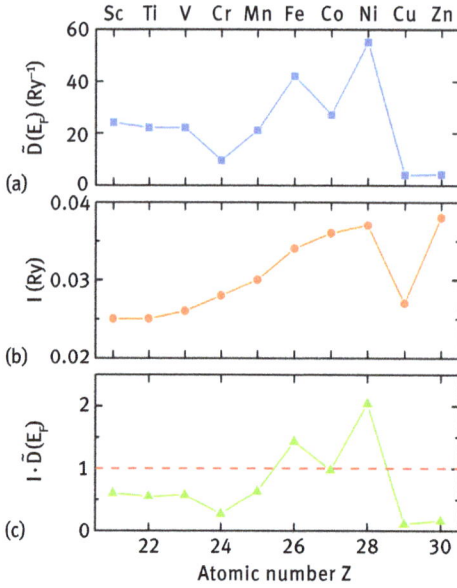

Figure 3.15: (a) Density of states $\tilde{D}(E_F)$ at the Fermi energy for 3d-transition metals with atomic number Z. (b) Corresponding Stoner parameter I. (c) Product $I\tilde{D}(E_F)$. Data taken from Janak [46].

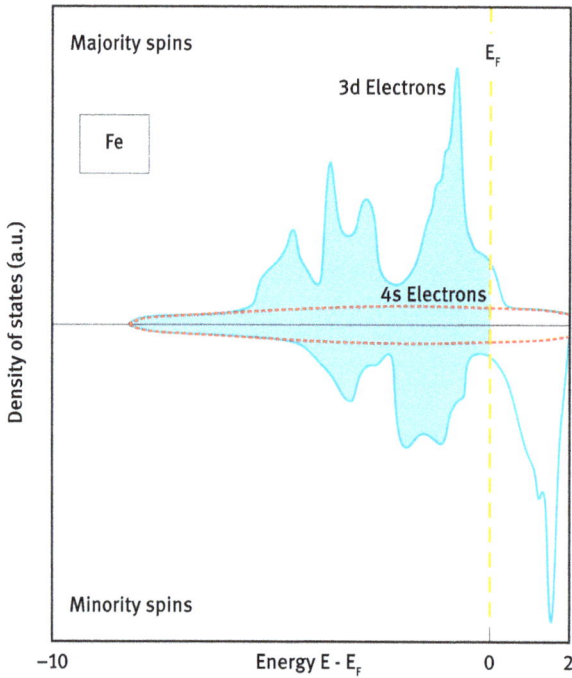

Figure 3.16: Density of states of the majority and minority spins of Fe. Figure adapted from Moruzzi et al. [47].

fraction correspond to minority spins. In case of the majority spins the band is shifted downwards so that all $3d$-bands are almost completely below the Fermi energy and thus almost fully occupied. In case of minority spins a large part of the $3d$-density of states is located above the Fermi energy, so that the states are only partly filled. As a consequence, the number of minority spins filled up to the Fermi energy is lower than for the majority spins. A net magnetization is observed.

3.5 Summary

- In paramagnetism, given magnetic moments are aligned along an outer magnetic field, while for diamagnetism counter-flowing currents are induced which shield the outer magnetic field.
- The paramagnetism of localized magnetic moments is determined by the magnetic moments due to the orbital angular momentum and the electron spin.
- Pauli paramagnetism regards the paramagnetic contribution owing to spin alignment of free electrons.
- The energy of spin-singlet and spin-triplet states is different due to exchange interaction. This results from the Pauli principle, i. e. the total wave function must be antisymmetric with regard to particle exchange.
- Ferromagnetism of $3d$-transition metals, i. e. Fe, Co, or Ni, can be explained within the Stoner model. Here, the increase of kinetic energy after spin polarization is compensated by the gain of exchange energy.

Exercises

Problem 3.1. Calculate the eigenstates of the Pauli spin matrix σ_y.

Problem 3.2. Confirm that the spinors

$$|\psi_+\rangle = \begin{pmatrix} \cos(\theta/2) \\ e^{i\phi}\sin(\theta/2) \end{pmatrix} \quad \text{and} \quad |\psi_-\rangle = \begin{pmatrix} -\sin(\theta/2) \\ e^{i\phi}\cos(\theta/2) \end{pmatrix} \tag{3.77}$$

are eigenfunctions of the generalized Zeeman Hamiltonian given by equation (3.29).

Problem 3.3. The classical definition of the magnetic moment is

$$\vec{\mu} = I\vec{F},$$

with I being the current and F being the encircled area. Show that this expression is equivalent to

$$\vec{\mu} = -\frac{e}{2m_0}\vec{r}\times\vec{p} = -\frac{e}{2m_0}\vec{L},$$

with $\vec{r}$, $\vec{p}$, and $\vec{L}$ being the position, momentum, and angular momentum vector, respectively.

Problem 3.4. In the Stoner model the gain of kinetic energy is compared to the decrease of potential energy. Calculate the relative gain of kinetic energy for a metallic two-dimensional system.

4 Diluted magnetic semiconductors

4.1 Overview

Diluted magnetic semiconductors (DMS) are alloys of nonmagnetic semiconductors doped with magnetic elements. The goal is to create materials which still have semi-conducting properties, i. e. doping, gate-operation, or sensitivity to light, but also possess magnetic properties at the same time. As illustrated in Figure 4.1, one can view a diluted magnetic semiconductor as a link between nonmagnetic semiconduc-tors, i. e. GaAs or InAs, and pure magnetic semiconductors, i. e. chalcogenides like EuO or spinels, e. g. $CdCr_2Se_4$. In a diluted ferromagnetic semiconductor, a small number of atoms in the semiconductor are replace by dopants, which possess a large magnetic moment. The doping is in the percentage range. Typical dopant atoms are $3d$-elements, such as Fe, Cr, or Mn, which result in diluted magnetic semiconductor materials such as

- II-VI semiconductors: $Cd_{1-x}Mn_xSe$, $Cd_{1-x}Mn_xTe$, $Zn_{1-x}Co_xTe$;
- IV-VI semiconductors: $(Pb,Sn,Mn)Te$;
- III-V semiconductors: $In_xMn_{1-x}As$, $Ga_xMn_{1-x}As$, $Ga_xMn_{1-x}N$.

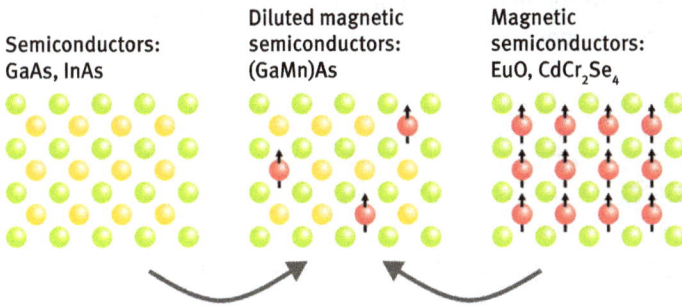

Figure 4.1: Diluted magnetic semiconductor can be seen as an intermediate material, between a pure semiconductor and a pure magnetic semiconductor.

The properties of diluted magnetic semiconductors fall into two categories: first, paramagnetic materials which possess a largely enhanced g-factor; As we will dis-cuss in detail below, these DMS materials can be used as a spin polarizer for spin injection; and second, ferromagnetic materials where the coupling between the mag-netic moments of the dopants is sufficiently strong, so that a ferromagnetic state is achieved.

https://doi.org/10.1515/9783110639001-004

4.2 II-VI diluted magnetic semiconductors

In order to obtain a II-VI diluted magnetic semiconductor, a $3d$-transition metal can be incorporated into a nonmagnetic semiconductor [48]. These magnetic atoms replace the group II elements in the crystal lattice. A typical II-VI diluted magnetic semiconductor is $Cd_{1-x}Mn_xSe$. The advantage of using $3d$-transition metals is the large local moment of the dopant atoms and the good solubility. Furthermore, no electrical doping, i. e. no p- or n-type doping, occurs, because a transition metal atom replaces a group II atom in the lattice. The magnetic properties of II-VI diluted magnetic semiconductor can be explained in terms of exchange interaction between electrons in the s- or p-orbitals and the d-orbitals. This sp-d exchange interaction can be interpreted as a very large Zeeman energy splitting with an effective factor g_{eff}:

$$\Delta E = g_{eff}\mu_B B, \tag{4.1}$$

with μ_B being the Bohr magneton and B the external magnetic field.

If $3d$-transition metals are incorporated in a II-VI semiconductor, the two electrons of the s-orbital are employed for the binding in the crystal lattice, since the transition metal is placed on a site of the group II elements. Owing to uncompensated spins of electrons in the $3d$-shell the magnetic moment of these atoms can be very high. Therefore, these atoms act as localized magnetic moments. As an example the electronic configuration of a Mn atom is shown in Figure 4.2. Owing to the binding in the crystal it is in a Mn^{2+} state.

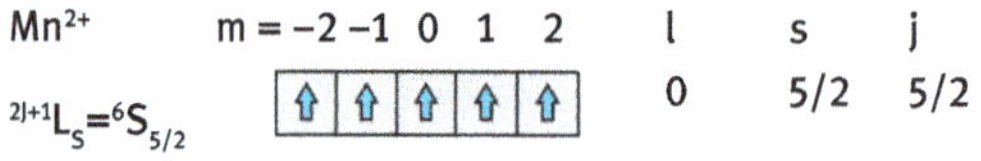

Figure 4.2: Electron configuration in the d-shell of Mn^{2+}. The total angular momentum is zero, while the spin quantum number s is $5/2$.

According to Hund's rule, all five electron spins in the d-shell are oriented in the same direction (cf. Section 3.3.2). The total angular momentum quantum number l is zero, while the total spin quantum number has the maximum value of $s = 5/2$.

In II-VI diluted magnetic semiconductors the free electrons in the s-orbitals (conduction band) or the holes in the p-orbitals (valence band) couple to the localized spins in the d-orbitals of the transition metals. As a consequence, the spins of the free carriers are aligned. The whole process is called sp-d exchange coupling. The resulting energy splitting of the free electrons can be much stronger than the bare Zeeman splitting in an external magnetic field. This finally results in a largely enhanced g-factor for the free electrons.

Similar to equation (3.66), for the *sp-d* exchange coupling the corresponding Heisenberg Hamiltonian can be expressed as

$$\hat{H}_{ex} = -\sum_{\vec{R}_i} J_{sp-d}(\vec{r} - \vec{R}_i)\vec{S}_i\vec{s}, \tag{4.2}$$

where J_{sp-d} is the coupling parameter, $\vec{r}$, $\vec{R}_i$ the position of the electron and Mn^{2+} atom, and $\vec{s}$, $\vec{S}_i$ the free electron spin and spin of the *i*-th localized Mn atom, respectively. The geometrical situation is illustrated in Figure 4.3.

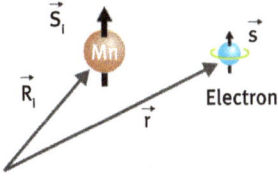

Figure 4.3: Manganese atom with a total spin of $\vec{S}_i$ at position $\vec{R}_i$ with a free electron at $\vec{r}$.

In order to theoretically describe the system more easily, two approximations are performed. First, the virtual crystal approximation is applied, where the few magnetic atoms are effectively homogeneously distributed over the lattice by preserving the total magnetic moment. The distribution is replaced by a continuum with the same average local moment density. Thus, for homogeneously distributed magnetic atoms, their individual magnetic moments are reduced accordingly. For the virtual crystal approximation the Heisenberg Hamiltonian is replaced by

$$-\sum_{\vec{R}_i} J_{sp-d}(\vec{r} - \vec{R}_i)\vec{S}_i\vec{s} \rightarrow -x \sum_{\text{all } \vec{R}} J_{sp-d}(\vec{r} - \vec{R})\vec{S}\vec{s}, \tag{4.3}$$

with x being the concentration of magnetic atoms. As one can see here, now the sum is performed over all lattice sites compared to the smaller number of sites of the actual magnetic atoms. The essence of the virtual crystal approximation is illustrated in Figure 4.4.

Secondly, the mean-field approximation, where the spin of the localized atoms is replaced by the thermal average of all localized spins $\vec{S}_i \rightarrow \langle S_z \rangle$. Without losing generality, we assumed the magnetization along the *z*-direction. Including both approximations, the total Hamiltonian can be written as

$$\hat{H}_{ex} = -s_z \langle S_z \rangle x \sum_{\vec{R}} J_{sp-d}(\vec{r} - \vec{R}). \tag{4.4}$$

The advantage of expressing the Hamiltonian in this way is that we can deal with a periodic system. As a consequence, the basis functions of the band structure can be

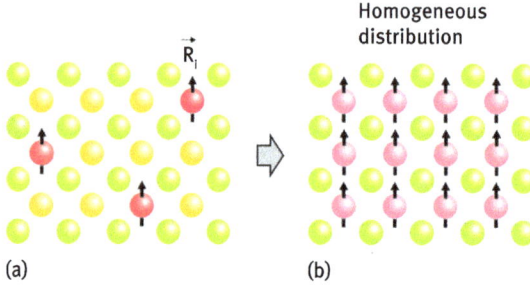

(a)   (b)

Figure 4.4: Schematic illustration of the virtual crystal approximation. (a) Diluted magnetic semiconductor with randomly distributed magnetic atoms. (b) Homogeneously distributed magnetic atoms with corresponding lower magnetic moments.

employed for calculating the energy eigenfunctions. The conduction band basis functions are composed from s-type atomic wave functions:

$$|\psi_s^\uparrow\rangle = \left|\frac{1}{2}, \frac{1}{2}\right\rangle, \quad |\psi_s^\downarrow\rangle = \left|\frac{1}{2}, -\frac{1}{2}\right\rangle. \tag{4.5}$$

Here, $|s^\uparrow\rangle$ and $|s^\downarrow\rangle$ are the states of the delocalized electrons. They should not be confused with the states of the localized electrons. The states $|\psi_s^\uparrow\rangle$ and $|\psi_s^\downarrow\rangle$ are eigenfunctions of the spin operator σ_z. Including the normal Zeeman effect into equation (4.4) the energy splitting in the conduction band can be expressed by the matrix elements

$$E^\uparrow = \langle\psi_s^\uparrow|\hat{H}_{ex}|\psi_s^\uparrow\rangle \quad \text{and} \quad E^\downarrow = \langle\psi_s^\downarrow|\hat{H}_{ex}|\psi_s^\downarrow\rangle. \tag{4.6}$$

Inserting the explicit expression for H_{ex} one arrives at the energy eigenstates

$$E^{\uparrow,\downarrow} = \pm\frac{1}{2}(g\mu_B B - N_0\alpha x\langle S_z\rangle), \tag{4.7}$$

with N_0 being the number of cations per unit volume and $\alpha = \langle\psi_s^\downarrow|J_{sp-d}|\psi_s^\downarrow\rangle$ the exchange integral. From the energy splitting described by equation (4.7) an effective g-factor can be defined:

$$g_{eff} = g - N_0\alpha x\langle S_z\rangle/\mu_B B \approx -N_0\alpha x\langle S_z\rangle/\mu_B B. \tag{4.8}$$

Often, the exchange contribution dominates, so that g can be neglected. For this case the energy splitting due to the coupling of the delocalized electrons and the localized electrons in the d-shell is given by

$$\Delta E_{sd} = -N_0\alpha x\langle S_z\rangle. \tag{4.9}$$

If one assumes that the localized moments are mainly responsible for the magnetization M, one can express the spin of the localized moments as

$$\langle S_z\rangle = \frac{M}{xN_0 g\mu_B}. \tag{4.10}$$

Furthermore, with the susceptibility $\chi_{Mn} = \mu_0 M/B$ the effective g-factor can be expressed as

$$|g_{eff}| = \frac{\chi_{Mn}\alpha}{g\mu_0\mu_B^2}. \tag{4.11}$$

Accordingly, for carriers in the valence band the exchange integral α needs to be replaced by the integral over the valence band functions $|\psi_X\rangle$:

$$\beta = \langle\psi_X|J_{sp-d}|\psi_X\rangle, \tag{4.12}$$

so that the energy splitting is given by

$$\Delta E_{pd} = -N_0\beta x\langle S_z\rangle. \tag{4.13}$$

Finally, the effective g-factor for holes in the valence band can be written as

$$|g_{eff}| = \frac{\chi_{Mn}\beta}{g\mu_0\mu_B^2}. \tag{4.14}$$

In Figure 4.5 the splitting of the s- and p-orbitals due to sp-d exchange interaction is illustrated. The experimental exchange constants of some II-IV diluted magnetic semiconductors are listed in Table 4.1.

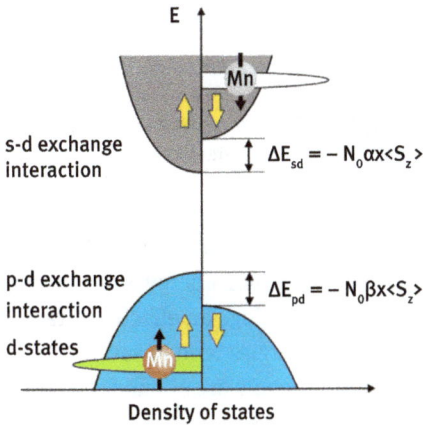

Figure 4.5: Schematics of the spin splittings of the s- and p-bands due to sp-d exchange interaction.

The strong Zeeman splitting in II-VI diluted magnetic semiconductors can be measured by optical means, e. g. by magneto-reflectance measurements. In Figure 4.6 the transitions are shown for circular polarized light (σ^+, σ^-), where the angular momentum is changed by $\pm\hbar$.

Table 4.1: Experimental sp-d exchange constants $N_0\alpha$ and $N_0\beta$ for II-VI diluted magnetic semiconductors in eV. Data taken from Furdyna [48].

Alloy	$N_0\alpha$	$N_0\beta$
$Zn_{1-x}Mn_xSe$	0.26	−1.11
$Zn_{1-x}Mn_xTe$	0.18	−1.05
$Cd_{1-x}Mn_xS$	0.22	−1.80
$Cd_{1-x}Mn_xSe$	0.26	−1.11
$Zn_{1-x}Mn_xTe$	0.22	−0.88
$Hg_{1-x}Mn_xSe$	0.4	−0.7
$Zn_{1-x}Mn_xTe$	0.4	−0.6

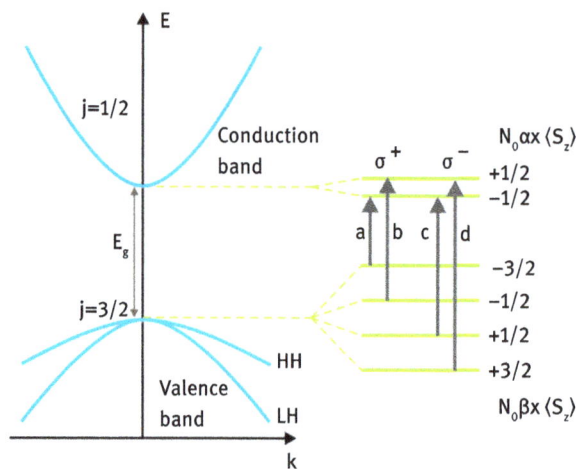

Figure 4.6: Schematic illustration of the transitions in $Zn_{0.95}Mn_{0.05}Te$ (after Furdyna et al. [48]). Compared to the situation shown in Figure 4.5 the external magnetic field is inversed.

The resulting measurement on a $Zn_{0.95}Mn_{0.05}Te$ crystal is shown in Figure 4.7 [49]. At low magnetic fields up to about 5 T a linear increase of the photon energy is found, corresponding to a linear energy splitting of the levels. In this range an effective g-factor can be defined. At larger magnetic fields the transition energies tend to saturate. The reason for this behavior is that all localized magnetic moments are aligned, and thus their magnetization is saturated. The small slope is due to the still remaining normal Zeeman effect of the electrons.

4.3 III-V diluted magnetic semiconductors

Compared to II-VI diluted magnetic semiconductors in III-V materials, the solubility of magnetic atoms is low. In order to incorporate magnetic atoms in III-V semiconduc-

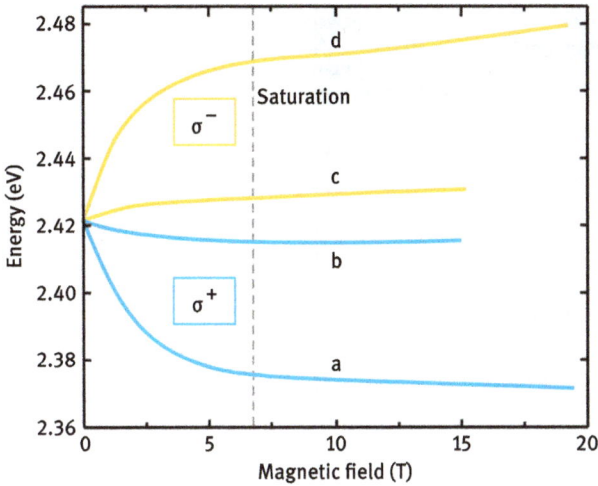

Figure 4.7: Optical transitions in $Zn_{0.95}Mn_{0.05}Te$ extracted from magneto-reflectance measurements as a function of magnetic field, with σ^+ and σ^- the two orientations of circularly polarized light. Graph adapted from Aggarwal et al. [49].

tors to form a diluted magnetic semiconductor, the crystal growth has to be performed far from equilibrium conditions. This can be achieved by using molecular beam epitaxy (MBE). Because of the low solubility, the layers are grown at lower temperatures compared to pure III-V semiconductor layers [50]. In Figure 4.8 a phase diagram of $Ga_{1-x}Mn_xAs$ is shown as a function of Mn composition and substrate temperature [51]. The range in the center corresponds to the crystalline metallic phase, which is the most interesting one for device applications. Here, the Mn atoms are incorporated on a cation-substitutional site. The area of the metallic phase is surrounded by an insulating phase. Here it comes to compensation effects, i. e. due to incorporation of transition metal atoms on interstitial sites. Above around 300 °C growth of a homogeneous $Ga_{1-x}Mn_xAs$ layer is inhibited, and MnAs nanocrystals in a GaAs matrix are formed. If the growth temperature is too low, i.e below about 180 °C, a rough layer or even a polycrystalline phase is formed.

A typical layer system with a $Ga_{1-x}Mn_xAs$ top layer is shown in Figure 4.9. The $Ga_{1-x}Mn_xAs$ layer is grown at a relatively low temperature of about 250 °C, while the buffer layer consisting either of GaAs or AlGaAs is grown at a much higher temperature of around 600 °C.

In Figure 4.10 a magnetization curve is shown for $Ga_{1-x}Mn_xAs$ with $x = 0.035$ [52]. The magnetization was measured by means of a SQUID (superconducting quantum interference device) magnetometer. The clear hysteresis is an indication of ferromagnetism. Owing to ferromagnetic properties a remnant magnetization M_r remains at zero field. The magnetization M_r diminishes at temperatures higher than the Curie temperature T_C, which is about 60 K for this particular sample. From the saturation

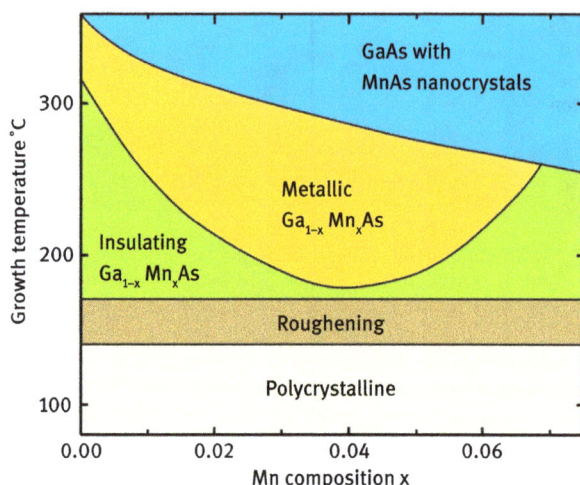

Figure 4.8: Phase diagram of $Ga_{1-x}Mn_xAs$ grown by molecular beam epitaxy. At high temperatures MnAs nanocrystals in a GaAs matrix are formed. Between 180 °C and about 300 °C $Ga_{1-x}Mn_xAs$ is grown. The distribution of Mn atoms in the crystal, i. e. on substitutional or interstitial sites, determines if the layer is metallic or insulating. Below about 180 °C the layers are either rough or polycrystalline. After Ohno et al. [51].

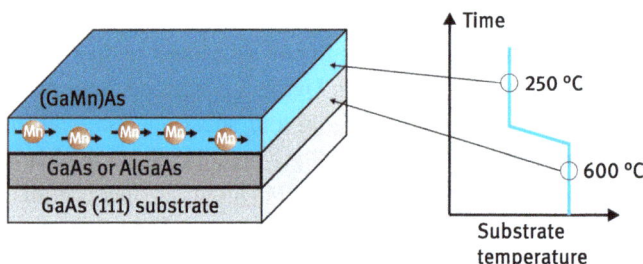

Figure 4.9: Typical layer system with a $Ga_{1-x}Mn_xAs$ top layer and a GaAs or AlGaAs buffer layer on top of a GaAs (001) substrate. On the right side the change of the growth temperatures is illustrated.

magnetization the magnetic moment of the localized Mn atoms can be determined. The saturation magnetization is given by

$$M_s = xN_0g\mu_B S, \tag{4.15}$$

with N_0 the concentration of cation sites. Experimentally one finds an average value of $S = 2.2$ [52], which indicates a mixture of Mn^{2+}: $S = 5/2$ and Mn^{3+}: $S = 2$ magnetic moments.

The origin of the two different states of Mn is explained in Figure 4.11. For the binding with an atom on a group III site three electrons are required; two electrons from the 4s-shell and one of the 3d-shell. Thus, the magnetic moment of the manganese

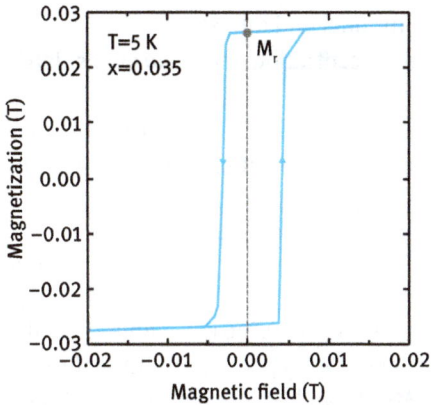

Figure 4.10: Magnetization of a $Ga_{1-x}Mn_xAs$ layer with $x = 0.035$ as a function of magnetic field. M_r is the remnant magnetization. Adapted from Ohno et al. [52].

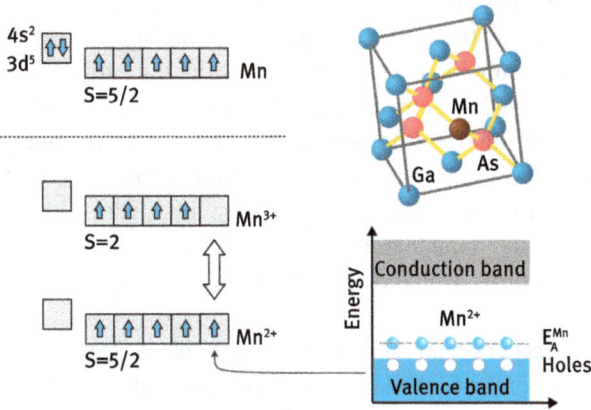

Figure 4.11: Incorporation of Mn in a III-V semiconductor lattice on a group III-site. The two electrons of the $4s^2$ states and one of the $3d$ electrons are taken for the binding to the neighboring atoms in the crystal. The manganese atom is in the Mn^{3+} state with a total spin of $S = 2$. The energy to refill the $3d$ state with an additional electron from the valence band is relatively low. In this case a hole is left in the valence band. The manganese atom is now in the Mn^{2+} state with a total spin of $S = 5/2$.

atom is $2\mu_B$. In a crystal the contribution of the orbital angular momentum can be neglected. The energy to refill the $3d$-shell with an additional electron from the valence band is relatively low. In that case the $3d$-shell is filled with five electrons so that the magnetic moment of the Mn atom is $5/2\mu_B$. At the same time, a free hole is created in the valence band. Hence, in III-V semiconductors a Mn atom on a substitutional site acts as an acceptor, i. e. the $Ga_{1-x}Mn_xAs$ layer is p-type. However, depending on the growth condition, a larger fraction of Mn atoms can also be incorporated on interstitial sites in the GaAs lattice. In that case the Mn atom becomes a double donor. As a consequence, the coexistence of Mn atoms on substitutional and interstitial sites leads to a self-compensation, i. e. a reduction of free hole concentration or even to an insulating state (cf. Figure 4.8).

By incorporating Mn into GaAs, the lattice constant changes [51]. The Mn atoms substitute the Ga atoms in the lattice. The increase of the lattice constant a with increasing Mn composition x is described by Vegard's law:

$$a = a_{GaAs}(1 - x) + a_{MnAs}x. \tag{4.16}$$

The corresponding lattice constants of GaAs is $a_{GaAs} = 0.566$ nm, while the value for a hypothetical zincblende MnAs crystal is about $a_{MnAs} = 0.6$ nm. In Figure 4.12 it is shown, how the lattice constant of $Ga_{1-x}Mn_xAs$ and $In_{1-x}Mn_xAs$ varies with Mn composition x. The lattice constants can be determined by x-ray diffraction. Up to the achieved maximum Mn content of about 9 % a linear increases with x following Vegard's law is observed for $Ga_{1-x}Mn_xAs$. The lattice constant of $In_{1-x}Mn_xAs$ as a function of Mn content is also shown in Figure 4.12. Since the lattice constant of InAs ($a_{InAs} = 0.606$) is larger than $a_{MnAs} = 0.6$ nm the lattice constant of $In_{1-x}Mn_xAs$ decreases linearly with x. For both diluted magnetic semiconductors the extrapolated lattice constant for zincblende MnAs (0.598 nm) are in good agreement with the MnAs lattice constant extrapolated from the (In,Mn)As side (0.601 nm) [51]. Interestingly, even for relatively large Mn contents no significant strain relaxation is observed, when the diluted magnetic semiconductors are grown on an undoped layer, with a higher or lower lattice constant. This is due to the fact, that the diluted magnetic semiconductors are grown at low temperatures, which results in an enhanced critical thickness.

The lattice constant has a direct effect on the magnetic properties. As can be seen in Figure 4.13 (a), for $Ga_{1-x}Mn_xAs$ on GaAs the larger lattice constant of $Ga_{1-x}Mn_xAs$ leads to compressive strain in the top layer. In this case the easy axis of the magnetization, i. e. the energetically preferred orientation of M, lies in-plane. In contrast for $In_{1-x}Mn_xAs$ on InAs the lattice constant of $In_{1-x}Mn_xAs$ is smaller than that of InAs, which leads to tensile strain in the top layer. Now, the easy axis of the magnetization is oriented out-of-plane, as illustrated in Figure 4.13 (b).

One possible theoretical description of the origin of ferromagnetism in III-V diluted magnetic semiconductors is based on the Zener model ($p-d$ exchange model) [53, 54, 55]. Here, the ferromagnetic state is driven by the exchange coupling of the free carriers, i. e. holes, and the localized spins (cf. Figure 4.14). Since the hole originates from exciting an electron into the Mn $3d$ spin-up state, as depicted in Figure 4.11, it is in a spin-down state.

The energetic situation is illustrated in Figure 4.15.

Owing to the exchange coupling a finite magnetization develops which leads to an increase of the free energy of the localized spin. At the same time the magnetization results in a lowering of the carrier energy, according to the Pauli paramagnetism

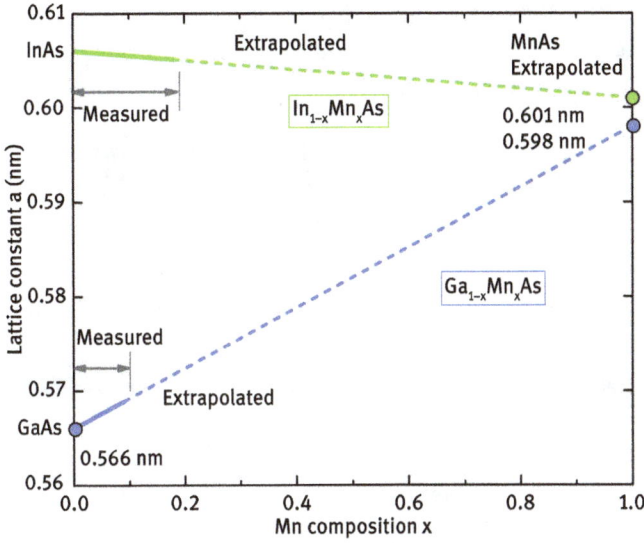

Figure 4.12: Lattice constants of $In_{1-x}Mn_xAs$ and $Ga_{1-x}Mn_xAs$ as a function of Mn composition x. The full lines represent the range measured by x-ray diffraction, while the dashed lines are extrapolated. The extrapolation leads to about 0.6 nm at $x = 1$, which is believed to be the lattice constant of hypothetical zincblende MnAs. Figure adapted from Ohno [51].

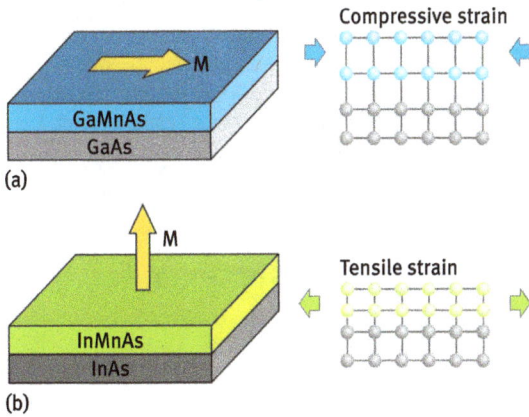

Figure 4.13: (a) and (b) Illustration of compressive and tensile strain in a $Ga_{1-x}Mn_xAs/GaAs$ and an $In_{1-x}Mn_xAs/InAs$ layer system, respectively. The orientation of the magnetization depends on the strain, i. e. in-plane for compressive and out-of-plain for tensile strain, respectively.

discussed in Section 3.3.3. At sufficiently low temperatures, lowering the free carrier energy overcompensates the increase of the free energy, caused by the polarization of localized spins. The total free energy of the system decreases. ΔF_{total} is the sum of the

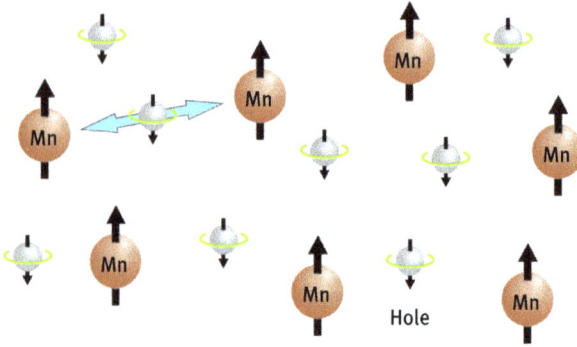

Figure 4.14: Mn atoms with localized magnetic moments surrounded by free holes. The holes mediate the magnetic coupling between the localized moments.

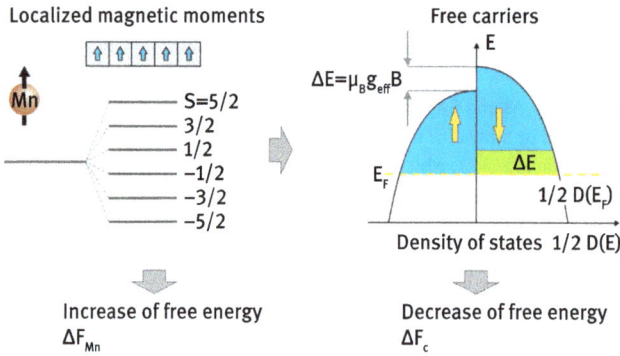

Figure 4.15: Illustration of the energetic contributions in a III-V diluted magnetic semiconductor. The free energy of the localized spins increases due to the magnetization. The spin polarization of the localized spins leads to a spin splitting of the bands of the delocalized holes.

contribution of the localized moments ΔF_{Mn} and that of the free carrier ΔF_{c}:

$$\Delta F_{\mathrm{total}} = \Delta F_{\mathrm{Mn}} + \Delta F_{\mathrm{c}} \tag{4.17}$$

$$\approx \frac{\mu_0 M^2}{2\chi_{\mathrm{Mn}}} - \frac{\chi_{\mathrm{c}}}{2\mu_0} B^2 \tag{4.18}$$

$$= \mu_0 \left(\frac{1}{2\chi_{\mathrm{Mn}}} - \frac{\chi_{\mathrm{c}}}{2\chi_{\mathrm{Mn}}^2} \right) M^2, \tag{4.19}$$

where χ_{Mn} and χ_{c} are the susceptibilities of the localized magnetic atoms and the free holes, respectively. Here, we made use of the fact, that the field contribution of the localized magnetic moments dominate, so that we can replace B by $\mu_0 M / \chi_{\mathrm{Mn}}$. At the ferromagnetic transition temperature both energy contributions balance. A further reduction of temperature gives rise to a spontaneous polarization of localized spins and spin splitting. Thus, at the Curie temperature T_C we have the condition

$$\left(\frac{1}{2\chi_{Mn}} - \frac{\chi_c}{2\chi_{Mn}^2} \right) = 0, \tag{4.20}$$

which corresponds to

$$\chi_{Mn} = \chi_c. \tag{4.21}$$

We first derive the expression for the susceptibility of the free carriers. As outlined in Section 4.2, the p–d exchange interaction leads to an effective g-factor for the holes:

$$g_{eff} = \frac{\chi_{Mn}\beta}{g\mu_0\mu_B^2}, \tag{4.22}$$

with g being the g-factor of the localized spins, which is 2 for Mn. In the next step, g_{eff} is inserted into the expression for the susceptibility of free carriers (cf. Pauli paramagnetism, equation (3.58)) given by

$$\chi_c = \mu_0 \frac{g_{eff}^2}{4} \mu_B^2 D(E_F), \tag{4.23}$$

so that one obtains for χ_c

$$\chi_c = \left(\frac{\chi_{Mn}^2 \beta^2}{g^2 \mu_0 \mu_B^2} \right) D(E_F)/4. \tag{4.24}$$

Here, $D(E_F)$ is the density of states for holes in the valence band at the Fermi energy. As for the II-VI diluted magnetic semiconductors discussed before, the contribution of the localized moments to the magnetization is much stronger than that of the delocalized carriers. Therefore, the latter contribution is neglected, so that the magnetization is only expressed by the localized magnetic moments (cf. equation (3.46)):

$$M = x N_0 g S \mu_B B_J(y), \tag{4.25}$$

where $y = g S \mu_B B / k_B T$ and $B_J(y)$ being the Brillouin function. For $y \ll 1$ one obtains the Curie law with the susceptibility inversely proportional to the temperature:

$$\chi_{Mn} = \mu_0 x N_0 \frac{(g\mu_B)^2}{3} \frac{S(S+1)}{k_B T}. \tag{4.26}$$

At the transition to the ferromagnetism ($T = T_C$) with $\chi_{Mn} = \chi_c$ the Curie temperature is given by [53]

$$T_C = \frac{x N_0 S(S+1) A_F D(E_F) \beta^2}{12 k_B}. \tag{4.27}$$

Here, the factor A_F expressing the enhancement by electron-electron interaction was added [54].

In Figure 4.16 the normalized ferromagnetic temperatures for $Ga_{1-x}Mn_xAs$ and $In_{1-x}Mn_xAs$ are shown as a function of hole concentration [53, 54]. One finds that T_C increases with increasing hole concentration and approximately follows the square-root concentration dependence of the density of states at the Fermi level.

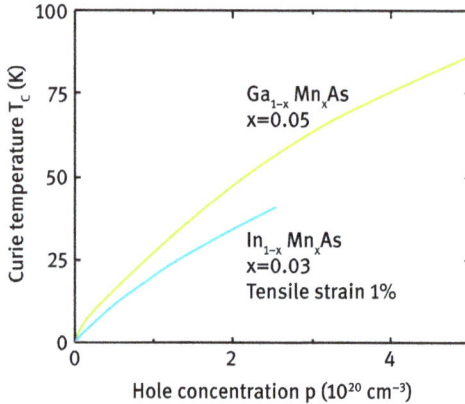

Figure 4.16: Calculated Curie temperature T_C as a function of hole concentration p for $Ga_xMn_{1-x}As$ an $In_{1-x}Mn_xAs$. Data taken from Dietl et al. [53] and Ohno [54].

Figure 4.17 (a) shows the experimentally determined Curie temperature of $Ga_xMn_{1-x}As$ as a function of Mn content x. The T_C values of the as-grown samples monotonously increase up to about $x = 0.05$, while for larger Mn contents the Curie temperature tends to saturate [56]. This can be explained by the fact, that at larger Mn contents a larger fraction of Mn is incorporated on interstitial sites, leading to a self-compensation. This is also confirmed by the hole concentration plotted in Figure 4.17 (b), which saturates above $x = 0.05$. By annealing the samples for about 2h at 180 °C the number of Mn atoms on interstitial sites can be reduced, and the hole concentration increases correspondingly. This results in a monotonous increase of T_C up to $x = 0.09$. For the highest Mn content a Curie temperature of more than 170 K was reported [56].

Using the Zener-model, i. e. the $p-d$ exchange model, the Curie temperatures were calculated for different II-VI and III-V diluted magnetic semiconductors [53]. For most diluted magnetic semiconductors, it was predicted that T_C is below room temperature. There were only two semiconductors, i. e. GaN and ZnO, with a T_C higher than 300 K. Alternatively, the magnetic properties of III-V diluted magnetic semiconductors can also be decribed by other theoretical models, i. e. density function theory [57]. Within that approach it is also possible to include the orbital character of the atoms in the lattice, and thus the angular dependence of the interaction can be considered. For some

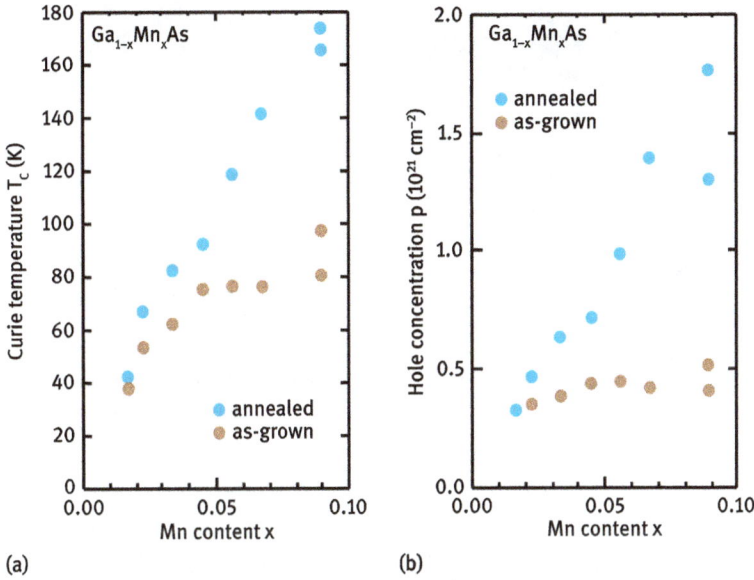

Figure 4.17: (a) Experimentally determined Curie temperature T_C as a function of Mn composition x for $Ga_x Mn_{1-x} As$. The data is shown for as-grown as well as for annealed samples. (b) Corresponding values of the hole concentration p. Data take from Jungwirth et al. [56].

materials, e. g. GaN, a lower Curie temperature, compared to the results of the p–d exchange model was found. The lower T_C was attributed to localization of energetically low lying donor states and to the resulting percolation effects.

4.4 Transport properties of III-V diluted magnetic semiconductors

The magnetic properties of diluted magnetic semiconductor layers can be investigated by transport measurements, e. g. by utilizing the Hall effect. Here, an external magnetic field $\vec{B}_0$ is applied perpendicularly to a semiconductor layer. The measurement set-up is depicted in Figure 4.18.

When a current flows through this conductor, it is observed that electrons or holes are accumulated on one side of the conductor. The reason of this accumulation is the Lorentz force:

$$\vec{F}_L = q\vec{v} \times \vec{B}_0,\tag{4.28}$$

which deflects the propagating carriers to one side. Here, $\vec{v}$ is the velocity of the carriers and q the charge. For a given current the flow of negatively charged electrons and positively charged holes is opposite. Therefore, both kind of carriers are deflected to the same side of the conductor. The accumulated carriers build up an electric field $\vec{\mathcal{E}}$

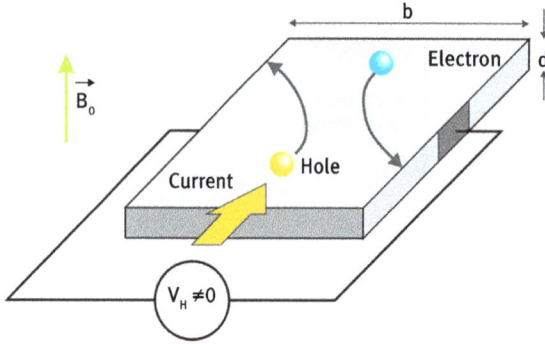

Figure 4.18: Schematic of the Hall effect. In the presence of a magnetic field electrons and holes are deflected in opposite directions by the Lorentz force. The Hall voltage V_H builds up due to the carrier accumulation at the sample edges.

perpendicularly to the current flow, so that the corresponding force $\vec{F}_e = q\vec{\mathcal{E}}$ compensates the deflection from the Lorentz force. The electric field results in a Hall voltage V_H across the conductor:

$$V_H = \mathcal{E}b, \tag{4.29}$$

with b being the width of the conductor. In equilibrium the Lorentz force and the force due to the electric field are compensating each other:

$$|\vec{F}_L| = |\vec{F}_e|, \tag{4.30}$$

which results in

$$V_H = vbB_0. \tag{4.31}$$

The current I through the sample is given by

$$I = jdb, \tag{4.32}$$

where d is the thickness of the layer, and j is the current density given by

$$j = qn_{3D}v, \tag{4.33}$$

with n_{3D} being the carrier concentration. One finally finds for the Hall voltage

$$V_H = \frac{bj}{qn_{3D}}B_0 = \frac{1}{qn_{3D}}\frac{IB_0}{d}. \tag{4.34}$$

Often the Hall resistance is used, which is defined by

$$R_H = \frac{V_H}{I} = \frac{r_H}{d}B_0, \tag{4.35}$$

where $r_{\mathrm{H}} = 1/(qn_{\mathrm{3D}})$ is the Hall coefficient. Depending on the type of carriers a positive or negative voltage is observed, i.e. a negative voltage for electrons $q = -e$ and a positive voltage for holes $q = +e$, with e the elementary charge. The possibility to find out what kind of carriers are present can be utilized in Hall effect measurements on diluted magnetic semiconductors, as will be explained below. Furthermore, the Hall voltage is inversely proportional to the carrier concentration. Thus, the measurement of the Hall voltage is a very important tool, especially for semiconducting materials, to gain information on the type of carrier and the carrier concentration.

Let us go back to the Hall effect measurements on diluted magnetic semiconductors. Here, a magnetic field $B_0 = \mu_0 H$ is applied by an external magnet coil perpendicularly to the epitaxial layers. The voltage drop is measured across the current leads. In diluted magnetic semiconductors the magnetization M due to polarization of the localized magnetic moments contributes to the total field $B = \mu_0(H + M)$ (cf. equation (3.8)). As a consequence, the Hall signal should be directly affected by M. Some care has to be taken regarding the preferred magnetization direction. In case of $Ga_{1-x}Mn_xAs$ grown on GaAs, the diluted magnetic semiconductor is compressively strained because of the larger lattice constant. As we learned in the previous section, for compressively strained layers the easy axis of the magnetization is in-plane (cf. Figure 4.13 (a)). However, for Hall effect measurements it is advantageous that the magnetization is parallel to the external perpendicularly oriented magnetic field. As illustrated in Figure 4.19 (a), in order to turn the easy axis from in-plane into out-of-plane direction, the $Ga_{1-x}Mn_xAs$ layer can be grown on an InGaAs layer having a larger lattice constant than GaAs. As a consequence, now there is tensile strain in the $Ga_{1-x}Mn_xAs$ layer and the easy axis points out-of-plane.

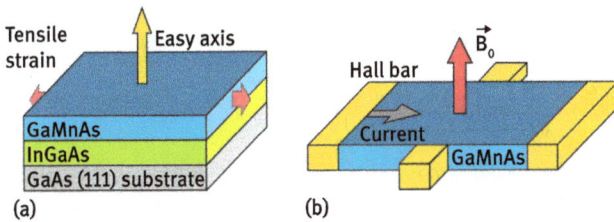

Figure 4.19: (a) Layer system with a $Ga_{1-x}Mn_xAs$ layer on an InGaAs buffer layer. Owing to the larger lattice constant of InGaAs compared to GaAs, the diluted magnetic semiconductor layer is tensile-strained. As a consequence, the easy axis points out-of-plane. (b) Schematic of a Hall bar sample.

As shown in Figure 4.19 (b), in a Hall measurement setup an external magnetic field is applied perpendicularly to the layer system.

The Hall voltage drop is measured across the current channel. As explained above, in nonmagnetic layers, the Hall resistance R_{Hall} is proportional to the external field $B_0 = \mu_0 H$. In diluted magnetic semiconductors the Hall resistance is also affected by the magnetization M of the layer, so that the total Hall resistance can be expressed as [58]

$$R_{\text{Hall}} = \frac{r_{\text{H}}}{d}\mu_0 H + \frac{r_s}{d}\mu_0 M. \tag{4.36}$$

Here d is the layer thickness and r_{H} the ordinary Hall coefficient. Furthermore, r_s is the anomalous Hall coefficient resulting from the additional contribution of M to the total field. As can be inferred from Figure 4.20 (a), the Hall voltage directly reflects the magnetization [59]. The Hall curve observed here, has a similar shape as the magnetization measured by a SQUID magnetometer (cf. Figure 4.10). In particular, the Hall signal at low temperatures, i. e. 5 K and 20 K, shows a hysteresis being characteristic for ferromagnetic materials. When the temperature is increased the corresponding coercive field becomes smaller. The coercive field is the field, where the magnetization is inversed under external magnetic field bias. At T = 60 K the ferromagnetic property is lost, no hysteresis is present. The material is in a paramagnetic state. Thus, the Hall measurements are an interesting option for gaining information on the magnetic properties, in addition to direct magnetization measurements using a SQUID magnetometer.

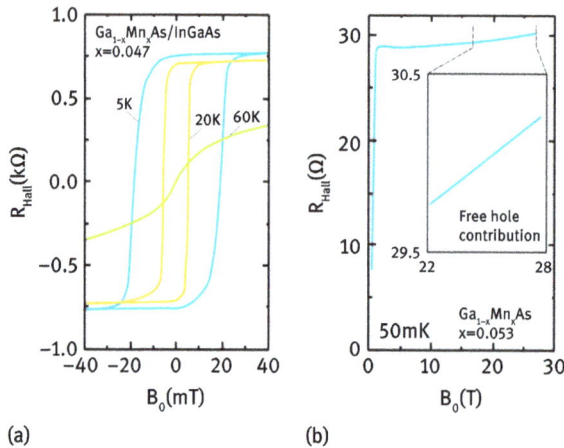

(a) (b)

Figure 4.20: (a) Hall resistance R_{Hall} as a function of an external magnetic field B_0 for $Ga_{1-x}Mn_xAs$ (x = 0.047) grown on InGaAs at different temperatures. Graph adapted from Chiba et al. [59]. (b) Hall resistance R_{Hall} as a function of an external magnetic field of $Ga_{1-x}Mn_xAs$ (x = 0.053). The linear slope at large magnetic fields shown in the inset gives information on the type of free carriers, i. e. holes in the present case. Graph adapted from Ohno et al. [60].

For nonmagnetic semiconductor layers, the sign of the Hall constant is directly related to the type of carriers, i. e. electrons or holes. As can be seen in Figure 4.20 (b), the magnetization of the diluted magnetic semiconductors dominate the Hall signal at low magnetic fields. However, by measuring the slope of the Hall resistance R_{Hall} at larger magnetic fields in the range of 22 to 28 T, where the magnetization M due to the localized magnetic moments is saturated, information on the type of carriers can be gained. This is illustrated in Figure 4.20 (b) (inset), where R_{Hall} is shown at very high magnetic fields. In this range a linear increase of R_{Hall} is found, which can be assigned to the free holes in the layer. From the slope a hole concentration of 3.5×10^{20} cm^{-3} was extracted.

A basic property of semiconductor materials is the possibility to control the carrier concentration by the field effect using a gate electrode. We also learned that ferromagnetism in III-V diluted magnetic semiconductors relies on the exchange coupling between localized moments and free holes. Thus, it is an interesting question, whether or not ferromagnetism in a diluted magnetic semiconductor can be controlled by a gate electrode. That this is indeed the case is shown in Figure 4.21. Here the ferromagnetism is controlled by a gate placed on top of an $In_{1-x}Mn_xAs$ layer [61]. At positive gate voltages $V_g > 0$, where the hole concentration is reduced, a small hysteresis and a small saturation magnetization is found in the R_{Hall} vs. B_0 curves indicating a weak ferromagnetism. In contrast, at zero and negative gate voltages $V_g > 0$, where the hole concentration is successively increased, the hysteresis and the saturation magnetization is increased. This clearly demonstrates that the hole concentration has a large impact on the formation of ferromagnetism.

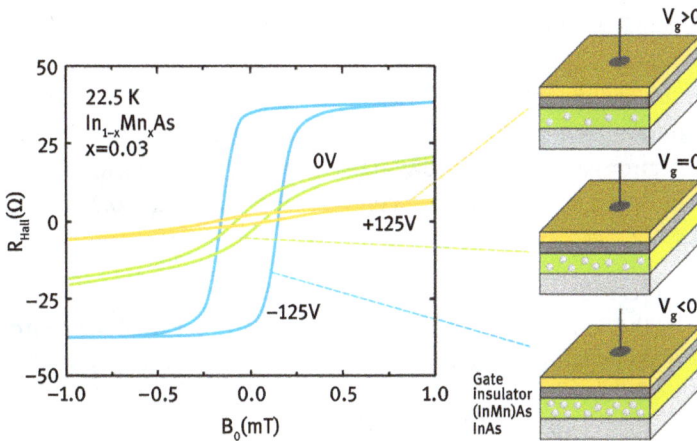

Figure 4.21: Hall resistance of an $In_{1-x}Mn_xAs$ layer ($x = 0.03$) as a function of an external magnetic field B_0 for three different gate voltages, i. e. +125, 0, and −125 V, respectively. On the right side the situation within the layer system at different gate biases is illustrated. At $V_g > 0$ the hole concentration is lower compared to the case at zero gate voltage, while for $V_g < 0$ the concentration is enhanced. Figure adapted from Ohno et al. [61].

It could also be demonstrated that the Curie temperature can be controlled by adjusting the hole concentration by means of a gate electrode [61]. This is in accordance with the theoretical model and the corresponding formula for T_C, i. e. equation (4.27), introduced above. Here, a higher T_C value is expected for larger density of states at the Fermi level $D(E_F)$, i. e. larger hole concentrations.

4.5 Summary

- Diluted magnetic II-VI semiconductors doped with $3d$-transition metals are paramagnetic. Free electrons and localized magnetic moments are coupled via sp-d exchange coupling. The magnetic behavior can be described by an enhanced effective g-factor.
- In III-V diluted magnetic semiconductors the Curie temperature is determined by the balance between the increase of free energy of the localized moments and the decrease of free energy of the free holes.
- The magnetic properties can be analyzed by means of the Hall effect.
- The magnetism of III-V diluted magnetic semiconductors can be controlled by means of a gate electrode.

Exercises

Problem 4.1. Let us consider a II-VI diluted magnetic semiconductor with spin 1/2-localized magnetic moments. The concentration of the magnetic electrons is given by x. Find an expression for the energy splitting ΔE_{sd} of the conductive electrons due to exchange as a function of temperature T and external magnetic field. The Zeeman splitting of the free electrons is neglected. How large is the energy splitting at 5 K at an external magnetic field of 1 T and 10 T for an exchange constant of $N_0\alpha = 0.25$ eV? We assume $x = 0.1$ and $g = 2$.

Problem 4.2. The Hall resistance of a diluted magnetic semiconductor is affected by two contributions, i. e. the contribution of carriers in a magnetic field and the contribution owing to the magnetization of the localized carriers:

$$R_{\text{Hall}} = \frac{r_H}{d} B_0 + \frac{r_s}{d} \mu_0 M,$$

with r_H and r_s being the normal and ordinary Hall coefficient, d the thickness of the samples, and B_0 the external magnetic field. Let us assume a diluted magnetic II-VI semiconductor with spin $S = 1/2$ localized moments at a temperature of $T = 5$ K. The g-factor is assumed to be 2. Calculate the magnetic field at which M is 90 % and 99 % of its saturation value M_0. Close to the saturation of M, the ordinary Hall effect can be resolved, since in the equation given above the second contribution saturates so that the slope of the first contribution can be measured.

Problem 4.3. Calculate the Curie temperature T_C using equation (4.27) for GaAs doped with Mn atoms. We assume a Fermi energy of the holes of 100 meV, with a hole effective mass of $m^* = 0.51m_0$. Furthermore, the doping concentration shall be 10 %. For βN_0 we take 1.2 eV with $N_0 = 2.2 \times 10^{28}$ m^{-3}.

5 Magnetic electrodes

5.1 Overview

In a ferromagnetic material at zero external magnetic field, the magnetization is usually observed to be subdivided into different areas, the so-called magnetic domains. As can be seen in Figure 5.1, in each domain the magnetization has a certain orientation, which differs from the neighboring ones. At the boundary of these areas a domain wall is formed. Here, the orientation of the magnetization is changed.

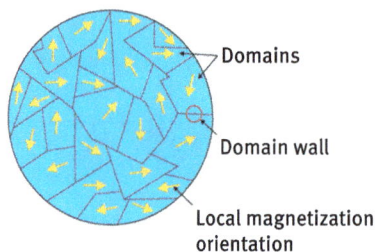

Figure 5.1: Schematic illustration of a ferromagnetic material with magnetic domains. In each domain the orientation of the magnetization is fixed. The domains are separated by domain walls.

The formation of magnetic domains has a big impact on applications in magneto- and spin-electronics. A prominent example is a hard-disc drive for large scale memory. Here, magnetic domains are formed by local magnetization using the magnetic field generated by the write head, so that the digital information 0 and 1 is encoded by two opposite magnetic orientations.

For spin electronic device applications, magnetic domains are relevant for two reasons. First, for spin injectors, e. g. as needed in a spin field-effect transistor, the magnetization of the injector electrode must be well defined. Here, the formation of different magnetic domains is usually counterproductive, since from each domain the spin is injected in a different orientation. For a proper design of an injector electrode, it is of utmost importance to gain a detailed knowledge on the magnetization in a ferromagnetic electrode. As a second application, magnetic domains are also used directly for information processing [11]. Using dedicated shapes of magnetic electrodes, magnetic domains can be trapped at certain positions, e. g. by dents in ferromagnetic stripes. By means of an electric current, these domains can be moved back and forth [62]. Even logic circuits can be realized solely based on ferromagnetic domains [63].

In this chapter, we will give a brief overview of the formation of magnetic domains and of the parameters which determine their orientation and size. Furthermore, different forms of domain walls will be introduced. Subsequently, the switching of magnetic domains under an external magnetic field is discussed. It will be shown how

https://doi.org/10.1515/9783110639001-005

the switching process can be measured. By performing micromagnetic simulations, the transitions of magnetic domains during the application of an external magnetic field can be visualized. Alternatively, magnetic domains can also be manipulated by an electrical current, which offers many opportunities for applications in information processing and data storage [64].

5.2 Formation of magnetic domains

There are a number of reasons why a certain configuration of domains is formed in a magnetic material instead of a uniform magnetization. The main reason for the formation of domains is the interplay between different energy contributions.

5.2.1 Magnetic stray field

One important reason for the formation of domains is the lowering of the magnetic energy due to the magnetic stray field. A sequence of different magnetic configurations is shown in Figure 5.2.

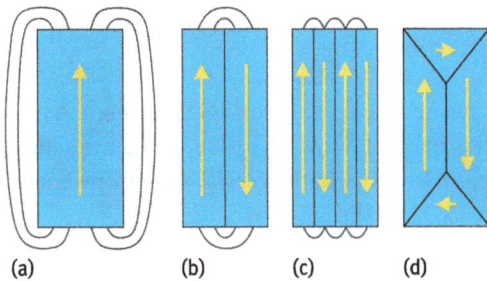

(a)   (b)   (c)   (d)

Figure 5.2: (a)–(d) Decrease of the total magnetic stray field energy by forming magnetic domains.

In Figure 5.2 (a) the material is homogeneously magnetized in only one direction. Here one finds a large number of magnetic field lines in the outer space around the ferromagnetic material. The energy contained in the outer space is given by the integral over the outer volume:

$$U_{\text{stray}} = \frac{1}{2\mu_0} \int B_0^2 \, dV. \tag{5.1}$$

As depicted in Figure 5.2 (b), by splitting up the magnetization in two opposite orientations, the field in the outer space and the corresponding energy can already be reduced significantly. Obviously, a further reduction is gained by subdividing the magnetization into smaller domains with opposite magnetization (cf. Figure 5.2 (c)). By forming domains where the magnetization is rotated in steps of 90 °, the field lines can be kept completely inside the magnetic material, as shown in Figure 5.2 (d). In this configuration the magnetic energy is minimized, so that in principle this kind of configuration is favored at zero external magnetic field. Of course the inside of the material contains some magnetostatic energy. Generally one can write for the magnetostatic energy density as follows:

$$u_{st} = \frac{1}{2\mu_0} B^2. \tag{5.2}$$

5.2.2 Crystal anisotropy

There are additional mechanisms which determine the final domain configuration in a ferromagnetic material. One of them is crystal anisotropy. Owing to the regular arrangement of the atoms in a crystal lattice their coupling, in particular their magnetic coupling, depends on the crystal orientation. This is illustrated in Figure 5.3 (a) for a Co crystal. The magnetic energy depends on the angle ϕ of the magnetization with respect to the c-axis. Usually there is at least one axis, the so-called easy axis, which is preferred energetically for the orientation of the magnetization. The energy density originating from the crystal anisotropy can be written as

$$u_{an} = K_u[1 - (\vec{m} \cdot \vec{e}_{ea})^2], \tag{5.3}$$

with K_u being the uniaxial anisotropy coefficient, $\vec{e}_{ea}$ the unit vector along the easy axis and the normalized magnetization given by $\vec{m} = \vec{M}/M_0$. Here, M_0 is the saturation magnetization. Alternatively, one can also express the crystal anisotropy energy density by the angle ϕ between the easy axis and the magnetization:

$$u_{an} = K_u \sin^2 \phi. \tag{5.4}$$

In some cases higher order terms are included with the corresponding anisotropy coefficients given by K_{u1} and K_{u2}:

$$u_{an} = K_{u1} \sin^2 \phi + K_{u2} \sin^4 \phi. \tag{5.5}$$

For the Co crystal a minimum energy is gained if the magnetization is oriented along the c-axis. As can be seen in Figure 5.3 (b), if an outer magnetic field is applied along the c-axis the field required to orient the magnetization along this direction is small. This axis is called easy axis. Thus the magnetization is saturated already at small external magnetic fields. In contrast, if the outer magnetic field is applied perpendicularly to the c-axis, according to equation (5.4) a large energy is required to reorient the magnetization. This kind of axis is called a hard axis. This is the reason, why the magnetization only increases slowly with increasing external field (cf. Figure 5.3 (b)).

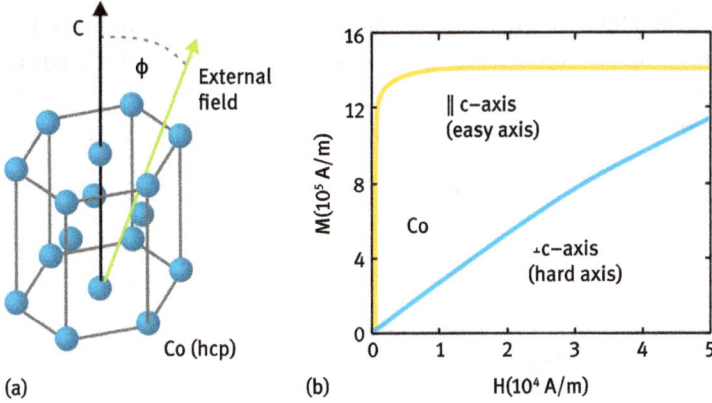

Figure 5.3: Crystal anisotropy in Co. (a) Crystal structure (hexagonal closed packed) and orientation of the outer magnetic field. The easy axis is along the c-axis of the crystal. Here the magnetization along this axis is driven to saturation by application of a small external field. In contrast, for the hard axis, which is oriented perpendicularly to the c-axis, the magnetization only slowly follows the external field. (b) Typical measurement of the magnetization of Co. For a magnetic field along the easy axis the magnetization switches at small magnetic fields, while for a field along the hard axis the magnetization is changed slowly.

The fundamental origin of the anisotropy contribution is the interaction of the electron spin with the orbital motion of the electrons by spin-orbit coupling. The corresponding charge distribution connected to this is nonisotropic. By changing the spin orientation, the overlap of the orbitals in the crystal and thus the exchange interaction is changed.

5.2.3 Form anisotropy contribution

Another contribution which determines the formation of domains is the form anisotropy. Here, for a given magnetic moment in a material, the magnetic dipole-dipole interaction with neighboring magnetic moments is considered. The energy related to the magnetic dipole-dipole interaction for the two magnetic moments at sites i and j is given by

$$U_{\text{dipol}} = -\vec{\mu}_j \vec{B}_i = \frac{\vec{\mu}_i \vec{\mu}_j}{r_{i,j}^3} - \frac{3(\vec{r}_{i,j}\vec{\mu}_i)(\vec{r}_{i,j}\vec{\mu}_j)}{r_{i,j}^5}, \tag{5.6}$$

with $\vec{\mu}_j$ being a magnetic moment in the magnetic field $\vec{B}_i$ created by a second magnetic moment $\vec{\mu}_i$. Each of these magnetic moments is a magnetic dipole which couples to the other one by the dipole-dipole interaction. The interaction depends on the distance $\vec{r}_{i,j}$ between the magnetic moments. According to the dipole-dipole interaction, the preferred orientation of the magnetization depends on the arrangement of neighboring magnetic moments. For a ferromagnetic bar, as shown in Figure 5.4 (a), one finds that the minimum energy is gained if the magnetization is aligned along the long axis of the bar. Thus, the easy axis corresponds to this direction. In case that the magnetic electrode has a square shape, one finds two easy axes because of the higher symmetry (cf. Figure 5.4 (b)).

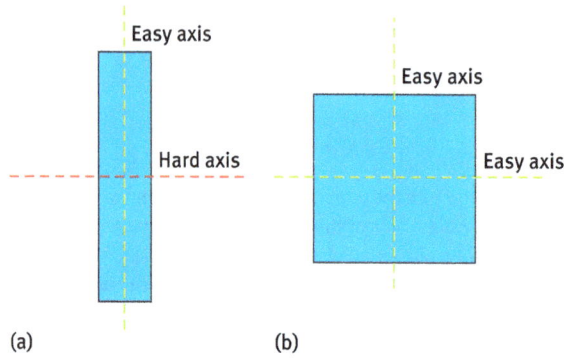

Easy axis

Easy axis

Hard axis

Easy axis

Easy axis

(a)

(b)

Figure 5.4: Form anisotropy: the easy axis is found to be along the long axis of the ferromagnetic bar. (a) Magnetic bar: the easy axis is aligned along the long axis, while the hard axis is along the short axis. (b) square shaped electrode: because of the symmetric geometry, here two perpendicular easy axes are present.

5.2.4 Exchange energy contribution

In Section 3.4.1 the exchange interaction was discussed. For a ferromagnetic material, this energy contribution expresses the tendency to keep neighboring magnetic moments aligned to each other. The energy density connected to the exchange interaction can be written as

$$u_{ex} = \sum_{i=x,y,z} A(\vec{\nabla} m_i)^2,$$

(5.7)

where A is the exchange stiffness constant. It is related to the previous expression for exchange coupling equation (3.66) by $A = JS^2 a$, with $J = J_{ij}$ the exchange coupling constant, S the spin, and a the lattice constant.

5.3 Domain walls

At the boundary between two magnetic domains the orientation of the magnetization is changed. This boundary is called a domain wall. Usually the orientation is not changed abruptly, but rather over many lattice planes. The formation of a domain wall is governed by the interplay between exchange and anisotropy energy.

As we will see later for a particular magnetic material, the boundary between two magnetic domains can be quite complex and depends on many details. However, there are two basic types of domain walls which can be considered. The first is the so-called Bloch wall, which is depicted in Figure 5.5.

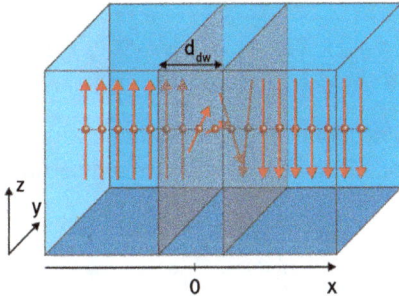

Figure 5.5: Rotation of the magnetization within a Bloch domain wall, with d_{dw} being the thickness of the domain wall.

Here the magnetization is rotated about the axis perpendicularly to the domain wall. Let us assume that the domain wall is perpendicular to the x-axis, and thus the magnetization rotates about that axis. We can use polar coordinates to describe the magnetization within the yz-plane:

$$m_z(x) = \cos \phi(x), \quad m_y(x) = \sin \phi(x), \tag{5.8}$$

with the boundary conditions given by

$$\phi(-\infty) = 0, \quad \phi(\infty) = \pi. \tag{5.9}$$

The magnetization within the domain wall can be determined by energy minimization. There are two competing mechanisms, i. e. the exchange energy and anisotropy energy. The exchange energy, given by equation (5.7), is minimized if the domain wall is wide. Thus, the misalignment between neighboring spins should be spread over a large distance along the x-direction. In contrast, the anisotropy energy contribution is small if the magnetization after rotation in the domain wall is aligned back along the easy axis over a distance as short as possible. The task is to minimize the total domain wall energy given by

$$\gamma_B = \int_{-\infty}^{\infty} \left[K_u \sin^2 \phi + A \left(\frac{d\phi}{dx} \right)^2 \right] dx, \tag{5.10}$$

with the first term representing the crystal anisotropy and the second one the exchange interaction. The latter term results by differentiating the magnetization given by equation (5.8). For the energy minimization δy_B needs to be zero:

$$\delta y_B = 2K_u \sin\phi \cos\phi - 2A\frac{d^2\phi}{dx^2} = 0. \tag{5.11}$$

Here, we made use of the Euler–Lagrange differential equation, which minimizes the integral of the general form:

$$\int f(x, y, y')\, dx \rightarrow \frac{\partial f}{\partial y} - \frac{d}{dx}\left(\frac{\partial f}{\partial y'}\right) = 0, \tag{5.12}$$

where $y' = dy/dx$. In order to obtain an expression connecting the distance x with the magnetization angle ϕ, we multiply the equation given above by $d\phi/dx$ and perform an integration:

$$\int\left(2K_u \sin\phi \cos\phi - 2A\frac{d^2\phi}{dx^2}\right)\frac{d\phi}{dx}\, dx = K_u \sin^2\phi - A\left(\frac{d\phi}{dx}\right)^2 + C = 0, \tag{5.13}$$

with C an integration constant. In the above expression we made use of

$$2A\frac{d\phi}{dx}\left(\frac{d^2\phi}{dx^2}\right) = \frac{d}{dx}\left[A\left(\frac{d\phi}{dx}\right)^2\right]. \tag{5.14}$$

Far from the domain wall, i.e. for $\phi = \pi$, the derivative $d\phi/dx$ vanishes, thus $C = 0$. From equation (5.13) we get the expression

$$dx = \pm\sqrt{\frac{A}{K_u}}\frac{d\phi}{\sin\phi}. \tag{5.15}$$

Integration gives

$$x = \pm\sqrt{\frac{A}{K_u}}\ln\left(\tan\frac{\phi}{2}\right). \tag{5.16}$$

This expression can finally be transformed into

$$\cos\phi = \tanh(x/\sqrt{(A/K_u)}) \tag{5.17}$$

by making use of $\cos\phi = \tanh(\tan\phi/2)$. The above expression can now be used to determine the domain wall width d_{dw}. It is defined by the condition, where the tangent at $x = 0$ crosses the $\phi = 0$ and $\phi = \pi$ line:

$$d_{dw} = \pi\sqrt{\frac{A}{K_u}}. \tag{5.18}$$

Usually the thickness of a Bloch domain wall is several tens of a nanometer.

Especially for thin layers the magnetization is not rotated around the axis perpendicular to the domain wall, but rather rotated within the plane of the layer. This domain wall, illustrated in Figure 5.6, is called a Néel wall. For thin layers the Néel domain wall is energetically favorable compared to the Bloch wall, because the magnetic field lines remain within the material.

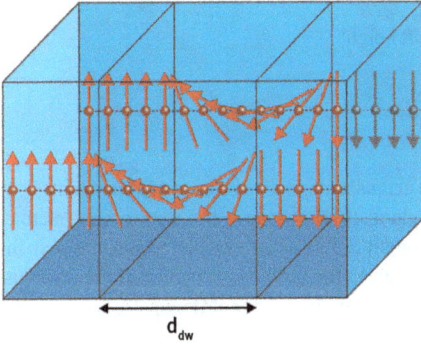

Figure 5.6: Reversal of the magnetization in a Néel domain wall.

For small ferromagnetic structures the magnetization reversal can be more complex, e. g. by means of a vortex domain wall shown in Figure 5.7. For even smaller widths a transverse domain wall is often preferred.

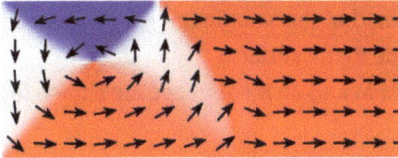

Figure 5.7: Example of a vortex domain wall in a ferromagnetic bar.

5.4 Ferromagnetic electrodes

In this section we will have a closer look at the formation of domains in magnetic electrodes. A very powerful tool to investigate domain structures is the scanning magnetic force microscope (MFM). A schematic of the setup is depicted in Figure 5.8 (a).

An atomic force microscope (AFM) tip is covered with a magnetized ferromagnetic material. The tip is scanned at a small distance across the surface of a magnetic material containing different magnetic domains. Depending on the orientation of the magnetization and the corresponding stray field, the magnetized tip is either slightly attracted or repelled, due to magnetic dipole-dipole interaction. The force on the tip is

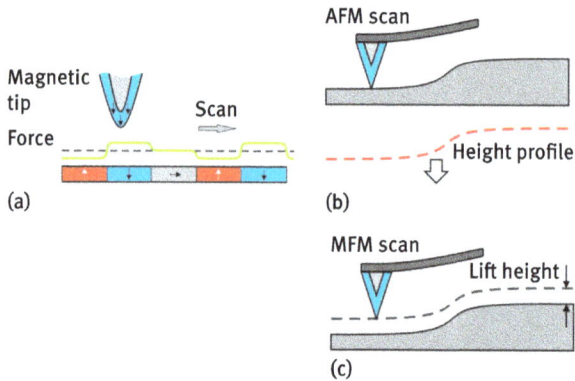

Figure 5.8: (a) Magnetic force microscope tip scanned across a magnetic surface with domains of different magnetization. The tip is magnetized out-of-plane. (b) First scan in the atomic force microscope mode, in order to obtained information on the height profile of the sample. (c) Second scan in the magnetic force microscope mode at a fixed distance using the profile information of the previous scan. Here, only the magnetic stray field changes the oscillation frequency of the tip.

employed to generate an image of the magnetic landscape by a change of the tip oscillation frequency. Details of the mode of operation are illustrated in Figure 5.8 (b) and (c). During the first scan the MFM runs in the atomic force microscope mode, where the tip is lowered close to the surface of the sample. In this mode van der Waals forces apply, and the height profile of the sample is obtained. In the second run this information is used to scan the surface at a fixed distance, following the surface profile. The distance is so large that only the stray field of the sample imposes a force on the tip, and no direct mechanical contact is made to the surface.

As an example, in Figure 5.9 (a) and (b) an AFM and an MFM image are shown for a Co bar, respectively.

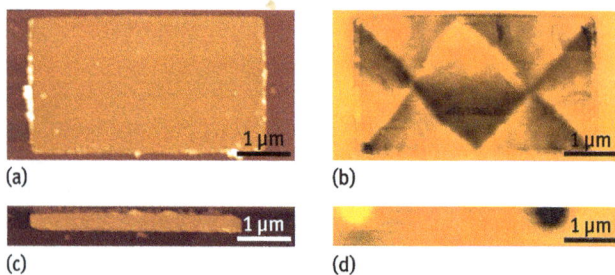

Figure 5.9: (a) Atomic force microscopy image of a Co bar. The width of the bar is 2.5 μm, while the length is 4.45 μm. (b) Corresponding magnetic force microscopy image. (c) Atomic force microscopy image of a bar with a smaller width of 250 nm. (d) Corresponding magnetic force microscopy image. Images provided by S. Heedt, Forschungszentrum Jülich.

The Co bar has a dimension of $2.5 \times 4.45\,\mu m^2$ and thus a relative small aspect ratio of about 1.8. The measurements were performed at zero external magnetic field. The height data of the AFM scan was utilized for the subsequent MFM scan. The MFM image clearly shows a formation of a domain structure. The domain walls cross the sample at an angle of about 45 °. As explained above, the formation of the domains ensures that the external magnetic field and thus the magnetic energy is minimized. Since the aspect ratio is relatively close to one, the form anisotropy does not play a big role. This is different for the sample depicted in Figure 5.9 (c) and (d). The Co bar analyzed here, has a width of only 250 nm and a length of $4.45\,\mu m$, resulting in a much larger aspect ratio of 17.8. As one can infer from the MFM image in Figure 5.9 (d), only a single domain is present. Only at the ends the magnetic field lines are entering and leaving the Co bar, indicated by the bright and dark spots. Owing to its elongated shape, the form anisotropy plays a major role, resulting in a single domain.

In Figure 5.10 a sequence of MFM images of a permalloy (Ni 80 %, Fe 20 %) bar is shown [65]. Different magnetic fields were applied parallel to the long axis of the bar. The permalloy bar has a dimension of $2 \times 4\,\mu m^2$, resulting in a relatively small aspect ratio of 2.

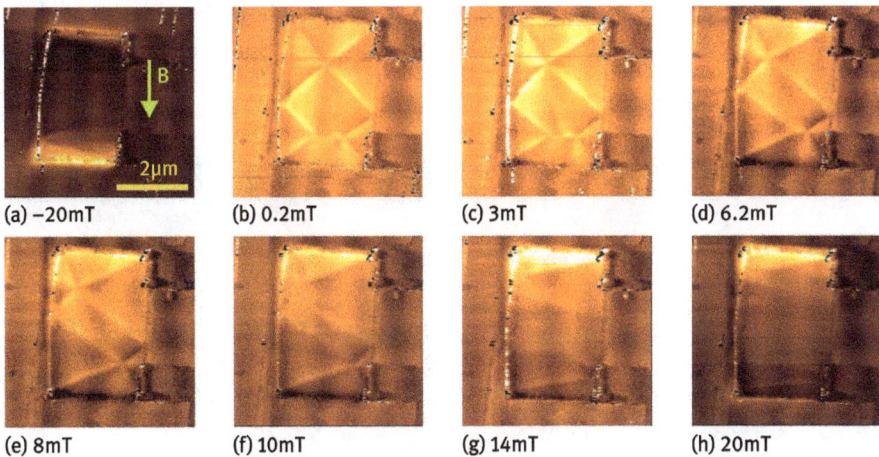

(a) −20mT   (b) 0.2mT   (c) 3mT   (d) 6.2mT

(e) 8mT   (f) 10mT   (g) 14mT   (h) 20mT

Figure 5.10: Magnetic force microscope images of a permalloy bar at different magnetic fields applied along the long axis of the bar. The bar has a size of $2 \times 4\,\mu m^2$. The external magnetic field is applied in-plane along the vertical direction.

As can be seen in Figure 5.10 (a), at sufficiently large external magnetic fields, i. e. −20 mT, a single domain is formed. Here, the external magnetic field polarizes the complete bar in one direction. Around zero magnetic field a more complex multidomain structure is found, as can be seen in Figure 5.10 (b). It resembles the domain structure of the Co bar previously shown in Figure 5.9 (b). The domain structure is basically symmetric along the two symmetry axes of the bar. In the sequence depicted in Figure 5.10 (b)–(h), the external magnetic field was stepwise increased up to +20 mT. As can be seen here, the symmetric domain structure found close to zero field is successively distorted by enlarging the domain areas with a magnetization along the external magnetic field. At the largest field of +20 mT, only a single domain is present, where the magnetic field lines are entering and leaving at the lower and upper side walls.

In Figure 5.11 a sequence of MFM images is shown for a narrow bar at various external magnetic fields. The bar has a width of only 420 nm and a length of 4 μm, corresponding to a much larger aspect ratio of 9.5, compared to the sample shown in Figure 5.10. As before, at a large external magnetic field of –20 mT a single domain is present (cf. Figure 5.11 (a)). At zero magnetic field, a single domain is still found, although the bright and dark areas at the terminals are broadened (cf. Figure 5.11 (b)). The direction of magnetization remains in the same direction as before at –20 mT. The magnetization is the remnant magnetization, left over after setting the field back to zero. As can be seen in the next image, at a field of +4 mT the magnetization is inversed. Here, the magnetic field is larger than the coercive field, i. e. the field required to reverse the magnetization. At +10 mT a full single domain is formed. In Figure 5.11 (e)–(h), the reversed sequence is shown, where the magnetic field is varied from –0.6 mT to –5.2 mT. Once again the magnetization is reversed at around $|B| = 4$ mT.

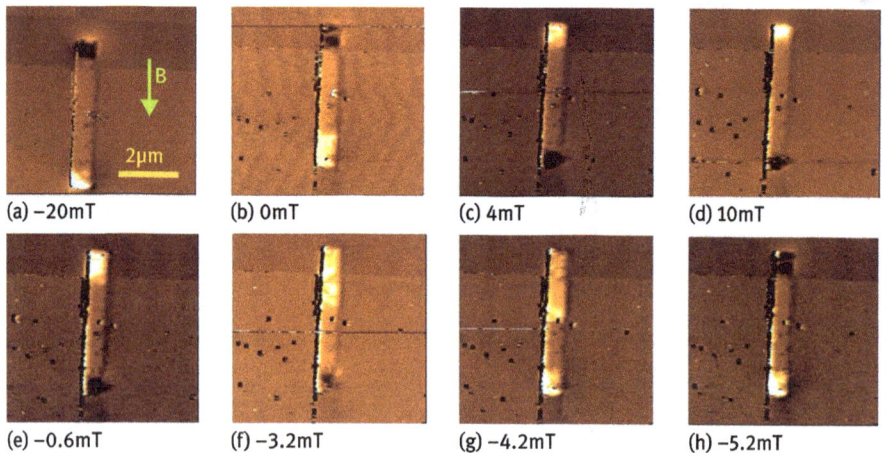

(a) –20mT   (b) 0mT   (c) 4mT   (d) 10mT

(e) –0.6mT   (f) –3.2mT   (g) –4.2mT   (h) –5.2mT

Figure 5.11: Magnetic force microscope images of a permalloy bar at different magnetic fields applied along the long axis of the bar. The width and length of the bar is 0.42 μm and 4 μm, respectively. The external magnetic field is applied in-plane along the vertical direction.

5.5 Local Hall effect measurements

We have learned in the previous section that the properties of magnetic electrodes can be analyzed by means of magnetic force microscopy. An alternative method to gain information on the magnetization is to use a superconducting quantum interference device (SQUID). However, in order to obtain a sufficiently large signal, a large array of identical magnetic electrodes needs to be measured. The magnetization of a single ferromagnetic bar can be investigated by means of local Hall effect measurements [66, 67, 65]. A typical sample is shown in Figure 5.12. Here, a Hall cross structure is defined in a semiconductor heterostructure by dry etching. The heterostructure contains a two-dimensional electron gas. The ferromagnetic bar is placed on top of the Hall bar so that the magnetic stray field of the bar penetrates the center of the Hall cross.

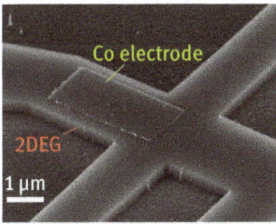

Figure 5.12: (a) Scanning electron micrograph of a Hall bar structure with a ferromagnetic Co electrode on top. The Hall bar is defined by dry etching. It consists of a two-dimensional electron gas in an InGaAs/InP heterostructure. Image provided by S. Heedt, Forschungszentrum Jülich.

In Figure 5.13 details of the measurement setup are shown. In the center of the Hall bar the perpendicular component of the stray field results in a Hall voltage V_H. A switching of the magnetization results in a change of V_H. The bias current I flows perpendicularly to the Hall voltage probes. In order to change the magnetization of the ferromagnetic bar, an in-plane magnetic field $\vec{B}_{ext}$ is applied parallel to the long side of the ferromagnetic electrode. While varying $\vec{B}_{ext}$ the local Hall voltage is recorded.

A typical example of a local Hall measurement is shown in Figure 5.14 [68].

Each time the magnetization is reversed, the Hall voltage V_H changes stepwise when the coercive field is reached. By scanning the external field in both directions a hysteretic curve is observed which directly reflects the magnetization of the ferromagnetic bar. As can be seen in Figure 5.14, the smaller the width of the magnetic bar, the larger is the coercive field. This is a consequence of the form anisotropy discussed in Section 5.2. Usually, the sample used for local Hall measurements is patterned in such a way that a larger number of Hall crosses are placed in a row. By that only a single current has to be applied for all Hall crosses. Each Hall cross is equipped with a ferromagnetic bar of different dimensions. Thus, with a single device many Hall bar structures can be measured simultaneously. This makes the local Hall effect a very

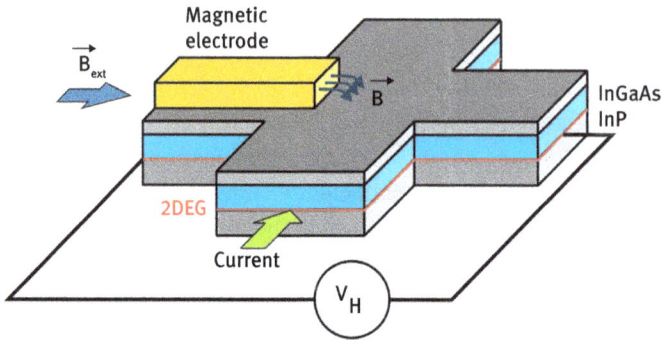

Figure 5.13: Schematic illustration of a sample used for local Hall measurements. The perpendicular component of the stray field of the ferromagnetic bar leads to a Hall voltage. The semiconductor consists of a two-dimensional electron gas (2DEG) in an InGaAs/InP heterostructure. The external magnetic field is applied in-plane along the ferromagnetic bar.

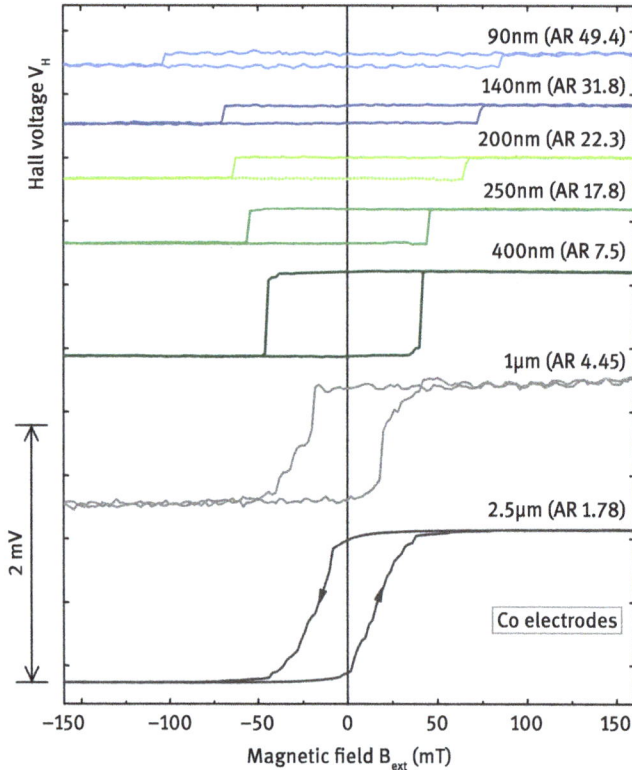

Figure 5.14: Measurement of the local Hall voltage V_H for ferromagnetic Co bars of different widths ranging from 90 nm to 2.5 μm. The corresponding aspect ratios (AR) are 49.4 down to 1.78. The measurements were taken at 4 K. Figure provided by S. Heedt, Forschungszentrum Jülich.

powerful tool to investigate the properties of ferromagnetic bars. This is important especially for investigations in connection with the spin field-effect transistor, since by the local Hall effect measurements the coercive fields of the two ferromagnetic bars used as injectors and emitters can be determined. By applying an external magnetic field a parallel and antiparallel configuration can be achieved between injector and detector. However, in contrast to MFM measurements, no details on the domain pattern are resolved by the local Hall effect measurements, because only the stray field at the terminals of the ferromagnetic bar is detected. Nevertheless, a closer look on the set of measurements plotted in Figure 5.14 can give at least some qualitative information. For the Co bar with the smallest aspect ratio, the transition is rather smooth. This can be attributed to a more complex domain structure, similar to the one shown in Figure 5.10. As one can see there, the widths of the domain at the terminal changes rather smoothly. In contrast, for Co bars with a width smaller than 400 nm a sharp step in the Hall voltage is observed. This indicates that only a single domain is present due to the large form anisotropy contribution.

5.6 Micromagnetic simulations

In addition to experimental techniques described above, the magnetic properties can also be calculated directly by utilizing micromagnetic simulation tools [69]. The calculations are based on the Landau–Lifshitz–Gilbert equation

$$\frac{\partial \vec{M}}{\partial t} = -|\gamma| \vec{M} \times \vec{H}_{\text{eff}} + \alpha \frac{\vec{M}}{M} \times \frac{\partial \vec{M}}{\partial t}, \tag{5.19}$$

which describes the change of the magnetization $\vec{M}$ with time by a precession and a damping term. The precession occurs if the magnetization has a component perpendicular to the effective field $\vec{H}_{\text{eff}}$. The effective field is a combination of the externally applied field and the field due to the magnetization. The parameter α in the second term is the phenomenological Gilbert damping parameter. In Figure 5.15 the evolution of the magnetization in the course of time is illustrated. Here, the vector $-\vec{M} \times \vec{H}_{\text{eff}}$ is responsible for the spin precession, while the vector $-\vec{M} \times (\vec{M} \times \vec{H}_{\text{eff}})$ points towards the precession axis and thus drags the magnetization towards $\vec{H}_{\text{eff}}$. Here we made use of the fact, that $\partial \vec{M}/\partial t$ is proportional to $-\vec{M} \times \vec{H}_{\text{eff}}$. The latter is part of the damping contribution in the Landau–Lifshitz equation. As can be seen in Figure 5.15, due to the damping contribution the magnetization precesses towards the effective magnetic field until both are aligned.

In Figure 5.16 a typical result of a micromagnetic simulation of a 200 nm wide and 4.45 μm long Co bar is shown [69]. The external magnetic field is applied parallel to the long axis of the bar.

At a large external magnetic fields of 300 mT the magnetization is completely aligned with the external field. The parallel magnetization remains even when the

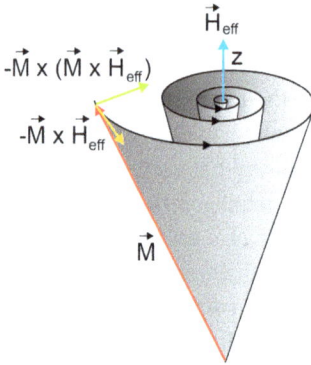

Figure 5.15: Precession of the magnetization $\vec{M}$ towards the effective magnetic field $\vec{H}_{eff}$ in the presence of a damping term. The vector $-\vec{M} \times \vec{H}_{eff}$ is responsible for the precession around $\vec{H}_{eff}$, while $-\vec{M} \times (\vec{M} \times \vec{H}_{eff})$ points towards the effective field axis and results in a damping.

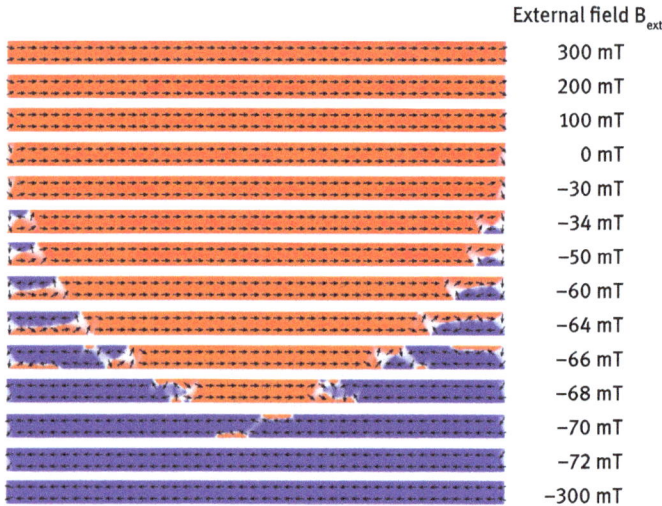

	External field B_{ext}
	300 mT
	200 mT
	100 mT
	0 mT
	−30 mT
	−34 mT
	−50 mT
	−60 mT
	−64 mT
	−66 mT
	−68 mT
	−70 mT
	−72 mT
	−300 mT

Figure 5.16: (a) A set of micromagnetic simulations of a Co electrode using the OOMMF software package [69] with various external magnetic fields applied. The width and length of the electrode are 200 nm and 4.45 μm, respectively. Figure provided by S. Heedt, Forschungszentrum Jülich.

external magnetic field is switched off. At −34 mT vortex domain walls are formed at the terminals. With increasing field strength these domain walls shift towards the center of the bar until they finally merge and get canceled at around −70 mT. Thus the magnetization is completely reversed.

In many spintronic device applications the spin is injected and detected by a pair of ferromagnetic electrodes. Often it is desirable to either switch both electrodes in a parallel or in an antiparallel configuration. This can be achieved by employing two electrodes with different aspect ratios and thus different coercive fields. In Figure 5.17 micromagnetic simulations of such a pair of electrodes are shown. At 300 mT both electrodes contain single domains, which are parallel aligned. At –30 mT a more complex domain structure develops in the lower electrode with the smaller aspect ratio, while the upper electrode remains in a single domain mode. At –40 mT the magnetization of the lower electrode is reversed, i. e. an antiparallel configuration is obtained. By further increasing the field strength, finally a parallel configuration is achieved which is reversed compared to the initial situation. A comparison with the simulations of the single bar shown in Figure 5.16 reveals that the mutual influence due to dipole-dipole interaction results in a lowering of the coercive field of the upper bar by about 3 mT.

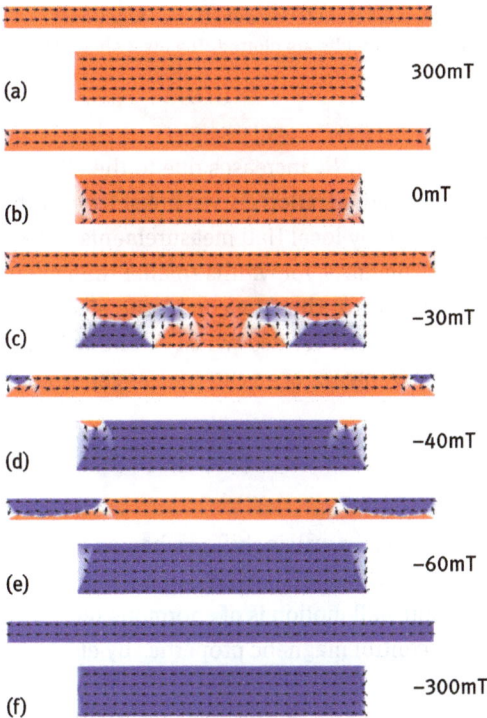

Figure 5.17: (a) A set of micromagnetic simulations of two Co electrode using OOMMF [69] with various external magnetic fields applied along the bar. The width and length of the bottom electrode are 500 nm and 3 µm, respectively. The top electrode is the same as in Figure 5.16 (200 nm × 4.45 µm). The electrodes are separated by 250 nm. Figure provided by S. Heedt, Forschungszentrum Jülich.

From the local Hall effect experiments on ferromagnetic bars with different geometry we found that the coercive field H_c increases with increasing aspect ratios. This is a consequence of the form anisotropy. This behavior is also confirmed by simulations, as shown in Figure 5.18.

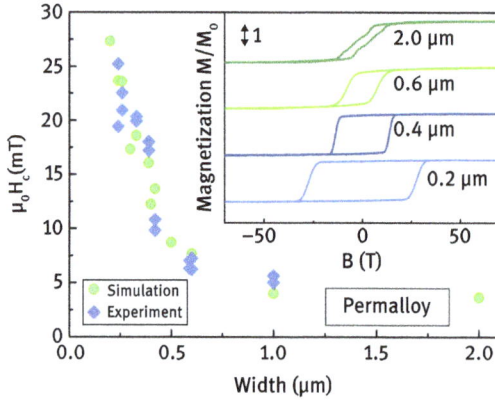

Figure 5.18: Coercive field $\mu_0 H_c$ as a function of width of different ferromagnetic permalloy bars (80 % Ni and 20 % Fe). Experimental as well as calculated results are plotted. The inset shows the corresponding simulated magnetization curves.

For decreasing width, thus increasing aspect ratio, H_c increases due to the effect of the form anisotropy. For selected widths the magnetization curves are shown in the inset. The experimental values of H_c determined by local Hall measurements fit very well to the corresponding simulated ones. Particularly, for widths smaller than about 500 nm a strong impact of the width on H_c is found. Thus, by combining electrodes with different aspect ratios, pairs of electrodes can be created which can be switched from a parallel to an antiparallel configuration by means of an external magnetic field.

5.7 Domain wall motion

From the previous sections we learned that a domain wall is shifted when an external magnetic field is applied. Here we will show that a domain wall can also be moved by an electrical current. Of course this domain wall motion is of enormous relevance for technical applications, since it allows to control magnetic properties by electrical means.

Let us consider a conductive metallic ferromagnetic stripe, which contains a domain wall, as depicted in Figure 5.19. The electrical current is assumed to flow from right to left. Due to the negative charge, the electrons propagate in opposite direction to the current. The electrons approaching the domain wall are spin polarized in the up-direction due to the magnetization of the localized magnetic moments in the left domain. Within the domain wall the magnetization of the localized magnetic moments is rotated. Here, a Néel domain wall is assumed. When the electrons cross the domain wall during propagation their spin is polarized along the localized moments, as shown in Figure 5.19 (a) and (c). After crossing the domain wall the spin of the mobile electrons is finally inversed. The spin is rotated from $\hbar/2$ to $-\hbar/2$ and thus changed

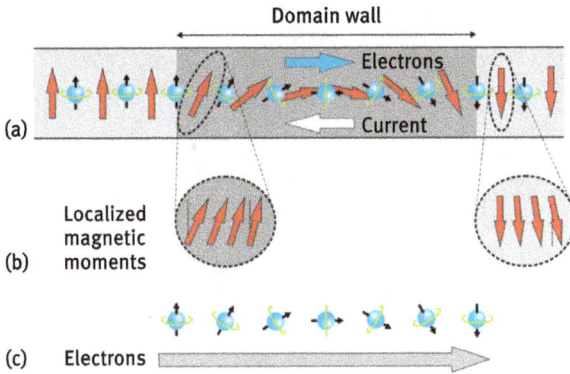

Figure 5.19: (a) Shift of a domain wall by an electrical current. (c) The spin orientation of the electrons, which are spin polarized by the left-hand side domain, is rotated by $-\hbar$ after crossing the domain wall. (b) Owing to the conservation of angular momentum the net spin of the localized moments has to be rotated by the opposite angular momentum. This effectively leads to a propagation of the domain wall towards the right-hand side.

by $-\hbar$. Owing to angular momentum conservation the spin of the localized magnetic moments has to be changed by $+\hbar$, accordingly. This effectively leads to a shift of the domain wall to the right side due to the gradual rotation of the magnetic moments in the domain wall, as illustrated in Figure 5.19 (b).

In Figure 5.20 a typical device with a working principle based on the formation and propagation of magnetic domains is shown.

Such a kind of structure can be used as a shift register [70] or as a magnetic memory [64]. It consists of a ferromagnetic stripe, with a width of about 200 nm and a length of several micrometers [70]. The ferromagnetic strip is crossed by two normal metal leads. Lead A is connected to two pulse generators PG1 and PG2 connected to the upper and lower terminal, respectively. The current flows between A and B. As shown in Figure 5.20 (a), by applying a negative pulse to PG1 a so-called tail-to-tail domain wall is formed. Tail-to-tail means, that the magnetization points outwards in opposite direction. Here we assumed that the ferromagnetic stripe is initially polarized with a magnetization pointing from left to right. The fringing field generated by a current pulse flowing through the crossing between lead A and the ferromagnetic stripe results in a magnetization pointing from right to left, forming the corresponding tail-to-tail domain wall. As we learned above, by driving a current through a ferromagnetic conductor, domain walls are moved. Thus, depending on the length of the pulse, the domain wall is shifted by a certain distance. In the situation depicted in Figure 5.20 (b), a negative pulse is applied from the lower side of lead A via PG2. Here we assumed that the stripe was initially magnetized in the negative direction. Due to the different bending of the current path a magnetization pointing from left to right is induced, resulting in a so-called head-to-head domain wall, where the magnetization

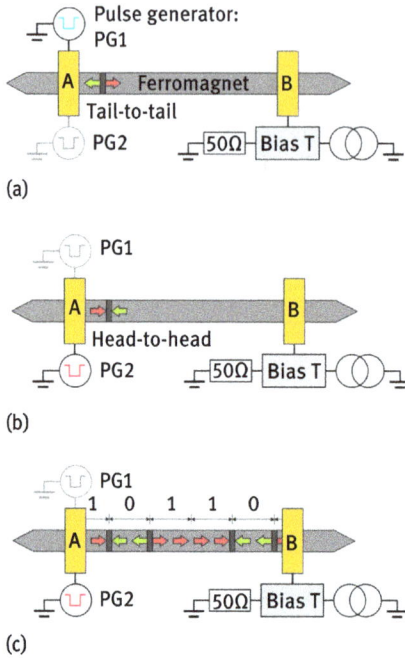

(a)

(b)

(c)

Figure 5.20: Ferromagnetic stripe crossed by two normal metal leads. Lead A is connected to two pulse generators PG1 and PG2. The electrical current flows through the ferromagnetic stripe between A and B. (a) A negative pulse applied from PG1 leads to a formation of a tail-to-tail domain wall. Depending on the pulse length the domain wall is shifted by a certain distance from A. (b) Applying a negative pulse generated by PG2 results in a head-to-head domain wall. (c) Example of a bit sequence after applying according pulses to PG1 and PG2.

points towards each other. By the length of the current pulse the distance of the domain wall propagated from lead A is determined. In the case where the magnetization underneath lead A is already pointing from left to right, no domain wall is formed after applying a pulse from PG2. However, all domain walls already present in the stripe are shifted by a certain distance because of the current flowing between lead A and B. In Figure 5.20 (c) a situation is shown where a number of subsequent pulses were applied to PG1 or PG2. After each pulse the domain walls are shifted by a fixed amount, so that a domain pattern is formed in the ferromagnetic stripe, which corresponds to a bit pattern. Thus, this device can be used as a shift register, where on the left side a bit sequence is transferred in the stripe which is read out at lead B [70]. The read-out can be done by measuring the electrical resistance of the ferromagnetic stripe. This is possible, since each domain wall in the stripe results in a stepwise change of the resistance. By using a longer ferromagnetic stripe and placing a read and write head at the center, a so-called racetrack memory can be realized [64]. For the read head a magnetic tunnel junction can be utilized, while for the write head the fringing fields of a domain wall in a nanowire crossing the storage ferromagnetic stripe can be used. By driving a current through the ferromagnetic stripe the domains wall pattern can be shifted back and forth across the read and write heads. By inserting small dents at the sidewall of the magnetic wires the magnetic domain walls can be pinned [64]. These dents are placed in regular distances, to assign each bit to a certain area.

5.8 Summary

- In magnetic materials the magnetization is often subdivided in magnetic domains. Each domain has a certain orientation of magnetization. Domains are formed in order to minimize the magnetic energy.
- Domain walls are found at the boundary between magnetic domains. The widths of a domain wall is determined by the competition between different energy contributions, e. g. crystal anisotropy and exchange energy contribution.
- The switching field, i. e. coercive field, of a magnetic bar is determined by its shape via the form anisotropy contribution.
- The properties of single magnetic electrodes can be analyzed by means of local Hall effect measurements.
- By micro-magnetic simulations based on the Landau–Lifshitz–Gilbert equation the domains in magnetic electrodes can be calculated.

Exercises

Problem 5.1. Calculate the domain wall width of Fe and permalloy using the parameters $A = 10 \, \text{pJm}^{-1}$ and $8 \, \text{pJm}^{-1}$ and $K_u = 0.15 \, \text{kJm}^{-3}$ and $0.50 \, \text{kJm}^{-3}$, respectively.

Problem 5.2. Find an expression the energy γ_B per unit area contained in a Bloch domain wall. Hint, use the expression for γ_B given by equation (5.10) in connection with equation (5.13).

Problem 5.3. Let us assume a Hall bar with a magnetic electrode on top. The perpendicular component of the stray field B of the magnetic electrode shall be 50 mT. The Hall bar is formed in a two-dimensional electron gas with a sheet electron concentration of $n_{2D} = 8 \times 10^{11} \, \text{cm}^{-2}$. We assume a measurement current of $I = 100 \, \mu\text{A}$. Calculated the expected Hall voltage.

6 Spin injection

6.1 Overview

A typical spin electronic device consists of three components: spin injector, spin manipulator, and spin detector. All these elements are used in the Datta–Das spin field-effect transistor [6]. As shown in Figure 6.1, first a spin-polarized current is provided. For this purpose, a spin injector consisting of magnetic material is utilized as a source electrode. Second, the spin orientation is manipulated by means of a gate electrode. For the spin manipulation, the Rashba effect is employed. The Rashba effect will be discussed in detail in the next chapter. Third, the spin orientation is detected by a magnetic drain contact. The transfer of electrons between source and drain electrodes and thus the current flow between both contacts is determined by the spin orientation of the impinging electrons with respect to the magnetization of the drain contact.

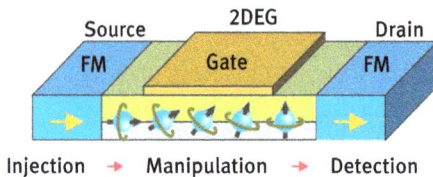

Figure 6.1: Datta–Das spin transistor as an example of a typical spin electronic device [6]. The spins are injected from a ferromagnetic (FM) source contact into a two-dimensional electron gas (2DEG). Here, the spin manipulation takes place by biasing the gate electrode. At the ferromagnetic drain contact the spin orientation is detected.

In this chapter, the properties of the spin injectors and spin detectors are discussed. For both components the same mechanism applies, just the current direction is reversed. In the following, we will only discuss the spin injector, keeping in mind that all findings are valid for spin detectors as well. For spin injectors and detectors various magnetic materials can be employed:

- metallic ferromagnetic contacts, which usually have a high Curie temperature, e. g. Fe, Co, or permalloy (80 % Ni, 20 % Fe);
- half-metallic ferromagnets as ideal spin-polarizers, e. g. Fe_3O_4;
- diluted magnetic semiconductors for good material matching, e. g. $In_{1-x}Mn_xAs$, $Ga_{1-x}Mn_xAs$, or $Cd_{1-x}Mn_xTe$;
- Heusler alloys, ferromagnetic alloys, even though the constituting elements are non-ferromagnetic, e. g. Co_2MnAl, Ni_2MnAl, or Ni_2MnIn.

https://doi.org/10.1515/9783110639001-006

6.2 Resistor model

The most simple way to describe the properties of a spin injector is the resistor model [71]. The underlying idea is that in a magnetic material each of the two spin orientations can be treated separately, i. e. spin scattering is neglected. As known for metallic ferromagnetic materials, e. g. Ni or Fe, the density of states at the Fermi energy differs for the majority and minority spin. Since the density of states is directly related to the conductivity, it is plausible that the conductivity differs as well. Thus, as illustrated in Figure 6.2, the ferromagnetic electrode can be treated as two resistors connected in parallel, one for each spin orientation.

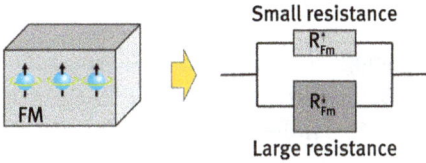

Figure 6.2: Ferromagnetic (FM) contact modeled by two resistance channels in parallel corresponding to the two spin orientations.

The difference in the conductivity of the two channels in the ferromagnet can be quantified by the spin polarization β, which is defined as

$$\beta = \frac{\sigma^\uparrow_{FM} - \sigma^\downarrow_{FM}}{\sigma^\uparrow_{FM} + \sigma^\downarrow_{FM}}, \tag{6.1}$$

with $\sigma^\uparrow_{FM}$ and $\sigma^\downarrow_{FM}$ the conductivity of spin-up and spin-down electrons in the ferromagnet. The spin polarization β can be used to calculate the resistances $R^\uparrow_{FM}$ and $R^\downarrow_{FM}$ in the ferromagnet connected to the spin-up and spin-down orientation:

$$R^\uparrow_{FM} = \frac{2R_{FM}}{1+\beta}, \quad R^\downarrow_{FM} = \frac{2R_{FM}}{1-\beta}, \tag{6.2}$$

with R_{FM} being the total resistance given by

$$\frac{1}{R_{FM}} = \frac{1}{R^\uparrow_{FM}} + \frac{1}{R^\downarrow_{FM}}. \tag{6.3}$$

In three dimensions the resistance R can be calculated from the conductivity σ by $R = d/(F\sigma)$, with F and d the cross section area and the thickness of the electrode, respectively. Of course the geometry is identical for both spin channels. The expressions for $R^\uparrow_{FM}$ and $R^\downarrow_{FM}$ follow from rewriting equation (6.1) to

$$\sigma^\uparrow_{FM} = \sigma^\downarrow_{FM} \frac{1+\beta}{1-\beta}, \tag{6.4}$$

which can be written as

$$\frac{1}{R_{FM}^{\uparrow}} = \frac{1}{R_{FM}^{\downarrow}} \frac{1+\beta}{1-\beta}. \tag{6.5}$$

For all-electrical measurements it is not sufficient to reverse the magnetization of a single ferromagnetic injector to observe a difference in the electrical current through a ferromagnet/semiconductor (FM/SC) structure. This is because when the magnetization is reversed, the resistance for a given spin orientation is changed from a large to a small value or vice versa. However, since the resistance change is exactly opposite for each spin orientation, the total resistance of the network remains the same. As a consequence, the total current through the ferromagnet/semiconductor structure is not changed after magnetization reversal.

The situation differs completely in a ferromagnet/semiconductor/ferromagnet (FM/SC/FM) structure. As illustrated in Figure 6.3, the current through the structure is larger if both ferromagnetic electrodes are aligned in parallel, while a smaller current flows if both electrodes are aligned antiparallel.

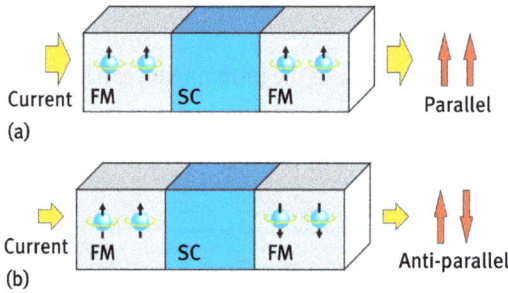

Figure 6.3: Structure with ferromagnetic (FM) contacts on both sides of the semiconductor (SC). (a) In the case where both ferromagnets are aligned in parallel a large current can flow through the structure. (b) For an antiparallel configuration, a smaller current flows. In both cases, the same voltage bias is assumed.

The reason for the difference in the resistance can be inferred from the resistance networks of both configurations depicted in Figure 6.4.

Here, each spin orientation is represented by a separate channel which does not intermix with the other one. Each channel is constituted by a series of three resistors, i. e. the spin-dependent resistors $R_{FM}^{\uparrow}$ and $R_{FM}^{\downarrow}$ of both ferromagnets and the spin-independent resistor of the semiconductor. Each spin channel in the semiconductor has a resistance of $2R_{SC}$, with R_{SC} the total resistance of the semiconductor. For the parallel configuration shown in Figure 6.4 (a), the spin-up channel has a low resistance while the spin-down channel has a high resistance. In contrast, for the antiparallel configuration shown in Figure 6.4 (b), both spin channels have the same resistance.

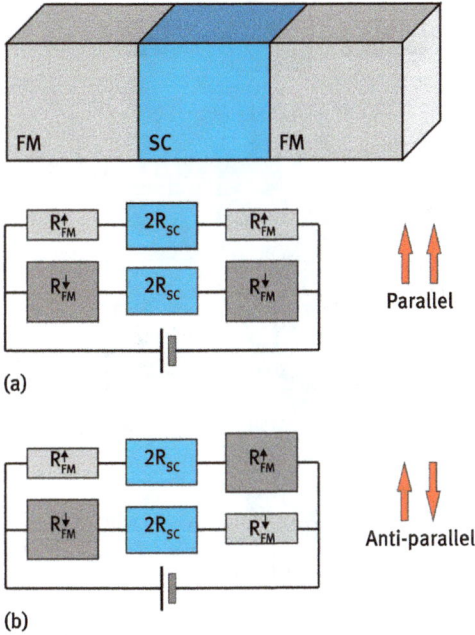

Figure 6.4: Resistance model for the parallel (a) and anti-parallel (b) configuration, respectively. For each spin orientation, a separate current path exists, which is not intermixed with the other. In the semiconductor channel, the resistance for each spin orientation is assumed to be the same, i. e. $2R_{SC}$, with R_{SC} the total resistance of the semiconductor.

By calculating the total resistance of the complete network, one finds that the parallel configuration has a lower resistance than the antiparallel configuration. This can be attributed to the dominance of the low resistance spin-up channel.

It is instructive to have a closer look at the voltage drops for the different spin channels in the parallel and anti-parallel configuration. As shown in Figure 6.5 (a), in the parallel configuration the voltage drop in the ferromagnetic electrodes is lower for the spin-up channel compared to the spin-down channel owing to the lower resistance of the former [71]. Since the total voltage drop is fixed, the voltage drop in the semiconductor is larger for spin-down carriers compared to spin-up carriers. Due to the fact that the resistance in the semiconductor is spin-independent, the current of spin-up carriers is larger than the one for spin-down carriers because of the larger voltage drop. Thus, in total, a spin-polarized current flows in the semiconductor. In contrast, for the antiparallel configuration, the voltage drop in the semiconductor is identical for both spin orientations. As a consequence, the net spin current is unpolarized.

The spin polarization of the current in the semiconductor for the parallel configuration is defined as

$$\alpha = \frac{j^{\uparrow} - j^{\downarrow}}{j^{\uparrow} + j^{\downarrow}}. \tag{6.6}$$

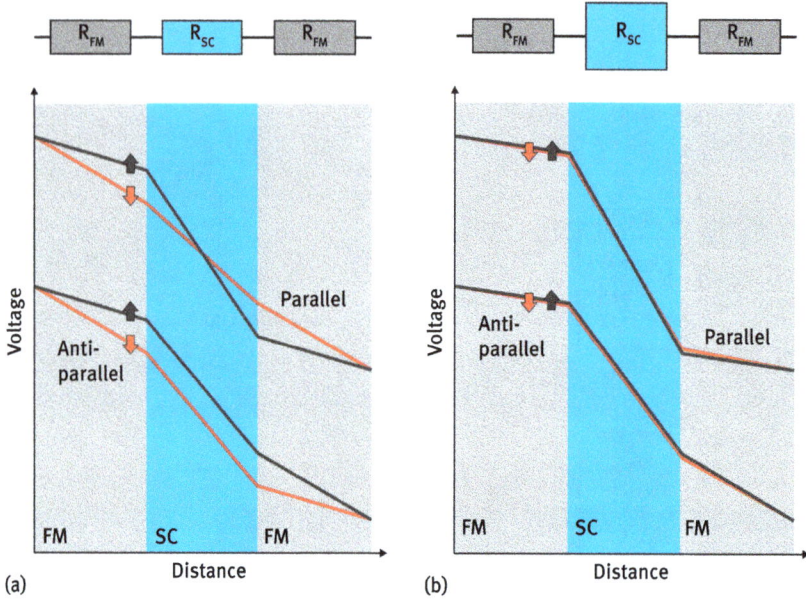

Figure 6.5: (a) Voltage drop as a function of distance for each spin channel in the parallel and anti-parallel configuration. In the parallel configuration, the voltage drop in the semiconductor is different for the two spin orientations. Therefore, a spin-polarized current flows in the semiconductor. In the antiparallel configuration, the voltage drop in the semiconductor is identical, therefore no net spin polarization occurs. (b) Corresponding situation for the case with a much larger resistance of the semiconductor compared to the one of the metal electrodes. Figure adapted from Schmidt and Molenkamp [71].

From the resistor network discussed above, the current spin polarization can be calculated. The resulting expression is

$$\alpha = \beta \frac{R_{FM}}{R_{SC}} \frac{2}{2\frac{R_{FM}}{R_{SC}} + (1 - \beta^2)},$$ (6.7)

with β given by equation (6.1) and R_{SC} and R_{FM} the resistances of the semiconductor segment and of each ferromagnet electrode, respectively. One finds that the spin polarization α in the semiconductor is increased for increasing spin polarization β in the ferromagnets. An interesting feature contained in equation (6.7) is that α scales with the ratio between the resistance of the ferromagnet and the semiconductor R_{FM}/R_{SC}. Especially in case of a metallic ferromagnet as a spin injector, R_{FM} is much smaller than the resistance R_{SC} of the semiconductor. Consequently, the spin-polarized current in the semiconductor is expected to be small. This is illustrated in Figure 6.6, where α is plotted for different ratios R_{FM}/R_{SC}. While for a ratio of 1 the dependence of α on β is almost linear, at $R_{FM}/R_{SC} = 0.01$ a noticeable spin polarization is only found for $|\beta| > 0.8$. However, often such large β values cannot be achieved in the ferromagnetic electrodes.

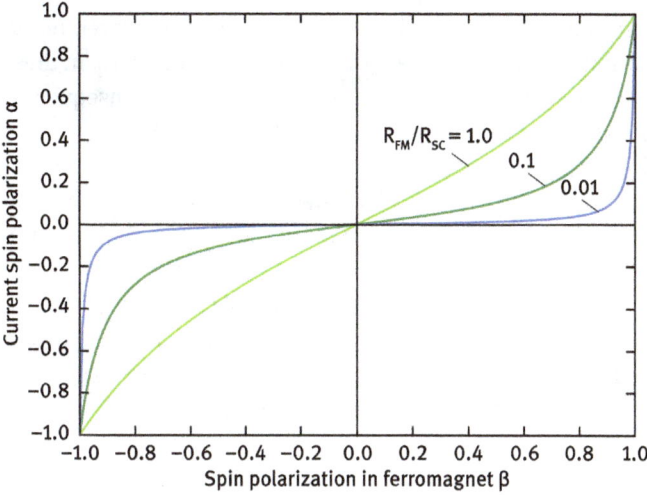

Figure 6.6: Current spin polarization α calculated according to equation (6.7) for different ratios R_{FM}/R_{SC} as a function of the spin polarization β in the ferromagnet.

For practical applications, the relative difference of the resistance between the parallel and anti-parallel configuration is of interest. A detailed calculation gives

$$\frac{\Delta R}{R_{par}} = \frac{\beta^2}{1-\beta^2} \frac{R_{FM}^2}{R_{SC}^2} \frac{4}{4\frac{R_{FM}^2}{R_{SC}^2} + 2\frac{R_{FM}}{R_{SC}} + (1-\beta^2)}. \tag{6.8}$$

Here, R_{par} is the total resistance in the parallel configuration. Once again, if R_{FM}/R_{SC} is small, $\Delta R/R_{par}$ is expected to be small as well. In fact, this is illustrated in Figure 6.5, where the case of comparable ferromagnet and semiconductor resistances is compared to the case where the semiconductor has a much larger resistance. As can be seen in Figure 6.5 (b), due to the much larger resistance in the semiconductor, the main voltage drop occurs there. As a consequence, in the parallel configuration the difference in the voltage drop for both spin orientations, and thus the spin-polarized current is very small.

All in all, one finds that a large mismatch between the resistances of the ferromagnet and the semiconductor, as is usually expected for highly conductive metallic ferromagnets, leads to a very small resistance change $\Delta R/R_{par}$ and to a very small current spin polarization α in the semiconductor. The consequence of the large conductivity mismatch was first pointed out by Schmidt et al. [72].

6.3 Local description of spin injection

In the resistor model described in the previous section, the interaction between the spin channels was neglected completely. In semiconductors this assumption often holds, especially if the contribution of spin-orbit coupling is small. However, in ferromagnetic materials, the spin-flip scattering length can be short. Therefore, in these materials the spin alters its orientation due to scattering events. As a consequence, the resistor model based on two separate uncoupled spin channels fails.

In the model described below, these shortcomings are avoided. This improved model includes spin scattering and is based on a local description of the electrochemical potential $\mu^{\uparrow,\downarrow}$ for each spin channel. An important assumption is that the spin scattering is much slower than all other electronic scattering processes. This still allows the use of the division into two separate spin channels. The properties of the two channels are described in linear response, i. e. restricting to one dimension the current density for the spin-up and spin-down carriers. The current density is proportional to the gradient of the corresponding electrochemical potential:

$$\frac{\partial \mu^{\uparrow,\downarrow}}{\partial x} = -\frac{ej^{\uparrow,\downarrow}}{\sigma^{\uparrow,\downarrow}}. \tag{6.9}$$

Here, the conductivity for each spin channel is given by the Einstein relation, cf. equation (2.33):

$$\sigma^{\uparrow,\downarrow} = D(E_F)^{\uparrow,\downarrow} e^2 \mathcal{D}, \tag{6.10}$$

with $D(E_F)^{\uparrow,\downarrow}$ being the density of states at the Fermi level and $\mathcal{D}$ the diffusion constant. The exchange of carriers with spin of opposite orientations is determined by the spin diffusion equation

$$\frac{\mu^{\uparrow} - \mu^{\downarrow}}{\tau_{sf}} = \mathcal{D}\frac{\partial^2(\mu^{\uparrow} - \mu^{\downarrow})}{\partial x^2}. \tag{6.11}$$

Here, τ_{sf} is the spin-flip scattering time. As long as there is an imbalance between $\mu^{\uparrow}$ and $\mu^{\downarrow}$, spin-flip scattering occurs at a rate τ_{sf}^{-1}. The spin-flip or spin diffusion length λ_{sf} is connected to τ_{sf} by

$$\lambda_{sf} = \sqrt{\tau_{sf}\mathcal{D}}. \tag{6.12}$$

It is the typical length scale on which the spin information is lost. In the ferromagnet, the spin polarization can be expressed as

$$\beta = \frac{D^{\uparrow}(E_F) - D^{\downarrow}(E_F)}{D^{\uparrow}(E_F) + D^{\downarrow}(E_F)}, \tag{6.13}$$

with $D^{\uparrow}(E_F)$ and $D^{\downarrow}(E_F)$ being the density of states of the spin-up and spin-down states at the Fermi energy, respectively. Using the Einstein relation for the conductance one can also write the spin polarization as

$$\beta = \frac{\sigma^{\uparrow}_{FM} - \sigma^{\downarrow}_{FM}}{\sigma^{\uparrow}_{FM} + \sigma^{\downarrow}_{FM}}. \tag{6.14}$$

By employing this expression, one obtains the conductivity of each spin channel as a function of β. These expressions have to be inserted in equation (6.9), in order to calculate the current densities and electrochemical potentials. In case of the semiconductor, the conductivity does not depend on the spin orientation. Therefore, one can simply write

$$\sigma_{SC}^{\uparrow,\downarrow} = \frac{\sigma_{SC}}{2}, \tag{6.15}$$

with σ_{SC} being the total conductivity of the semiconductor.

Let us discuss the profile of the electrochemical potentials in a ferromagnet/semiconductor structure close to the interface. The general solutions of the diffusion equation on the ferromagnet and semiconductor side can be written as

$$\mu^{\uparrow,\downarrow}(x) = \mu_0 + ax + c^{\uparrow,\downarrow}e^{x/\lambda_{FM}} \quad \text{for } x \leq 0 \tag{6.16}$$

and

$$\mu^{\uparrow,\downarrow}(x) = \tilde{\mu}_0 + bx + d^{\uparrow,\downarrow}e^{-x/\lambda_{SC}} \quad \text{for } x > 0, \tag{6.17}$$

respectively. Here, λ_{FM} and λ_{SC} are the spin-flip lengths in the ferromagnet and in the semiconductor, respectively, while μ_0 and $\tilde{\mu}_0$ are the offsets of the electrochemical potentials. Without losing generality, one can set $\mu_0 = 0$.

Next we have to find expressions for the different coefficients. The relation between $c^{\uparrow}$ and $c^{\downarrow}$ can be obtained from the current conservation within the ferromagnet, i. e. the total current density has to be constant $j(x) = j^{\uparrow}(x) + j^{\downarrow}(x)$. The current density is obtained from equation (6.9) by inserting equation (6.16), which results in

$$j^{\uparrow,\downarrow} = -\frac{\sigma_{FM}^{\uparrow,\downarrow}}{e}\left(a + \frac{c^{\uparrow,\downarrow}}{\lambda_{FM}}e^{x/\lambda_{FM}}\right). \tag{6.18}$$

By assuming that the total current density at $x \to -\infty$ and at $x = 0$ is the same one finds

$$c^{\uparrow} = -c^{\downarrow}\frac{\sigma_{FM}^{\downarrow}}{\sigma_{FM}^{\uparrow}}. \tag{6.19}$$

In the same manner one obtains for the corresponding coefficients in the semiconductor:

$$d^{\uparrow} = -d^{\downarrow}. \tag{6.20}$$

The continuity of the electrochemical potential at $x = 0$ gives

$$c^{\uparrow} = \tilde{\mu}_0 + d^{\uparrow}, \tag{6.21}$$

$$c^{\downarrow} = \tilde{\mu}_0 + d^{\downarrow}, \tag{6.22}$$

which results in

$$\tilde{\mu}_0 = \frac{c^{\uparrow} + c^{\downarrow}}{2} \tag{6.23}$$

and

$$d^{\uparrow} = \frac{c^{\uparrow} - c^{\downarrow}}{2}. \tag{6.24}$$

In the limits of $x \to -\infty$ and $x \to +\infty$ for the ferromagnet and semiconductor, respectively, both spin channels are in equilibrium, and therefore one can write

$$a = -\frac{ej}{\sigma_{FM}}, \tag{6.25}$$

$$b = -\frac{ej}{\sigma_{SC}}. \tag{6.26}$$

From the continuity of the current $j^{\uparrow}(x)$ at $x = 0$ we obtain the relation

$$j^{\uparrow}(0) = -\frac{\sigma_{FM}^{\uparrow}}{e}\left(a + \frac{c^{\uparrow}}{\lambda_{FM}}\right) = -\frac{\sigma_{SC}/2}{e}\left(b - \frac{d^{\uparrow}}{\lambda_{SC}}\right). \tag{6.27}$$

Inserting equation (6.24) results in

$$c^{\uparrow} = \left[\frac{\sigma_{SC}}{2e\lambda_{SC}}\frac{1}{2}\left(1 + \frac{\sigma_{FM}^{\uparrow}}{\sigma_{FM}^{\downarrow}}\right) + \frac{\sigma_{FM}^{\uparrow}}{e\lambda_{FM}}\right] = j(0)\left(\frac{1}{2} - \frac{\sigma_{FM}^{\uparrow}}{\sigma_{FM}}\right). \tag{6.28}$$

By using

$$\frac{1}{2} - \frac{\sigma_{FM}^{\uparrow}}{\sigma_{FM}} = -\frac{1}{2}\beta \tag{6.29}$$

and

$$1 + \frac{\sigma_{FM}^{\uparrow}}{\sigma_{FM}^{\downarrow}} = \frac{2}{1 - \beta}, \tag{6.30}$$

derived from equation (6.14), one finally obtains for the coefficients $c^{\uparrow}$, $c^{\downarrow}$, $d^{\uparrow}$, $d^{\downarrow}$, and $\tilde{\mu}_0$ in equations (6.16) and (6.17)

$$c^{\uparrow} = -\frac{\lambda_{SC}}{\sigma_{SC}}\frac{ej\beta(1 - \beta)}{1 + \frac{\lambda_{SC}}{\lambda_{FM}}\frac{\sigma_{FM}}{\sigma_{SC}}(1 - \beta^2)}, \tag{6.31}$$

$$c^{\downarrow} = +\frac{\lambda_{SC}}{\sigma_{SC}}\frac{ej\beta(1 + \beta)}{1 + \frac{\lambda_{SC}}{\lambda_{FM}}\frac{\sigma_{FM}}{\sigma_{SC}}(1 - \beta^2)}, \tag{6.32}$$

$$d^{\uparrow} = -d^{\downarrow} = -\frac{\lambda_{SC}}{\sigma_{SC}} \frac{ej\beta}{1 + \frac{\lambda_{SC}}{\lambda_{FM}} \frac{\sigma_{FM}}{\sigma_{SC}}(1 - \beta^2)}, \tag{6.33}$$

$$\tilde{\mu}_0 = +\frac{\lambda_{SC}}{\sigma_{SC}} \frac{ej\beta^2}{1 + \frac{\lambda_{SC}}{\lambda_{FM}} \frac{\sigma_{FM}}{\sigma_{SC}}(1 - \beta^2)}. \tag{6.34}$$

Using these coefficients, the profile of electrochemical potentials for spin-up and spin down electrons in the ferromagnet and in the semiconductor can be calculated.

In Figure 6.7, typical profiles of the electrochemical potentials at a ferromagnet/semiconductor interface are shown for $\sigma_{FM} = \sigma_{SC}$ and $\sigma_{FM} = 10\sigma_{SC}$, respectively. For the spin polarization in the ferromagnet, $\beta = 0.5$ was assumed.

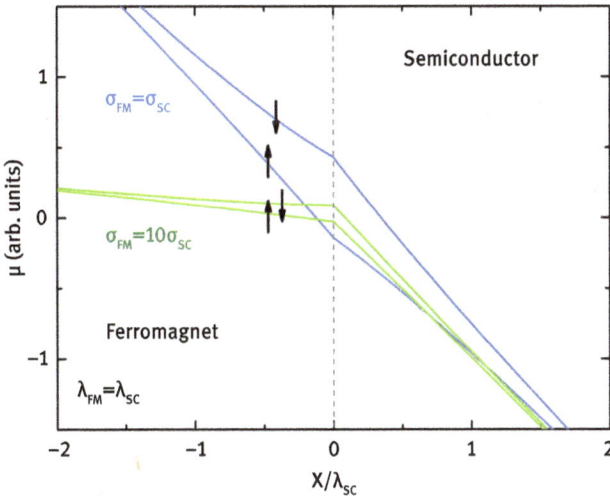

Figure 6.7: Electrochemical potential for spin-up and spin-down carriers at a ferromagnet/semiconductor structure. For the spin-flip lengths $\lambda_{FM} = \lambda_{SC}$ was assumed, while for the conductivities $\sigma_{FM} = \sigma_{SC}$ (blue curve) and $\sigma_{FM} = 10\sigma_{SC}$ (green curve) were taken. The spin polarization β was set to 0.5 in both cases.

Far from the interface, the electrochemical potentials of both spin orientations are the same due to the effect of spin scattering. Close to the interface, the electrochemical potentials are split. This gives an imbalance of the spin current at the interface and thus in a spin polarization. As one can see in Figure 6.7, a conductivity mismatch ($\sigma_{FM} = 10\sigma_{SC}$) results in a smaller splitting between $\mu^{\uparrow}(x)$ and $\mu^{\downarrow}(x)$ and thus in a smaller difference of the spin currents. Owing to the larger conductivity in the ferromagnet, the voltage drop is smaller, i. e. the slope of $\mu^{\uparrow}(x)$ and $\mu^{\downarrow}(x)$ is smaller.

Using the approach described above, the current spin polarization in the semiconductor α can be calculated for a ferromagnet/semiconductor/ferromagnet structure [71]. From the diffusion equations together with the proper boundary conditions one obtains

$$\alpha = \beta \frac{\lambda_{FM}}{x_{SC}} \frac{\sigma_{SC}}{\sigma_{FM}} \frac{2}{2\frac{\lambda_{FM}\sigma_{SC}}{x_{SC}\sigma_{FM}} + (1 - \beta^2)}, \tag{6.35}$$

with x_{SC} as the length of the semiconductor section. In order to keep the formula relatively simple, spin-flip scattering in the semiconductor was neglected, i. e. $\lambda_{SC} \rightarrow \infty$. The functional dependency is the same as in equation (6.7) if one replaces R_{FM}/R_{SC} by σ_{SC}/σ_{FM} and includes the additional prefactor λ_{FM}/x_0. Thus, we expect a decreased spin polarization in the semiconductor with increasing σ_{SC}/σ_{FM}, as already illustrated in Figure 6.6. The spin polarization for a fixed ratio $\sigma_{SC}/\sigma_{FM} = 0.1$ is plotted in Figure 6.8 (a) for various ratios λ_{FM}/x_0. On finds that for increasing spin scattering in the ferromagnet, thus decreasing λ_{FM}, the spin polarization decreases. The underlying reason can be inferred from Figure 6.7. For decreasing λ_{FM} the splitting of the electrochemical potentials $\mu_\uparrow(x)$ and $\mu_\downarrow(x)$ at the interface becomes smaller. This reduces the achievable spin polarization right away. In Figure 6.8 (b) it is shown, how fast the spin polarization in the semiconductor drops with increasing length x_0 of the semiconductor segment. Here $\lambda_{FM} = 10$ nm and $\sigma_{SC}/\sigma_{FM} = 0.1$ were assumed. For a very large polarization in the ferromagnet, i. e. $\beta = 0.98$ more than 30 % spin polarization can be achieved for $x_0 = \lambda_{FM}$. However, for smaller values of β only very small values of α are expected.

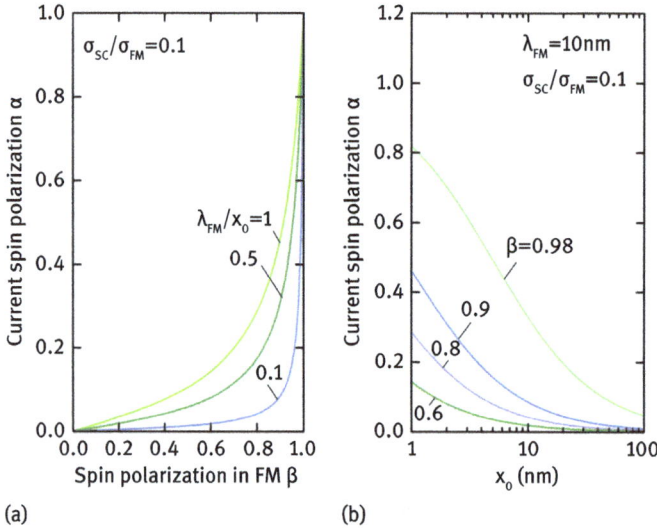

(a)   (b)

Figure 6.8: (a) Current spin polarization in the semiconductor α as a function of the spin polarization in the ferromagnet β for different λ_{FM}/x_0 assuming a fixed ratio $\sigma_{SC}/\sigma_{FM} = 0.1$. (b) Current spin polarization as a function of the length x_0 of the semiconductor layer for different values of β. Here we assumed $\lambda_{FM} = 10$ nm, and $\sigma_{SC}/\sigma_{FM} = 0.1$. For both graphs spin scattering in the semiconductor was neglected.

Based on equation (6.35), the relative change in the resistance $\Delta R/R_{par}$ between the parallel and antiparallel configuration can be calculated [71]:

$$\frac{\Delta R}{R_{par}} = \frac{\beta^2}{1-\beta^2} \frac{\lambda_{FM}^2}{\sigma_{FM}^2} \frac{\sigma_{SC}^2}{x_{SC}^2} \frac{4}{4\frac{\lambda_{FM}^2 \sigma_{SC}^2}{x_{SC}^2 \sigma_{FM}^2} + 2\frac{\lambda_{FM}\sigma_{SC}}{x_{SC}\sigma_{FM}} + (1-\beta^2)}, \qquad (6.36)$$

with R_{par} being the resistance of the parallel configuration. From this formula we can once again infer that for small ratios σ_{SC}/σ_{FM}, as usually occurs when a semiconductor is combined with a metallic ferromagnet, a very small resistance change $\Delta R/R_{par}$ is expected.

This can be seen in Figure 6.9. With decreasing σ_{SC}/σ_{FM} the curves of $\Delta R/R_{par}$ vs. β are shifted downwards by about two orders of magnitude if the conductance ratio is reduced by one order of magnitude.

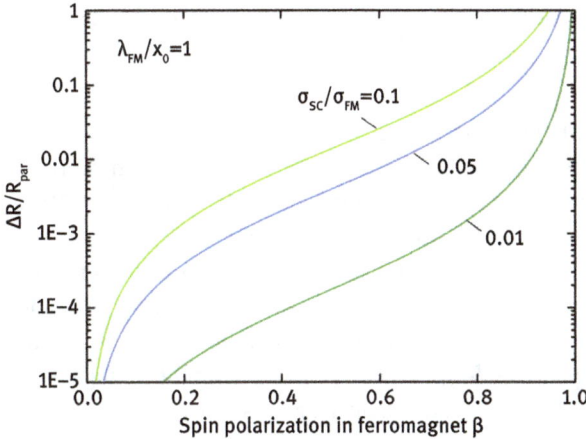

Figure 6.9: Dependence of $\Delta R/R_{par}$ on the spin polarization β in the ferromagnet for various ratios $\sigma_{SC}/\sigma_{FM} = 0.1, 0.05,$ and 0.01. Here, $\lambda_{FM}/x_0 = 1$ was assumed.

We will now have a look at how the electrochemical potentials for both spin orientations depend on the spatial position in the FM/SC/FM structure for various spin scattering lengths λ_{SC} in the semiconductor. In Figure 6.10 (a), λ_{SC} is comparable to the length of the semiconductor section x_{SC}.

The electrochemical potentials are bent towards each other, due to the finite spin scattering length. If the spin-flip length in the semiconductor is much smaller than its length $\lambda_{SC} \ll x_{SC}$, the electrochemical potentials at the center of the semiconductor in the antiparallel magnetization configuration as well as in the parallel magnetization configuration are equilibrated (cf. Figure 6.10 (b)). Thus, $\Delta R/R_{par}$ is expected to be literally zero. Only at the two interfaces, the electrochemical potentials are split, indicating a spin imbalance in the spin currents. In Figure 6.10 (c), the case $\lambda_{SC} \to \infty$

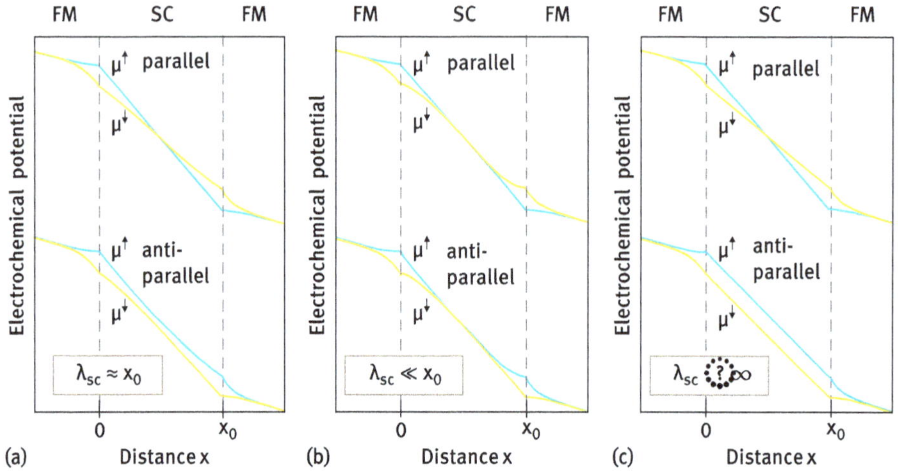

Figure 6.10: Electrochemical potential in a ferromagnet/semiconductor/ferromagnet structure for the parallel and anti-parallel configuration. (a) $\lambda_{SC} \approx x_0$, (b) $\lambda_{SC} \ll x_0$, (c) $\lambda_{SC} \to \infty$. After Schmidt and Molenkamp [71].

is illustrated (cf. equation (6.36)). Since no spin-flip scattering occurs in the semiconductor, the electrochemical potential decreases linearly. Due to the different slopes of $\mu^\uparrow(x)$ and $\mu^\downarrow(x)$ in the parallel case, there is a current imbalance (cf. equation (6.9)) between both spin directions, whereas for the antiparallel case the gradient of $\mu^\uparrow(x)$ and $\mu^\downarrow(x)$ is the same, so that the current density for each spin direction is the same, i. e. there is no spin polarized current. As discussed in the previous section, the situations shown in Figure 6.10 are idealized, since the voltage drops in the semiconductor are comparable to the voltage drops in the ferromagnet. In most cases the voltage drops in the semiconductor are much larger, so that we once again arrive at the conclusion that the spin polarization is very low.

6.4 Optical detection of spin-polarized carriers

In the previous sections, we learned that the electrical injection and detection efficiency can be rather low. As given in Figure 6.11 (a), typical values for the injection efficiency are around 1 %. In an all-electrical device, e. g. a ferromagnet/semiconductor/ferromagnet structure, the spin-polarized carriers have to be detected as well. Since the detection process is just the reversed injection process, the detection efficiency is also about 1 %. Thus, the total efficiency is around 0.01 %. As a matter of fact, this value is very low, and this makes it extremely difficult to detect any spin signal by electrical means and also to get any information about spin injection and detection efficiency in general.

However, there is an alternative method to gain information on spin injection, i. e. by optically detecting the spin injection into the semiconductor, as illustrated in

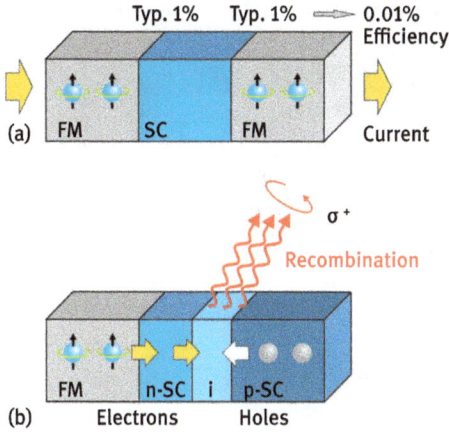

Figure 6.11: (a) Typical values of injection and detection efficiency in a ferromagnet/semiconductor/ferromagnet (FM/SC/FM) device are 1% each, which results in a total efficiency of 0.01%. (b) Schematic of the optical detection of the spin polarization by means of circularly polarized light using a light emitting diode. Here spin-polarized electrons are injected into the n-type area of the semiconductor. These electrons recombine in the intrinsic area with the unpolarized holes from the other side.

Figure 6.11 (b). This method relies on a *pin* light emitting diode (LED), as sketched in Figure 6.12. A *pin* diode is a *pn*-diode with an intrinsic (*i*) layer inserted in between. Holes and electrons are transferred from a *p*-type and an *n*-type layer, respectively, into the intrinsic quantum well. The quantum well layer consists of a semiconductor material with a lower band gap than the ones of the *p*- and *n*-type layers. Electrons and holes are gathered in the quantum well and recombine by emitting a photon with an energy corresponding to the energy difference between the electron and the hole state.

The process of electron-hole recombination and the corresponding emission of photons can be used to gain information on the spin polarization of holes or electrons. The underlying reason is that the spin polarization of the recombined carriers is directly connected to the circular polarization of the emitted light. Thus, by injecting spin-polarized carriers into the semiconductor and analyzing the degree of circular polarization, information on the spin injection efficiency can be gained. In principle, there are two possibilities to realize such kind of spin-LED. As illustrated in Figure 6.12 (a), spin-polarized electrons can be injected using either a II-VI diluted magnetic semiconductor or a metallic ferromagnetic contact. In these structures, the spins of the holes supplied for the recombination are unpolarized, since the corresponding contact is made of a non-magnetic material.

In Figure 6.12 (b), a spin-LED is depicted, where spin-polarized holes are injected, while the spin of the electrons is unpolarized. Spin-polarized holes can be generated

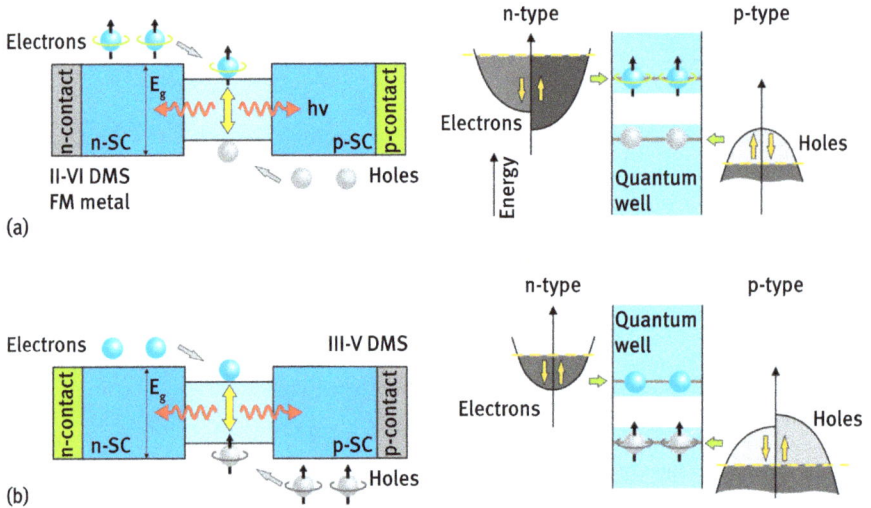

Figure 6.12: Spin *pin* light-emitting diode: (a) Electrons and holes are supplied from the *n*- and *p*-type sections and gathered in the intrinsic center quantum well. The quantum well is formed by a semiconductor with a smaller band gap. The recombination of an electron and a hole results in the emission of a photon. The energy of the emitted photon *hv* corresponds to the energetic difference between the electron and hole states in the quantum well. In the *n*-type semiconductors, spin-polarized electrons are supplied by a spin-injector i. e. a II-VI diluted magnetic semiconductor or a ferromagnetic metal. In the sketch on the right-hand-side, the density of states of both electrodes and the occupation in the quantum well are depicted. (b) Schematic of a spin-LED with a *p*-type spin injector, e. g. by using a III-V diluted magnetic semiconductor (DMS). The electrons recombining with the holes in the quantum well are non-spin-polarized. The sketch on the right-hand-side illustrates the density of states of the electrodes and the occupation of carriers in the quantum well.

by using a III-V diluted magnetic semiconductor, which is a *p*-type semiconductor (cf. Section 4.3).

In order to understand why the recombination of spin-polarized electrons or holes results in a circular polarization of the emitted photons, we have to take a look on the band structure of a typical III-V semiconductor, as schematically illustrated in Figure 6.13 (a). To obtain any optical transition, the semiconductor has to be a direct band gap semiconductor, i. e. the maximum of the valence band has to be at the same position in momentum space as the minimum of the conduction band. The fact that the minimum of the conduction band and the maximum of the valence band are located at the same position in momentum space, i. e. the Γ-point in the Brillouin zone, ensures that for the recombination of electrons from the conduction band and holes from the valence band no momentum is transferred. Thus, a photon with zero momentum can be emitted directly.

For the optical transitions, the angular momentum has to be conserved. As depicted in Figure 6.13 (a), the conduction band is formed by *s*-type atomic orbitals with orbital angular momentum quantum number $l = 0$, so that the total angular momen-

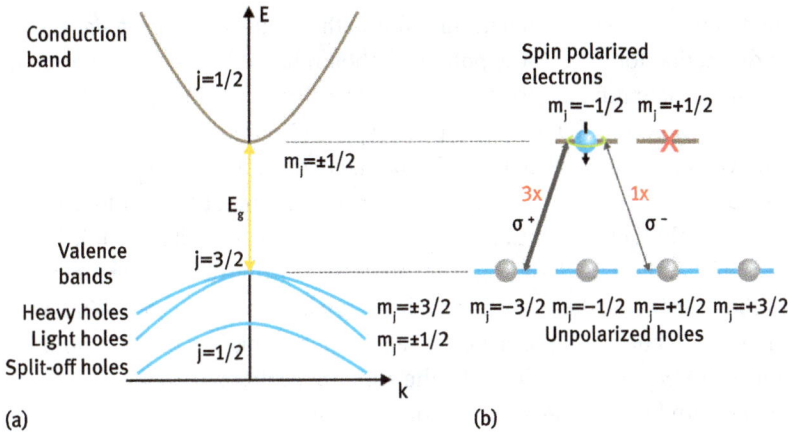

Figure 6.13: (a) Schematic band structure of a typical III-V semiconductor. The states of the conduction band are formed by s-type orbitals, with spin $\pm 1/2$ and orbital angular momentum $l = 0$. Thus, the total angular momentum quantum number is $j = 1/2$. The valence band states are formed by p-type orbitals with angular momentum $l = 1$, so that the total angular momentum of the states are either $j = 3/2$ or $1/2$. The latter is the split-off band with the magnetic quantum numbers $m_j = \pm 1/2$. The former set of bands is subdivided in the heavy and light hole bands. (b) Optical transitions between the conduction and the valence band. Due to conservation of the angular momentum and the fixed momentum of $\Delta m = \pm 1$ of the emitted circularly polarized photons, only certain transitions are possible. In the scheme, we assumed that only spin-down polarized carriers are injected into the conduction band while the corresponding spin-up state is empty.

tum is $j = 1/2$. The valence band states are formed by p-type atomic orbitals with orbital angular momentum $l = 1$. Thus, owing to spin-orbit coupling the total angular momentum can either be $j = 1/2$ or $3/2$. The band belonging to $j = 1/2$ is the split-off band, which is separated at the Γ-point from the upper valence bands by Δ_{so}. The upper valence bands with $j = 3/2$ are subdivided in the heavy hole and light hole band corresponding to $m_j = \pm 3/2$ and $\pm 1/2$, respectively.

In Figure 6.13 (b), the transitions between the conduction and valence bands are sketched. Here, we disregard transitions connected to the split-off band. The angular momentum of the emitted circularly polarized photon can be

$$\Delta m = \pm 1. \tag{6.37}$$

For a right-handed circularly polarized photon σ^+, one gets

$$\Delta m = +1 \leftrightarrow \sigma^+,$$

while for a left-handed circularly polarized photon σ^-, the angular momentum is

$$\Delta m = -1 \leftrightarrow \sigma^-.$$

Owing to the given total angular momentum value in the conduction and valence band and the condition that for a circularly polarized photon $\Delta m = \pm 1$, the conservation of the total angular momentum allows only certain transitions (cf. Figure 6.13 (b)). For example the transition of a spin down electron $m_j = -1/2$ in the conduction band to a state in the valence band is only possible for valence band states $m_j = -3/2$ and $+1/2$, by the constraints connected to the angular momentum of the photon. In the first transition, a right-handed circularly polarized photon σ^+ is emitted, while in the second one, a left-handed circularly polarized photon σ^-. Owing to the different matrix elements of the transitions, the transition to the heavy hole state, i. e. $m_j = -3/2$, with an emission of a σ^+ photon is 3 times more likely than the emission of a σ^- photon to the light hole state ($m_j = +1/2$). Similarly, the spin-up electrons with $m_j = +1/2$ can only recombine with holes in the $m_j = -1/2$ or $+3/2$ states. In this case, the transition to the heavy hole state, i. e. $m_j = +3/2$, with the emission of a σ^- photon is 3 times more likely than the transition to the light hole state by emitting a σ^+ photon.

If all states in the conduction band are equally occupied, i. e. no spin polarization of the carriers, the same number of σ^+ and σ^- photons is emitted. Thus, in total no circularly polarized light is observed. However, in case that spin-polarized electrons are present in the conduction band, circularly polarized light is emitted. This is illustrated in Figure 6.13 (b), where spin-down electrons are assumed to be injected into the conduction band by a spin injector electrode, while the spin-up states remain empty. In principle, right- and left-polarized photons are emitted, however, since the emission of a σ^+ photon is 3 times more likely than the emission of a σ^- photon [73], and in total right-handed circularly polarized light is observed. Correspondingly, for spin injection of spin-up electrons, left-handed circularly polarized light would be expected. Thus, the observation of circularly polarized light is a direct indication of spin polarized carriers in the semiconductor.

The degree of circularly polarized light P_{opt} can be directly related to the occupation of spin-up and spin-down states, $n^\uparrow$ and $n^\downarrow$, respectively, by accounting for the different transition probabilities:

$$P_{opt} = \frac{\sigma^+ - \sigma^-}{\sigma^+ + \sigma^-} = \frac{(3n^\uparrow + n^\downarrow) - (3n^\downarrow + n^\uparrow)}{(3n^\uparrow + n^\downarrow) + (3n^\downarrow + n^\uparrow)} = \frac{1}{2}\frac{n^\uparrow - n^\downarrow}{n^\uparrow + n^\downarrow}. \tag{6.38}$$

6.5 Experiments on optical detection of spin polarization

In Figure 6.14, a typical setup for measuring a spin polarization in a semiconductor is shown. The circularly polarized light emitted from the sample is converted into linearly polarized light by means of a $\lambda/4$-plate. The orientation of the linearly polarized light is then analyzed by a linear polarizer. The spectral distribution of the emitted light is measured by a spectrometer. In most cases, the sample is placed in a magnet so that the magnetization of the spin injector can be changed.

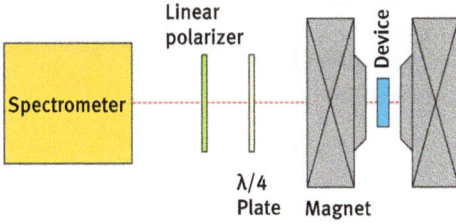

Figure 6.14: Setup to measure the spin polarization in a semiconductor by optical means. The circularly polarized light emitted from the sample is transformed into linearly polarized light by a $\lambda/4$-plate. By using a linear polarizer, the polarization orientation is determined. The spectrometer is used to measure the emission as a function of wavelength. The magnet coils are employed to apply an external magnetic field in order to polarize the injector.

A typical spin-LED is shown in Figure 6.15 [74]. Here, a II-VI diluted magnetic semiconductor, i. e. BeMnZnSe, is used for injecting spin-polarized electrons. The LED itself consists of an intrinsic GaAs quantum well and p- and n-type AlGaAs barrier layers. The spin-polarized electrons are injected into the n-type AlGaAs layer, while the unpolarized holes are supplied by the p-type layer. In order to obtain spin-polarized electrons, an external magnetic field has to be applied. This is necessary, since II-VI diluted magnetic semiconductors are not ferromagnetic. The magnetic field is oriented perpendicularly to the layer system. The circularly polarized light is also collected along this direction.

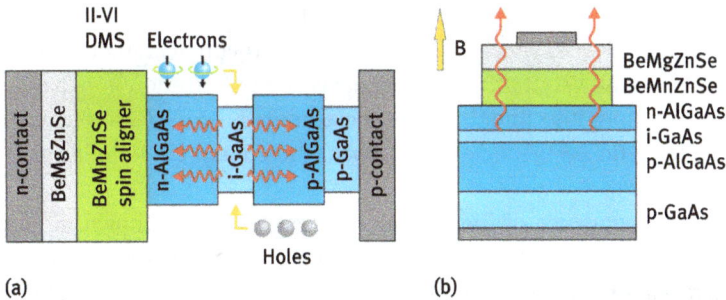

Figure 6.15: (a) Layer structure of a spin-LED based on a BeMnZnSe II-VI diluted magnetic semiconductor spin injector. For the LED, an AlGaAs/GaAs/AlGaAs layer system is used. The spin-polarized electrons and the unpolarized holes recombine in the GaAs quantum well. (b) Cross section of the spin LED device. The magnetic field for spin alignment is oriented perpendicularly to the layer system. After Fiederling et al. [74].

In Figure 6.16 (a), the degree of optical polarization P_{opt} of the emitted photons is plotted as a function of the applied magnetic field [74]. At zero field no polarization is observed, since no spin-polarized carriers are present in the $Be_{0.07}Mn_{0.03}Zn_{0.9}Se$ layer. This is due to the fact that this material is a strong paramagnet and not a ferro-

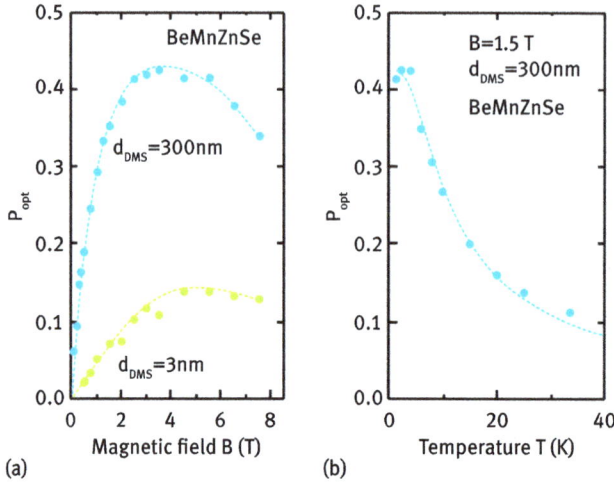

Figure 6.16: (a) Optical polarization P_{opt} as function of magnetic field of a spin LED with a 300 nm and 3 nm thick $Be_{0.07}Mn_{0.03}Zn_{0.9}Se$ spin injector, respectively. The temperature was below 4 K. (b) Degree of optical polarization P_{opt} as a function of temperature for the sample with a 300 nm thick $Be_{0.07}Mn_{0.03}Zn_{0.9}Se$ spin aligner. The magnetic field was fixed at 1.5 T. Graphs adapted from Fiederling et al. [74].

magnet. If the magnetic field is increased, first a linear dependence is found, which can be explained in the frame of the giant Zeeman effect (cf. Section 4.2). Above 3 T a saturation is observed which is attributed to a saturation of the magnetization of the localized magnetic moments. For the 300 nm thick BeMnZnSe layer a maximum polarization of 43 % is achieved, which is very close to the theoretical limit of 50 % (cf. equation (6.38)). For the very thin BeMnZnSe layer the polarization is reduced, owing to an incomplete spin injection. In order to exclude any artifacts, a reference sample with a nonmagnetic BeMgZnSe electrode was studied as well. Here, no spin polarization was found [74].

The temperature dependence of the optical polarization P_{opt} is shown in Figure 6.16(b). As can be seen here, P_{opt} and thus the spin polarization decreases significantly with increasing temperature. Around 30 K, P_{opt} is only one quarter of the initial value. The reason for this strong decrease of P_{opt} is the decrease of the effective g-factor, leading to a smaller spin splitting of the energy levels. If the energy splitting is smaller, the probability of thermally exciting the upper level with the opposite spin orientation is higher. Thus, the net spin polarization is reduced.

A layer of a III-V diluted magnetic semiconductor, e. g. GaMnAs, can also be employed to inject spin-polarized carriers into an LED structure [58]. In contrast to the case discussed above, here the spin injector is p-type, and thus spin-polarized holes are injected. Furthermore, the spin injector is ferromagnetic. A schematic of a spin-LED based on a III-V diluted magnetic semiconductor as spin injector is depicted in Figure 6.17. The p-type GaMnAs spin polarizer is magnetized in-plane; therefore the

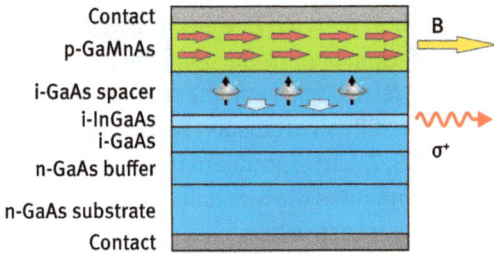

Figure 6.17: Schematics of a spin-LED with a *p*-type GaMnAs spin injector. The LED layer stack consists of an InGaAs quantum well with GaAs barriers. From the bottom contact unpolarized electrons are supplied. Figure adapted from Ohno et al. [58].

light emitted from the LED structure has to be collected from the side, in order to detect the circular polarization correctly.

Due to the ferromagnetic properties of the GaMnAs spin polarizer, the measured degree of optical polarization P_{opt} shows a clear hysteresis in the corresponding experiments [58]. By increasing the temperature the change in polarization decreases as well as the coercive field, i. e. the width of the hysteresis. Above 52 K, the hysteresis in P_{opt} vanished, since the Curie temperature was exceeded. Thus the optical response directly reflects the magnetic properties of the GaMnAs spin polarizer.

6.6 Injection through a barrier

In the previous Sections 6.2 and 6.3, we discussed that the large difference in the conductivity between the semiconductor and the metallic spin injector and detector results in a considerable lowering of the resistance change $\Delta R/R_{par}$ between the parallel and antiparallel magnetization of the magnetic electrodes. As mentioned already, this so-called conductivity mismatch problem was first pointed out by Schmidt et al. [72]. However, very soon after, Rashba [75] suggested that this problem can be avoided by introducing a barrier layer between the magnetic electrode and the semiconductor, as illustrated in Figure 6.18.

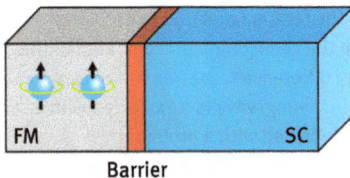

Figure 6.18: Layer system with a barrier placed between the ferromagnetic spin injector and the semiconductor.

Below we will discuss two theoretical approaches: first, the free electron approximation, where the electron transport through the ferromagnet/semiconductor interface is
described by plane waves [76]: scattering in the ferromagnet and the semiconductor is
neglected; in the second part we will discuss the diffusive transport regime, including
finite spin diffusion lengths in both materials [75, 77]. Irrespective of both approaches,
one finds that introducing an interface barrier enhances the resistance change considerably in a ferromagnet/semiconductor/ferromagnet structure.

6.6.1 Free electron approximation

A relatively simple theoretical approach to describing the transport from a ferromagnet to a semiconductor through a barrier is to use the free electron approximation in
the ballistic limit [76]. As illustrated in Figure 6.19 (a), in the ferromagnetic layer the
spin states are energetically shifted by the exchange energy Δ_{ex}. For the semiconductor we assume a two-dimensional electron gas (2DEG), since this is the basis for the
spin field-effect transistor. The conductivity mismatch between the ferromagnet and
the semiconductor is mainly due to the different electron concentrations. This can be
quantified by the energetic offset Γ between the lower conduction band of the ferromagnet and the semiconductor. For the tunneling barrier the most simple description
using a δ-shaped barrier is employed, with the barrier height quantified by U_0. The
electronic states are filled up to the common Fermi energy E_F.

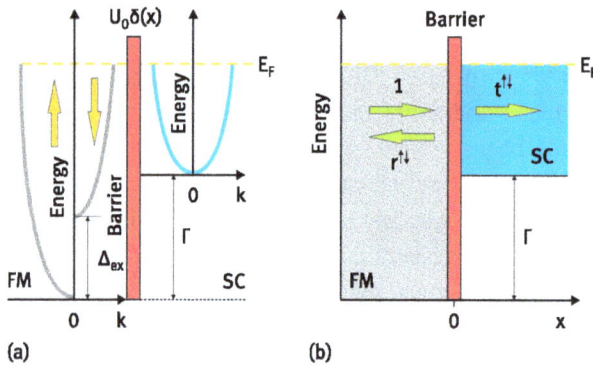

Figure 6.19: (a) Ferromagnet/semiconductor structure with a δ-shaped barrier inserted at the interface in the free electron approximation. The energy shift in the ferromagnet due to exchange
interaction is given by Δ_{ex}. The lower electron concentration in the semiconductor compared to the
metallic ferromagnet is incorporated by introducing the energy difference Γ. (b) Transfer of carriers
at a tunneling barrier described by plane waves. A part of the incoming wave is transferred with a
probability $t^{\uparrow\downarrow}$ and the remaining part of the incoming wave is reflected with a probability $r^{\uparrow\downarrow}$.

As depicted in Figure 6.19 (b), the charge carrier transport through that tunneling barrier at $x = 0$ is described by quantum mechanical reflection and transmission. Here

we employ the free electron approximation, where the electronic states are described by plane waves. Scattering within the ferromagnet and the 2DEG is neglected. In order to obtain the plane wave solutions in the ferromagnet ($x < 0$) and the semiconductor ($x > 0$), we have to solve the Schrödinger equation in each section. The corresponding single electron Hamiltonian is given by

$$\hat{H} = -\frac{\hbar^2}{2m(x)}\Delta + V(x) + U_0\delta(x) - \vec{h}(x)\cdot\hat{\vec{\sigma}}, \tag{6.39}$$

where $m(x)$ equals the free electron mass m_0 in the ferromagnet and the effective electron mass m^* in the semiconductor, respectively. Owing to the lower electron mass and lower electron concentration in the semiconductor, a potential step of $V(x) = \Gamma$ is assumed for $x > 0$. In the ferromagnet $V(x) = 0$. The last term represents the exchange splitting in the ferromagnet $\vec{h}(x) = \pm h_0\vec{e}_y$, with $\Delta_{\text{ex}} = 2h_0$ the exchange splitting energy (cf. Figure 6.19 (a)). Within the semiconductor $\vec{h}(x) = 0$. In the ferromagnet ($x < 0$), the energy eigenvalues of the Schrödinger equation are given by

$$E_{\text{FM}}^\uparrow = \frac{\hbar^2}{2m_0}(\vec{k}_{\text{FM}}^\uparrow)^2, \tag{6.40}$$

$$E_{\text{FM}}^\downarrow = \frac{\hbar^2}{2m_0}(\vec{k}_{\text{FM}}^\downarrow)^2 + \Delta_{\text{ex}}, \tag{6.41}$$

while the eigenvalues in the semiconductor are

$$E_{\text{SC}}^\sigma = \frac{\hbar^2}{2m^*}(\vec{k}_{\text{SC}}^\sigma)^2 + \Gamma. \tag{6.42}$$

Here, σ denotes the spin direction, which can be either spin-up ($\uparrow$) or spin-down ($\downarrow$). Since we assumed a two-dimensional system for the semiconductor, the wave vectors have only two components $\vec{k}_{\text{FM}}^\sigma = (k_{\text{FM},x}^\sigma, k_{\text{FM},y}^\sigma)$. The plane wave solutions in the different regions for spin-up electrons are

$$\Psi_{\text{FM}}^\uparrow(\vec{r}) = \begin{pmatrix} 1 \\ 0 \end{pmatrix} e^{i\vec{k}_{\text{FM}}^\uparrow \vec{r}} + r^\uparrow \begin{pmatrix} 1 \\ 0 \end{pmatrix} e^{-i\vec{k}_{\text{FM}}^\uparrow \vec{r}}, \tag{6.43}$$

$$\Psi_{\text{SC}}^\uparrow(\vec{r}) = t^\uparrow \begin{pmatrix} 1 \\ 0 \end{pmatrix} e^{i\vec{k}_{\text{SC}}^\uparrow \vec{r}}, \tag{6.44}$$

which corresponds to a situation that a plane wave with a wave vector $\vec{k}_{\text{FM}}^\uparrow$ impinges at the barrier and is either reflected with a reflection amplitude $r^\uparrow$ or transmitted into the semiconductor with a transmission amplitude $t^\uparrow$. The reflection and transmission amplitudes are determined by the boundary conditions. For the spin-down state, similar expressions are obtained. Since at a given energy in the ferromagnet the wave vectors for spin-up and spin-down are different, the corresponding reflection and transmission amplitudes differ as well. This will finally lead to an imbalance of the spin currents through the barrier. At quasi-equilibrium the transport takes place at the Fermi energy $E_{FM}^\sigma = E_{SC}^\sigma = E_F$. Under these conditions the absolute values of the wave vectors in the ferromagnet and semiconductor are

$$k_{FM}^\uparrow = \frac{1}{\hbar} \sqrt{2m_0 E_F}, \tag{6.45}$$

$$k_{FM}^\downarrow = \frac{1}{\hbar} \sqrt{2m_0 (E_F - \Delta_{ex})}, \tag{6.46}$$

$$k^{SC} = \sqrt{2\pi n_{2D}}, \tag{6.47}$$

with n_{2D} being the electron concentration in the two-dimensional electron gas. Since the potential is only varied along the x direction, k_y^σ is the same in both materials because of momentum conservation. With ϕ the angle of incidence at the interface on the ferromagnetic side, the wave vector components in x-direction can be expressed as

$$k_{FM,x}^\sigma(\phi) = k_{FM}^\sigma \cos\phi, \tag{6.48}$$

$$k_{SC,x}^\sigma(\phi) = \sqrt{(k_{SC}^\sigma)^2 - (k_{FM,y}^\sigma(\phi))^2}. \tag{6.49}$$

Using these wave vectors, the boundary conditions for δ-potentials [78] at $x = 0$ are applied for the wave functions:

$$\Psi_{FM}^\sigma(x = 0, y) = \Psi_{SC}^\sigma(x = 0, y) \tag{6.50}$$

as well as for the derivatives:

$$\frac{\hbar^2}{2m^*} \frac{\partial \Psi_{SM}^\sigma(\vec{r})}{\partial x}\bigg|_{x=0+} - \frac{\hbar^2}{2m_0} \frac{\partial \Psi_{FM}^\sigma(\vec{r})}{\partial x}\bigg|_{x=0-} = U_0 \Psi_{FM}^\sigma(x = 0, y). \tag{6.51}$$

The last expression is an appropriate derivative boundary condition for δ-shaped potentials [78]. By applying the boundary conditions, the spin-dependent transmission coefficient can be determined:

$$T^\sigma(\phi, Z) = \frac{v_{SC,x}^\sigma}{v_{FM,x}^\sigma} |t^\sigma|^2 = \frac{4 v_{SC,x}^\sigma v_{FMx}^\sigma}{4(v_{FM}^\uparrow Z)^2 + (v_{SC,x}^\sigma + v_{FM,x}^\sigma)^2}. \tag{6.52}$$

Here, the group velocities $v_{FM,x}^\sigma = \hbar k_{FM,x}^\sigma / m_0$ and $v_{SC,x}^\sigma = \hbar k_{SC,x}^\sigma / m^*$ were inserted. For the sake of better comparison the dimensionless Z parameter for the barrier strength given by

$$Z = \frac{U_0}{\hbar v_{FM}^\uparrow} \tag{6.53}$$

was included in equation (6.52). The transmission coefficient is the parameter, which determines the current from the ferromagnet to the semiconductor. Since $v_{FM,x}^\uparrow$ and

$v_{\text{FM},x}^{\uparrow}$ differ for finite exchange energies Δ_{ex}, the transmission coefficient at the Fermi energy is also different for both spin directions. Using the expression for the transmission coefficient, the spin-dependent current density can be calculated explicitly

$$j^{\sigma} = \frac{e^2 V}{(2\pi)^2} \int\limits_{k_{\text{FM},x}^{\sigma} > 0} d^2 \vec{k}_{\text{FM}}^{\sigma} \delta(E_{\sigma} - E_{\text{F}}) T^{\sigma} v_{\text{FM},x}^{\sigma}. \tag{6.54}$$

Here, we assumed that the transport takes place at the Fermi energy, which is valid at small bias voltages V. The current density is determined by the group velocity and the corresponding transmission coefficient. The integration is performed over positive wave vectors only. Using equation (6.48) the integration can be transformed to an integration over the angle of incidence:

$$j^{\sigma} = \frac{e^2 V k_{\sigma}^{\text{FM}}}{h\pi} \int\limits_{0}^{\phi_{\sigma}^c} d\phi T^{\sigma}(\phi, Z) \cos\phi, \tag{6.55}$$

with ϕ_c^{σ} being the critical angle of incidence. For angles larger than ϕ_c^{σ}, total reflection occurs, since the kinetic energy in the x-direction is too small to overcome the potential step Γ at the interface. Note that due to the different group velocities the current density is different for spin-up and spin-down. The current densities for both spin directions can be employed to calculate the current polarization $\alpha = (j^{\uparrow} - j^{\downarrow})/(j^{\uparrow} + j^{\downarrow})$, with $j = j^{\uparrow} + j^{\downarrow}$ the total current density.

In Figure 6.20 the effect of an interface barrier on the current polarization α is illustrated.

Within the given range of the electron concentration $n_{2\text{D}}$ the current polarization increases substantial with increasing barrier strength Z. Note that for $Z = 0$, thus no δ-barrier, the current polarization inverses when $n_{2\text{D}}$ is increased. This is an effect of the decreasing potential step Γ with increasing electron concentration. The larger Z is the smaller is the effect of $n_{2\text{D}}$ on the current polarization. The carrier transfer is solely governed by the δ-barrier.

In order to verify spin polarization by an electrical measurement we have to measure the current through an ferromagnet/semiconductor/ferromagnet sandwich, where the magnetization of the ferromagnetic electrodes can be switched between a parallel and an antiparallel configuration. The corresponding potential profile and energy-momentum dispersions are illustrated in Figure 6.21. Here two interfaces have to be considered resulting in total transmission probabilities for the whole structure [76]. Using these quantities the relative resistance change $\Delta R/R_{\text{par}}$ between the parallel and antiparallel configuration can be calculated, with R_{par} being the resistance for the parallel configuration.

The corresponding results from the calculation of $\Delta R/R_{\text{par}}$ as a function of the barrier height Z for different potential step heights Γ/E_{F} are shown in Figure 6.22 (a).

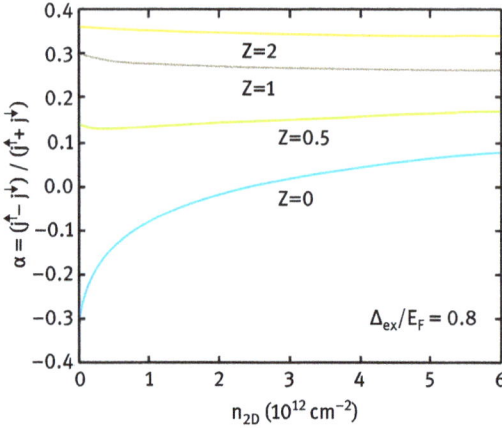

Figure 6.20: Current polarization α as a function of the electron density in the two-dimensional electron gas for various values of the barrier strength Z with $\Delta_{ex}/E_F = 0.8$. For E_F a value of 5 eV was assumed. The effective mass was assumed to be $m^* = 0.04m_0$. Figure adapted from Heersche et al. [76].

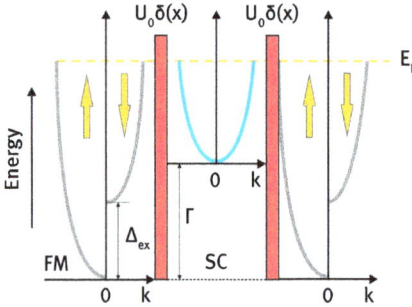

Figure 6.21: Potential landscape and energy-momentum dispersions of a ferromagnet/semiconductor/ferromagnet structure with a δ-shaped barrier at each interface. Here the configuration with a parallel orientation of the magnetization of both ferromagnets is shown.

For Z larger than about 0.5 the resistance change $\Delta R/R_{par}$ is generally increasing with Z. For smaller Γ/E_F values, thus larger electron concentrations in the two-dimensional electron gas, $\Delta R/R_{par}$ is larger. This can be explained by the improved conductance matching between the ferromagnet and semiconductor. Depending on the specific value of Γ an increase of more than a factor of 10 can be achieved. In the calculations performed for $\Delta_{ex}/E_F = 0.6$, the resistance change is modulated for $Z < 0.5$. This can be attributed to the sign change of current polarization α (cf. Figure 6.20). In Figure 6.22 (b), the resistance change $\Delta R/R_{par}$ is plotted as a function of Z and various values of the exchange splitting in the ferromagnet. Once again one finds that the resistance change improves with increasing Z, apart from some modulations at smaller values of Z. As expected, $\Delta R/R_{par}$ increases for increasing Δ_{ex}/E_F. Although

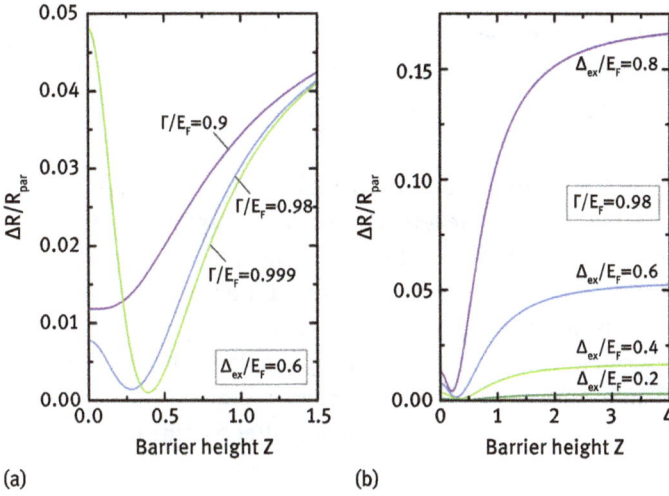

Figure 6.22: (a) Normalized resistance difference between the parallel and anti-parallel magnetization configuration $\Delta R/R_{par}$ in a ferromagnet/semiconductor/ferromagnet structure as a function of barrier height Z for various values of the step height Γ/E_F. The exchange splitting was fixed to Δ_{ex}/E_F, assuming $E_F = 5$ eV. (b) Corresponding plot for various values of Δ_{ex}/E_F with Γ/E_F fixed to 0.98.

the resistance change improves for increasing values Z one should keep in mind that increasing Z, goes along with a corresponding increase of the total resistance.

6.6.2 Diffusive transport regime

The transport in a ferromagnet/semiconductor/ferromagnet structure with interface barriers can also be treated theoretically in the diffusive transport regime [75, 77]. A schematic of the system is shown in Figure 6.24 (inset). The semiconductor is coupled on both sides to the ferromagnetic electrodes by a barrier layer. In contrast to the previous case, a bulk semiconductor is assumed. The approach discussed here is an extension of the model introduced in Section 6.3. As an additional feature a spin-dependent interface resistance per unit cross section area

$$r_\pm = 2r_B(1 \pm \gamma)$$

is included, with r_B being the total interface resistance and γ a parameter quantifying the spin dependent tunneling. The resistances per unit cross section of the ferromagnet and semiconductor are r_{FM} and r_{SM}, respectively. These values are calculated by multiplying the resistivities ρ_{FM} and ρ_{SC} with the respective spin diffusion lengths λ_{FM} and λ_{SC}. This procedure is plausible, since for distances larger than the spin diffusion length all spin imbalances are basically compensated. By solving the diffusion equation under the appropriate boundary conditions the current spin polarization at the ferromagnet/semiconductor interface is obtained:

$$\alpha = \frac{j^{\uparrow} - j^{\downarrow}}{j^{\uparrow} + j^{\downarrow}} = \frac{\beta r_{FM} + \gamma r_B}{r_{FM} + r_{SC} + r_B}, \tag{6.56}$$

with β being the spin polarization of the ferromagnet (cf. equation (6.13)). If one ne-
glects the interface resistance in the above formula, the current spin polarization reduces to $\alpha = \beta/(1 + r_{SC}/r_{FM})$, which corresponds to the previously discussed case where
for $r_{SC} \gg r_{FM}$ the current polarization is largely suppressed. However, if an interface
barrier is included, one finds that α remains large as long as $r_B > r_{SC}$.

In Figure 6.23, the spatial dependence of current spin polarization for three different scenarios is shown. In case of a metallic ferromagnet, e. g. Co, combined with
a metallic Cu layer, a noticeable spin polarization remains after passing the interface,
even without introducing an interface barrier. However, by combining the ferromagnetic Co electrode (FM) directly with a semiconductor (SC), i. e. $r_B = 0$, basically no
spin polarization is left in the semiconductor. Only if an interface barrier is introduced
which is in the order of or larger than the resistance of the semiconductor $r_B \geq r_{SC}$,
a noticeable current spin polarization is expected in the semiconductor section.

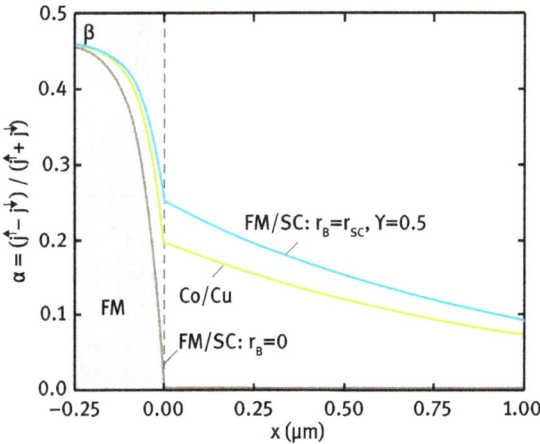

Figure 6.23: Current polarization α at a ferromagnet/semiconductor interface. It can be clearly seen
that the introduction of an interface barrier ($r_B \neq 0$) leads to a noticeable polarization of current
in the semiconductor. In the case where no interface barrier is present the current polarization is
zero in the semiconductor. As a reference, the result for an all-metal interface, i. e. Co/Cu, is also
shown. The following parameters were used: $\beta = 0.46$, $\lambda_{FM} = 60$ nm, $r_{FM} = 4.5 \times 10^{-15}$ Ωm^2 for Co;
$r_{Cu} = 6 \times 10^{-15}$ Ωm^2, $\lambda_{Cu} = 1$ µm for Cu; and $r_{SC} = 4 \times 10^{-9}$ Ωm^2, and $\lambda_{SC} = 2$ µm for the semiconductor.
Figure adapted from Fert and Jaffrès [77].

Let us once again move over to a ferromagnet/semiconductor/ferromagnet structure
including a barrier at each interface, as depicted in Figure 6.24. In this case, the resis-

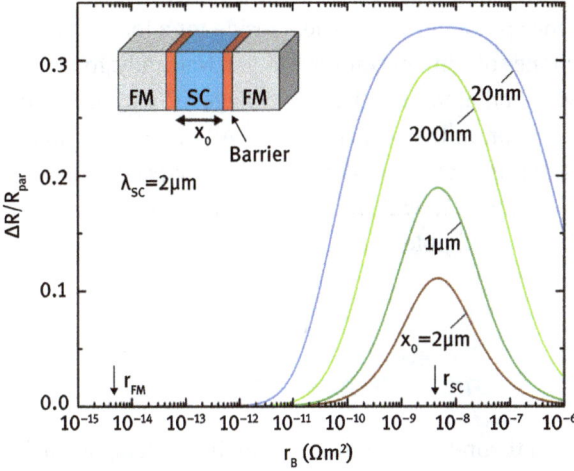

Figure 6.24: Relative resistance change $\Delta R/R_{par}$ as a function of interface resistance r_B for different lengths x_0 of the semiconductor. For the spin diffusion length λ_{SC} is assumed to be 2 µm. The inset shows a schematic of the sample layout. Figure adapted from Fert and Jaffrès [77].

tance change between the parallel and antiparallel configuration is given by [77]

$$\Delta R = \frac{1}{F} \frac{2(\beta r_{FM} + \gamma r_B)^2}{(r_B + r_{FM})\cosh(\frac{x_{SC}}{\lambda_{SC}}) + \frac{r_{SC}}{2}[1 + (\frac{r_B}{r_{SC}})^2]\sinh(\frac{x_{SC}}{\lambda_{SC}})}, \qquad (6.57)$$

with x_{SC} the thickness of the semiconductor layer and F the cross section area. In the case of a small thickness of the semiconductor layer compared to its spin diffusion length, $x_0 \ll \lambda_{SC}$, the expression given above simplifies to

$$\Delta R = \frac{1}{F} \frac{2(\beta r_{FM} + \gamma r_B)^2}{(r_B + r_{FM}) + \frac{r_{SC}}{2}[1 + (\frac{r_B}{r_{SC}})^2]\frac{x_{SC}}{\lambda_{SC}}}. \qquad (6.58)$$

In order to obtain the relative change of resistance $\Delta R/R_{par}$, we need to know the expression for the resistance for the parallel configuration, which is given by [79, 77]

$$R_{par} = \frac{1}{F}\left[2(1-\beta^2)r_{FM} + r_{SC}\frac{x_{SC}}{\lambda_{SC}} + 2(1-\gamma^2)r_B \right.$$
$$\left. + 2\frac{(\beta-\gamma)^2 r_{FM}r_B + r_{SC}(\beta^2 r_{FM} + \gamma^2 r_B)\tanh(\frac{x_{SC}}{2\lambda_{SC}})}{(r_{FM} + r_B) + r_{SC}\tanh(\frac{x_{SC}}{2\lambda_{SC}})} \right]. \qquad (6.59)$$

The resulting change of the resistance $\Delta R/R_{par}$ as a function of interface resistance r_B is shown in Figure 6.24. It can be seen here that first $\Delta R/R_{par}$ increases when the interface barrier is increased. Further on, when r_B is approximately as large as the resistance of the semiconductor, r_{SC}, a maximum is reached. For even larger interface resistances the resistance difference $\Delta R/R_{par}$ drops. This can be explained by the fact that for large interface barriers the spin-polarized carriers reside for a longer time in the semiconductor, so that the chance of spin relaxation is larger. Naturally, for a thinner semiconductor layer the highest obtainable resistance change $\Delta R/R_{par}$ increases, because of the reduced spin relaxation in the semiconductor. All in all, one can say that introducing an interface barrier indeed results in a largely enhanced resistance change. As can be inferred from Figure 6.24, while for values of r_B much smaller than r_{SC} the resistance change $\Delta R/R_{par}$ is negligible, for r_B in the order of r_{SC} ratios up to 30 % can be achieved.

6.7 Experiments on spin injectors with interface barriers

In the previous two sections it was theoretically shown that by introducing a barrier between the highly conductive metallic ferromagnet and the semiconductor the resistance change $\Delta R/R_{par}$ can be largely enhanced. In this section, experiments are introduced where a barrier layer is inserted between the ferromagnet and the light emitting diode (LED) structure used for spin detection. There are a number of possibilities to insert a barrier layer. One option is to use a dielectric oxide barrier, e. g. Al_2O_3 or MgO. Another possibility is to use the intrinsic properties of the semiconductor, i. e. a Schottky barrier, which is formed at a metal/semiconductor interface.

A typical cross-section of a spin-LED with a ferromagnetic injector and an oxide interface barrier layer is shown in Figure 6.25 [80]. Here, a 2 nm thick Al_2O_3 layer is used as a barrier. As a ferromagnetic electrode, either Co, Ni, or permalloy is employed. Spin-polarized electrons are injected from the ferromagnet into the semiconductor through the Al_2O_3 barrier and recombine with the unpolarized holes in the GaAs quantum well. The orientation of the magnetization in the ferromagnetic film is defined by an external magnetic field perpendicular to the layer system. The polarized light is also collected in this direction.

The electroluminescence measurements are performed at opposite magnetic fields [80]. The measurement setup is shown in Figure 6.14. The difference of the detected circularly polarized light is a measure for the degree of spin polarization if the measurement is performed with the same settings of the polarizers. For a proper evaluation of the spin polarization, the effect of magnetic circular dichroism has to be subtracted. Information on the latter can be obtained from photoluminescence measurements. The total spin polarization was found to be around 1 % [80].

Alternatively, a Schottky barrier can also be used to increase the spin injection efficiency [81, 82]. The corresponding spin-LED layer structure and potential profile are depicted in Figure 6.26 (a) and (b), respectively. The height of the Schottky barrier is fixed because it is given by the material combination, in this case Fe/AlGaAs. However, the width of the barrier and thus the transparency can be tuned by doping, which determines the width of the space charge layer. The spin-polarized electrons

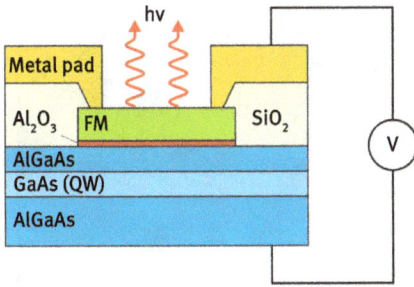

Figure 6.25: Cross section of an AlGaAs/GaAs/AlGaAs spin-LED with an Fe, Ni, or permalloy ($Ni_{80}Fe_{20}$) layer as a ferromagnetic (FM) injector. The barrier layer is made of Al_2O_3. The ferromagnetic layer is sufficiently thin that the light emitted from the quantum well can be transmitted. The spin-polarized electrons recombine in the GaAs quantum well (QW) with the unpolarized holes. The holes are supplied from the bottom layer. Schematics adapted from Manago and Akinaga [80].

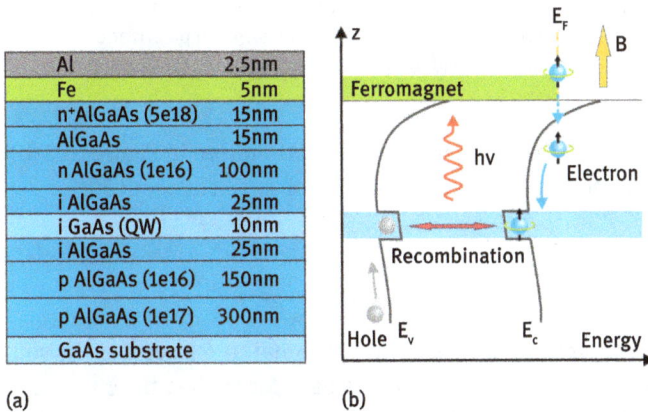

(a)   (b)

Figure 6.26: (a) Layer sequence of a spin-LED using a Schottky barrier to enhance the spin injection efficiency. The width of the Schottky barrier is shaped by the doping profile of the AlGaAs layers. The electron and hole pairs recombine in the GaAs quantum well. The doping is given in cm^{-3}. (b) Schematics of the band diagram of the spin-LED structure.

are injected through the Schottky barrier and are collected in the GaAs quantum well. Here the electrons recombine with the unpolarized holes.

In Figure 6.27 (a) an electroluminescence measurement is shown for 0 and 6 T external magnetic field. At zero field no difference is detected for the two orientations, σ^+ and σ^-, of circularly polarized light. Here the remnant magnetization is in-plane, so that no difference is expected. With increasing magnetic field the magnetization of the ferromagnetic layer is rotated out of plane. As a consequence, the difference between σ^+ and σ^- increases systematically, which can be attributed to the recombination of more and more spin-polarized electrons with unpolarized holes. At a field of 6 T a large difference in the electroluminescence signal is observed. The circular opti-

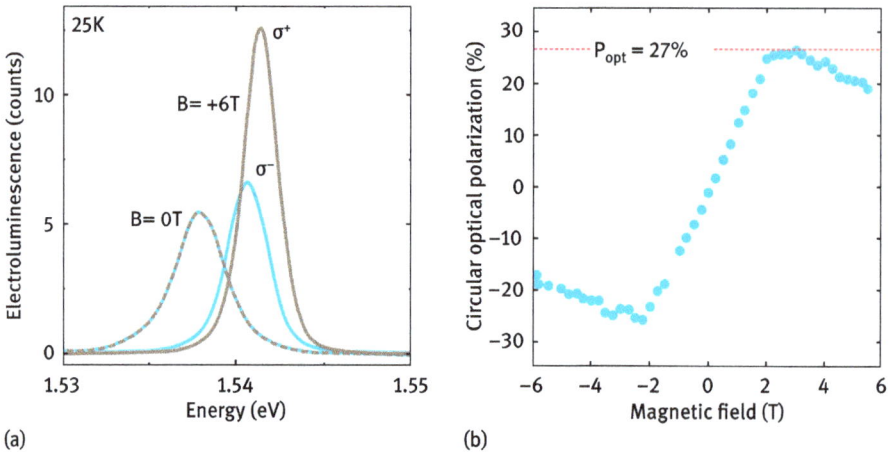

(a)   (b)

Figure 6.27: Electroluminescence at 0 and 6 T for an AlGaAs/GaAs/AlGaAs spin-LED with a Schottky barrier. An Fe layer was used as a spin injector. The layer sequence is given in Figure 6.26. σ^+ and σ^- are the different orientations of circularly polarized light. (b) Circular optical polarization as a function of the external magnetic field. Figure provided by B. Beschoten, RWTH Aachen University.

cal polarization as a function of external magnetic field is depicted in Figure 6.27 (b). As can be seen here, a maximum polarization of 27 % is observed at about 3 T.

6.8 Nonlocal spin injection

The devices we discussed so far were realized as two-terminal structures, i. e. the voltage drop is measured at the same pair of contacts as used for feeding the current. In these so-called spin valve structures, constituted of a normal metal or semiconductor sandwiched between two ferromagnetic electrodes, the spin signal is measured locally. The two-terminal configuration of course also implies that some side-effects might contribute to the measured result. The most obvious is the contact resistance, which results in an additional voltage drop. Furthermore, in magnetic structures other effects can contribute to spurious effects, i. e. the impact of local stray fields from the ferromagnetic metal electrodes or the anisotropic magnetoresistance (AMR) of the injector and detector. Therefore, these measurements are often considered to be unreliable for proving spin injection [83, 84]. In order to avoid the problems mentioned above, so-called nonlocal measurements of spin imbalance are performed [85]. In this setup a four-terminal configuration is used.

In Figure 6.28 (a), typical sample used for nonlocal measurements is shown [68].

It consists of an InN nanowire which is covered at the center by two Co electrodes, which are used for spin injection and detection. The Co electrodes have a different width, in order to obtain different switching fields (cf. Section 5.4). A detail of the InN nanowire crossed by these two ferromagnetic electrodes is depicted in Figure 6.29.

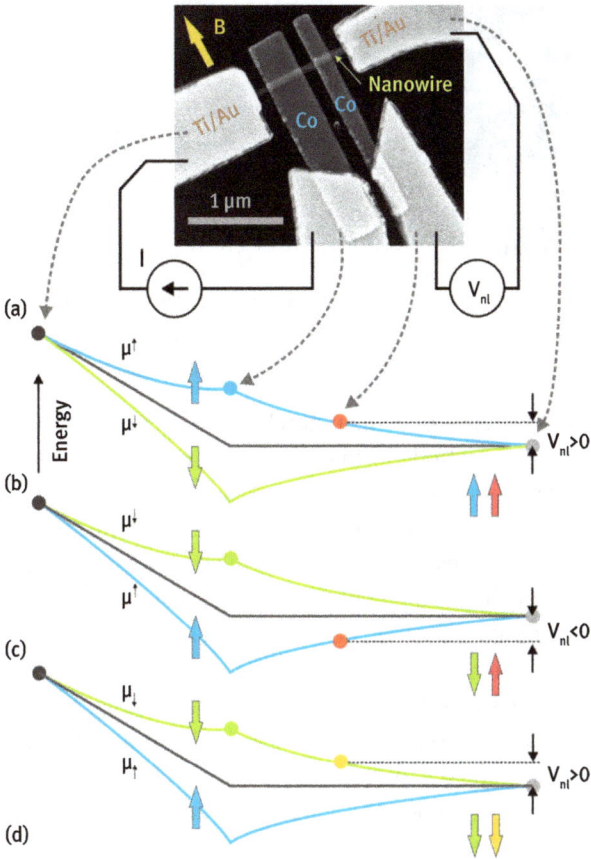

Figure 6.28: (a) Top view of an InN nanowire crossed by two ferromagnetic Co electrodes of different widths. Both ends of the nanowire are contacted by nonmagnetic Ti/Au electrodes. The external
magnetic field is applied along the Co electrodes. (b) Electrochemical potential for injection from
the left Co electrode. The voltage is applied between the left normal metal contact and the left Co
electrode. The bias voltage induces a current flow between these two electrodes. The voltage drop
is measured between the right Co electrode and the right normal metal contact. For the situation depicted here, both Co electrodes are in spin-up, i. e. parallel configuration. Owing to its polarization,
the right Co electrode picks up the electrochemical potential of the spin-up electrons. A nonlocal
voltage $V_{nl} > 0$ is measured between the two right contacts. (c) After applying a magnetic field opposite to the magnetization, first the wider Co electrode reversed its magnetization, and thus an
antiparallel configuration is obtained. The majority spins are spin-down. The right ferromagnetic
electrode still picks up the electrochemical potential of the spin-down electrodes, thus a negative
voltage is measured. (d) Upon further increasing the external magnetic field, the right Co electrode
inverses its magnetization as well, so that the parallel configuration is achieved. Now a positive
voltage is detected once again.

The magnetization direction of the Co electrodes is controlled by applying an external
in-plane magnetic field oriented along the stripes. As can be seen in Figure 6.28 (a),

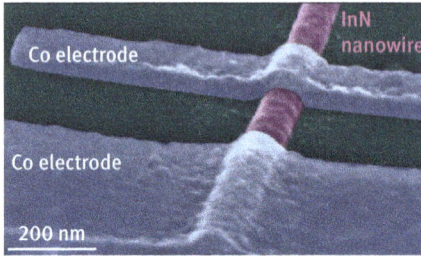

Figure 6.29: Scanning electron beam micrograph of an InN nanowire crossed by two Co electrodes of different widths. Image provided by S. Heedt, Forschungszentrum Jülich.

each end of the InN nanowire is covered with a Ti/Au normal metal contact. By driving a bias current I from the left Co electrode to the left normal contact, spins are injected into the InN nanowire. This leads to an imbalance of the spin occupation in the nanowire. Figure 6.28 (b) shows the situation where spin-up electrons are injected from the left Co electrode and propagating diffusively within the nanowire. As a consequence, the corresponding electrochemical potential $\mu^\uparrow$ is higher than $\mu^\downarrow$ for spin-down electrons. Towards the left normal contact, $\mu^\uparrow$ and $\mu^\downarrow$ approach each other owing to the finite spin diffusion length λ_{SC} in the semiconductor. Because of the voltage drop resulting from driving the current between these two contacts, the average electrochemical potential has a finite slope. The voltage drop is measured between the right Co detector electrode and the right normal contact. As illustrated in Figure 6.28 (b), owing to its magnetization parallel to the injector Co electrode, the electrochemical potential of the spin-up electrons is detected. Whereas the right normal metal contact is non-spin-selective and picks up the average electrochemical potential. As a consequence, a positive voltage drop $V_{nl} > 0$ is expected. Here, the open-circuit voltage drop is detected, and thus no carrier transport is involved. Figure 6.28 (c) corresponds to the situation where the magnetization of the injector electrode is inverted. This switching can be achieved by applying a sufficiently large external magnetic field in opposite direction, which exceeds the coercive field of that electrode. As long as the external field is smaller than the coercive field of the second narrower Co electrode, this one remains in its initial magnetization orientation. Thus, here both electrodes are in an anti-parallel configuration. For the case shown in Figure 6.28 (c), the majority carriers are spin-down electrons. However, owing to its inverse magnetization the Co spin detector electrode picks up the electrochemical potential $\mu^\uparrow$ of the spin-up minority electrons. This is the reason why a negative voltage is measured. Thus, it is expected that the switching between parallel and anti-parallel configuration is accompanied by a reversal of the measured nonlocal voltage V_{nl}. Of course this will only happen if a spin imbalance is induced in the semiconductor by injection spin-polarized electrons from the left Co electrode. Thus, the reversal of the nonlocal voltage is a clear indication of a successful spin injection. If the external magnetic field is further increased, eventually the magnetization of the detector electrode aligns to the injector electrode, so that once again a parallel configuration is obtained (cf. Figure 6.28 (d)). Now, the detector senses the majority spin-up electrochemical potential $\mu^{\uparrow}$. Hence, the expected voltage drop is positive.

The corresponding set of measurements is shown in Figure 6.30.

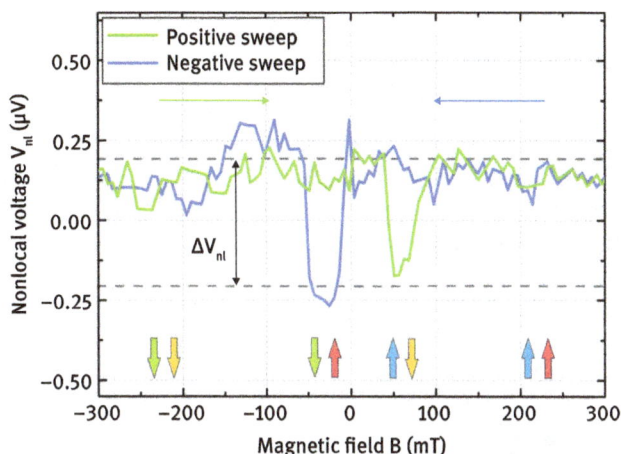

Figure 6.30: Nonlocal spin valve measurements on an InN nanowire at a temperature of 4 K. The magnitude of the effect can be quantified by the change of the non-local voltage ΔV_{nl}. The bias current was 2.5 µA. The arrows indicate the direction of magnetization of the two Co electrodes, i. e. parallel or antiparallel configuration. Graph provided by S. Heedt, Forschungszentrum Jülich.

The spin-polarized carriers are injected from a Co electrode into an InN nanowire, while the nonlocal voltage V_{nl} probing the spin-related signal is measured between the second narrower Co electrode and an adjacent normal metal contact. The respective sample image and measurement scheme are shown in Figure 6.28 (a). The nonlocal voltage curves are obtained during a magnetic field sweep along the easy axis of the Co contacts. A bias current of a few µA was driven between the spin injector and the right normal electrode. As one can infer from Figure 6.30, when sweeping the magnetic field between ±300 mT, one observes pronounced changes in the nonlocal voltage ΔV_{nl} at certain field values. Due to the different widths of the Co electrodes, the magnetization reversal occurs at different coercive fields. In the regions of negative values of V_{nl}, the spin injector and detector are oppositely magnetized. At high fields the switch of narrower Co electrode brings the system to a parallel state and thus V_{nl} back to positive values. As expected for very narrow magnetic leads, abrupt resistance changes associated with transitions between the two magnetization states of both Co electrodes occur at +41 mT and at −52 mT for the wider and the narrower magnetic electrode, respectively.

6.9 Optical spin generation

Although electrical spin injection can be used to generate spin polarized carriers in a semiconductor, its efficiency is rather low, as we could infer from the previous sections. Anyway, there is an alternative way based on optically absorption.

6.9.1 Optical absorption

Let us assume a circularly polarized beam of light which is focused on a semiconductor. As we learned in Section 6.4, circularly polarized photons possess an angular momentum of $\pm\hbar$ for right and left circular polarization, respectively. When a circularly polarized beam of light with a photon energy larger than the semiconductor band gap is focused on the semiconductor, electron-hole pairs are generated during photon absorption. The optical excitations are illustrated in Figure 6.31.

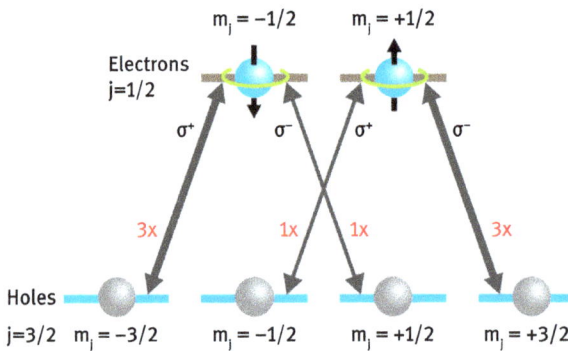

Figure 6.31: Optical selection rules between holes in the valence band with total angular momentum $j = 3/2$ and electrons in the conduction band with $j = 1/2$. The probability for heavy hole transitions ($m_j = 3/2$) is three times as large as for light hole transitions $m_j = 1/2$. Note that the angular momentum quantum numbers m_j refer to the occupied electron states.

In total four different absorption processes are allowed, two of them involve heavy hole states with $m_j = \pm3/2$, whereas the other involve light hole states with $m_j = \pm1/2$. As can be seen in Figure 6.31, one possible absorption process for a σ^+ photon is the excitation of a $m_j = -3/2$ valence band electron into a $m_j = -1/2$ state in the conduction band. During that process the photon transfered an angular momentum of $\hbar$ to the electron and thus increase m_j by +1. Alternatively, a σ^+ photon can also excite a valence band electron with $m_j = -1/2$ to a conduction band state with $m_j = +1/2$, also increasing m_j by +1. However, this process is three times less likely than the previous one. Thus, absorption of σ^+ photons will create a net spin polarization of 50 % spin-down electrons. As can be inferred from Figure 6.31, absorption of σ^- photons will result in a net spin polarization of spin-up electrons. At the same time the holes in the valence band are polarized as well. However, owing to the much stronger spin-orbit coupling in the valence band compared to the conduction band, the hole spin polarization is lost much more quickly.

An electron excited from the conduction band leaves a hole state with opposite spin. Based on that, the absorption processes can also be viewed from a different perspective. An electron excited from the $m_j = -3/2$ state into a $-1/2$ state by a σ^+ photon leaves a heavy hole with $m_{j,HH} = +3/2$. In total the electron-hole pair created by the σ^+ photon has a total z angular momentum of

$$m_{j,HH} + m_j = +\frac{3}{2} - \frac{1}{2} = +1, \tag{6.60}$$

which is equal to that of the photon. Thus, the angular momentum of the photon is transferred to the electron-hole pair by the absorption process.

6.9.2 Spin coherence and spin dephasing

As we will discuss below, by optical excitation with circularly polarized light using ultra-short laser pulses an ensemble of coherent nonequilibrium spins can be created. There are two mechanisms which subsequently destroy the coherence and relax the ensemble to equilibrium. Its time evolution is described by the times T_1 an T_2, the longitudinal and transversal spin coherence times, respectively. In order to explain both contributions, we are considering electrons exposed to a magnetic field in z-direction, as shown in Figure 6.32 (a). The Zeeman effect results in a spin splitting. Assuming a negative effective g-factor, e. g. $g = -0.44$ in GaAs, the spin-down state corresponds to the ground state when the magnetic field points downwards. By optical excitation a nonequilibrium ensemble is created with electrons lifted from the spin-down ground state to the spin-up excited state. As illustrated in Figure 6.32 (b), one possible mechanism for regaining the equilibrium state is a relaxation from the excited state to the ground one. Here, we used the representation of the spin states on the surface of the Bloch sphere where the upper and lower poles represent the $|\uparrow\rangle$ and $|\downarrow\rangle$ states, respectively. For the relaxation an energy transfer corresponding to the Zeeman energy is required, i. e. a transfer to the lattice by phonons. The energy relaxation occurs with a certain probability within the time T_1 being the characteristic time for the longitudinal spin magnetization to reach the equilibrium state. In contrast, the transversal decoherence time T_2 describes the loss of phase relation between the two eigenstates. In the example shown in Figure 6.32, the superposition of the spin-down and spin-up states connected by a phase factor $e^{i\phi}$ is changed by a random phase shift of $\Delta\phi$. The transversal decoherence time T_2 is the characteristic time in which the phase relation between the two spin eigenstates is lost. This process occurs without any energy transfer or change of occupation. Note, that in the set-up considered here, the spin

(a) (b) (c)

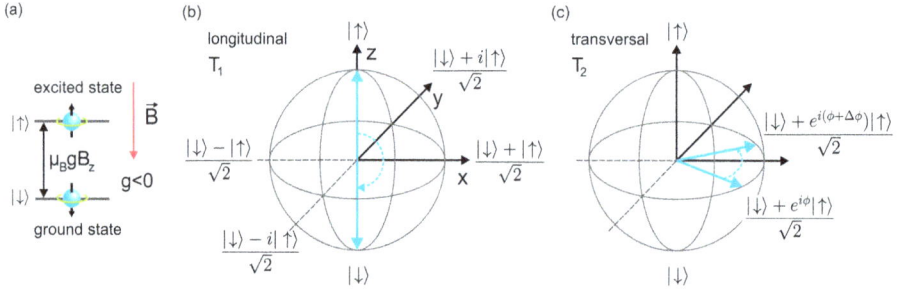

Figure 6.32: (a) Energy levels of an electron spin due to the Zeeman effect in a magnetic field pointing downwards along the z-axis. We assumed a negative g-factor, as it is common for many III-V semiconductors. (b) Bloch sphere with an initial $|\uparrow\rangle$ state which relaxes to the $|\downarrow\rangle$ ground state. (c) Dephasing where the phase of the state $|\downarrow\rangle + e^{i\phi}|\uparrow\rangle$ is shifted by $\Delta\phi$.

is precessing about the magnetic field, i. e. the phase ϕ is changing constantly and well-determined in the course of time, thus, the coherence is preserved. In a two-level system the energy relaxation connected to the longitudinal decoherence is often suppressed due to energy conservation. This generally leads to $T_1 \gg T_2$ [86].

So far we only considered the decoherence of a single spin characterized by T_1 and T_2. However, in optical experiments a whole ensemble amounting to the order of 10^{15} of electrons is coherently excited [87]. Only in the case that all conduction electrons are identical and noninteracting, the transverse decay of the net spin magnetization would correspond to the intrinsic decoherence time T_2 of the individual electrons. However, due to variations of the local magnetic field or the electron g-factor, there might be inhomogeneous effects in addition to the intrinsic homogeneous decoherence. In such an environment each spin precess at different rates resulting in an additional dephasing of the spin polarization. Even when all spins are evolving coherently this leads to a spreading of the relative spin orientations. Nevertheless, in principle there is a possibility to distinguish between both contributions by employing so-called spin echo techniques. Here, the spin orientation is rotated by 180° about the z-axis at certain moments, so that the phase shifts due to inhomogeneous effects develop inversely in time. However, spin echo is rarely used in time-resolved optical experiments so that the inhomogeneous effects cannot be singled out. Therefore, the measured transverse spin decay is connected to the so-called transverse spin dephasing time T_2^*. The homogeneous and inhomogeneous contributions to T_2^* can be expressed by

$$\frac{1}{T_2^*} = \frac{1}{T_2} + \frac{1}{T_2^{\text{inh}}}. \tag{6.61}$$

One finds that T_2^* is a lower bound to the transverse spin coherence time T_2. The transverse spin dephasing time is often also named spin lifetime.

6.9.3 Optical detection of magnetization

Optical methods can also be used to probe the net magnetization of electron spins in a semiconductor. The underlying mechanism is the magneto-optical Kerr effect. An illustration of the set-up is given in Figure 6.33. A linearly polarized laser beam impinges on the surface of a magnetized material. If the material is transparent, part of the light is transmitted, as illustrated in Figure 6.33 (a). While the beam passes the material the polarization axis is rotated. Thus, the polarization of the transmitted beam is rotated by a certain angle with respect to the incoming beam. This effect is called the Faraday effect, with the corresponding rotation angle of the polarization being the Faraday angle θ_F. Part of the incoming laser beam is reflected. As can be inferred from Figure 6.33 (b), for a magnetized material the polarization axis of the reflected beam is also turned with respect to the polarization of the incoming beam. This effect is called the magneto-optical Kerr effect, or abbreviated as MOKE. The corresponding rotation angle of the polarization, i. e. the Kerr angle θ_K, depends on the magnetization $\vec{M}$.

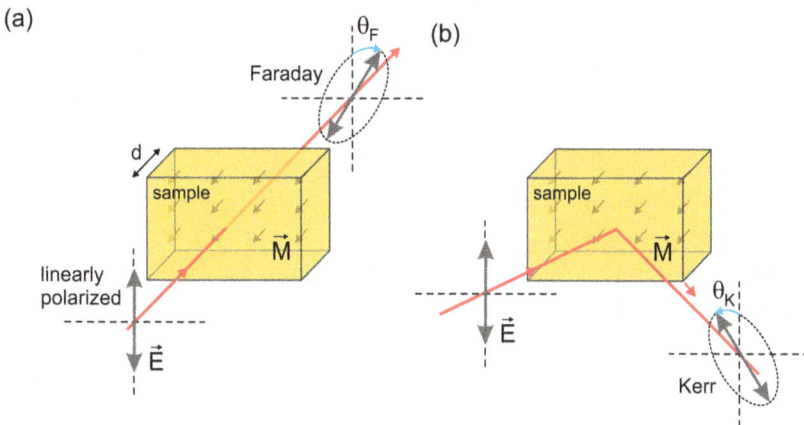

Figure 6.33: Illustration of the magneto-optical Kerr and Faraday effects. A linearly polarized laser beam is focused on the sample. (a) Geometrical situation of the Faraday effect. The magnetization of the sample causes a rotation of the plane of polarization and a slight ellipticity. (b) Corresponding geometry of the Kerr effect.

The Faraday as well as the Kerr rotation have their origin in the energetic splitting of spin-up and spin-down electrons in the presence of a magnetic field. The strength of the polarization rotation is largely enhanced in the vicinity of optically allowed band-to-band transitions, e. g. transitions as shown in Figure 6.31. However, due to the presence of the magnetic field, the electron and hole states experience an additional two-fold and four-fold Zeeman splitting, respectively. As a consequence, the optical transitions associated with right- and left-circularly polarized light will split in energy. For the sake of simplicity we assume two Zeeman split transitions, which result in two absorption resonances $\alpha_\pm$ for right- and left-circularly polarized light, respectively, as depicted in Figure 6.34. The difference of the absorption is related to a corresponding difference of the real parts of the refractive indices $n_\pm$ (cf. Figure 6.34), which in principle can be gained from the Kramers–Kronig relation between $\alpha_\pm$ and $n_\pm$. Regarding the Faraday effect the incoming linearly polarized light passing the material can be decomposed in a superposition of two opposite circularly polarized light components. The difference of the refractive indices results in a difference of phase velocities between the right- and left-circularly polarized component. Neglecting the damping due to absorption we can express the electric field oscillations of right- and left-circular components after passing the material of thickness d by

$$\mathcal{E}_\pm = \mathcal{E}_0 e^{i(k_\pm d - \omega t)}. \tag{6.62}$$

Here, $\mathcal{E}_0$ is the field amplitude at $z = 0$, ω is the oscillation frequency, and $k_\pm$ are the corresponding wave vectors. By making use of the relation between the wave vector and the oscillation frequency $k_\pm = n_\pm \omega / c_0$, with c_0 the speed of light, one can write

$$\mathcal{E}_\pm = \mathcal{E}_0 e^{i(n_\pm \omega d / c_0 - \omega t)}. \tag{6.63}$$

The phase shift $\Delta\phi$ between both components after passing the material of thickness d amounts to

$$\Delta\phi = \frac{\omega d}{c_0}(n_+ - n_-). \tag{6.64}$$

The phase shift of the two circularly polarized components results in a corresponding Faraday rotation angle of the linearly polarized light

$$\theta_F = \frac{\omega d}{2c_0}(n_+ - n_-). \tag{6.65}$$

Thus, the magnitude of the Faraday rotation angle increases with the difference of the refractive indices originating from the Zeeman splitting, and the thickness of the material. A schematics of the dependence of θ_F on ω is shown in Figure 6.34 (c).

The Zeeman splitting usually results in a magnetization of the material, so that the degree of Faraday rotation is at the end a measure for the magnetization of the material. Often the Faraday rotation is expressed in terms of the Verdet constant $V(\lambda, T)$, which depends on the photon wavelength λ and the temperature

$$\theta_F = V(\lambda, T) M d. \tag{6.66}$$

Here, M is the projection of the magnetization on the direction of the incoming laser beam. In a similar fashion the Kerr angle θ_K also depends on the projection of the magnetization. A closer look on both effects reveals that the transmitted and reflected

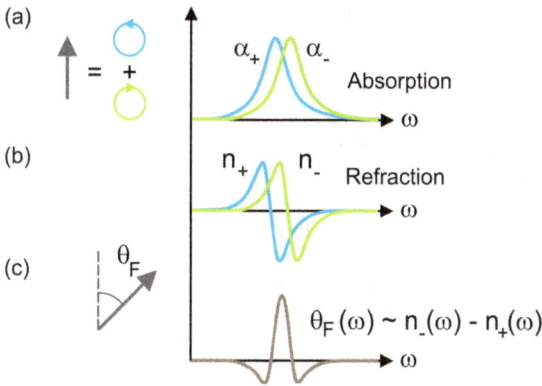

Figure 6.34: (a) Idealized Zeeman-split absorption resonances α_+, α_- corresponding to right- and left-circularly polarized light, respectively. The linearly polarized light is decomposed into a super-position of right- and left-circularly polarized light components. (b) Corresponding real parts of the refractive indices n_+ and n_-. (c) Resonant Faraday rotation by an angle θ_F. The Faraday rotation is only strong in the vicinity of the resonance.

beams have a slight ellipticity. Up to now we assumed that the magnetization was induced by means of an external magnetic field. However, as we will see in the next section, it is also possible to generate spin-polarized electrons, and thus a magnetization in the material is possible by exposing the sample to circularly polarized light by employing the process described in Section 6.9.1.

6.9.4 Pump-probe experiments

One can combine the optical spin generation with an optical measurement of the spin orientation. These experiments are particularly interesting when information on the spin dynamics shall be gained. In these so-called pump-probe experiments a short circularly polarized laser pulse is focused on the sample, in order to generate spin-polarized carriers. The typical beam arrangement is depicted in Figure 6.35. The pump and the probe beams are at almost normal incidence to the sample surface. Spin precession is achieved by applying a transversal magnetic field (Voigt geometry). The projection of the magnetization along the probe beam is measured through the Faraday θ_F or Kerr angle θ_K at different time delays Δt.

The complete optical set-up is depicted in Figure 6.36. The source of the fs-pulses is a laser system with an Ar-ion pump laser and a Ti:sapphire laser. The pulse has a length of typically 100 fs. The central wavelength is about 800 nm, corresponding to a photon energy of 1.55 eV. A beam splitter divides the primary pulse into the pump and probe pulses. The time delay Δt between the pump and probe pulse is adjusted by a delay line, which mechanically changes the path length of the pump beam, with

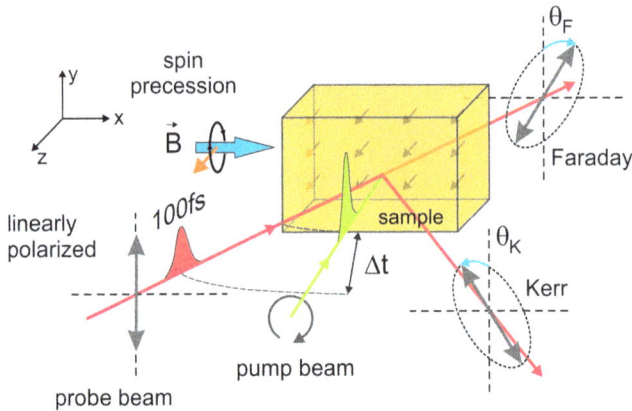

Figure 6.35: Beam arrangement of a pump-probe experiment. A circularly polarized laser pulse with a pulse width of typically 100 fs is focused on the sample. Spin-polarized carriers are generated in the semiconductor. The corresponding magnetization is detected by the time-resolved measuring of the Faraday or Kerr rotation angle. A transversal magnetic field $\vec{B}$ is applied to let the electron spins precess. The linearly polarized probe pulse is delayed by a time Δt with respect to the pump pulse.

respect to the probe beam. Both the pump and probe beams are focused on the sample located in a magneto-optical cryostat. The circularly polarized photons of the pump pulse generate spin-polarized electrons in the conduction band. Owing to the strong spin-orbit coupling in the valence band the holes are depolarized very quickly. The spin-polarized electrons in the semiconductor are probed by measuring the reflection of a linearly polarized laser pulse, i. e. the Kerr rotation angle is determined. A diode bridge is used to measure the polarization rotation of the probe beam. Here, the linear polarization of the probe beam is split into two orthogonal linear beams of the same intensity. This is achieved by first rotating the probe polarization to 45° from verti-cal, and then passing through a polarization beam splitter. The two emerging beams are focused on a pair of Si photo-diodes, where the difference between the two diode currents is a measure of the Kerr rotation angle. In order to suppress the effect of the pump on the probe and to obtain a better signal-to-noise ratio, lock-in techniques are used. Here, the probe beam is modulated by an optical chopper, while the circular polarization of the pump beam is modulated by a photoelastic modulator.

By using a successively longer delay Δt of the probe pulse the dynamics of spins generated by the pulse beam can be measured. Since by the pump pulse an ensem-ble of spin-polarized electrons is generated, the measurement of the Kerr or Faraday angle corresponds to the properties of a large number of electrons, i. e. generally with averaged spin orientations. As long as no external magnetic field is applied, it can be expected that the spin polarization decays exponentially with the characteristic decay time given by the transverse dephasing time T_2^*. However, as illustrated in Figure 6.35, in most cases a fixed external magnetic field is applied, which is aligned perpendic-

1 beam splitter
2 delay line
3 polarizer
4 photoelastic modulator
5 filter wheel
6 neutral density filter
7 mechanical chopper
8 λ/2 retardation plate
9 polarizer
10 focusing lens
11 sample
12 magneto-optical cryostat
13 spatial filter
14 λ/2 retardation plate
15 polarizing beam-splitter
16 diode bridge

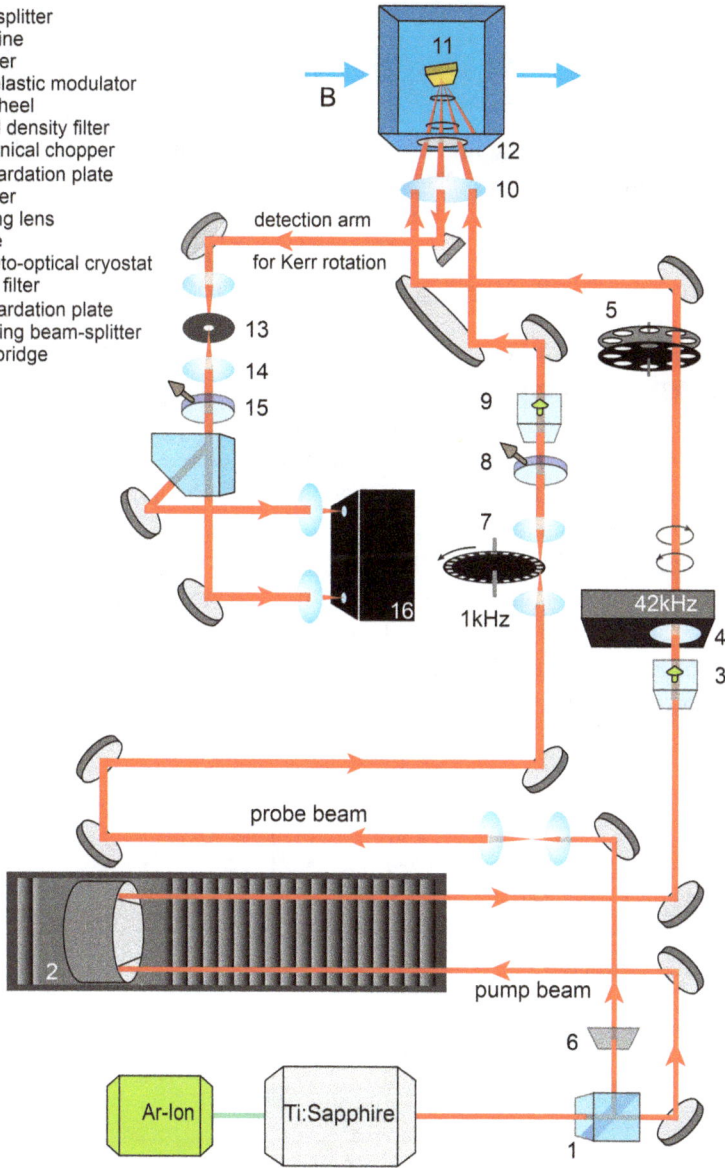

B

11

12

10

detection arm
for Kerr rotation

13

14

15

5

9

8

7

42kHz

4

3

1kHz

16

probe beam

pump beam

6

Ar-Ion

Ti:Sapphire

2

1

Figure 6.36: Optical setup for time-resolved Kerr rotation experiments. The different components are label by numbers. Figure provide by Bernd Beschoten, RWTH Aachen.

ularly to the orientation of the optically generated spins. This results in a spin precession about axis of the external magnetic field with the Larmor frequency given by $\omega_L = g\mu_B B/\hbar$. The Larmor precession results in an oscillation of the spin component along the probe beam. The corresponding expectation value along that direction is given by

$$\langle S_z \rangle = \frac{1}{2}\hbar \cos(\omega_L \Delta t). \tag{6.67}$$

As a consequence, the Faraday and Kerr rotation angle oscillates accordingly. Including the exponential dephasing one finds for the Faraday rotation angle as a function of the delay time

$$\theta_F(\Delta t) = \theta_{max} \exp\left(-\frac{\Delta t}{T_2^*}\right) \cos(\omega_L \Delta t). \tag{6.68}$$

with θ_{max} is the angle at $\Delta t = 0$.

Figure 6.37 shows a typical measurement of the Faraday rotation angle θ_F of an InGaAs layer as a function of the delay time Δt [88]. At $\Delta t = 0$ the pump and probe beam coincide. The circularly polarized pump beam generates spin-polarized carriers, which results in a finite Faraday rotation angle. The in-plane magnetic field leads to a spin precession and thus to an oscillation of $\theta_F(\Delta t)$. In the course of time the envelope of the oscillation amplitude decreases, following an exponential dependence, as described by equation (6.68). From that exponential decay the dephasing time T_2^* can be extracted, while from the oscillation period the g-factor can be determined. In Figure 6.37 the Faraday angle is shown for excitation with σ^+ and σ^-. Since the corresponding orientations of the initially generated spins are opposite, the time-dependence of the Faraday rotation angles are inversed to each other.

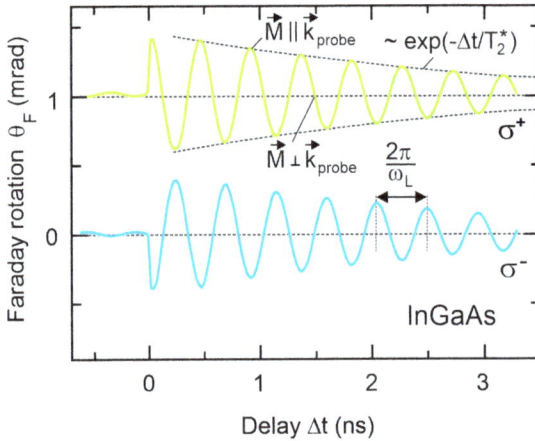

Figure 6.37: Measurement outcome of a pump-probe experiment on a 500 nm thick $In_x Ga_{1-x}As$ layer with $x = 0.049$: Time-resolved Faraday rotation after excitation by circularly polarized pump pulses with σ^+ and σ^- polarization, respectively. The data was taken at $T = 30$ K in a transverse magnetic field of 0.5 T [88]. The σ^+ curve was shifted for clarity. As shown for the σ^+ curve, the envelope can be described by an exponential decrease. The cases where the magnetization $\vec{M}$ is parallel and perpendicular to the wave vector of the probe beam $\vec{k}_{probe}$ are indicated, respectively. The oscillation period is given by $2\pi/\omega_L$. Figure provided by Bernd Beschoten, RWTH Aachen.

For many spintronic applications it is desired that the spin state be preserved as long as possible. This can be directly translated to a long spin dephasing time T_2^*. Indeed, in n-type GaAs very long dephasing times exceeding 100 ns are observed [87]. Figure 6.38 illustrates how the spin coherence and spin precession in Si-doped n-type GaAs changes when the doping concentration is varied [89]. For a carrier concentration of $n = 2 \times 10^{15}$ cm^{-3} determined at room temperature, the time-resolved Kerr signal expressed by θ_K shows a damped oscillation, very similar to the one depicted in Figure 6.37. From this measurement a g-factor of -0.4315 and T_2^* of about 2 ns were extracted. Obviously, upon increasing the doping concentration by a factor of ten, i. e. to 2×10^{16} cm^{-3}, the precession frequency and thus the g-factor changes. Furthermore, the coherence time is enhanced by a factor of five. When the doping concentration is increased further, a decrease of the g-factor and of T_2^* is observed.

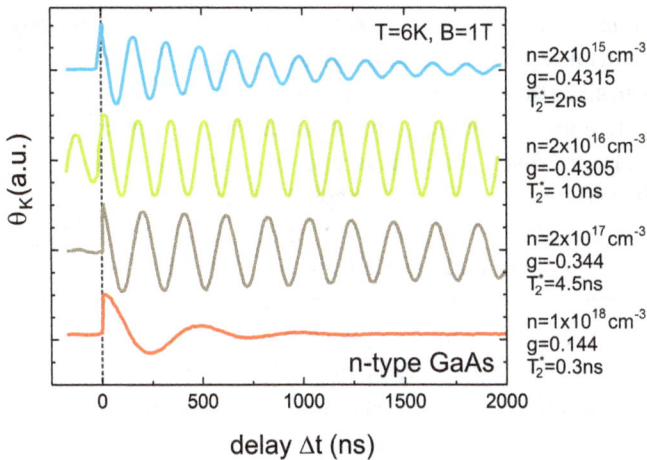

Figure 6.38: Time-resolved Kerr rotation θ_K for Si-doped n-type bulk GaAs samples at a temperature of 6 K in a magnetic field of 1 T. The given carrier densities are determined at room temperature. The g-factors are determined from the oscillation period. The measurements were taken at a laser energy of 1.4938 eV. The repetition interval T_{rep} was about 12.5 ns. Figure adapted from Heidkamp [89].

The sample with the longest spin dephasing time has a doping concentration in the vicinity of the critical doping concentration n_c of the metal–insulator transition [90]. For doping concentrations lower than n_c the sample is a so-called Mott insulator [91]. Here, the isolated shallow Si donor states are not thermally ionized at the low measurement temperatures so that transport is only possible when the electrons are hopping from one localized Si donor site to a neighboring one. This process is called variable range hopping. Close to but below n_c neighboring donor states start to overlap so that an impurity band is formed. Due to the overlap of the donor orbitals the electronic state is delocalized and the sample enters the metallic regime.

The rather long spin lifetimes are not directly expected, when one would assume the carrier recombination time as a limiting factor. In order to explain the different mechanisms which finally results in such long spin lifetimes we refer to Figure 6.39. First of all, we have to assume a degenerate n-type doped semiconductor with the Fermi level located within the conduction band. The system is in equilibrium, no spin polarization and thus no magnetization is present. By excitation with a circularly polarized femtosecond pump pulse coherent spin polarized electrons and holes are generated. The imbalance is due to the optical selection rules introduced in Section 6.9.1 which result in an excitation three times greater for one spin orientation than for the other. In the valence band a rapid dephasing and relaxation of the hole states occurs on timescales of a few picoseconds [92]. The former is due to the degeneracy of the light-hole and heavy-hole band at the Γ-point where a strong valence band mixing occurs, while the latter is caused by the strong spin-orbit coupling of the valence bands in GaAs. In contrast, the spin distribution of the electrons remains unbalanced, since the spin-orbit coupling is much weaker in the conduction band. Simultaneously, electron-hole recombination takes place so that the valence band is refilled completely within about 1 ns. However, since the spins of hole states in the conduction band were equally distributed, the same number of electrons of both spin orientations recombine. As a result, the unbalanced distribution of electrons remains in the conduction band, which finally dephases within the rather long spin lifetime T_2^*.

A closer look at Figure 6.38 reveals that for the sample with the very long spin dephasing time T_2^* ($n = 2 \times 10^{16}$ cm^{-3}) a large Kerr rotation signal is observed at a negative time delay. This signal originates from the excitation of the previous pump pulse. Thus, the Kerr rotation signal does not completely die off within the repetition interval of $T_{\text{rep}} \approx 12.5$ ns. In order to determine T_2^* precisely a different technique is required, the resonant spin amplification, which is described in detail below.

6.9.5 Resonant spin amplification

We saw in the previous section that if the spin lifetime is significantly longer than the laser repetition interval T_{rep} the ensemble excited by the previous pump pulse is not completely dephased before the new pump pulse arrives. In that case a different techniques has to be employed to gain information on the spin lifetime. The working principle of the measurement scheme called resonant spin amplification is illustrated in Figure 6.40. In case that an integer multiple of the Larmor precession period given by $2\pi/\omega_L$ fits into the repetition interval T_{rep}, the subsequently injected spins can interfere constructively with the existing spins. In the example given in Figure 6.40 this corresponds to the trace for $T_{\text{rep}} = 5 \times 2\pi/\omega_L$. Since the constructive interference occurs not only for the latest spin ensemble, but also for its predecessors, all these spins are in resonance. However, if the spin precession does not exactly fit into the repetition interval, this resonance is greatly suppressed. This is because even the most recently

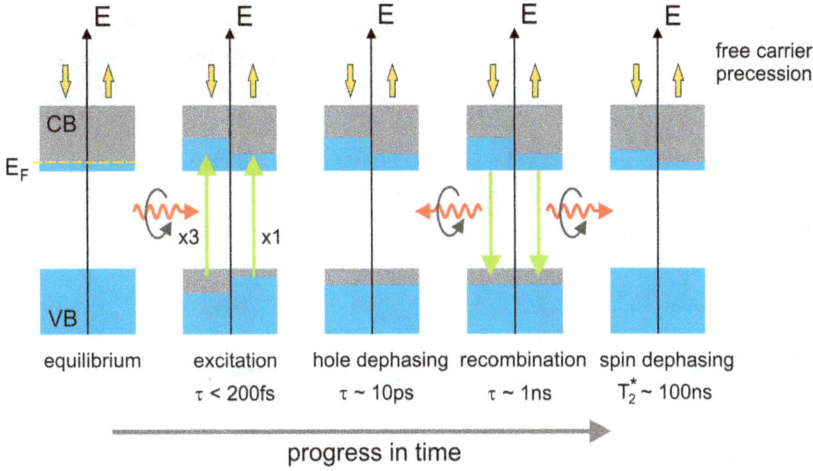

Figure 6.39: Sequence of mechanisms during optical pumping of electron spins. In equilibrium the Fermi level E_F is located in the conduction band. By excitation within a time of less than 200 fs by circularly polarized light spin polarized electrons and holes are created in the conduction (CB) and valence band (VB), respectively. Due to the strong spin-orbit coupling in the valence band, the holes dephase rapidly within about 10 ps, leaving a balanced number of hole states in the valence band. By recombination, the valence band is refilled completely within about 1 ns. As a result, an unbalanced distribution of electrons remains in the conduction band, which leads to a net precession of the magnetization when an in-plane magnetic field is applied. Figure adapted from Heidkamp [89].

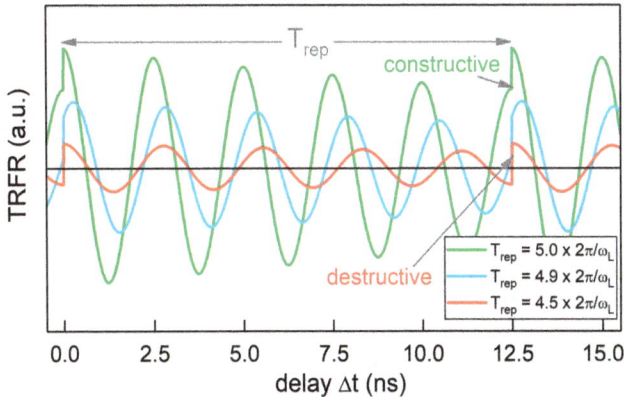

Figure 6.40: Interference of the precessing spins injected at $\Delta t = 0$ in a time-resolved Faraday rotation (TRFR) experiment. Three different exemplary simulations of delay line scans are shown assuming $g = -0.42$, $T_2^* = 30$, and $T_{rep} = 12.5$ ns. In case that the magnetic field is adjusted in a way that five Larmor precessions with a period of $2\pi/\omega_L$ fit within the repetition interval, constructive interference occurs. Upon decreasing the magnetic field, i. e. decreasing ω_L, the constructive interference is lost. In case of half-numbered precessions, i. e. $T_{rep} = 4.5 \times 2\pi/\omega_L$, the interference is destructive. Figure provided by Bernd Beschoten, RWTH Aachen.

created spin ensemble is only slightly off resonance, its predecessor is already twice off, and so on. As a consequence, the net magnetization is considerably reduced. For the special case that half-numbered precessions fit into the repetition interval, the newly injected spins are always opposite to the previous ones, so that in total a destructive interference occurs, as can be seen in Figure 6.40 (red curve).

The resulting magnetization M_s can be calculated by adding up all contributing exponentially damped Larmor precessions

$$M_s(\Delta t, \omega_L) = \sum_{n=1}^{\infty} M_0 e^{-(\Delta t + n T_{\text{rep}})/T_2^*} \cos[\omega_L(\Delta t + n T_{\text{rep}})], \tag{6.69}$$

with Δt within the interval $[-T_{\text{rep}}; 0)$. This formula can be written in a closed form [89]

$$M_s(\Delta t, \omega_L L) = \frac{M_0}{2} e^{-\frac{\Delta t + T_{\text{rep}}}{T_2^*}} \frac{\cos(\omega_L \Delta t) - e^{T_{\text{rep}}/T_2^*} \cos[\omega_L(\Delta t + T_{\text{rep}})]}{\cos(\omega_L T_{\text{rep}}) - \cosh(T_{\text{rep}}/T_2^*)}. \tag{6.70}$$

In resonant spin amplification experiments the repetition rate is usually fixed while ω_L is adjusted by tuning the magnetic field. Whenever the resonance condition is fulfilled a sharp peak is observed in the time-resolved Kerr rotation signal. This can be seen in Figure 6.41, where the Kerr rotation is measured at a fixed pump-probe delay of $\Delta t = -50$ ps. The peak positions can be easily identified with condition that the absolute value of the denominator of equation (6.70) has its local minimum, i. e. where the cosine term is equal to one. The spin dephasing time can be extracted by fitting each peak to equation (6.70). The decrease of the peak height with increasing magnetic field found in Figure 6.41 reflects a decrease of the spin dephasing time.

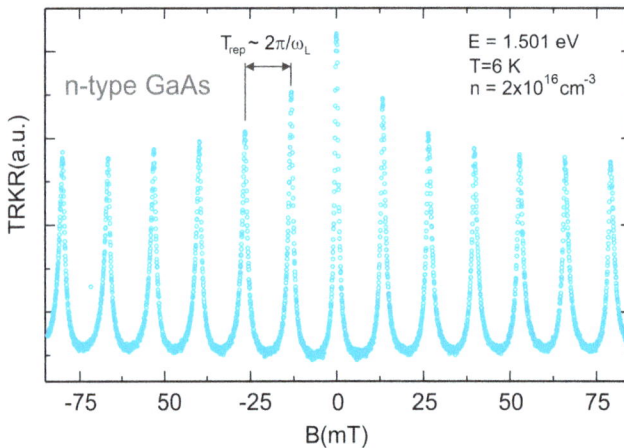

Figure 6.41: Resonant spin amplification scan on n-type GaAs at $T = 6$ K taken by time-resolved Kerr rotation (TRKR) by sweeping an in-plane magnetic field. The pump-probe delay was fixed at $\Delta t = 50$ ps. The peak distance corresponds to one Larmor precession during the laser repetition interval. Figure provided by B. Beschoten, RWTH Aachen.

6.10 Summary

- The resistor model can be used to describe the current polarization in a ferromagnetic/semiconductor/ferromagnet structure. Spin scattering is neglected.
- By a local description of spin injection based on the spin diffusion equation spin scattering can be included.
- Due to the conductivity mismatch between metallic ferromagnets and semiconductors the spin efficiency is very low. The injection efficiency can enhanced by introducing a barrier layer.
- Spin injection into a semiconductor can be measured by optical means, i. e. by means of a spin light emitting diode.
- Spin polarized carriers can be generated by circularly polarized light. The magnetization can be measured by means of the magneto-optical Kerr effect.

Exercises

Problem 6.1. Let us assume an electron in an external magnetic field $\vec{B}$. The corresponding Zeeman Hamiltonian is given by

$$\hat{H}_Z = \frac{g\mu_B}{\hbar} \hat{\vec{S}}\vec{B},$$

with $\hat{\vec{S}}$ being the spin operator $\frac{\hbar}{2}\hat{\vec{\sigma}}$. From the Heisenberg picture the time development of the spin operator is expressed by

$$i\hbar \frac{d}{dt}\hat{\vec{S}}_i = [\hat{\vec{S}}_i, \hat{H}_Z].$$

Show that inserting the Hamiltonian in the equation above results in

$$\frac{d}{dt}\langle\vec{S}\rangle = -\frac{g\mu_B}{\hbar}(\langle\vec{S}\rangle \times \vec{B}),$$

with $\langle\vec{S}\rangle$ being the expectation value of $\hat{\vec{S}}$. The equation describes the precession of optically generated electrons in an external magnetic field.

Problem 6.2. Confirm the expression for the current spin polarization α in the resistor model given by

$$\alpha = \beta \frac{R_{FM}}{R_{SC}} \frac{2}{2\frac{R_{FM}}{R_{SC}} + (1 - \beta^2)},$$

with β being the spin polarization in the ferromagnetic and R_{FM} and R_{SC} the resistances of the ferromagnet and semiconductor, respectively.

7 Spin transistor

7.1 Overview

In this chapter we will discuss the mechanism which allows the manipulation of the spin orientation in a spin electronic device by electrical means [13]. In Figure 7.1 a schematic of the spin field-effect transistor is shown [6, 93]. The spin transistor consists of three components: a spin injector as the source electrode, a semiconducting area where the spin orientation is changed by controlled spin precession, and a spin detector as the drain electrode. The requirements for the injector and detector were already explained in detail in the previous chapter. We assume that both magnetic electrodes are ideal ferromagnets. Here, we will focus on the properties of the semiconductor channel located in between the magnetic contacts. In Figure 7.1 (a) the spin precession in a spin transistor is shown when no voltage is applied to the gate electrode on top of the semiconductor channel. In the semiconductor channel the spin precesses owing to the presence of Rashba spin-orbit coupling. For the sake of simplicity we assumed that the electrons arrive at the drain contact with a spin orientation opposite to the spin orientation in the detector electrode. In this case the electrons cannot enter the detector, since no states are available there which the incoming electrons can occupy. As a consequence, the current through the spin transistor is blocked. As explained in detail below, by applying a gate voltage the strength of the Rashba effect can be enhanced. This leads to a stronger spin precession so that the spin orientation of the electrons arriving at the drain electrode fits to the magnetization of the detector electrodes. As illustrated in Figure 7.1 (b), now a current can flow in the structure, since the electrons can enter the drain electrode.

In the following, we first discuss the special type of heterostructure most suitable for observing a strong Rashba spin-orbit coupling. Subsequently, the physical background of spin-orbit coupling in a semiconductor heterostructure is explained in detail. Furthermore, methods are introduced to determine the strength of spin-orbit coupling by transport experiments.

7.2 InAs-based two-dimensional electron gases

For reasons which will become clear in the next section, semiconductor heterostructures based on InAs or its alloys are particularly suited for observing a strong spin-orbit coupling. A typical conduction band profile of a modulation doped InGaAs/InP heterostructure with a strained $In_{0.77}Ga_{0.23}As$ layer is depicted in Figure 7.2 (a) [29]. The lattice constant of the $In_{0.77}Ga_{0.23}As$ layer is larger than that of InP. Compared to $In_{0.53}Ga_{0.47}As$, which is lattice matched to InP, the higher In content of 77 % leads to a smaller band gap and a smaller effective electron mass. In order to prevent lattice relaxation, the $In_{0.77}Ga_{0.23}As$ layer thickness has to be kept below a critical thickness

https://doi.org/10.1515/9783110639001-007

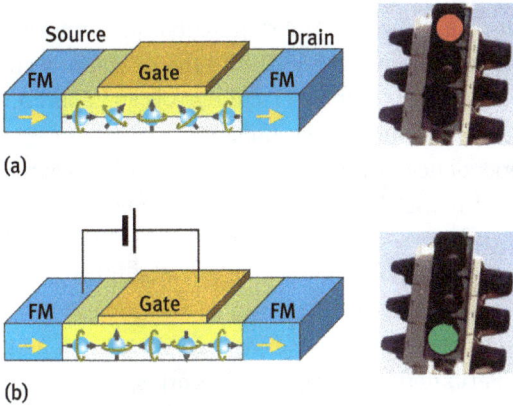

(a)

(b)

Figure 7.1: Schematics of a spin transistor: (a) Spin-polarized electrons are injected from the ferromagnetic source contact into the semiconductor. No gate voltage is applied. The spin precession due to the presence of the Rashba effect leads to a spin orientation which is opposite to the magnetization of the ferromagnetic drain electrode. No current can flow in this case, corresponding to a red traffic light. (b) A gate voltage is applied, which enhances the Rashba effect. Because of the stronger spin precession, now the spin orientation of the electrons arriving at the drain electrode fits to its magnetization. A current can flow, corresponding to a green traffic light.

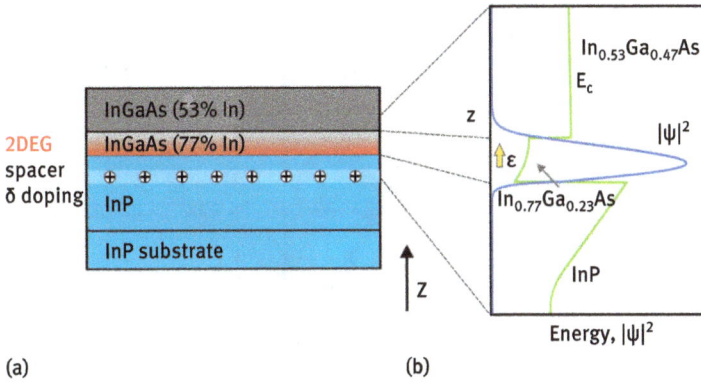

(a)   (b)

Figure 7.2: (a) Layer sequence of an $In_xGa_{1-x}As/InP$ heterostructure. The 2DEG is located in a strained $In_{0.77}Ga_{0.23}As$ layer at the boundary to the InP spacer. (b) Conduction band profile E_c and squared amplitude $|\psi|^2$ of the electron wave function in an InGaAs/InP heterostructures with an inserted strained $In_{0.77}Ga_{0.23}As$ layer. Due to the bending of the conduction band an electric field $\mathcal{E}$ is present in the quantum well.

which is about 10 nm. On top this layer is capped by an $In_{0.53}Ga_{0.47}As$ layer which is lattice matched to InP. The electrons in the 2DEG are provided by the n-type doped InP layer, which is separated from the 2DEG by a spacer layer. The whole structure is grown on a semi-insulating InP wafer. The electron wave function $\psi(z)$ is mainly located in the strained $In_{0.77}Ga_{0.23}As$ layer. Owing to the different barrier materials and the doping at only one side of the $In_{0.77}Ga_{0.23}As$ well, the rectangular quantum well is tilted. The tilted potential profile results in an electric field in the quantum well. The presence of an electric field $\mathcal{E}$ in the two-dimensional electron gas is an important prerequisite for the Rashba effect discussed in the next section.

In Figure 7.3 the temperature dependence of the mobility of a two-dimensional electron gas in an InGaAs/InP heterostructure is shown. Down to temperatures of around 40 K the mobility is limited by phonon scattering. At temperatures below 40 K the mobility saturates, which can be attributed to alloy scattering. Mobilities above $100\,000\ cm^2/Vs$ can be achieved. In contrast to a two-dimensional electron gas in an AlGaAs/GaAs heterostructure, the electron gas is located in a ternary material, where the In and Ga atoms are statistically distributed in the crystal. The corresponding disorder potential results in an additional scattering contribution, which limits the mobility at low temperatures. By introducing a strained $In_{0.77}Ga_{0.23}As$ layer the alloy scattering contribution is lowered, owing to the smaller disorder potential [29]. As an alternative barrier material $In_{0.52}Al_{0.48}As$ lattice matched to $In_{0.53}Ga_{0.47}As$ can also be used [94], especially when the layer system is grown by MBE, since InP layers are difficult to grow by this method.

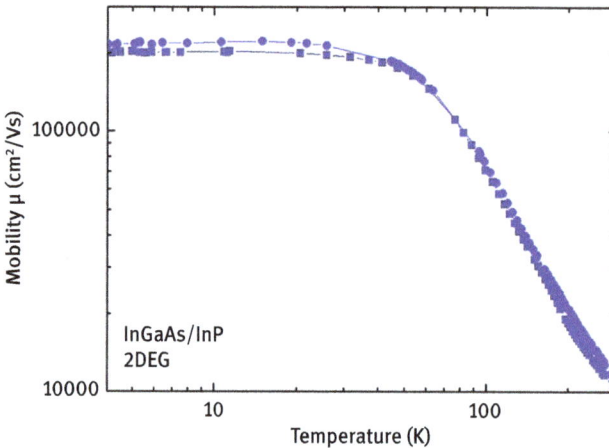

Figure 7.3: Temperature dependence of the mobility of a two-dimensional electron gas in an In-GaAs/InP heterostructure. At low temperatures the mobility is limited by alloy scattering, while at higher temperatures phonon scattering is relevant. The measurements of two Hall bars of the same wafer are shown.

7.3 The Rashba effect

We have seen in the previous section that, owing to the different materials forming the barriers of the $In_{0.77}Ga_{0.23}As$ quantum well, the potential profile is asymmetric.

This tilted energy potential V is connected to an electric field $\vec{\mathcal{E}} = (1/e)\vec{\nabla}V$ for an electron with charge $-e$ ($e > 0$), which is oriented perpendicularly to the plane of the two-dimensional electron gas. In general, in their own frame of reference electrons propagating in an electric field $\vec{\mathcal{E}}$ experience an effective magnetic field $\vec{B}$ given by the Lorentz transformation

$$\vec{B} = -\frac{1}{c^2}\vec{v} \times \vec{\mathcal{E}}, \tag{7.1}$$

with c being the velocity of light. As illustrated in Figure 7.4 (a), this magnetic field is oriented perpendicularly to the direction of propagation and to the electric field $\vec{\mathcal{E}}$.

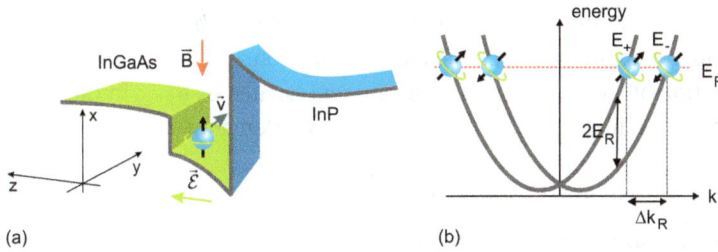

Figure 7.4: (a) Schematic illustration of the Rashba spin-orbit coupling in a two-dimensional electron gas. (b) Energy-momentum dispersion with an energy splitting due to the Rashba effect.

In the nonrelativistic approximation the Hamiltonian responsible for the spin-orbit coupling of a free electron propagation in an electric field can be directly derived from the Dirac equation by a power expansion in $(v/c)^2$ [95]:

$$\hat{H}_{so} = -\frac{\hbar}{4m_0^2 c^2}\hat{\vec{\sigma}}(\hat{\vec{p}} \times \vec{\nabla}V), \tag{7.2}$$

with $\vec{p}$ with the momentum and $\vec{\sigma}$ the Pauli spin matrices. This contribution to the total Hamiltonian is called Pauli spin-orbit term. We can also be more specific and assume a radially symmetric potential V so that $\vec{\mathcal{E}} = (1/e)\vec{\nabla}V = (1/e)(\vec{r}/r)dV/dr$. Inserting this into equation (7.2) results in the well-known spin-orbit coupling Hamiltonian in atomic physics:

$$\hat{H}_{so} = \frac{1}{2m_0^2 c^2}(\hat{\vec{S}} \cdot \hat{\vec{L}})\frac{1}{r}\frac{dV}{dr}. \tag{7.3}$$

Here, we inserted the expressions for the spin $\vec{S} = (\hbar/2)\vec{\sigma}$ and the orbital angular momentum $\vec{L} = \vec{r} \times \vec{p}$. As a result of the spin-orbit coupling term, at a given $\vec{L}$ an energy splitting occurs between spin-up and spin-down electrons.

In a two-dimensional electron gas, as shown in Figure 7.4 (a), the electrons are moving in an electric field, which originates from the tilt of the potential profile, i. e.

the potential gradient $\vec{\nabla} V$ is nonzero. As pointed out by Bychkov and Rashba [13], this electric field leads to an energy splitting between spin-up and spin down electrons, similar to the spin-orbit coupling contribution in atoms. In analogy to equation (7.2), the corresponding Hamiltonian, i. e. the so-called Rashba Hamiltonian, is given by

$$\hat{H}_R = \alpha_R \vec{e}_z (\hat{\vec{\sigma}} \times \vec{k}), \tag{7.4}$$

with the wave vector of the electrons in the two-dimensional electron gas given by $\vec{k} = \vec{p}/\hbar$ and $\vec{e}_z$ the unit vector in z-direction. We assumed that the electric field is oriented along the z-direction, perpendicularly to the plane of the two-dimensional electron gas. The parameter α_R is the Rashba parameter which is a measure for the strength of the spin-orbit coupling. Beside other parameters, which will be discussed later, it also contains the strength of electric field, i. e. the strength of $(1/e)\vec{\nabla} V$. Thus, the Rashba Hamiltonian expresses the same mechanisms contained in equation (7.2), where the contribution of $\vec{\nabla} V$ and the prefactors are represented by $\alpha_R \vec{e}_z$. In the given frame of reference the Rashba Hamiltonian can also be written more explicitly as

$$\hat{H}_R = \alpha_R (\hat{\sigma}_x k_y - \hat{\sigma}_y k_x). \tag{7.5}$$

More stringent calculations show, that the Rashba parameter is considerably larger than naïvely expected for free electrons in a two-dimensional electron gas propagating in an electric field $\vec{\mathcal{E}}$ [95]. The reason is that the electrons not only experience the electric field $\vec{\mathcal{E}}$ due to the macroscopic potential in the quantum well, but also the potential gradients due to the electron orbitals of the atoms forming the crystal. This is also the reason why the Rashba effect is particularly strong for crystals containing atoms with a large atomic number, i. e. In or Sb, since here the potential modulations are strong. As illustrated in Figure 7.4 (b), the Rashba effect results in an energy splitting. For electrons with two spin orientations being perpendicular to the electric field and to the direction of motion, the energy splitting for a given value of $|\vec{k}|$ can be expressed as

$$E_R = \pm \alpha_R |\vec{k}|. \tag{7.6}$$

For a fixed energy, e. g. the Fermi energy E_F, the wave vector difference Δk_R can be evaluated analytically (cf. Figure 7.4 (b)):

$$\Delta k_R = \frac{2m^* \alpha_R}{\hbar^2}. \tag{7.7}$$

Here a parabolic band relation with a constant effective mass m^* was assumed. The spins of the corresponding eigenstates are oriented perpendicularly to the direction of motion in the plane of the two-dimensional electron gas. As for free electrons aligned parallel or antiparallel to an external magnetic field, one can also define an effective magnetic field [96] as

$$\vec{B}_{\text{eff}} = \Delta k_R \langle \vec{S} \rangle_+, \tag{7.8}$$

with the spin expectation value

$$\langle \vec{S} \rangle_+ = \langle \psi_+^R | \hat{\vec{S}} | \psi_+^R \rangle \qquad (7.9)$$

belonging to the upper of the two energy branches E_+ and E_- shown in Figure 7.4 (b). The eigenfunctions for each branch are $\psi_\pm^R$ and can be written as

$$|\psi_\pm^R(\vec{k}_\parallel)\rangle = \frac{e^{i\vec{k}_\parallel \vec{r}_\parallel}}{2\pi} \xi_{\vec{k}_\parallel}(z) \frac{1}{\sqrt{2}} \begin{pmatrix} 1 \\ \mp i e^{i\varphi} \end{pmatrix}, \qquad (7.10)$$

with $\vec{k}_\parallel = k_\parallel (\cos\varphi, \sin\varphi, 0)$ being the in-plane k vector described in polar coordinates and $\vec{r}_\parallel = (x, y, 0)$. The in-plane component of the wave function is described by a plane wave $\exp(i\vec{k}_\parallel \vec{r}_\parallel)$, while the so-called envelope function of the confined state of the quantum well is given by $\xi_{\vec{k}_\parallel}(z)$.

According to the definition in equation (7.8), the effective magnetic field has the same orientation as $\langle \vec{S} \rangle_+$. In Figure 7.5 (a) the energy-momentum paraboloids are shown for the two Rashba spin-split branches. As can be seen here, the effective magnetic field is always perpendicular to the direction of the k-vector. The spin eigenvalues of the upper and lower branches are aligned parallel and antiparallel to B_{eff}, respectively. In contrast to the Zeeman effect, where an external magnetic field is applied, here the orientations of the spin eigenvalues are not fixed for all k-values. The spin eigenvalues are rather perpendicular to the k-vectors. The spins are aligned perpendicular to the effective magnetic field B_{eff}, which is always perpendicular to the direction of motion. As can be seen in Figure 7.5 (b), B_{eff} rotates in a cycle around $k = 0$.

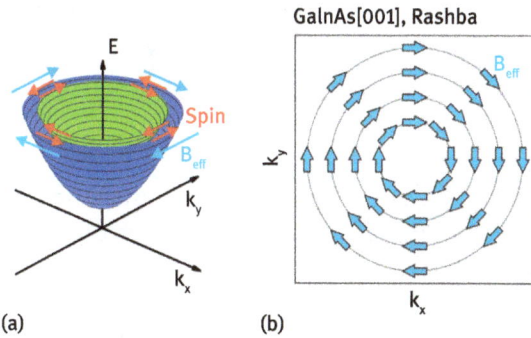

Figure 7.5: (a) Energy-momentum dispersion for electrons in a two-dimensional electron gas including the Rashba effect. The effective magnetic field is oriented perpendicularly to the corresponding k-vector. The spin eigenvalues are arranged clockwise and counter-clockwise around the energy axis. (b) Orientation of the effective magnetic field in k-space in the presence of the Rashba effect.

In case that for a given k-vector the spin state is not an eigenstate the spin is precessing around the effective magnetic field. This is illustrated in Figure 7.6 (a). This spin

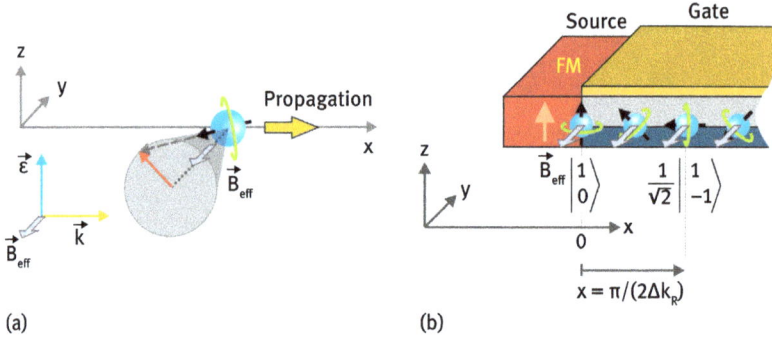

Figure 7.6: (a) Spin precession of the electron spin with the wave vector $\vec{k}$ around an effective magnetic field $\vec{B}_{\text{eff}}$. (b) Spin precession in a spin field-effect transistor. The magnetization of the source electrode is along the z-direction. The effective magnetic field $\vec{B}_{\text{eff}}$ is assumed to be along the $-y$-direction.

precession is the same as for free electrons with a magnetic moment $\vec{\mu}$ in an external magnetic field $\vec{B}$, where the torque leading to the spin precession is given by $\vec{\tau} = \vec{\mu} \times \vec{B}$. Generally, the spin precession from an initial state $|\psi\rangle$ to a final state $|\psi'\rangle$ after a certain propagation of the electron can be described by a 2×2 rotation matrix U_r

$$|\psi'\rangle = U_r|\psi\rangle. \tag{7.11}$$

To be more specific, let us consider the situation in a spin field-effect transistor. In Figure 7.6 (b) the area around the ferromagnetic source contact used as a spin injector is depicted. The electrons in the source contact are assumed to be polarized along the z-direction. Neglecting the contribution from the envelope function $\xi_{\vec{k}_\parallel}(z)$, the wave function at $x = 0$ is given by

$$|\psi(0)\rangle = \begin{pmatrix} 1 \\ 0 \end{pmatrix} = \frac{1}{2}\left[\begin{pmatrix} 1 \\ -i \end{pmatrix} + \begin{pmatrix} 1 \\ i \end{pmatrix}\right]. \tag{7.12}$$

Here the spin state along the z-direction is decomposed into the eigenstates for the electron propagation along the x-direction. These two basis states, which propagate with the wave vectors $k_x \mp \Delta k_R/2$, respectively, are

$$|\psi(x)\rangle = \frac{1}{2}\left[\exp\left[i(k_x - \Delta k_R/2)x\right]\begin{pmatrix} 1 \\ -i \end{pmatrix} + \exp\left[i(k_x + \Delta k_R/2)x\right]\begin{pmatrix} 1 \\ i \end{pmatrix}\right]. \tag{7.13}$$

Owing to the different phase accumulations $(k_x \mp \Delta k_R/2)x$, after a distance x the spin expectation value vector is given by

$$\langle \vec{S}(x)\rangle = \begin{pmatrix} \sin(-\Delta k_R x) \\ 0 \\ \cos(\Delta k_R x) \end{pmatrix}. \tag{7.14}$$

As illustrated in Figure 7.6b, the spin precesses around the effective magnetic field $\vec{B}_{\text{eff}} = (0, B_y, 0)$. After a distance $x = \pi/(2\Delta k_{\text{R}})$ the spin is rotated by 90 ° about $\vec{B}_{\text{eff}}$. We can define the spin precession length

$$l_{\text{R}} = \frac{\pi \hbar^2}{m^* \alpha_{\text{R}}}, \tag{7.15}$$

which is the length on which the electron precesses a full cycle. Below, we will explain how the Rashba parameter α_{R} and thus the spin precession length l_{R} can be tuned by means of a gate voltage. This allows to adjust the accumulated precession angle between the source and drain contact and by that control the device resistance.

7.4 Strength of the Rashba spin-orbit coupling

In this section, it will be sketched how the strength of the Rashba coupling parameter α_{R} can be calculated. We begin by introducing the $k \cdot p$ model, which can be utilized to calculated the energy bands and wave function in the vicinity of some important points in the Brillouin zone. In our case it is usually the Γ-point ($k = 0$). However, the $k \cdot p$ theory only applies to an ideal periodic crystal with the electronic states. In order to find an expression for the Rashba coefficient, an extension is necessary, the so-called envelope function approximation. Here confined states, such as two-dimensional states in a semiconductor quantum well, can be treated.

7.4.1 The $k \cdot p$ method

The $k \cdot p$ method allows us to calculate the band structure around some distinct wave vectors in the Brillouin zone [97]. For direct band gap semiconductors such as GaAs, InP, or InAs, we are interested in the Γ-point, since here one finds the minimum of the conduction band. Using the $k \cdot p$ method the energy band and corresponding wave functions can be calculated using the second order perturbation method. As a basis, the energies and wave function at the Γ-point are taken. The wave functions in the vicinity of the Γ-point are described by superpositions of these basis functions. This means that the bands, e. g. valence and conduction bands are coupled. In practice symmetry arguments, deduced from the particular symmetry of the crystal lattice, can be utilized to reduced the number of parameters considerably [97, 95].

In a periodic crystal the wave functions can be expressed by Bloch functions:

$$\varphi_{v\vec{k}}(\vec{r}) = e^{i\vec{k}\cdot\vec{r}} u_{v\vec{k}}(\vec{r}). \tag{7.16}$$

Here the first factor is a plane wave, while the second represents the lattice periodic part $u_{v\vec{k}}(\vec{r})$ of the Bloch function in the periodic crystal potential $V_0(\vec{r})$. The parameter

ν is the band index. In a III-V semiconductor crystal the $u_{\nu\vec{k}}(\vec{r})$ of the valence band, often written as $|X\rangle$, $|Y\rangle$, and $|Z\rangle$, are constituted from the p-orbital wave functions, whereas the lattice periodic part $|S\rangle$ of the conduction band Bloch function originates from s-orbitals. The Schrödinger equation for the Bloch states in a crystal with the lattice periodic potential $V_0(\vec{r})$ can be written as

$$\left[\frac{\hat{p}^2}{2m_0} + V_0(\vec{r})\right]e^{i\vec{k}\vec{r}}u_{\nu\vec{k}} = E_\nu(\vec{k})e^{i\vec{k}\vec{r}}u_{\nu\vec{k}}, \tag{7.17}$$

with m_0 being the free electron mass. The effect of the kinetic energy operator on the plane wave part of $\varphi_{\nu\vec{k}}(\vec{r})$ can be evaluated directly:

$$\left[\frac{\hat{p}^2}{2m_0} + V_0 + \frac{\hbar^2 k^2}{2m_0} + \frac{\hbar}{m_0}\vec{k}\cdot\hat{p}\right]|\nu\vec{k}\rangle = E_\nu(\vec{k})|\nu\vec{k}\rangle. \tag{7.18}$$

Here we expressed the lattice periodic part by $|\nu\vec{k}\rangle$ which is an expansion in terms of band edge functions

$$|\nu\vec{k}\rangle = \sum_{\nu'} c_{\nu\nu'}(\vec{k})|\nu'0\rangle. \tag{7.19}$$

The name $k \cdot p$ theory originates from the corresponding factor in the Hamiltonian. By applying a second order perturbation theory the energy of the νth band at finite values of $\vec{k}$ can be obtained:

$$E_\nu(\vec{k}) = E_\nu(0) + \frac{\hbar^2 k^2}{2m_\nu^*}, \tag{7.20}$$

where

$$\frac{m_0}{m_\nu^*} = 1 + \frac{2}{m_0}\sum_{\nu'}\frac{|\langle\nu 0|\hat{p}|\nu'0\rangle|^2}{E_\nu(0) - E_{\nu'}(0)}. \tag{7.21}$$

With $E_\nu(0)$ and $E_{\nu'}(0)$ the corresponding band energies at the Γ-point. Thus, the coupling of adjacent bands ν' expressed by $\langle\nu 0|\hat{p}|\nu'0\rangle$ results in an expression for the effective electron mass m_ν^* in band ν.

If Pauli spin-orbit coupling (cf. equation (7.2)) is included, the corresponding Schrödinger equation is given by

$$\left[\frac{\hat{p}^2}{2m_0} + V_0 + \frac{\hbar^2 k^2}{2m_0} + \frac{\hbar}{m_0}\vec{k}\cdot\left(\hat{p} + \frac{\hbar}{4m_0 c^2}\hat{\vec{\sigma}}\times\vec{\nabla}V_0\right) + \frac{\hbar}{4m_0^2 c^2}\hat{p}\cdot\hat{\vec{\sigma}}\times(\vec{\nabla}V_0)\right]|n\vec{k}\rangle$$
$$= E_n(\vec{k})|n\vec{k}\rangle. \tag{7.22}$$

Similar to equation (7.19), the wave functions $|n\vec{k}\rangle$ are expanded by the wave functions at the Γ-point $|\nu'0\rangle$, but now including the spin eigenstate $|\sigma\rangle$:

$$|n\vec{k}\rangle = \sum_{\nu',\sigma'=\uparrow,\downarrow} c_{n\nu'\sigma'}(\vec{k})|\nu'\sigma'\rangle. \tag{7.23}$$

Here, we abbreviated $|v'0\rangle \otimes |\sigma'\rangle$ by $|v'\sigma'\rangle$. Multiplication of the Schrödinger equation, equation (7.22), with $\langle v\sigma|$ from the left, results in a set of equations for the energy dispersion $E_n(\vec{k})$:

$$\sum_{v',\sigma'} \left\{ \left[E_{v'}(0) + \frac{\hbar^2 k^2}{2m_0} \right] \delta_{vv'} \delta_{\sigma\sigma'} + \frac{\hbar}{m_0} \vec{k} \cdot \vec{P}_{vv'\sigma\sigma'} + \Delta_{vv'\sigma\sigma'} \right\} c_{nv'\sigma'}(\vec{k})$$

$$= E_n(\vec{k}) c_{nv\sigma}(\vec{k}), \tag{7.24}$$

with

$$\vec{P}_{vv'\sigma\sigma'} = \langle v\sigma| \left[\hat{\vec{p}} + \frac{\hbar}{4m_0 c} \hat{\vec{\sigma}} \times \vec{\nabla} V_0 \right] |v'\sigma'\rangle, \tag{7.25}$$

and

$$\Delta_{vv'\sigma\sigma'} = \frac{\hbar}{4m_0^2 c^2} \langle v\sigma| [\hat{\vec{p}} \cdot \hat{\vec{\sigma}} \times \vec{\nabla} V_0] |v'\sigma'\rangle. \tag{7.26}$$

Similar to the previously discussed term $\langle v0|\hat{\vec{p}}|v'0\rangle$, here $\vec{P}_{vv'\sigma\sigma'}$ results in a mixing of band edge states $|v0\rangle$. The matrix elements of the spin-orbit coupling $\Delta_{vv'\sigma\sigma'}$ induce a splitting of degenerate energy levels.

For a III-V semiconductor it is often sufficient to consider the three valence band periodic functions $|X\rangle$, $|Y\rangle$, and $|Z\rangle$ and the conduction band $|S\rangle$ function. In addition, the symmetry of the crystal lattice has to be included to gain the relevant matrix elements. If spin-orbit coupling is neglected in equation (7.25) the matrix element which couples the valence band to the conduction band can be expressed as [95]

$$P = \langle S|\hat{p}_x|X\rangle. \tag{7.27}$$

As we know from equation (7.21), this contributes to the expression for the effective mass. Furthermore, the spin-orbit term given by equation (7.26) can be expressed as

$$\Delta_{so} = -\frac{3i\hbar}{4m_0^2 c^2} \langle X| [(\vec{\nabla} V_0) \times \hat{\vec{p}}]_y |Z\rangle. \tag{7.28}$$

As shown in Figure 7.7, this term is responsible for the energy splitting in the valence band between the heavy and light hole bands and the split-off band.

7.4.2 Envelope function approach

By utilizing the $k \cdot p$ method information on the bulk band structure is obtained, including spin-orbit coupling effects. However, this approach is insufficient to describe the Rashba effect. The reason is that the Rashba effect originates from local potential

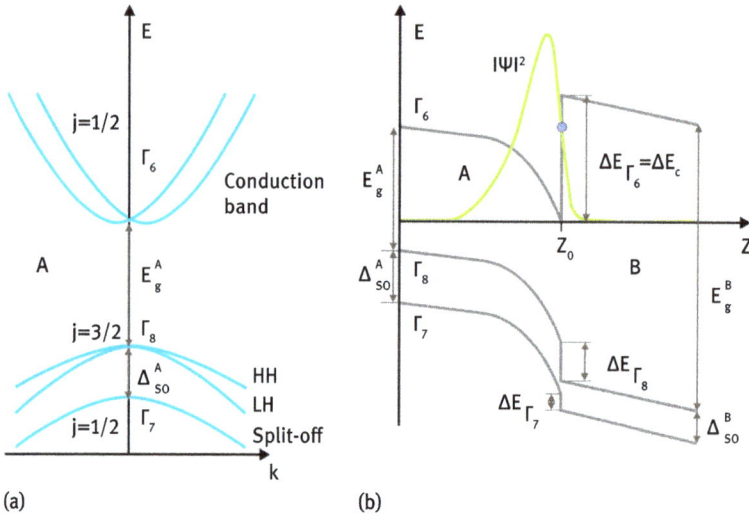

(a) (b)

Figure 7.7: (a) Schematic band structure of a III-V semiconductor (material A), with an energy band gap E_g^A. The split-off band is separated by Δ_{so}^A from the light and heavy hole band. (b) Profile of the conduction and valence band at the interface of a semiconductor with a smaller (A) and larger (B) band gap. E_g^A and E_g^B denote the band gaps of semiconductor A and B, while ΔE_c and ΔE_v denote the conduction and valence band offsets, respectively. Also shown is $|\psi|^2$ of the wave function in the quantum well.

gradients within a semiconductor heterostructure. Thus, the total potential lacks the required periodicity for the $k \cdot p$ method. A suitable approach to solve that problem is the envelope function approximation, which allows us to describe electron states in the presence of electric fields that vary slowly on the length scale of the lattice parameter. The envelope function can be constructed in the following way:

$$\Psi(\vec{r}) = \sum_{v'\sigma'} \psi_{v'\sigma'}(\vec{r})|v'\sigma'\rangle. \tag{7.29}$$

It is expressed in a similar fashion as the expansion of the wave function for the $k \cdot p$ method given by equation (7.23). However, in contrast to the coefficients $c_{nv'\sigma'}(\vec{k})$, which are spatially independent, in the envelope function approach we introduced position dependent coefficients $\psi_{v'\sigma'}(\vec{r})$. By these coefficients the Bloch functions are modulated on a length scale larger than the lattice parameter. In Figure 7.8 a sketch of a wave function in the envelope function approximation is shown. The envelope function itself corresponds to the wave function, which is obtained from a Schrödinger–Poisson solver in the free electron approximation, i. e. as schematically shown in Figure 7.2 (b). However, one should keep in mind that the wave function on a microscopic scale is much more modulated given by the lattice periodic part of the Bloch function. In most cases this fact is neglected, since it in a way already included by the effective mass approach, i. e. equation (7.21) directly shows that the coupling of the Bloch

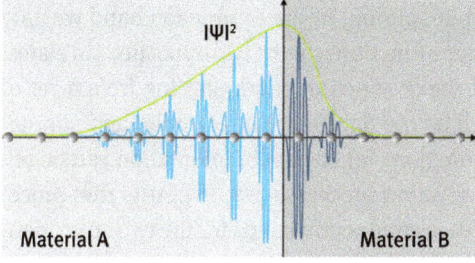

Figure 7.8: Schematic illustration of a wave function at a heterointerface between material A and B in the envelope function approximation.

function determines the effective mass. Yet for the expression of Rashba coupling parameter it is essential to include the rapidly varying lattice periodic part of the Bloch function explicitly. The reason is that these potential variations result in strong electric field modulations, which contribute to the spin-orbit coupling. Only by including this mechanism theoretical values of the Rashba coefficient are obtained which match the experimental values.

In the envelope function approximation the Schrödinger equation at $B = 0$ reads

$$\left[\frac{(-i\hbar\vec{\nabla})^2}{2m_0} + V_0(\vec{r}) + \frac{\hbar}{4m_0c^2}(-i\hbar\vec{\nabla}) \cdot \hat{\vec{\sigma}} \times \vec{\nabla}V_0 + V(\vec{r}) \right] \Psi(\vec{r}) = E\Psi(\vec{r}). \qquad (7.30)$$

In addition to the lattice periodic potential $V_0(\vec{r})$, the Hamiltonian also contains the slowly varying potential $V(\vec{r})$ of the quantum well. Furthermore, Pauli spin-orbit coupling is included. The wave function $\Psi(\vec{r})$ is defined by equation (7.29). By multiplying the Schrödinger equation with $\langle v\sigma|$ from the left and integrating over one unit cell of the lattice one obtains

$$\sum_{v'\sigma'} \hat{H}_{v,v'\sigma\sigma'} \psi_{v'\sigma'}(\vec{r}) = E\psi_{v\sigma}(\vec{r}), \qquad (7.31)$$

with

$$\hat{H}_{v,v'\sigma\sigma'} = \left[-\frac{\hbar^2}{2m_0}\Delta + E_{v'}(0) + V(\vec{r}) \right]\delta_{vv'\sigma\sigma'}$$

$$+ \frac{\hbar}{im_0}\vec{\nabla} \cdot \vec{P}_{vv'\sigma\sigma'} + \vec{\Delta}_{vv'\sigma\sigma'}. \qquad (7.32)$$

Here we made use of the fact the $V(\vec{r})$ and $\psi_{v\sigma}(\vec{r})$ are slowly varying within one unit cell, so that they can be taken out of the integral. Furthermore, $\vec{P}_{vv'\sigma\sigma'}$ and $\vec{\Delta}_{vv'\sigma\sigma'}$ are defined by equations (7.25) and (7.26), respectively. The system given above is a set of coupled differential equations, with the eigenfunctions given by the set of envelope functions. Since we are in particular interested in spin-orbit coupling effects, these envelope functions need to be spinors.

For the calculation of the Rashba spin-splitting in the conduction band we only care about the Γ_6-conduction band envelope functions for the two opposite spin states. The term Γ_6 relates to the symmetry of the conduction band. Starting from a set of bands, i. e. the conduction band as well as the light hole, heavy hole, and split-off bands, the conduction band envelope functions for each spin orientation can be obtained by performing a Löwdin renormalization procedure [98, 99]. After that procedure, the Hamiltonian only contains the two conduction bands for the two spin orientations instead of all eight bands.

We are interested in a heterostructure of two semiconductor materials, where material A is the small band gap material, e. g. InGaAs, and material B the large band gap material, e. g. InP. The band profile of such a heterojunction is depicted in Figure 7.7 (b). The conduction band, or more precisely the Γ_6 conduction band, is represented by a set of functions obeying the symmetry of that band. The Γ_8-band represents the light and heavy hole bands, whereas the Γ_7-valence band corresponds to the split-off band. Beside the smooth slope of the conduction and valence band edges one also has to include the abrupt band offsets, i. e. ΔE_{Γ_6} and ΔE_{Γ_8} for the conduction and valence bands, respectively. Furthermore, the offset in the split-off band ΔE_{Γ_7} has to be considered. After performing the Löwdin renormalization, the Hamiltonian for the conduction band envelope functions can be written as [100]

$$\hat{H} = \frac{(-i\hbar\vec{\nabla})^2}{2m^*(\vec{r})} + \Theta_B(z)\Delta E_c + V(\vec{r}) + \frac{1}{\hbar}\hat{\vec{\alpha}}_R\left(\hat{\vec{\sigma}} \times \frac{\hbar}{i}\vec{\nabla}\right). \tag{7.33}$$

Here, the step function $\Theta_B(z)$ with 1 inside material B and 0 otherwise takes care of including the conduction band offset. Epitaxial growth along the z-direction was assumed. The operator for the Rashba contribution is given by

$$\hat{\vec{\alpha}}_R(\vec{r}) = \frac{\hbar^2}{6m_0}\vec{\nabla}\left(\frac{E_p}{E - E_{\Gamma_7}^A - \Theta_B(z)\Delta E_{\Gamma_7} - V(\vec{r})} - \frac{E_p}{E - E_{\Gamma_8}^A - \Theta_B(z)\Delta E_{\Gamma_8} - V(\vec{r})}\right), \tag{7.34}$$

with E_p being the $k \cdot p$ interaction parameter describing the interaction between the conduction and the valence bands [101].

For heterostructure grown along the z-direction, the conduction envelope function $\psi(z)$ as well as the potential $V(z)$ depend only on the z-direction. This envelope function of the confined state at the heterointerface can be computed by means of a Schrödinger–Poisson solver. Subsequently, the Rashba parameter α_R is obtained by calculating the expectation value $\langle\psi(z)|\hat{\alpha}_R|\psi(z)\rangle$ using the following expression [100]:

$$\alpha_R = \frac{\hbar^2 E_p}{6m_0}\vec{e}_z\langle\psi(z)|\left[\left(\frac{1}{(E_g^A + \Delta_{so}^A)^2} - \frac{1}{(E_g^A)^2}\right)(1 - \Theta_B(z))V'(z)\right.$$
$$\left. + \left(\frac{1}{(E_g^B - \Delta E_c + \Delta_{so}^B)^2} - \frac{1}{(E_g^B - \Delta E_c)^2}\right)\Theta_B(z)V'(z)\right]|\psi(z)\rangle$$

$$+ \frac{\hbar^2 E_p}{6m_0} \bar{e}_z \left(\frac{\Delta_{so}^B - \Delta_{so}^A}{2(E_g^B - \Delta E_c + \Delta_{so}^B)^2} + \frac{\Delta_{so}^B - \Delta_{so}^A}{2(E_g^A + \Delta_{so}^A)^2} \right.$$

$$\left. - \frac{\Delta E_c}{2(E_g^B - \Delta E_c)^2} - \frac{\Delta E_c}{2(E_g^A)^2} \right) |\psi(z_0)|^2. \tag{7.35}$$

Here it was assumed that for small confinement energies E in equation (7.34) can be approximated by the conduction band edge in the low band gap material $E_c^A = E_{\Gamma_6}^A$. In that case $E - E_{\Gamma_7}^A$ corresponds to the band gap E_g^A. All the other energies are obtained accordingly and can be found in Figure 7.7 (b). The first two terms in equation (7.35) are simply proportional to the expectation values of the electric field $\mathcal{E}(z) = V'(z)/e$ in materials A and B, respectively. The abrupt change of the potential at the heterointerface results in an additional contribution, which is proportional to $|\psi(z_0)|^2$.

Although equation (7.35) looks rather complicate, some general requirements can be formulated to gain a large value of α_R. First of all, due to the inverse of the squared energies, a small band gap is advantageous. Furthermore, in order to prevent that the differences in equation (7.35) from being canceled out, $\Delta_{so}^{A,B}$ should be as large as possible. Thus, the Rashba spin-orbit coupling parameter is directly connected to the spin-orbit coupling in the valence band. This reflects that the Rashba effect originates from the coupling of the upper valence band and the split-off valence band to the conduction band. The smaller the energy differences between the valence band and the conduction band, the stronger the coupling is. The coupling itself is quantified by the interaction parameter E_p.

It is interesting to note that the expectation value $\langle \psi(z)|\mathcal{E}_c(z)|\psi(z)\rangle$, i. e. the electric field calculated from the gradient of the conduction band profile, is zero. Here, the spikes of the electric field at the interface owing to the conduction band offsets have to be included. The fact that $\langle \psi(z)|\mathcal{E}_c(z)|\psi(z)\rangle$ is zero is a consequence of the Ehrenfest theorem, i. e. the expectation value of the electric field of a bound state must strictly be zero [102]. However, since the band offsets in the valence band are different, the expectation value $\langle \psi(z)|\mathcal{E}_v(z)|\psi(z)\rangle$ is non-zero, with $\mathcal{E}_v(z)$ the gradient of the valence band profile. Thus, the naïve assumption that the strength of the Rashba coefficient is determined by the expectation value of the electric field in the conduction band is wrong. It is rather a consequence of the valence band profile. This once again shows, that the Rashba effect is mainly determined by the valence band which couples to the conduction band.

7.5 Magnetoresistance measurements

The strength of the Rashba parameter α_R can be determined by means of magneto-transport measurements. One possibility is to make use of the characteristic beating pattern observed in the oscillating magnetoresistance. These oscillations originate from the Landau quantization in a two-dimensional electron gas, when a magnetic field which is oriented perpendicularly is applied. The Landau quantization was described in Section 2.8.1. In the presence of the Rashba effect the Landau energy spectrum is modified.

7.5.1 Beating patterns due to the Rashba effect

If Rashba spin-orbit coupling is present, the Hamiltonian describing a two/dimensional system in a magnetic field (cf. equation (2.55)) acquires the following additional term:

$$\hat{H}_R = \frac{\alpha_R}{\hbar}\vec{e}_z[\hat{\vec{\sigma}} \times (\hat{\vec{p}} + e\vec{A})],\tag{7.36}$$

where $\vec{k}$ in equation (7.4) has been replaced by $(\hat{\vec{p}} + e\vec{A})/\hbar$. The total Hamiltonian can thus be expressed as

$$\hat{H} = \begin{pmatrix} \frac{\hat{p}_x^2}{2m^*} + \frac{(\hat{p}_y+eBx)^2}{2m^*} + \frac{g\mu_B B}{2} & -\alpha_R(i\frac{\partial}{\partial y} - \frac{eB}{\hbar}x - \frac{\partial}{\partial x}) \\ -\alpha_R(i\frac{\partial}{\partial y} - \frac{eB}{\hbar}x + \frac{\partial}{\partial x}) & \frac{\hat{p}_x^2}{2m^*} + \frac{(\hat{p}_y+eBx)^2}{2m^*} - \frac{g\mu_B B}{2} \end{pmatrix}.\tag{7.37}$$

The resulting energy spectrum is given by [13]

$$E_{n,\pm} = \begin{cases} \hbar\omega_c[n \pm \sqrt{(\frac{1}{2} - \frac{gm^*}{2m_0})^2 + 2\frac{\alpha_R m^*}{\hbar^3 \omega_c}n}], & n = 1, 2, \ldots \\ \hbar\omega_c(\frac{1}{2} - \frac{gm^*}{2m_0}), & n = 0, \end{cases}\tag{7.38}$$

with $\omega_c = eB/m^*$ being the cyclotron frequency. In Figure 7.9 the corresponding energy levels are plotted for $\alpha_R = 7.5 \times 10^{-12}$ eVm. Due to the inclusion of the Rashba effect the energy levels at a fixed magnetic field are no longer equally spaced. Owing to the irregular spacing of the Landau levels at the Fermi energy, a beating pattern is observed in the oscillatory magnetoresistance. Each time the spacing of the energy levels at the Fermi level is about the same in a certain magnetic field range, a node in the beating pattern is observed.

The strength of the spin-orbit coupling can be determined by analyzing the beating pattern in the Shubnikov–de Haas effect measurements. As we know from equation (2.83), the period of the Shubnikov–de Haas oscillations in $1/B$ is directly related to the sheet electron concentration. Looking at Figure 7.10 (a) we realize that the Rashba spin-orbit coupling leads to a splitting of the energy-momentum relation in two paraboloids.

Due to the different dispersion, the number of carriers filled up to the Fermi energy is slightly different in both spin-split subbands. Consequently, a superposition of two oscillations with slightly different frequencies is expected to result in a beating pattern in the Shubnikov–de Haas effect measurements.

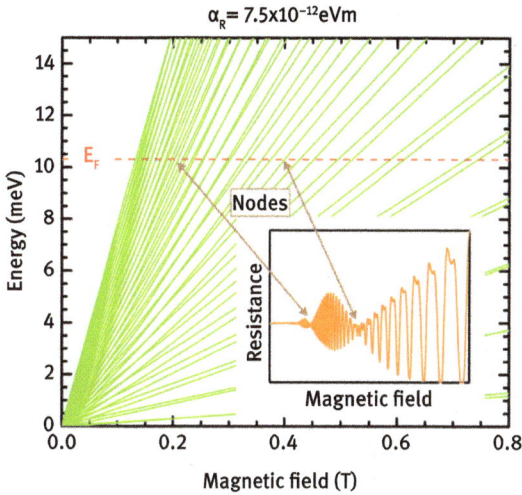

Figure 7.9: Landau level spectrum including the Rashba effect. The oscillatory magnetoresistance shows a beating pattern. The nodes in the beating pattern appear when the Landau levels at the Fermi energy E_F are equally spaced.

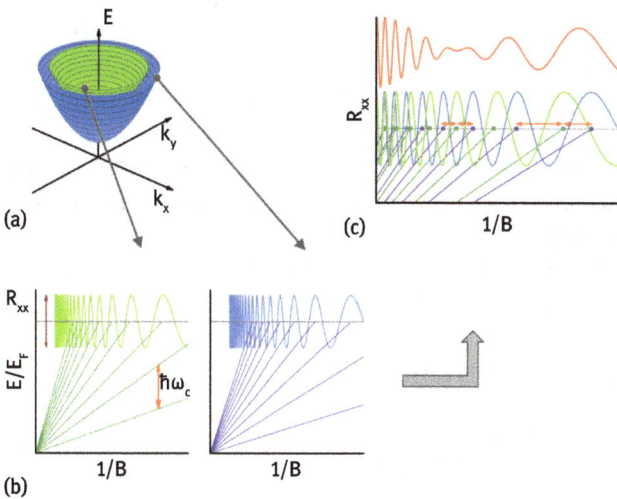

Figure 7.10: Origin of the beating pattern in the Shubnikov–de Haas oscillations. (a) The Rashba effect splits the energy momentum dispersion in two paraboloids of different size. Each paraboloid corresponds to a certain electron concentration. (b) Owing to the different electron concentrations, the oscillations in R_{xx} have a different period in $1/B$. (c) The superposition of both contributions leads to a beating pattern in the magnetoresistance oscillations.

That this is indeed the case can be seen in Figure 7.11. Due to the superposition of two oscillations slightly different frequencies a characteristic beating pattern is observed.

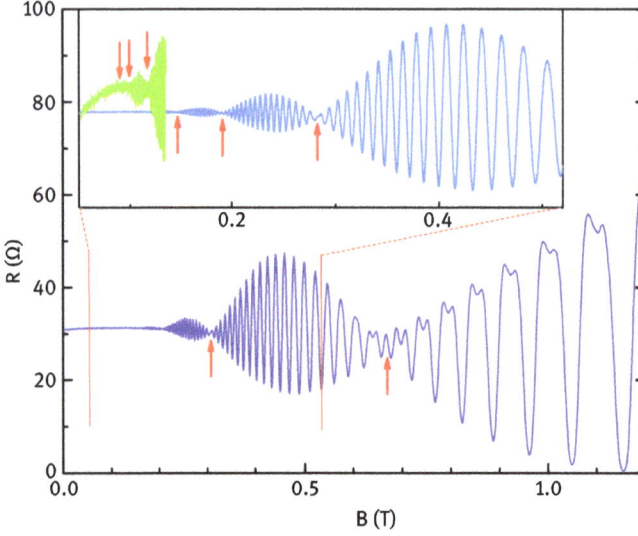

Figure 7.11: Shubnikov–de Haas measurement of a two-dimensional electron gas in an InGaAs/InP heterostructure. The Rashba spin-orbit coupling leads to a beating pattern. The measurements were performed at 50 mK. The red arrows indicate the nodes of the beating pattern. The inset shows a detail of the low field range.

According to equation (2.83) the sheet electron concentrations n_1, n_2 of each spin-split subband can be extracted by performing a fast Fourier transform in $1/B$. Note that the electron concentration determined from equation (2.83) has to be divided by two, due to the lifted spin degeneracy. As can be seen in Figure 7.12, the different electron concentrations result in a peak splitting in the Fourier transform.

The difference $\Delta n_{2D} = n_1 - n_2$ is a measure of the strength of the Rashba effect. By using a modified expression for the density of states, the following formula can be derived to calculate the Rashba spin-orbit coupling parameter from the electron concentrations of the spin-split subbands [103, 100]:

$$\alpha_R = \frac{\Delta n_{2D}\hbar^2}{m^*}\sqrt{\frac{\pi}{2(n_{2D} - \Delta n_{2D})}}. \tag{7.39}$$

Here, $n_{2D} = n_1 + n_2$ is the total sheet electron concentration. For the measurement shown in Figure 7.11 a coupling constant of $\alpha_R = 6.97 \times 10^{-12}$ eVm was extracted. Often it is more convenient to determine the value of n_{2D} directly by means of a Hall effect measurement.

7.5.2 Gate-control of the Rashba effect

So far it was shown that the Rashba effect leads to a characteristic beating pattern in the Shubnikov–de Haas oscillations. From the peak splitting found in the Fourier

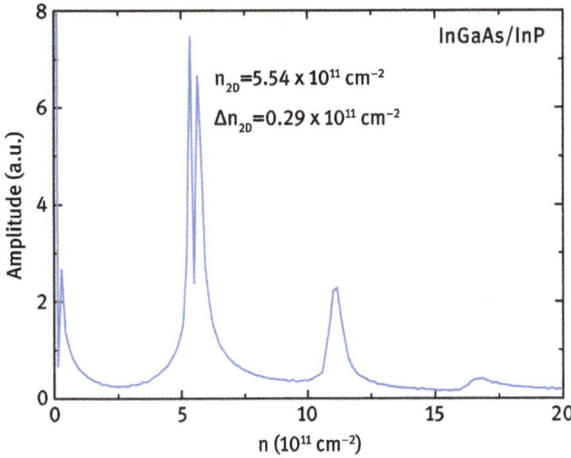

Figure 7.12: Fourier transform of the measurement shown in Figure 7.11. The Rashba effects results in a splitting of the first peak. The second peak belongs to a higher harmonic contribution to the oscillations.

transform of the oscillations the Rashba coupling coefficient can be extracted. The next thing to show is that α_R can be adjusted by applying a gate voltage. In Figure 7.13 (a), a set of magnetoresistance measurements of a two-dimensional electron gas in an InGaAs/InP heterostructure at various gate voltages is shown. By applying a more negative gate voltage, R_{xx} increases due to the lowering of the electron concentration. Furthermore, one finds that the node between 0.6 and 0.7 T shifts towards larger fields for more negative gate voltages. By performing a Fourier transform, α_R can be determined. It can be seen in Figure 7.13 (b) that for more negative gate voltages α_R increases. This clearly demonstrates that the Rashba coupling coefficient can be controlled by applying a gate voltage.

In Figure 7.14 the conduction band profile of a two-dimensional electron gas in an InGaAs/InP heterostructure is shown for two different gate voltages. It can be seen that for a the negative gate voltage of $-6\,\text{V}$ the potential profile is more tilted. This results in a larger contribution of the net electric field and leads to a larger value of α_R. Furthermore, for a more negative gate voltage $|\psi|^2$ is shifted towards the right-hand side interface.

In Figure 7.15 the measured Rashba coefficient α_R is plotted as a function of electron concentration for two different samples [100]. Once again, for lower electron concentrations α_R increases, which can be attributed to the larger tilting of the potential profile. As can be seen in the graph, the agreement between the experimental values of α_R and the calculated ones is good. However, it is important to note that for a proper calculation the contribution of the potential offsets at the interfaces has to be included. If only the macroscopic field is considered the calculated values of α_R are too small. In the experiment described above, the change of α_R is directly connected with

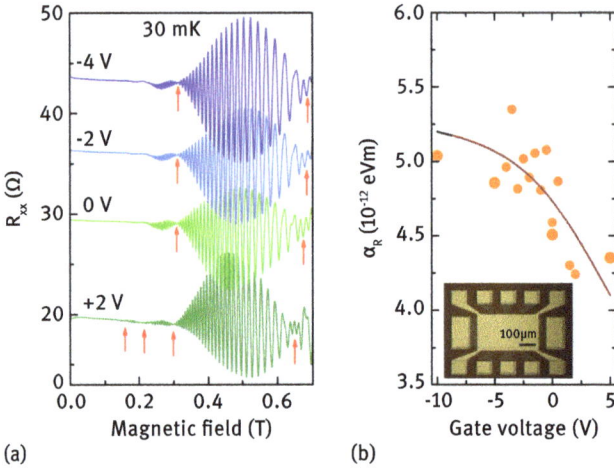

Figure 7.13: (a) Magnetoresistance R_{xx} of a two-dimensional electron gas in an InGaAs/InP hetero-structure for various gate voltages. The red arrows indicated the nodes of the beating pattern. (b) Rashba coupling parameter α_R as a function of gate voltage. The inset shows a Hall bar structure covered by a gate electrode.

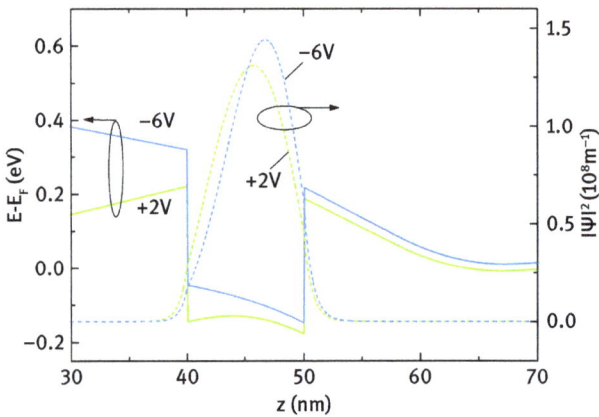

Figure 7.14: Conduction band profile of a two-dimensional electron gas in an InGaAs/InP hetero-structure for gate voltages of +2 and −6 V, respectively. Also shown are the corresponding probability densities $|\psi|^2$ of the wave functions.

a change in the electron concentration. Interestingly, one can also find a condition, where the electron concentration is kept constant but the symmetry of the quantum well is changed, i. e. the tilt of the quantum well potential is changed [104]. This can be realized by employing a top and back gate. Here, it could be shown that by balancing both gate voltages differently by keeping the concentration constant the Rashba coupling constant can be changed.

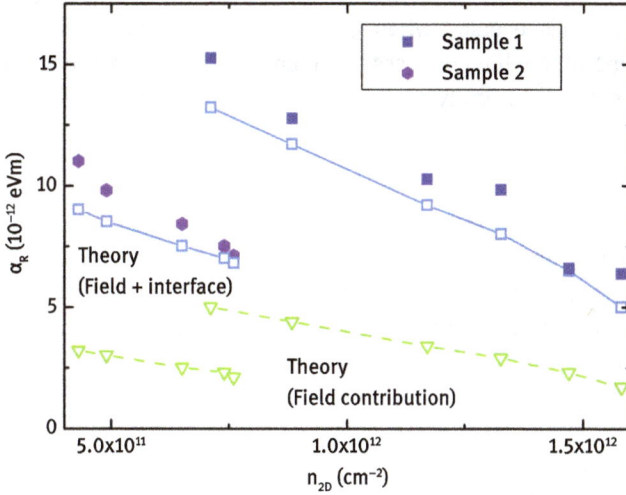

Figure 7.15: Measured Rashba coupling parameter α_R as a function of the electron concentration for two different InGaAs/InP samples. The experimental values are compared to the calculated values of α_R. The calculated values merely accounting for the contribution of the macroscopic electric field are too small to explain the measurements. It is crucial to include the contributions of the hetero-interfaces.

7.6 Bulk inversion asymmetry

Besides the Rashba spin-orbit coupling, there is another contribution, the so-called bulk inversion asymmetry term, which also results in a spin-splitting of the conduction band. The origin of this effect lies in the symmetry of the crystal. In this section we will first discuss two symmetry relations, i. e. time-reversal symmetry and spatial-inversion symmetry. Considering these symmetries gives information about the degeneracy of a state.

7.6.1 Time-reversal symmetry

In simple terms time-reversal symmetry can be imagined by winding a movie of a propagating particle backwards. The particle takes the same path but with opposite momentum and opposite spin. If that holds under time reversal, the particle will have the same energy. Time-reversal symmetry is broken if an external magnetic field is applied. Because of the vector product in the term of the Lorentz force, a charged particle is deflected in opposite direction, if the momentum is inversed. We can express the time-reversal symmetry in a more formal manner. For that we first define the time-reversal operator $\hat{T}$, which reverses spin and momentum:

$$\hat{T} = i\hat{\sigma}_y \hat{\mathcal{K}}. \tag{7.40}$$

Here, $i\hat{\sigma}_y$ takes care that the spin is reversed, when applied to a spinor. The operator $\hat{\mathcal{K}}$ is the complex conjugate operator, which reverses the momentum, e. g. a plane wave $\exp(i\vec{k}\vec{x})$ is transferred to $\exp(-i\vec{k}\vec{x})$. Applying $\hat{\mathcal{T}}$ twice to an electron with spin-1/2 results in

$$\hat{\mathcal{T}}^2 = -1, \tag{7.41}$$

i. e. $\hat{\mathcal{T}}$ is an antiunitarian operator $\hat{\mathcal{T}}^{-1} = -\hat{\mathcal{T}}$. If time-reversal symmetry holds, we should get the same energy for the initial state $|\psi\rangle$ and the time-reversed state $|\hat{\mathcal{T}}\psi\rangle$. This has consequences for the Hamilton operator $\hat{H}$, in order to maintain time-reversal symmetry. It should hold that

$$\hat{\mathcal{T}}\hat{H}\hat{\mathcal{T}}^{-1} = \hat{H}, \tag{7.42}$$

which is equivalent to the commutator relation

$$[\hat{H}, \hat{\mathcal{T}}] = 0. \tag{7.43}$$

Generally, symmetries can be expressed most compactly by a commutator relation, as done above. One can easily show that if the Hamiltonian is time-reversal symmetric one obtains the same value for the energy for $|\psi\rangle$ and $|\hat{\mathcal{T}}\psi\rangle$, i. e.

$$E = \langle\hat{\mathcal{T}}\psi|\hat{\mathcal{T}}\hat{H}\hat{\mathcal{T}}^{-1}|\hat{\mathcal{T}}\psi\rangle = \langle\psi|\hat{\mathcal{T}}^\dagger\hat{\mathcal{T}}\hat{H}\hat{\mathcal{T}}^{-1}\hat{\mathcal{T}}|\psi\rangle = \langle\psi|\hat{H}|\psi\rangle, \tag{7.44}$$

where we made use of the fact that $\hat{\mathcal{T}}^\dagger = \hat{\mathcal{T}}^{-1}$.

For a spin-1/2 particle one can directly prove that a time reversal invariant Hamiltonian has at least twofold degenerate eigenvalues. Let us assume that for the operator $\hat{\mathcal{T}}$ the equation $\hat{\mathcal{T}}|\psi\rangle = c|\psi\rangle$ does have a nondegenerate eigenvalue c. Applying $\hat{\mathcal{T}}$ once again leads to $\hat{\mathcal{T}}^2|\psi\rangle = \hat{\mathcal{T}}c|\psi\rangle = cc^*|\psi\rangle$. However, with $\hat{\mathcal{T}}^2 = -1$ this leads to a contradiction, since $cc^* = |c|^2 \neq -1$. This implies that for a system which maintains time reversal symmetry, the eigenvalues are at least twofold degenerate. This is the so-called Kramers degeneracy, which gives for the energy eigenvalues

$$E_+(\vec{k}) = E_-(-\vec{k}), \tag{7.45}$$

with $+/-$ indicating the spin orientation. Thus, time-reversed states, i. e. with inversed momentum $\vec{k} \to -\vec{k}$ and reversed spin, have the same energy.

7.6.2 Spatial inversion symmetry

Let us consider a crystal structure with an inversion center. A typical lattice is the diamond lattice, as shown in Figure 7.16 (a). As indicated here, the crystal possesses an inversion center, and thus the crystal is spatial inversion symmetric.

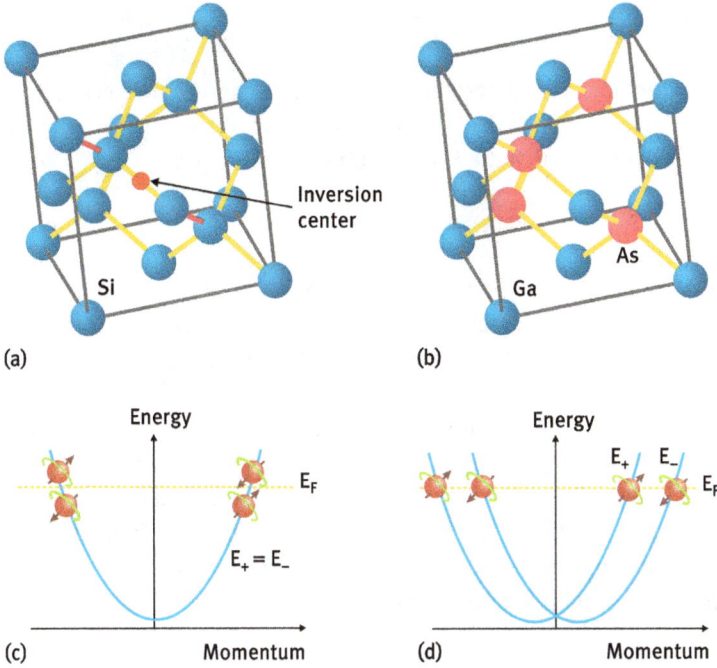

Figure 7.16: (a) Diamond lattice with an inversion center located between two atoms in the lattice. The red bars indicate the corresponding inversion symmetric bonds. A typical semiconductor with a diamond lattice is Si. (b) In the zincblende crystal no inversion center is found, because every second site in the lattice contains a different atom. Typical semiconductors with a zincblende configuration are GaAs or InAs. (c) Energy-momentum relation in case of inversion symmetry, where the states are spin degenerate. (d) Energy-momentum relation in case of no inversion symmetry, where the spin degeneracy is lifted.

Owing to this spatial inversion symmetry of the lattice potential in the Hamiltonian the electron energy must be the same under inversion of the k-vector:

$$E_{\pm}(\vec{k}) = E_{\pm}(-\vec{k}). \tag{7.46}$$

If we combine this with time-inversion symmetry (Kramers degeneracy) expressed by equation (7.45) we arrive at

$$E_{+}(\vec{k}) = E_{-}(\vec{k}). \tag{7.47}$$

Thus, the energies of the two spin orientation for any k-vector are degenerate. No spin-splitting of the energy levels can occur due to symmetry reasons (cf. Figure 7.16 (c)).

7.6.3 Dresselhaus term

In a crystal lattice, which does not possesses a spatial inversion center, e. g. a zinc blende lattice, as it is found for most III-V semiconductors (cf. Figure 7.16 (b)), the energies for a particle under reversal of the k-vector are no longer equal:

$$E_\pm(\vec{k}) \neq E_\pm(-\vec{k}). \tag{7.48}$$

Thus, together with the time inversion symmetry equation (7.45), which still holds, one finds that in general $E_+(\vec{k})$ is not the same as $E_-(\vec{k})$. Therefore, the spin degeneracy in the conduction band is lifted, as illustrated in Figure 7.16 (d). Only at $k = 0$ are the states degenerate.

It follows from the general considerations given above, in a lattice which does not possess an inversion center another spin-splitting contribution is found in addition to the Rashba effect. This contribution is due to the internal electric field in the crystal, which is not canceled out because of the lack of inversion symmetry. This results in an additional spin-orbit coupling term. For zinc blende crystals this contribution is called the Dresselhaus term, or bulk inversion asymmetry term [105]. The Hamiltonian consists of cubic terms

$$\hat{H}_D = \eta_D [\hat{\sigma}_x k_x (k_y^2 - k_z^2) + \hat{\sigma}_y k_y (k_z^2 - k_x^2) + \hat{\sigma}_z k_z (k_x^2 - k_y^2)]. \tag{7.49}$$

Similarly to α_R for the Rashba effect, the parameter η_D is a measure for the strength of the spin-orbit coupling due to bulk inversion asymmetry. For a two-dimensional electron gas in the xy-plane the corresponding Hamiltonian is given by

$$\hat{H}_D = \eta_D [\hat{\sigma}_x k_x (k_y^2 - \langle k_z^2 \rangle) + \hat{\sigma}_y k_y (\langle k_z^2 \rangle - k_x^2)]. \tag{7.50}$$

Since the Dresselhaus term contains k^3-contributions, it is also called the cubic spin-orbit contribution. In case of a two-dimensional electron gas, the expectation value of $\langle k_z^2 \rangle$ has to be inserted for the confinement direction. While $\langle k_z \rangle$ is zero the $\langle k_z^2 \rangle$ contribution is nonzero. For strong confinement one can assume $\langle k_z^2 \rangle \gg k_x^2, k_y^2$. In that case equation (7.50) can be approximated by a linearized form:

$$\hat{H}_D = \beta_D (\hat{\sigma}_x k_x - \hat{\sigma}_y k_y), \tag{7.51}$$

where $\beta_D = -\eta_D \langle k_z^2 \rangle$. Similarly to the Rashba effect, we can also assign an effective magnetic field B_{eff} to a given wave vector. This is illustrated in Figure 7.17 (a).

As shown in Figure 7.17 (a), in case of Dresselhaus spin-orbit coupling B_{eff} can also be parallel to $\vec{k}$. This is in contrast to the Rashba effect, where the effective magnetic field vector is always perpendicular to the k-vector (cf. Figure 7.17 (b)). A very interesting situation occurs, if the strength of the Dresselhaus spin-orbit coupling is just as big as the Rashba effect, i. e. $\beta_D = \alpha_R$. The resulting effective magnetic field is depicted

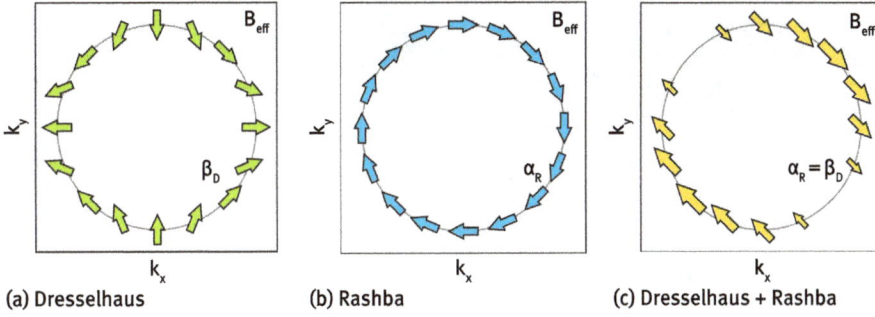

Figure 7.17: (a) Effective magnetic field B_{eff} for the linear Dresselhaus effect in a two-dimensional system. (b) Corresponding effective field in the presence of the Rashba effect. (c) Combined effective field in case that $\alpha_R = \beta_D$; the strength of Rashba and Dresselhaus effect are the same.

in Figure 7.17 (c). For k-vectors along the $(1\bar{1}0)$ direction one finds that the effective magnetic field diminishes. Thus, for wave vectors along that direction, no spin precession occurs. Moreover, the effective magnetic field is aligned only along the $[1\bar{1}0]$ axis. Based on this peculiar behavior a special type of spin field-effect transistor can be realized [8, 10]. In the most ideal case the one-dimensional channel between source and drain is aligned along the $[1\bar{1}0]$ direction. Let us assume that for a certain gate voltage $\beta_D = \alpha_R$. No spin precession occurs, since the effective magnetic field is zero. Thus the spin orientation of the electrons is preserved while they propagate between source and drain. The transistor is in the on-state, when assuming the same magnetization for the source and drain contact. By changing the gate voltage the Rashba coefficient can be changed so that $\beta_D \neq \alpha_R$. In that case a finite effective magnetic field appears, so that spin-precession occurs. Thus, except for a full rotation of the spin, the spin orientation of the electrons arriving at the drain contact does not match to the magnetization of the drain electrode.

In was realized that the case for $\beta_D = \alpha_R$ has SU(2) symmetry, a special unitary group of degree 2 represented by 2×2 matrices [106]. It was shown theoretically that a so-called persistent spin helix mode is formed. As illustrated in Figure 7.18 (a), assuming ballistic transport, and starting from a given point, the spin precession angle varies linearly with the distance propagated along the [110] direction.

Thus for a given distance moved along the [110] direction, the spin orientation is the same, no matter what the angle of propagation has been. A full spin rotation is achieved after propagating a distance half the Rashba spin precession length:

$$\frac{l_R}{2} = \frac{\pi \hbar^2}{2 m^* \alpha_R}. \tag{7.52}$$

The factor 1/2 appears, since the effective magnetic field for a propagation along [110] is twice as large as the Rashba field alone due the additional contribution of the Dresselhaus effective field. Even for the diffusive case, this has important consequences.

(a)

(b)

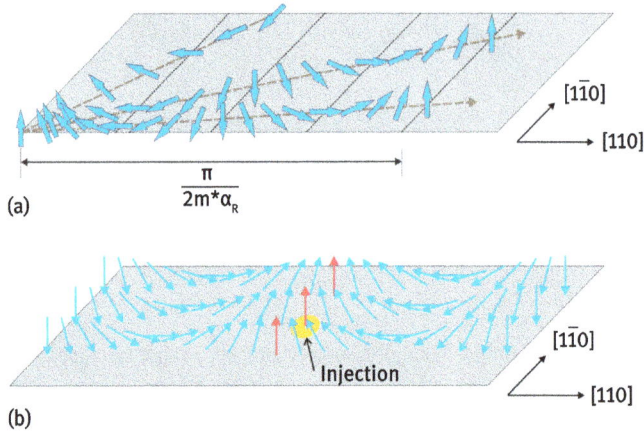

Figure 7.18: (a) Spin configuration of electrons starting at the same spot but moving ballistically in different directions. The spin precession only depends on the distance propagated along the [110] axis. (b) Diffusive evolution of spins injected in spin-up direction at the spot at the center. The spin polarization evolves into a persistent spin helix mode.

As illustrated in Figure 7.18 (b), injecting spins at a spot a spatially regular spin structure forms after diffusive evolution of the carriers in the semiconductor. This structure is called persistent spin helix. Along the [110] axis a well-defined spatially and temporarily fixed spin-orientation is found. The evolution angle of the spatial spin precession depends on the direction, since the effective magnetic field is opposite for opposite directions of motions parallel to the [110], as depicted in Figure 7.17 (c). As shown in Figure 7.18 (b), for diffusion along the [1$\bar{1}$0] direction the spin lifetime is infinitive, because the spin does not change its direction at all. This is in stark contrast to all other cases, where the effective magnetic field has non-vanishing components in all directions, e. g. for pure Rashba or Dresselhaus spin-orbit coupling as shown in Figure 7.17 (a) and (b). As a consequence the spin orientation decays in all directions.

The persistent spin helix could indeed be measured by optical means [107, 108]. By using a circularly polarized laser beam, electrons can be locally excited in a GaAs/AlGaAs quantum well, which possesses a well-defined spin orientation, i. e. pointing out of plane [108]. The spin orientation after diffusion in different directions can be locally measured by means of the magneto-optical Kerr effect by scanning a linearly polarized laser beam across the surface of the sample. Indeed, it could be confirmed that a persistent spin helix is formed. Another consequence of this special symmetry is the vanishing of the spin-dependent quantum interference [109], i. e. the so-called weak antilocalization effect, discussed in the next chapter, is suppressed.

7.7 Rashba effect in quasi one-dimensional structures

So far, we have only considered spin-orbit coupling in two-dimensional electron systems. However, it was already pointed out by Datta and Das [6] that a better signal modulation is expected if the transport is restricted to only one dimension. In addition, many concepts of spin electronic devices rely on one-dimensional transport channels. We first consider a planar one-dimensional system with a parabolic confinement potential. Here, the Hamilton matrix is deduced for the most general case, when an external magnetic field is applied. Based on this, we discuss the special case at zero field. As long as the geometrical confinement is not too strong a beating effect in the magnetoresistance is expected, as it is observed for the two-dimensional case. However, we will show that the node positions are shifted.

7.7.1 Rashba effect in planar quasi one-dimensional structures

One possible way to include spin-orbit coupling in a one-dimensional structure is to assume a two-dimensional electron gas which is additionally confined by a harmonic confinement potential:

$$V(x) = \frac{1}{2} m^* \omega_0^2 x^2, \tag{7.53}$$

where ω_0 is the characteristic oscillator frequency. By including the additional confinement $V(x)$ into the Schrödinger equation for a two-dimensional system equation (2.59) one arrives at

$$\left[-\frac{\hbar^2}{2m^*} \frac{\partial^2}{\partial x^2} + \frac{m^*}{2} \omega_c^2 (x - x_0)^2 + \frac{m^*}{2} \omega_0^2 x^2 + \frac{1}{2} g \mu_B \hat{\sigma}_z \right] \varphi_{nk}(x) = E \varphi_{nk}(x). \tag{7.54}$$

Here, we also added the Zeeman contribution, and thus $\varphi_{nk}(x)$ is a two-component spinor. By redefining the center coordinate x_0 and inserting an effective total frequency ω we can write

$$\left[-\frac{\hbar^2}{2m^*} \frac{\partial^2}{\partial x^2} + \frac{m^*}{2} \omega^2 (x - \tilde{x}_0)^2 + \frac{\omega_0^2}{\omega^2} \frac{\hbar^2 k_y^2}{2m^*} + \frac{1}{2} g \mu_B \hat{\sigma}_z \right] \varphi_{nk}(x) = E \varphi_{nk}(x), \tag{7.55}$$

with

$$\tilde{x}_0 = \frac{\omega_c^2}{\omega^2} x_0, \quad x_0 = -\frac{\hbar k_y}{eB} \quad \text{and} \quad \omega^2 = \omega_c^2 + \omega_0^2. \tag{7.56}$$

This Schrödinger equation represents the case of a harmonic oscillator of an effective frequency ω and a free motion along the y-direction. Similarly to equation (2.77), the eigenfunctions can be written as

$$\varphi_{n,\pm}(x) = \frac{1}{\sqrt{b}} \frac{\pi^{-1/4}}{\sqrt{2^n n!}} H_n\left(\frac{x - \tilde{x}_0}{b} \right) \exp\left(-\frac{(x - \tilde{x}_0)^2}{2b^2} \right) \chi_{\pm}, \quad n = 0, 1, 2, \ldots, \tag{7.57}$$

with $n = 0, 1, 2, \ldots$, $b = \sqrt{\hbar/m^*\omega}$ being the oscillator length of the parabolic confinement, and $H_n(x)$ the Hermite polynomials of integer order n. We added the spinors $\chi_\pm$ for spin up and down projected to the z-direction because of the Zeeman contribution. The corresponding eigen energies are

$$E_{n,\pm} = \hbar\omega\left(n + \frac{1}{2}\right) + \frac{\omega_0^2}{\omega^2}\frac{\hbar^2}{2m^*}k_y^2 \pm \frac{1}{2}g\mu_B B. \tag{7.58}$$

At zero magnetic field the $\omega = \omega_0$, and thus the energy splitting is simply determined by the geometric confinement energy $\hbar\omega_0$, as can be seen in Figure 7.19. With increasing magnetic field the effective energy splitting increases. At large magnetic fields with $\omega_c \gg \omega_0$ the energy levels approach the Landau levels of a two-dimensional system. This is plausible, since at large magnetic fields the cyclotron radius becomes smaller than the width of the geometrical confinement potential.

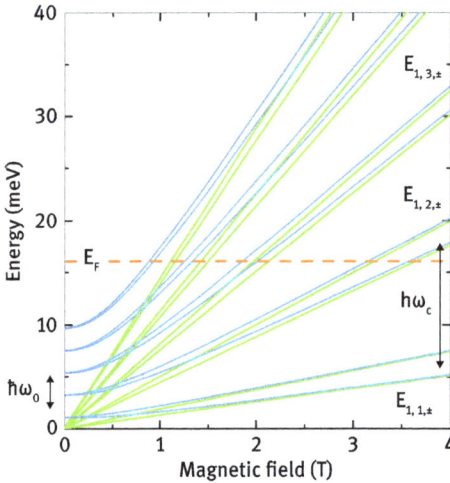

Figure 7.19: Magnetic field dependent dispersion of the energy levels of a one-dimensional electron gas with a parabolic confinement potential (blue curves), with $\hbar\omega_0 = 2\,\text{meV}$, $m^* = 0.037m_0$, and $g = -10$. As a reference, the energy levels of a two-dimensional electron gas is also shown (green lines). The parabolic confinement potential results in an additional energy splitting at $B = 0$.

In the presence of Rashba spin-orbit coupling, we need to add the following part to the Hamiltonian in equation (7.55):

$$\hat{H}_R = \alpha_R\left[\hat{\sigma}_x\left(k_y + \frac{eB}{\hbar}x\right) + i\hat{\sigma}_y\frac{d}{dx}\right]. \tag{7.59}$$

If we take $\varphi_{n,\pm}$ as basis functions, the energy eigenvalues can be calculated by diagonalizing the Hamilton matrix. The matrix elements concerning the Rashba Hamiltonian are given by

$$\langle \varphi_{n,\pm} | \hat{H}_R | \varphi_{n,\mp} \rangle = -\alpha_R \left(\frac{\omega_c^2}{\omega^2} - 1 \right) k_y, \tag{7.60}$$

$$\langle \varphi_{0,\pm} | \hat{H}_R | \varphi_{1,\mp} \rangle = \frac{\alpha_R}{b\sqrt{2}} \left(\frac{\omega_c}{\omega} \pm 1 \right), \tag{7.61}$$

$$\langle \varphi_{n,\pm} | \hat{H}_R | \varphi_{m,\mp} \rangle = \frac{\alpha_R}{b} \left(\frac{\omega_c}{\omega} \pm 1 \right) \sqrt{\frac{n+1}{2}} \delta_{n,m-1}$$
$$+ \frac{\alpha_R}{b} \left(\frac{\omega_c}{\omega} \mp 1 \right) \sqrt{\frac{n}{2}} \delta_{n,m+1}, \quad n \geq 1. \tag{7.62}$$

Written explicitly, the Hamilton matrix has the following form:

$$\begin{pmatrix} E_{0,+} & -\alpha_R(\frac{\omega_c^2}{\omega^2} - 1)k_y & 0 & -\frac{\alpha_R}{\sqrt{2}b}(\frac{\omega_c}{\omega} - 1) & \cdots \\ -\alpha_R(\frac{\omega_c^2}{\omega^2} - 1)k_y & E_{0,-} & -\frac{\alpha_R}{\sqrt{2}b}(\frac{\omega_c}{\omega} + 1) & 0 \\ 0 & -\frac{\alpha_R}{\sqrt{2}b}(\frac{\omega_c}{\omega} + 1) & E_{1,+} & -\alpha_R(\frac{\omega_c^2}{\omega^2} - 1)k_y \\ -\frac{\alpha_R}{\sqrt{2}b}(\frac{\omega_c}{\omega} - 1) & 0 & -\alpha_R(\frac{\omega_c^2}{\omega^2} - 1)k_y & E_{1,-} \\ \vdots & & & & \ddots \end{pmatrix}. \tag{7.63}$$

From the structure of the matrix we immediately see which states couple due to the Rashba effect. Let us begin with $\omega_0 = 0$, e. g. no confinement. In this case $\omega_c/\omega = 1$, so that $\langle \varphi_{0,\pm} | \hat{H}_R | \varphi_{1,\mp} \rangle = 0$ as well as the elements containing $\omega_c/\omega - 1$. The eigenvalue spectrum reduced to the results of a two-dimensional electron gas in the presence of the Rashba effect given by equation (7.38). We can also consider the case where no external magnetic field is applied, so that $\omega_c = 0$ [110]. The Hamilton matrix reduces to

$$\begin{pmatrix} E_0 & \alpha_R k_y & 0 & \frac{\alpha_R}{\sqrt{2}b} & \cdots \\ \alpha_R k_y & E_0 & -\frac{\alpha_R}{\sqrt{2}b} & 0 \\ 0 & -\frac{\alpha_R}{\sqrt{2}b} & E_1 & \alpha_R k_y \\ \frac{\alpha_R}{\sqrt{2}b} & 0 & \alpha_R k_y & E_1 \\ \vdots & & & & \ddots \end{pmatrix}. \tag{7.64}$$

Here, the terms $(\alpha_R/b)\sqrt{(n+1)/2}$, $n = 0, 1, 2, \ldots$ couple neighboring one/dimensional subbands. In order to get a feeling of how strong this coupling is, one has to compare the Rasbha energy contribution to the level splitting $\hbar\omega_0$. From diagonalizing the Hamilton matrix one finds that the relative strength of Rashba coupling with respect to the confinement is expressed by the ratio

$$\frac{\frac{1}{2b^2}\alpha_R^2}{(2E_0)^2} = \frac{\frac{m^*}{2\hbar^2}\alpha_R^2}{\hbar\omega_0} = \frac{\Delta_R}{\hbar\omega_0}, \tag{7.65}$$

where we defined the characteristic Rashba spin-orbit coupling energy by

$$\Delta_R = \frac{m^*}{2\hbar^2}\alpha_R^2. \tag{7.66}$$

In Figure 7.20 typical energy-momentum dispersions of one-dimensional systems at $B = 0$ with a harmonic confinement potential are shown.

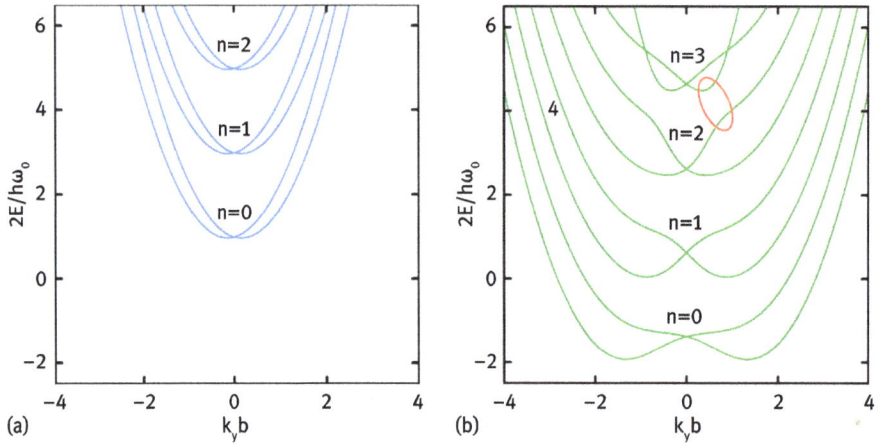

Figure 7.20: Energy-momentum dispersion relation for a one-dimensional conductor with a parabolic confinement potential along the x-direction. The k_y vector is normalized to the oscillator length $b = \sqrt{\hbar/m^*\omega_0}$, while the energy is normalized to $\hbar\omega_0/2$. (a) Weak coupling limit for $\Delta_R/\hbar\omega_0 = 0.01$. (b) Strong coupling limit for $\Delta_R/\hbar\omega_0 = 1$. The anticrossing between the $n = 2$ and 3 bands is marked.

In Figure 7.20 (a), the Rashba coupling energy is small compared to the confinement energy, i. e. $\Delta_R/\hbar\omega_0 = 0.01$. Here one finds a Rashba splitting of each subband. No significant coupling between neighboring bands is observed. The situation differs for a stronger Rashba contribution $\Delta_R/\hbar\omega_0 = 1$. In this case an anticrossing of bands is observed. In this range a superposition of the two anticrossing spin states occurs. Thus, the spin orientations do not correspond to the usual spin orientations of the bands far away from the anticrossing area.

Finally, we address the case where in the presence of finite Rashba spin-orbit coupling an external magnetic field is applied along the z-axis [111]. As can be seen in Figure 7.21 (a), for a strong confinement, i. e. narrow wire structure, the subband spectrum at $k_y = 0$ as a function of magnetic field resembles the one without Rashba spin-orbit coupling (cf. Figure 7.19).

At zero field the level splitting is given by $\hbar\omega_0$, while at increasing magnetic field the levels approach the linear dispersion of the Landau levels in a two-dimensional system. No clear signatures originating from the Rashba effect are found, because Δ_R is too small compared to $\hbar\omega_0$. Furthermore, as can be deduced from the almost equally

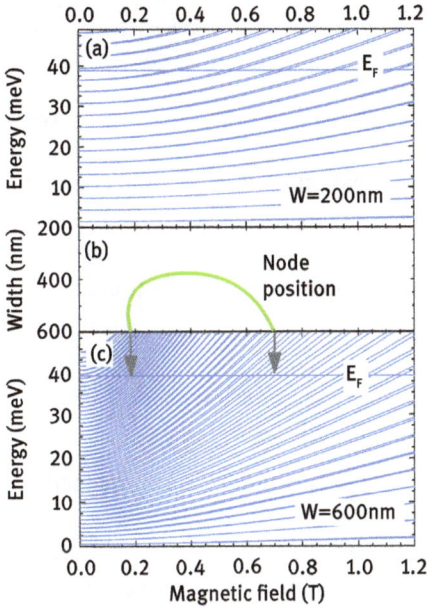

Figure 7.21: (a) and (c) Subband spectra at $k_y = 0$ as a function of the magnetic field for a 600 nm and 200 nm wide InGaAs/InP wire, respectively. A Rashba coupling parameter of $\alpha_R = 7.5 \times 10^{-12}$ eVm was assumed. (b) The position of the nodes in the magnetoresistance as a function of the wire width [111].

spacing between the crossing of the levels with E_F, no beating pattern is expected in the Shubnikov–de Haas oscillations. For weaker confinement, i. e. wider wire, the situation changes, as can be seen in Figure 7.21 (c). Here, the confinement energy $\hbar\omega_0$ is small compared to the previous case. Now the Rashba effect results in the variation of distance between the level crossings at E_F. As a result, a beating pattern is expected in the magneto-transport. For the present case, two nodes are expected. In Figure 7.21 (b) the node positions are plotted as a function of wire width, i. e. confinement energy. It can be seen that for decreasing wire width the two nodes approach each other. Above a width of about 400 nm no nodes are expected.

7.7.2 Helical energy gap

The combination of a strong effective Rashba field B_{eff}, cf. equation (7.8), with an external magnetic field B can lead to a partial spin-orbit gap in the energy momentum dispersion [112, 113]. As a consequence, spin and momentum are locked so that the transport is helical. In order to illustrate this feature we consider a one-dimensional system where the spin-orbit field is perpendicular to the external magnetic field. The corresponding Hamiltonian can be expressed by

$$\hat{H} = \frac{\hbar^2 k_y^2}{2m^*} + \alpha_R \sigma_x k_y + \frac{1}{2} g \mu_B \sigma_z B, \tag{7.67}$$

which contains the kinetic energy for the propagation along the y-direction, the Rashba term, and a Zeeman term with an external magnetic field B in z-direction. The effective spin-orbit field is along the x-direction, thus perpendicular to B. We neglected the effect of the vector potential on the momentum. The energy eigenvalue can be determined by exact diagonalization

$$E_{n,\pm} = E_n + \frac{\hbar^2 k_y^2}{2m^*} \pm \sqrt{\left(\frac{1}{2} g \mu_B B\right)^2 + (\alpha_R k_y)^2}, \tag{7.68}$$

where we also included a quantized energy levels E_n due to the confinement in x-direction.

In Figure 7.22 the evolution of the spin-orbit gap is illustrated. At zero external magnetic field we just have the two shifted dispersions for the two spin orientations owing to the presence of the Rashba effect. The spin is degenerate at $k = 0$ (cf. Figure 7.22 (a)). Upon applying an external magnetic field, the spin-degeneracy is lifted. Around $k = 0$ the Zeeman effect leads to a mixing of spin states and results in an avoided crossing, i.e. a so-called helical gap opens up, as shown in Figure 7.22 (b). The gap is partial, since it only develops at the center, while the outer branches of the dispersion are not interrupted. In the regime $g\mu_B B/2 \leq \alpha_R k_R$, with $k_R = m^* \alpha_R/\hbar^2$, the electronic transport within the gap has a helical nature. Here, spin and momentum are locked and the direction of the group velocity is directly connected with the spin orientation. If the Fermi energy is placed in the gap region, for left moving electrons on the outer left branch with $k < 0$ the electron spin basically points to the opposite direction than the spin of the right moving electrons in the outer right branch with $k > 0$. Due to the presence of the external magnetic field, both spins are oriented slightly in the direction of B. If the Zeeman energy dominates over the spin-orbit contribution, the partial gap is still present, as shown in Figure 7.22 (c). However, for the left and the right branches of the dispersion, with $k < 0$ and $k > 0$, respectively, the spin orientation is governed by the external magnetic field only. Thus, the helical nature is lost.

The presence of a partial energy gap can be revealed by performing quantized conductance measurements in a split-gate point contact (cf. Section 2.6.3). By changing the gate voltage, the Fermi level is shifted through the energy-momentum dispersion. The conductance is governed by the number of bands crossing the Fermi level. In case that Rasbha spin-orbit coupling is present but no external magnetic field is applied the dispersion is the one depicted in Figure 7.22 (a). Here, the conductance is quantized in steps of $2e^2/h$, as illustrated in Figure 7.23 (a). Although the energy momentum dispersion consists of two shifted non spin-degenerate parabolas, the conductance quantization in the presence of Rashba spin-orbit coupling is still in steps of $2e^2/h$.

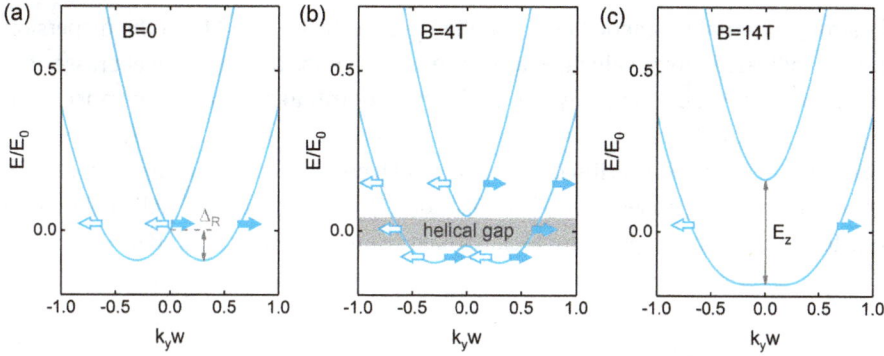

Figure 7.22: (a) Normalized energy-momentum relation for $B = 0$, with $E_0 = \hbar^2/(2m^*w^2)$ and w the width of the conductor. We assumed $w = 25\,\text{nm}$, $m^* = 0.037m_0$, and $\alpha_R = 1.2 \times 10^{-11}$ eVm. The two spinful subbands are shifted by $\pm m^*\alpha_R/\hbar^2$ and lowered in energy by $\Delta_R = m^*\alpha_R^2/2\hbar^2$. The arrows indicate the directions of the group velocities. (b) Corresponding dispersion, when a magnetic field of 4 T is applied. A partial helical gap is formed. We assumed a g-factor of $g = 4$. (c) Dispersion for $B = 14$ T. Here, the Zeeman energy $E_Z = g\mu_B B$ dominates over Δ_R.

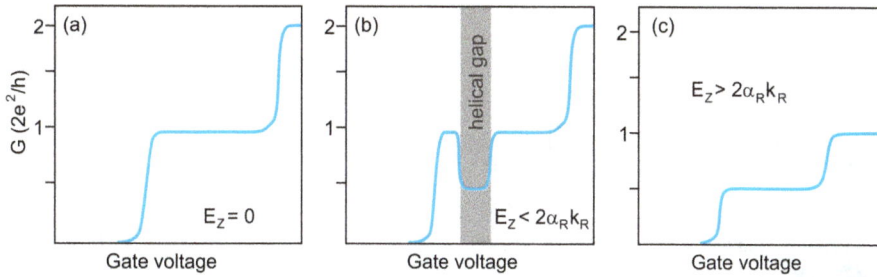

Figure 7.23: Schematic illustrations of the quantized conductance resulting from the three situations shown in Figure 7.22. (a) Quantized conductance in a one-dimensional conductor at zero magnetic field $E_Z = g\mu_B B = 0$. (b) Conductance in the presence of a helical gap, i. e. $E_Z < 2\alpha_R k_R$. While the Fermi energy is shifted by the gate voltage through the helical gap the conductance G drops from $2e^2/h$ to e^2/h. (b) If the Zeeman energy is sufficiently large, i. e. $E_Z > 2\alpha_R k_R$, the Zeeman splitting dominates, thus the conductance increases in steps of e^2/h.

The reason is, that the Fermi level crosses the bottom of the two branches simultaneously, i. e. the current is carried by a net electron flow in the two branches with a group velocity in the same direction, as explained in detail in Sect. 2.6.3. If an external magnetic field is applied, a helical gap is formed. Referring to Figure 7.22 (b), as long as the Fermi level is below the helical gap, the current is carrier by two branches. Consequently, the conductance is quantized at $2e^2/h$, as illustrated in Figure 7.23 (b). As soon as the Fermi level is located within the partial gap, only one branch contributes, thus the conductance is reduced to e^2/h. Upon further increasing the Fermi energy beyond the partial gap, the conductance value of $2e^2/h$ is recovered. The situation changes if a strong external magnetic field is applied. As can be deduced from the dispersion and the indicated group velocities in Figure 7.22 (c), once the Fermi level crosses the dispersion from the bottom, only a single branch contributes to the conductance. Consequently, the conductance is quantized to e^2/h right away, as illustrated in Figure 7.23 (c). Only if the Fermi level is lifted that much that it crosses the upper spin-split branch the conductance increases to $2e^2/h$. Indications of the presence of a helical gap have been found in quantum point contacts formed in an InAs nanowire [114].

7.7.3 Rashba effect in tubular structures

In Section 2.6.2 we introduced semiconductor nanowires, which were produced by a bottom-up approach. Usually, these nanowires have a hexagonal cross section, because of the faceted growth of the sidewalls. However, as an approximation, one can simplify the description by assuming a circular cross section. In this case one can use a cylindrical coordinate system, as defined in Figure 7.24 (a).

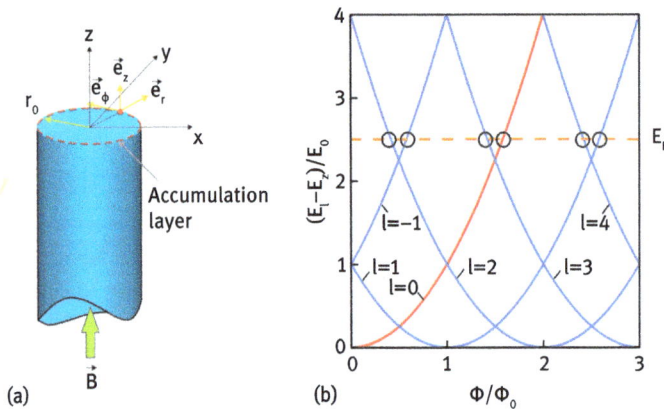

Figure 7.24: (a) Cylindrical nanowire with an accumulation layer at the surface. The system is described by using a cylindrical coordinate system, with the unit vectors $\vec{e}_r$, $\vec{e}_\phi$, and $\vec{e}_z$. (b) Level spectrum for a tubular conductor as a function of the normalized magnetic flux Φ/Φ_0 penetrating the cross section of the nanowire. The spectrum is periodic with the magnetic flux quantum $\Phi_0 = h/e$. The angular momentum quantum numbers l are labeled. The states are plotted for $k = 0$ and normalized to $E_0 = \hbar^2/2m^*r_0^2$.

In this framework the unit vectors are given by

$$\vec{e}_r = \begin{pmatrix} \cos\phi \\ \sin\phi \\ 0 \end{pmatrix}, \quad \vec{e}_\phi = \begin{pmatrix} -\sin\phi \\ \cos\phi \\ 0 \end{pmatrix}, \quad \vec{e}_z = \begin{pmatrix} 0 \\ 0 \\ 1 \end{pmatrix}. \tag{7.69}$$

For low band gap semiconductors, such as InAs or InN, an accumulation layer is formed at the surface of the nanowire (cf. Figure 2.19 (a)). As a simplification, we assume that the electrons are only located at the surface of the tube, while the inner part is insulating. The radius of the tubular conductor is fixed at $r = r_0$. We further assume that a magnetic field is applied along the z-direction $\vec{B} = (0,0,B)$. In the cylindrical coordinate system the corresponding vector potential can be chosen as

$$\vec{A} = \frac{1}{2}Br_0\vec{e}_\phi. \tag{7.70}$$

We are interested in the energy levels of this system in the presence of a magnetic field. For this we have to consider the kinetic moment operator $\hat{\pi}_\phi$, where the vector potential is included (cf. Section 2.8.1). The Hamiltonian describing this system is given by

$$\hat{H} = \frac{1}{2m^*}\left(-i\hbar\frac{1}{r_0}\frac{\partial}{\partial\phi} + \frac{1}{2}eBr_0\right)^2 - \frac{\hbar^2}{2m^*}\frac{\partial^2}{\partial z^2}. \tag{7.71}$$

An appropriate set of eigenfunctions solving the Schrödinger equation is

$$\psi_{lk} = e^{il\phi}e^{ikz}, \tag{7.72}$$

where the first part describes angular momentum states with the quantum number $l = 0, \pm1, \pm2, \ldots$ The second part corresponds to the free motion along the z-direction. The resulting energy eigenvalues are

$$E_{lk} = \frac{\hbar^2}{2m^*r_0^2}\left(l + \frac{\Phi}{\Phi_0}\right)^2 + \frac{\hbar^2k^2}{2m^*}, \tag{7.73}$$

with $\Phi = B\pi r_0^2$ being the magnetic flux penetrating the cross section area πr_0^2 and $\Phi_0 = h/e$ the magnetic flux quantum. In Figure 7.24 (b), the energy spectrum is shown as a function of the normalized flux Φ/Φ_0. Obviously, the spectrum is periodic with a period of a single magnetic flux quantum. Indeed, such a period was found in magnetotransport experiments, where a magnetic field was applied along the axis of the nanowire [115, 116]. Here, the magnetic field leads to a flux periodic population and depopulation of the upper energy level at the Fermi level. Assuming ideal ballistic transport in a one-dimensional system, the periodic change of the number of occupied channels results in a periodic conductance variation of $2e^2/h$ with a period of Φ_0.

As a next step we will take care of the Rashba effect in a cylindrical conductor. In an accumulation layer the confining potential will vary in radial direction, and thus the electric field $\vec{\mathcal{E}}$ will only have a radial component. This situation is depicted in Figure 7.25 (a), where $\vec{\mathcal{E}}$ points towards the center of the wire. To express the Rashba Hamiltonian given by equation (7.4) in cylindrical coordinates, we have to convert the Pauli spin matrices:

$$\sigma_r = \cos\phi\,\sigma_x + \sin\phi\,\sigma_y = \begin{pmatrix} 0 & e^{-i\phi} \\ e^{i\phi} & 0 \end{pmatrix}, \tag{7.74}$$

$$\sigma_\phi = -\sin\phi\sigma_x + \cos\phi\sigma_y = \begin{pmatrix} 0 & -ie^{-i\phi} \\ ie^{i\phi} & 0 \end{pmatrix}, \tag{7.75}$$

while σ_z remains. With the momentum operator given by

$$\hat{p} = \frac{\hbar}{i}\vec{\nabla} = \frac{\hbar}{i}\left(\frac{\partial}{\partial r}\vec{e}_r + \frac{1}{r}\frac{\partial}{\partial\phi}\vec{e}_\phi + \frac{\partial}{\partial z}\vec{e}_z\right), \tag{7.76}$$

we obtain for the Hamiltonian at zero external field:

$$\hat{H} = -\frac{\hbar^2}{2m^*r_0^2}\frac{\partial^2}{\partial\phi^2} + \alpha_R\left(\hat{\sigma}_\phi\frac{1}{i}\frac{\partial}{\partial z} - \hat{\sigma}_z\frac{1}{ir_0}\frac{\partial}{\partial\phi}\right). \tag{7.77}$$

To solve the Schrödinger equation, we can use the following ansatz for the wave functions:

$$|\psi\rangle = e^{ikz}e^{il\phi}\begin{pmatrix} \chi_1 \\ ie^{i\phi}\chi_2 \end{pmatrix}. \tag{7.78}$$

This form of the spinor is chosen, because it represents a state with a fixed total angular momentum. For the spin-down component the orbital angular momentum is larger by one compared to the spin-up state, due to the additional $i\exp(i\phi)$ factor. In the presence of spin-orbit coupling, the total angular momentum is conserved, and thus we can specify the state by the quantum number of the total angular momentum $j = l \pm 1/2$. Inserting the ansatz given by equation (7.78) into the Schrödinger equation we obtain the following coupled equations:

$$\left(\frac{\hbar^2}{2m^*r_0^2}l^2 - \alpha_R\frac{l}{r_0} - \tilde{E}\right)\chi_1 = -\alpha_R k\chi_2, \tag{7.79}$$

$$\left(\frac{\hbar^2}{2m^*r_0^2}(l+1)^2 + \alpha_R\frac{l+1}{r_0} - \tilde{E}\right)\chi_2 = -\alpha_R k\chi_1, \tag{7.80}$$

with $\tilde{E} = E - (\hbar k)^2/(2m^*)$ being the energy without the axial kinetic energy. From these equations the energy eigenvalues and the components χ_1 and χ_2 can be determined. At $k = 0$ the equations are decoupled, so that the energy eigenvalues for spin-up and spin-down for a given l are given by

$$E_{\pm,l} = \frac{\hbar^2}{2m^*r_0^2}l^2 - \frac{2\alpha_R}{r_0}(l\cdot s), \tag{7.81}$$

with the spin quantum number $s = \pm1/2$. We have written the energy in a form where the second term resembles the spin-orbit Hamiltonian for a radially symmetric potential given by equation (7.3). The spin-orbit contribution is opposite if the orbital angular momentum and spin are parallel or antiparallel, i. e. for a total angular momentum $j = \pm(|l| + |s|)$ or $j = \pm(|l| - |s|)$, respectively, with $l = \pm1, \pm2, \ldots$. Furthermore, for $l = 0$ we have $j = \pm 1/2$. In Figure 7.25 (b) the two parallel configurations belonging to $j = |l| + 1/2$ and $-|l| - 1/2$ are shown. For both configurations the spin-orbit energy contribution is the same.

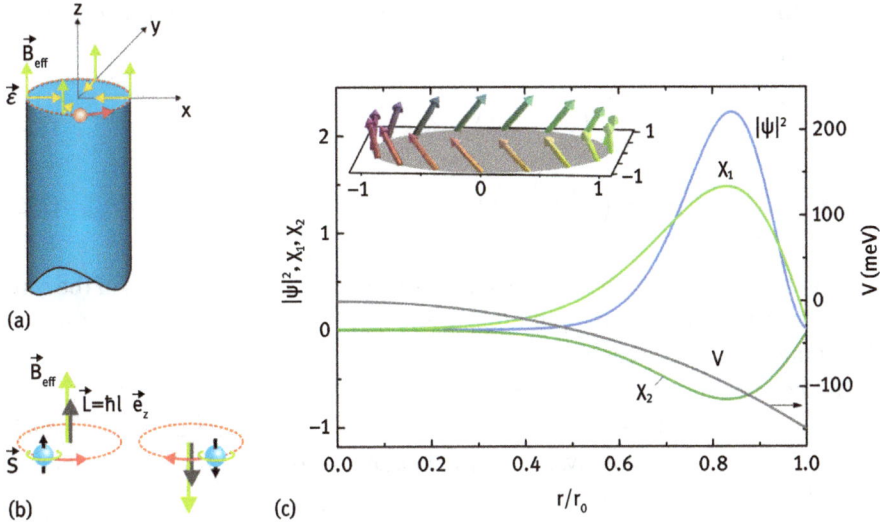

Figure 7.25: (a) Schematic illustration of the electric field $\vec{\mathcal{E}}$ and the resulting effective magnetic field $\vec{B}_{eff}$ of an electron circling around in the accumulation layer of the nanowire. (b) Two configurations, where the spin $\vec{S}$ is aligned in the same direction as the angular momentum $\vec{L}$. The effective magnetic field $\vec{B}_{eff}$ points along $\vec{L}$. The energy related to the spin-orbit coupling is the same for both configurations. (c) Simulations for an InAs nanowire: squared amplitude of the wave function $|\psi|^2$ with the corresponding spinor components χ_1 and χ_2 and the conduction band profile V as a function of the normalized radius r/r_0 for $j = 1/2$. The outer radius r_0 of the nanowire is assumed to be 50 nm. The inset shows the spin orientation along the circumference for $j = 1/2$ [117].

For $k \neq 0$ the two differential equations (7.79) and (7.80) are coupled. We omit the explicit solution here and discuss a more general situation below, where a finite width of the conductive channel is allowed.

In order to describe the physical situation more precisely, one can assume that the wave function in the accumulation layer extends into the nanowire. In that case we have to replace the Rashba coefficient α_R in equation (7.77) by $V'(r)y$, where $V'(r)$ is the derivative of potential $V(r)$ with respect to r. The material specific prefactor y is determined from the band structure [95]. The electric field is connected to the potential by $\vec{\mathcal{E}} = \vec{\nabla}V/e$. As can be seen Figure 7.25 (a), the electric field points towards the axis of the nanowire. Due to the circular motion of the electrons in the accumulation layer the electrons experience an effective magnetic field $\vec{B}_{eff}$, which is oriented along the nanowire axis. The components χ_1 and χ_2 of the spinor given in equation (7.78) become *r* dependent distribution functions

$$|\psi\rangle = e^{ikz} e^{il\phi} \begin{pmatrix} \chi_1(r) \\ ie^{i\phi}\chi_2(r) \end{pmatrix}. \tag{7.82}$$

Inserting this ansatz into the Schrödinger equation leads to the following coupled differential equations for χ_1 and χ_2 [117]:

$$-\frac{\hbar^2}{2m^*}\left(\chi_1'' + \frac{1}{r}\chi_1'\right) + (\widetilde{V}_{l,+} - \widetilde{E})\chi_1 = k\gamma V'\chi_2,$$

$$-\frac{\hbar^2}{2m^*}\left(\chi_2'' + \frac{1}{r}\chi_2'\right) + (\widetilde{V}_{l+1,-} - \widetilde{E})\chi_2 = k\gamma V'\chi_1. \tag{7.83}$$

The potential $\widetilde{V}_{l,\pm} = (\hbar l)^2/(2m^*r^2) + V \mp \gamma V'l/r$ contains the contributions of the centrifugal force and the diagonal spin-orbit term, while $\widetilde{E} = E - (\hbar k)^2/(2m^*)$ is the energy without the axial kinetic energy.

The components $\chi_1(r)$ and $\chi_2(r)$ can be calculated numerically [117]. In Figure 7.25 (c), a typical potential profile $V(r)$ of an InAs nanowire is shown, which was calculated self-consistently. At the wire boundary a barrier of infinite height was assumed. Also shown are the calculated squared amplitude $|\psi|^2$ for $j = 1/2$ at the Fermi energy together with the components $\chi_1(r)$ and $\chi_2(r)$ of the spinor given by equation (7.82).

By knowing $\chi_1(r)$ and $\chi_2(r)$, the spin expectation values in different orientations can be determined by integrating over the corresponding spin densities. One finds that the spin density in radial direction is zero. The tangential component of the spin density is given by

$$s_\phi(r) = \langle\psi|\hat{\sigma}_\phi|\psi\rangle = 2\chi_1(r)\chi_2(r), \tag{7.84}$$

while the component along the wire axis is

$$s_z(r) = \langle\psi|\hat{\sigma}_z|\psi\rangle = \chi_1(r)^2 - \chi_2(r)^2. \tag{7.85}$$

In Figure 7.25 (c) (inset) the spin orientation along the circumference is shown for $j = 1/2$. It can be seen that the spin only has a tangential and axial component. Averaging over the circumference leads to a net spin pointing along the z-axis. The tangential and axial spin densities s_ϕ and s_z for different values of the total angular momentum j are depicted in Figure 7.26. With increasing angular momentum, the spin points more and more towards the axis. This behavior can be explained by the fact that the effective magnetic field $\vec{B}_{\text{eff}}$ increases with increasing orbital angular momentum number l. The stronger the effective magnetic field is the stronger is the spin polarization towards the axis.

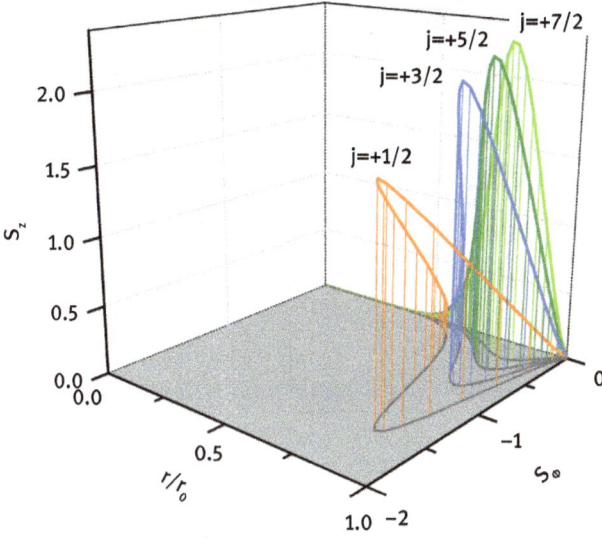

Figure 7.26: Spin density (s_ϕ, s_z) of the lower energy states for total angular momenta j = 1/2, 3/2, 5/2, and 7/2. The spin has only tangential and axial components [117].

In Figure 7.27 the energy $\tilde{E}$ vs k is shown for several j-values. The axial kinetic energy was left out in this plot. At $k = 0$ the coupling between l and $l + 1$ vanishes (cf. equation (7.83)). Therefore, a classification with respect to l is possible. The splitting between the second and third band ($l = \pm 1$) is caused by the diagonal part of $\hat{H}$. As discussed already in conjunction with equation (7.81), the states where the orbital angular momentum and the spin are parallel, i. e. $j = +3/2 = 1 + 1/2$ and $j = -3/2 = -1 - 1/2$, are lowered compared to the case where the orbital angular momentum and the spin are antiparallel, i. e. $j = \pm 1/2$. All bands are twofold degenerate, because states with a reversed angular momentum and spin have the same energy. The energy splitting increases proportional to l for the higher states.

7.8 Summary

- The spin precession of electrons in a two-dimensional electron gas can be controlled via the Rashba effect. The strength of the Rashba effect can be controlled by a gate electrode.
- The Rashba effect originates from the macroscopic electric field in a semiconductor heterostructures. The Rashba coupling parameter α_R increase for decreasing band gap and increasing split-off energy in the valence band.
- The Rashba coefficient can be extracted from the characteristic beating pattern in the Shubnikov–de Haas oscillations.

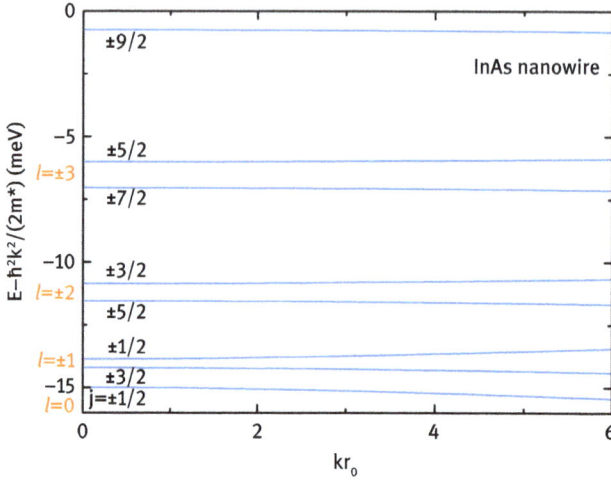

Figure 7.27: Energy vs normalized k dispersion, with r_0 the nanowire radius. The axial kinetic energy is left out. All bands are twofold degenerate. The bands can be labeled by the total angular momentum j [117].

- In crystals lacking an inversion center spin-orbit coupling due to bulk inversion asymmetry contributes.
- In wire structures the energy spectrum is determined by the quantization energy and the energy splitting due to the Rashba effect.

Exercises

Problem 7.1. Show that the spin-orbit Hamiltonian

$$\hat{H}_{so} = \frac{\hbar}{4m_0^2 c^2} \hat{\vec{\sigma}}(\hat{\vec{p}} \times \vec{\nabla} V),$$

with $\vec{p}$ the momentum, $\vec{\sigma}$ the Pauli spin matrices and V the potential, can be motivated by the magnetic energy

$$U = -\vec{\mu} \cdot \vec{B}.$$

Here $\vec{\mu}$ is the magnetic moment of the electron and $\vec{B}$ the effective magnetic field in the frame of reference of a charged particle in an electric field $\vec{\mathcal{E}}$.

Problem 7.2. Show explicitly for a spinor $|\psi\rangle$ that $\langle \hat{\mathcal{T}}\psi | \hat{\mathcal{T}}\psi \rangle = \langle \psi | \psi \rangle$, with $\hat{\mathcal{T}} = i\sigma_y \hat{\mathcal{K}}$ the time-reversal operator. Here, σ_y is the Pauli spin matrix for the y-direction and $\hat{\mathcal{K}}$ is the operator of the complex conjugation.

Problem 7.3. Show that the time-reversal operator $\hat{\mathcal{T}}$ is an operator which is proportional to the complex conjugate operator $\hat{\mathcal{K}}$.

Problem 7.4. The rotation of a spin-1/2 particle about an angle α around the y-axis is generally described by the operator

$$e^{-i\alpha\sigma_y/2}.$$

Show that for a spin-1/2 particle the following identity holds:

$$e^{-i\pi\sigma_y/2} = -i\sigma_y.$$

Here, the left side corresponds to a spin rotation of π about the y-axis.

Problem 7.5. Show that the Hamiltonian for a wire with a parabolic confinement potential

$$\hat{H} = -\frac{\hbar^2}{2m^*}\frac{\partial^2}{\partial x^2} + \frac{m^*}{2}w_c^2(x - x_0)^2 + \frac{m^*}{2}w_0^2 x^2,$$

can be written as

$$\hat{H} = -\frac{\hbar^2}{2m^*}\frac{\partial^2}{\partial x^2} + \frac{m^*}{2}w_c^2(x - \tilde{x}_0)^2 + \frac{w_0^2}{w^2}\frac{\hbar^2 k_y^2}{2m^*},$$

with

$$\tilde{x}_0 = \frac{w_c^2}{w^2}x_0, \quad x_0 = \frac{\hbar k_y}{eB}, \quad \text{and} \quad w^2 = w_c^2 + w_0^2.$$

Problem 7.6. Find an expression for the density of states $D(E)$ in the presence of Rashba spin-orbit coupling.

8 Spin interference

8.1 Overview

For most of the transport properties we discussed so far we referred to the particle property of the electrons. However, from quantum mechanics we also know that electrons have wave properties as well. In optics the interference of light waves play a major role, e. g. in a double slit experiment, where the light intensity at a screen is spatially modulated because of the interference of phase-shifted partial waves passing through one or the other slit. An important prerequisite for observing this interference effect is that the coherence of the light waves is preserved. Going back to electron waves in solid state materials, preserving electron coherence and observing interference effects is a major challenge. This is due to the fact that inelastic scattering processes break phase coherence. A typical example for an inelastic scattering event is phonon scattering, which is abundant at room temperature. The reason for the loss of phase coherence is immediately clear. Phonon scattering is connected to lattice vibrations. An electron wave is exposed to an effective scattering potential which varies in time, and therefore the well-determined phase is lost. As a consequence the fixed phase relation for a superposition of two electron waves is lost which is required to observe interference effects. In order to suppress electron-phonon scattering, the temperature needs to be lowered. Indeed at low temperatures of a few Kelvin and lower, transport phenomena can be observed which are due to electron interference. If spin-orbit coupling is present, or the scattering events not only change the momentum but also the spin orientation, the spin contribution has to be included in the interference process. A well-known example is the weak antilocalization effect. Here, a distinct conductance peak at zero magnetic field is observed. Spin interference effects are not only of fundamental interest, they are also a very versatile tool to get information on spin transport in semiconductors, e. g. the weak antilocalization effect can be employed to gain information on the spin-orbit scattering length. Beside effects based on diffusive transport, such as the aforementioned, interference effects can also be observed by guiding the electron partial waves along well-defined paths. A famous example for electron interference in ring-shaped conductors is the Aharonov–Bohm effect [118], where the interfere is controlled by a magnetic flux. In the presence of spin-orbit coupling, the internal magnetic fields results in a modification of the interference. Similar to the spin field-effect transistor, the interference can be controlled via the Rashba effect by means of a gate.

8.2 Electron interference effects

We will first discuss the underlying concept used to describe electron interference in metal and semiconductor structures. Typical phenomena are weak localization or uni-

https://doi.org/10.1515/9783110639001-008

versal conductance fluctuations. The phase coherence length l_φ of electron waves in a conductor is the relevant length scale, which determines under which conditions phase coherent transport phenomena can be observed. In most cases the electron transport is diffusive. Diffusive transport is characterized by the diffusion constant $\mathcal{D}$, which is defined as

$$\mathcal{D} = \frac{1}{d} v_F^2 \tau_e,$$

(8.1)

with $d = 1, 2, 3$ being the dimension of the system, v_F the Fermi velocity, and τ_e the elastic scattering time. The product of v_F and τ_e corresponds to the elastic mean free path

$$l_e = v_F \tau_e.$$

(8.2)

At low temperatures the phase coherence length l_φ is usually longer than the elastic mean free path l_e. This is because elastic scattering processes, e. g. ionized impurity scattering, are not phase-breaking. The electrons do not undergo an energy change, so that their wave length is preserved. Nevertheless, the direction of propagation is changed, and due to the influence of the scattering potential the phase of the electron wave is shifted. This phase shift happens in a controlled fashion, which is the same for repeated scattering events, and thus coherence is preserved. As mentioned above, in many cases $l_\varphi \gg l_e$, which means that the electron wave undergoes many elastic scattering events before the phase coherence is broken. Furthermore, all descriptions of the interference effects discussed here are based on the quasi classical theory, which means that one assumes that $\lambda_F \ll l_e$. Below we will first introduce the concept of how to describe phase coherent transport in semiconductors. Subsequently, phase-coherent transport in samples with a well-defined geometry will be discussed. Finally, diffusive phases coherent-transport, i. e. the weak localization effect is explained.

8.2.1 Electron interference effects

Let us assume a diffusive conductor, where the electron partial waves propagate from a starting point A to a second point Q along all possible paths (cf. Figure 8.1). These paths are determined by the distribution of elastic scattering centers. It is assumed that phase coherence is preserved along these paths.

According to Feynman one can describe each path j by a complex amplitude A_j defined by

$$A_j = C_j e^{i\varphi_j},$$

(8.3)

where φ_j is the phase shift the electron acquires on its way from A to Q along path j. For free electron propagation along path j, excluding scattering events, the corresponding

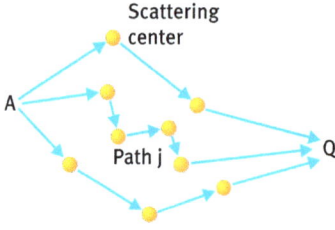

Figure 8.1: A set of electron trajectories for wave propagation from point A to Q. The transport is diffusive because of the presence of scattering centers in between A and Q.

phase accumulation $\widetilde{\varphi}_j$ can be determined from the action S_j by

$$\widetilde{\varphi}_j = \frac{1}{\hbar} S_j, \tag{8.4}$$

with the nonrelativistic action defined by

$$S_j = \int_{t_A}^{t_Q} L(\vec{v}, \vec{r}, t)\, dt = \int_{t_A}^{t_Q} \frac{m}{2} \vec{v}^2\, dt = \int_{\text{path } j} \frac{1}{2} \vec{p}\, d\vec{r}, \tag{8.5}$$

Here, $L(\vec{v}, \vec{r}, t)$ is the Lagrangian function of a free propagating electron, with $\vec{v} = \dot{\vec{r}}$ the time derivative of $\vec{r}$. Furthermore, t_A is the time when the electron starts at A and t_Q the time when it arrives at Q. As mentioned above, in addition to the phase shift accumulated during free propagation, the electrons also acquire a phase shift due to elastic scattering events. Consequently, the total phase shift φ_j is the sum of both contributions. The total probability P_{AQ} for an electron to be transported from A to Q is given by the square of the total amplitude along all possible paths:

$$P_{AQ} = \left| \sum_j A_j \right|^2 = \left| \sum_j C_j e^{i\varphi_j} \right|^2. \tag{8.6}$$

This expression is the key formula for all electron interference effects discussed below.

8.2.2 Aharonov–Bohm effect

We first start with the Aharonov–Bohm effect, which can be observed in samples with a ring-shaped geometry. The Aharonov–Bohm effect was theoretically predicted in 1959 [118]. The essence of this effect is that the vector potential $\vec{A}$ affects the interference of the electron partial waves propagating along the two branches of a conductive ring penetrated by a magnetic flux even when the magnetic field $\vec{B}$ within the conductor itself is zero. Since the electrons are not exposed to a magnetic field, classically no measurable effect should occur when the flux within the ring is changed. However, as we will see, the vector potential $\vec{A}$, which is nonzero in the conductor in contrast to $\vec{B}$, will induce a phase shift of the electron wave and thus affect the electron transport.

A schematic of the setup is depicted in Figure 8.2. The magnetic field B is restricted to an area within the ring structure, i. e. the area with magnetic flux Φ in Figure 8.2 (a). The magnetic field is zero within the conducting ring.

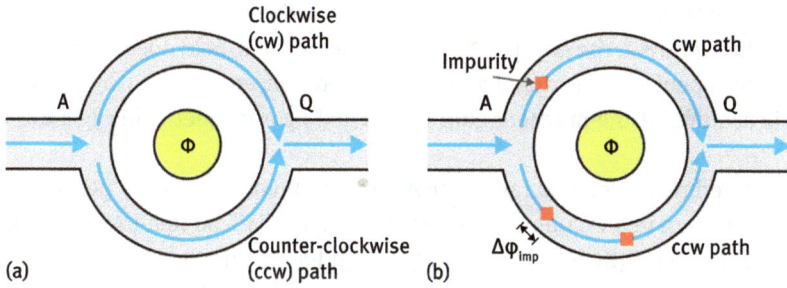

Figure 8.2: (a) Idealized ring-shaped conductor. At A the electron wave is split into two partial waves which propagate along a counter-clockwise (ccw) and clockwise (cw) path, respectively. At Q both partial waves interfere. (b) Ring-shaped conductor with impurities acting as scattering centers. The scattering event leads to an additional phase shift $\Delta\varphi_{imp}$.

In the presence of a vector potential the phase accumulation contains an additional contribution, since the Lagrangian function L of an electron with charge $q = -e \, (e > 0)$ is now given by

$$L(\vec{v}, \vec{r}, t) = \frac{m}{2}\vec{v}^2 - e\vec{v} \cdot \vec{A}. \tag{8.7}$$

Inserting L into equation (8.5), the phase difference of two electron partial waves emerging from A, propagating along the upper and the lower branch of the ring, and finally interfering at Q, is given by

$$\Delta\varphi = \chi_{ccw} - \chi_{cw} - \frac{e}{\hbar} \int_{ccw} \vec{A} \, d\vec{l} + \frac{e}{\hbar} \int_{cw} \vec{A} \, d\vec{l}$$

$$= \Delta\chi - \frac{e}{\hbar} \oint \vec{A} \, d\vec{l}. \tag{8.8}$$

The upper and lower path paths, i. e. clockwise (cw) and counter-clockwise (ccw), respectively, are depicted in Figure 8.2. In the formula given above, χ_{ccw} and χ_{cw} are the phases the electron waves acquire during the propagation along the counter-clockwise and clockwise paths, respectively, at zero magnetic field in the interior of the ring. Since the impurity configurations on both branches usually differ, the accumulated phases are different in both branches. In Figure 8.2 (b) a ring-shaped conductor including some scattering centers is shown. Each scattering center leads to an additional phase shift $\Delta\varphi_{\text{imp}}$ which contributes to χ_{ccw} and χ_{cw}. By making use of $\vec{\nabla} \times \vec{A} = \vec{B}$, equation (8.8) results in

$$\Delta\varphi = \Delta\chi - \frac{e}{\hbar} \int \vec{B}\, d\vec{f} \tag{8.9}$$

$$= \Delta\chi - 2\pi \frac{\Phi}{\Phi_0} . \tag{8.10}$$

The surface integral over $\vec{B}$ corresponds to the magnetic flux Φ penetrating the ring. As illustrated in Figure 8.2, the area penetrated by the magnetic field does not need to be as large as the opening of the ring. As can be inferred from equation (8.10), a phase shift of 2π is acquired if the magnetic flux is changed by a magnetic flux quantum $\Phi_0 = h/e$.

The first experiments demonstrating the Aharonov–Bohm effect were performed with an electron beam in a vacuum and a shielded magnet coil, so that the electrons were not exposed to a magnetic field [119, 120]. In solid-state samples the Aharonov–Bohm effect was demonstrated first in Au rings. The diameter of the ring structure was less than one micrometer with a wire width of a few tens of nanometers [121]. In these structures, it cannot usually be prevented that the magnetic field penetrates the conductor. Nevertheless, the vector potential $\vec{A}$ is still responsible for the effect on the electron interference pattern. A typical ring structure defined in an AlGaAs/GaAs semiconductor heterostructure is shown in Figure 8.3 [122].

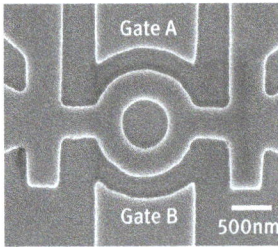

Figure 8.3: Scanning electron beam micrograph of a ring-shaped conductor. The structure was defined by wet chemical etching of an AlGaAs/GaAs heterostructure. The in-plane gates A and B can be used to change the electron concentration in the adjacent arms of the ring.

In Aharonov–Bohm experiments on nanoscaled structures, the electrons are usually scattered many times within each branch of the ring. This is due to the fact that mostly the elastic mean free path is smaller than the ring size, so that the transport is diffusive. Furthermore, often l_φ is comparable to the ring diameter. As a consequence, many electrons lose their phase memory while propagating through the ring. This is the reason why the oscillation amplitude is often considerably smaller than the total resistance of the structure.

As shown in Figure 8.4, pronounced Aharonov–Bohm oscillations are observed in ring structures based on two-dimensional electron gases located in an $In_{0.77}Ga_{0.23}As/$ InP heterostructure [123].

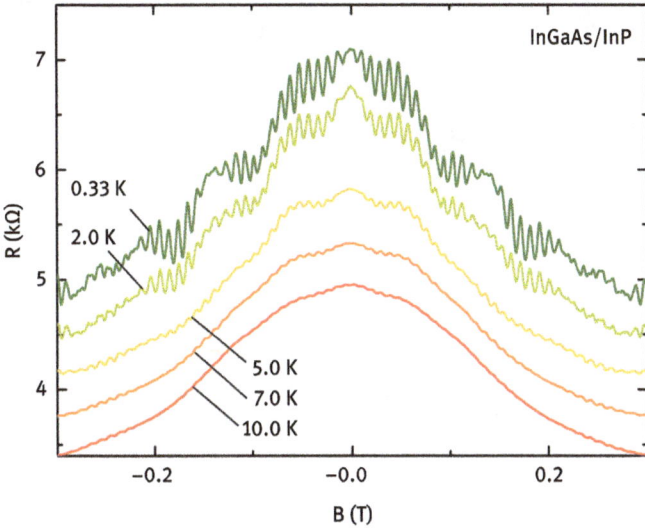

Figure 8.4: Aharonov–Bohm oscillations measured on a ring structure defined in an $In_{0.77}Ga_{0.23}As/InP$ heterostructure. The ring has an outer diameter of 700 nm with a width of about 85 nm. The measurements were performed at temperatures from 0.33 K up to 10 K. Figure adapted from Appenzeller et al. [123].

A comparison of the enclosed area of the ring confirmed that the oscillation period corresponds to a magnetic flux quantum Φ_0. Due to the low effective electron mass and because of the high electron mobility, the phase coherence length exceeds one micrometer at temperatures below 1 K. Consequently, large oscillation amplitudes up to 12 % are achieved.

In case of the Aharonov–Bohm effect, the interference at the branching point is determined by the two different paths, i. e. the ccw and cw paths, along the two branches of the ring, as illustrated in Figure 8.2. In the ideal case of fully symmetric paths, constructive interference and thus a maximum conductance/minimum resistance is expected at zero magnetic field, since the phase accumulation on both paths is the same ($\Delta\varphi = 0$). However, in reality the phase shift in each branch of the ring depends to a large extent on the distribution of scattering centers in the ring and the corresponding phase shifts imposed by the scattering events. Therefore, at zero magnetic field the phase shift $\Delta\varphi$ is finite, and no resistance minimum is observed at $B = 0$. Usually the magnetoresistance measurements are performed in an effective two-terminal configuration, which implies that the curve is symmetric with respect to magnetic field inversion (Onsager relation). The symmetry of the magnetoresistance can clearly be seen in Figure 8.4.

In addition to the control of the electron interference by a magnetic flux, the oscillation pattern can also be changed electrostatically by means of a gate electrode. A typical AlGaAs/GaAs sample with two in-plane gates is shown in Figure 8.3. By applying a voltage to one of the gates the electron concentration in the corresponding branch of the ring is varied. A change of the carrier concentration goes along with a change of the electron wavelength. As a result, the phase accumulated in this branch of the ring is changed. Of course this immediately affects the interference pattern, as can be seen in Figure 8.5 [122].

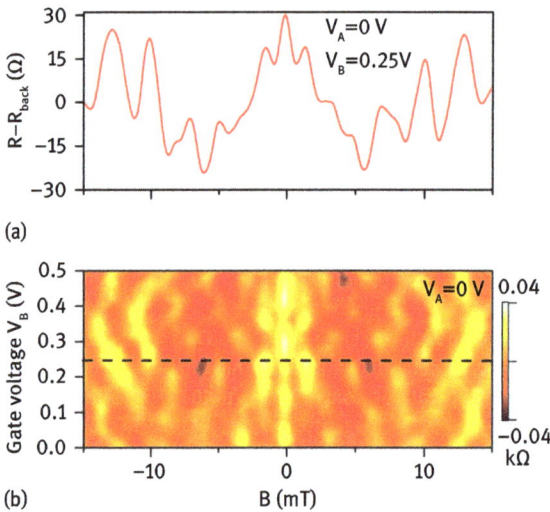

Figure 8.5: (a) Aharonov–Bohm magnetoresistance oscillations measured on a structure depicted in Figure 8.3. The slowly varying background resistance R_{back} was subtracted. (b) Normalized resistance oscillation as a function of magnetic field B and gate voltage V_B. The voltage at gate A was set to zero. Once again, the background resistance was subtracted from the total resistance. All measurements were performed at a temperature of 50 mK (Figure adapted from Krafft et al. [122]).

8.2.3 Altshuler–Aronov–Spivak oscillations

For the Aharonov–Bohm effect we considered the propagation and interference of two partial waves along two branches of a ring-shaped structure. Each partial wave propagates a distance corresponding to half the circumference of the ring. In the case where the phase-coherence length l_φ is sufficiently long, i. e. l_φ is in the order of the circumference of the ring, the electron partial waves might also propagate a full round, as shown in Figure 8.6.

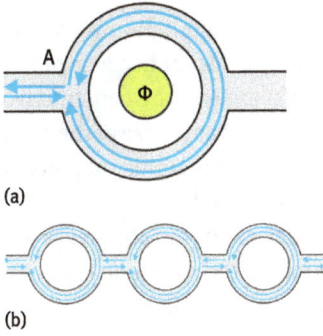

(a)

(b)

Figure 8.6: (a) Ring-shaped conductor, with two counter-propagating trajectories of electron partial waves starting at point A and interfering at A after a full round. (b) A series of interconnecting ring structures.

Let us assume an electron wave reaching point A from the left. Here it splits into two partial waves, propagation clockwise and counterclockwise around the ring. Returning to point A after a full circle both partial wave interfere. As long as no magnetic field is applied, time inversion symmetry holds, i. e. the counterclockwise propagation is the time-inversed one of the clockwise propagation. According to equation (8.5) the phase accumulation on both paths are the same, since for the counterclockwise propagation the integration path and the momentum are inversed with respect to the clockwise propagation. Thus, the phase accumulation for the propagation is identical. Being more specific, the electron partial waves propagate in two opposite directions with the corresponding complex amplitudes $A_{1,2} = C_{1,2} \exp(i\varphi_{1,2})$. According to equation (8.6) the total probability for an electron transfered from point A back to point A is given by

$$P_{AA} = |A_1 + A_2|^2 = |C_1|^2 + |C_2|^2 + 2\operatorname{Re}\left(C_1^* e^{-i\varphi_1} C_2 e^{i\varphi_2}\right), \tag{8.11}$$

with $\operatorname{Re}(z)$ the real part of z. Since for time reversed paths $C_1 = C_2$ and $\varphi_1 = \varphi_2$ holds, we obtain

$$|A_1 + A_2|^2 = 4|C_1|^2. \tag{8.12}$$

For the classical nonphase coherent transport regime the probability would simply be $|C_1|^2 + |C_2|^2$, which is a factor of two smaller than for phase coherent transport. The interference of the two partial waves at point A is always constructive. A constructive interference implies that the probability of the electron wave propagating back into the left lead is enhanced. Since the propagation is in the counter-direction of the incoming electron wave this implies that the resistance is enhanced. This property is fundamentally different from the Aharonov–Bohm effect, where the electron partial waves could accumulated different phases, since the paths were not identical, e. g. because of a different impurity configuration in each branch. For electron partial waves propagation a full round on time-reversed paths, the phase accumulation is exactly the same for both directions of propagation even if phase shifts occur because of impurity scattering. Furthermore, because of time inversion symmetry the constructive interference of the closed loops does not depend on the exact shape or size of the ring, presuming that phase-coherence is maintained. We will see later that this property is the basis for the weak localization effect discussed in the next section.

When the ring is penetrated by a magnetic field the phases of the counter propagating partial waves are shifted. The phase shift acquired by the magnetic field can be calculated by integrating the corresponding vector potential $\vec{A}$ along each path, similarly to equation (8.8). Since these trajectories are closed loops the phase shift is directly connected to the enclosed flux Φ. However, owing to the reversed propagation, the phase shift is opposite for both partial waves.

In the presence of a magnetic field, the phase accumulation along a closed loop with area F where the electron propagates in the counter-clockwise direction can thus be expressed by

$$\varphi_{ccw} \rightarrow \varphi_{ccw} - \frac{e}{\hbar} \oint_{ccw} \vec{A}\,d\vec{l} = \varphi_{ccw} - 2\pi \frac{\Phi}{\Phi_0}, \tag{8.13}$$

with Φ being the magnetic flux penetrating the loop. For a propagation in the opposite direction one obtains

$$\varphi_{cw} \rightarrow \varphi_{cw} - \frac{e}{\hbar} \oint_{cw} \vec{A}\,d\vec{l} = \varphi_{cw} + 2\pi \frac{\Phi}{\Phi_0}. \tag{8.14}$$

The phase difference between both trajectories is therefore

$$\Delta\varphi = -2\pi \frac{\Phi}{(\Phi_0/2)}. \tag{8.15}$$

Correspondingly, the total probability to return to point A, previously expressed by equation (8.11), is modified to

$$\begin{aligned} P_{AA} &= \left| C_{ccw} e^{i\varphi_{ccw}} e^{-i2\pi\Phi/\Phi_0} + C_{cw} e^{i\varphi_{cw}} e^{i2\pi\Phi/\Phi_0} \right|^2, \\ &= |C_{ccw}|^2 [1 + \cos(4\pi\Phi/\Phi_0)]. \end{aligned} \tag{8.16}$$

Here, we assumed $C_{ccw} = C_{cw}$ and $\varphi_{ccw} = \varphi_{cw}$, because of time-reversal symmetry at $B = 0$. From the equation given above one can conclude that the resistance is expected to oscillate with a period of $\Phi_0/2$. The magnetoresistance oscillations described above are named after Altshuler, Aronov, and Spivak [124], who predicted this effect theoretically. On a single ring-shaped structure, the Altshuler–Aronov–Spivak oscillations are often masked by the Aharonov–Bohm oscillations. This is because for the latter the required phase-coherent paths are shorter; therefore the interference is more likely and stronger. However, by connecting a number of rings in series, as depicted in Figure 8.6 (b), the Aharonov–Bohm effect can be suppressed. The reason is that for each ring the phase shift $\Delta\chi$ in equation (8.10) between each branch is different, since for each ring the impurity configuration is different. Thus, each ring is out of phase with the others. As a result, the contributions of the Aharonov–Bohm oscillations are averaged out. For the Altshuler–Aronov–Spivak oscillations the situation is different, since here the oscillations always start with a resistance maximum at zero field. If the dimensions of each ring are identical the oscillation periods are also the same. Thus all rings oscillate in phase.

The first experimental demonstration was provided by Sharvin und Sharvin [125]. In their experiment a thin Mg film was evaporated on the surface of a quartz filament. The magnetic field was applied in axial orientation with respect to the filament while the current was flowing through the Mg film along the filament. This setup corresponds to a situation where the rings shown in Figure 8.6 (b) are effectively stacked up to form a tube-like conductor. A comparison with the cross-section of the filament confirmed, that the resistance oscillations indeed had a period of $\Phi_0/2$. Beside cylindrical samples, Altshuler–Aronov–Spivak oscillations can also be observed in planar quantum wire networks [126]. Pronounced oscillations were also found in chain as well as in mesh ring structures. Depending on the type of material the magnetoresistance shows either a maximum or minimum at $B = 0$. As we will learn later, a minimum can be attributed to the presence of spin-orbit coupling.

8.2.4 Weak localization

So far we have discussed interference effects where the path of the interfering partial waves is well-defined by the sample geometry. Yet, interference effects can also be observed in disordered diffusive samples. Once again, closed loops are required, but this time these closed loops result from elastic scattering within the sample. An important prerequisite to fulfil this condition is that the elastic mean free path is smaller than l_φ, so that the electron can undergo several elastic scatterings before phase breaking occurs. This corresponding situation is illustrated in Figure 8.7 (a).

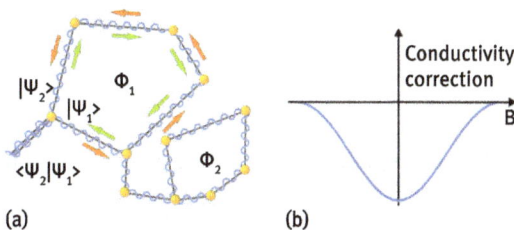

(a)   (b)

Figure 8.7: (a) Electron interference in closed loops leading to the weak localization effect. (b) Conductivity correction due to weak localization: The conductivity at zero magnetic field is reduced.

While the electron is propagating diffusively through the conductor, some closed loop trajectories are present. As for the Altshuler–Aronov–Spivak oscillations, the electron partial waves can propagate within these loops in clockwise and counterclockwise direction. As we already know from the discussion in Section 8.2.3, at zero magnetic field the interference is constructive, leading to an increased return probability. Due to the statistical distribution of scattering centers, the loops in the conductor differ in their size, as shown in Figure 8.7 (a). However, constructive interference occurs for all loops at $B = 0$ no matter what size they have. As a result, the overall resistance is increased compared to a classical conductor. This phenomenon is called the weak localization effect.

In order to quantitatively describe the weak localization effect, we only consider those scattering processes where the electrons return to their starting points. Thus, instead of equation (8.6), where the probability was expressed for a propagation from point A to Q, we now restrict ourself to closed loops

$$P_{\text{closed}} = \left| \sum_j A_j \right|^2 = \left| \sum_j C_j e^{i\varphi_j} \right|^2, \tag{8.17}$$

to calculate the return probability. For a diffusive two-dimensional system one obtains for the return probability per area $1/(4\pi Dt)$ [127]. For the total return probability one has to take care that the phase of the electrons is preserved, which gives us a factor $\exp(-t/\tau_\varphi)$, with τ_φ the phase-coherence time. The condition that the electron has to be scattered at least once, is included by the factor $[1 - \exp(-t/\tau_e)]$. Considering all components mentioned above the correction to the conductivity $\delta\sigma_{\text{loc}}^{2D}$ with respect to the classical value σ_{2D} at $B = 0$ can be expressed as [127, 128]

$$\delta\sigma_{\text{loc}}^{2D}(0) = -\frac{2\hbar}{m^*}\sigma_{2D} \int_0^\infty dt \frac{1}{4\pi Dt}(1 - e^{-t/\tau_e})e^{-t/\tau_\varphi}. \tag{8.18}$$

Here a prefactor $2\hbar/m^*$ resulting from the path integral formalism was included. In the expression given above the two-dimensional conductance given by $\sigma_{2D} = e^2 m^* D/(\pi\hbar^2)$ was inserted. Solving the integral results in

$$\delta\sigma_{\text{loc}}^{2D}(0) = -\frac{e^2}{2\pi^2\hbar} \ln\left(1 + \frac{\tau_\varphi}{\tau_e}\right). \tag{8.19}$$

The weak localization vanishes, if the phase coherence time τ_φ is smaller than τ_e, since then the logarithmic factor tends towards zero.

For a quasi one-dimensional structure where l_φ exceeds the width W the diffusion is effectively reduced to one dimension, so that the return probability per unit are can now be expressed by $W^{-1}(4\pi Dt)^{-1/2}$. Thus the corresponding correction to the conductivity is given by [128]

$$\delta\sigma_{1D}(0) = -\frac{2\hbar}{m^*}\sigma_{2D} \int_0^\infty dt W^{-1}(4\pi Dt)^{-\frac{1}{2}}(1 - e^{-t/\tau_e})e^{-t/\tau_\varphi}. \tag{8.20}$$

Note, that we consider a diffusive two-dimensional stripe of width W (quasi one-dimensional system), and therefore the conductivity for a two-dimensional conductor has to be inserted. Performing the integration in the expression given above results in

$$\delta\sigma_{1D}(0) = -\frac{e^2}{\pi\hbar}\frac{\sqrt{\mathcal{D}}}{W}\left[\sqrt{\tau_\varphi} - \left(\frac{1}{\tau_\varphi} + \frac{1}{\tau_e}\right)^{-\frac{1}{2}}\right]. \qquad (8.21)$$

In the case where the phase coherence time is much larger than the elastic scattering time $\tau_\varphi \gg \tau_e$ one can simplify this expression to

$$\delta\sigma_{1D}(0) = -\frac{e^2}{\pi\hbar}\frac{l_\varphi}{W}, \qquad (8.22)$$

with $l_\varphi = \sqrt{\mathcal{D}\tau_\varphi}$ being the phase coherence length. For the following discussion we will restrict ourselves to this approximation. The conductance correction $\delta G_{1D}(0)$ for a wire of width W can be calculated by using $G_{1D} = (W/L)\sigma_{2D}$, with L the length of the wire. This results in

$$\delta G_{1D}(0) = -\frac{e^2}{\pi\hbar}\frac{l_\varphi}{L}. \qquad (8.23)$$

A direct comparison of the one- and two-dimensional case reveals that the correction to the conductivity is much larger for the one-dimensional conductor.

If the sample is exposed to a magnetic field a phase shift according to equation (8.15) occurs for each closed loop. Since each closed loop encircles a different area, the magnetic flux and thus the phase shift is different for each loop. Thus, by applying a magnetic field the constructive interference, which occurs for all loops at $B = 0$, is lost on average. As a result, the conductance gradually decreases with increasing magnetic field. This is schematically illustrated in Figure 8.7 (b), where the conductivity has a minimum at zero field, while it increases monotonously when a magnetic field is applied.

In order to theoretically describe the dependence of the conductance on the magnetic field, we need to introduce the magnetic relaxation time τ_B. This quantity can be motivated by the following arguments. The phase shift in a closed loop is given by

$$\Delta\varphi = 2\pi\frac{\Phi}{\Phi_0} = \frac{F}{l_m^2}, \qquad (8.24)$$

with Φ the flux in a closed loop and l_m the magnetic length defined by equation (2.58). In order to account for this contribution, we need to insert an additional factor $\langle\langle\mathcal{P}_B(t)\rangle\rangle$ in equations (8.18) and (8.20). Here, $\langle\langle\ldots\rangle\rangle$ represent the average overall classical paths that close after time t and $\mathcal{P}_B(t)$ given by

$$\mathcal{P}_B(t) = \exp\left(i\Delta\varphi\right) = \exp\left(i\frac{2e}{\hbar}\oint_{l(t)}\vec{A}d\vec{l}\right). \qquad (8.25)$$

The phase shift $\Delta\varphi$ corresponds to the phase shift acquired after propagation in two time reversed paths (cf. equation (8.15)). For the two-dimensional system the conductance correction is then given by [127, 128]

$$\delta\sigma_{2D}(B) = -\frac{2\hbar}{m}\sigma_{2D}\int_0^\infty dt \frac{1}{4\pi\mathcal{D}t}(1 - e^{-t/\tau_e})e^{-t/\tau_\varphi}\langle\!\langle\mathcal{P}_B\rangle\!\rangle. \tag{8.26}$$

For diffusive motion the average over all classical paths decays exponentially with time:

$$\langle\!\langle\mathcal{P}_B(t)\rangle\!\rangle = e^{-t/\tau_B}, \tag{8.27}$$

with τ_B being the magnetic relaxation time. Generally, τ_B depends on the type of conductor, i. e. its dimensionality or the kind of boundary scattering and thus has to be determined for each situation separately [129]. Let us find an expression for τ_B for a diffusive two-dimensional system. The average length a particle diffuses within a time t is given by $l = \sqrt{\mathcal{D}t}$, which corresponds to an area $F = \mathcal{D}t$. The magnetic relaxation time τ_B is defined by the diffusion time connected to an area penetrated by a flux quantum Φ_0. Thus, in equation (8.24) the area F has to be replaced by $\mathcal{D}\tau_B$. By convention, in the usual definition of the magnetic relaxation length $l_m^2/2$ is inserted in equation (8.24) instead of l_m^2, so that one finally arrives at

$$\tau_B = \frac{l_m^2}{2\mathcal{D}}. \tag{8.28}$$

Using the magnetic scattering time τ_B an expression for the conductivity change with magnetic field can be derived [130, 131]. Since the derivation is quite elaborate, we only give the final expression, which is given by

$$\Delta\sigma_{2D}(B) = \delta\sigma_{2D}(B) - \delta\sigma_{2D}(0)$$

$$= \frac{e^2}{2\pi^2\hbar}\left[\Psi\left(\frac{1}{2} + \frac{\tau_B}{2\tau_\varphi}\right) - \Psi\left(\frac{1}{2} + \frac{\tau_B}{2\tau_e}\right) + \ln\left(\frac{\tau_\varphi}{\tau_e}\right)\right]. \tag{8.29}$$

Here $\Psi(x)$ is the digamma function defined by $\Gamma'(x)/\Gamma(x)$ with $\Gamma(x)$ the gamma function

$$\Gamma(t) = \int_0^\infty x^{t-1}e^{-x}\,dx. \tag{8.30}$$

Generally, the conductance increases compared to the zero field case, since the localization is lifted with increasing magnetic field. In order to compare magnetoconductance measurements to the model given above, it is often convenient to define characteristic magnetic fields connected to elastic scattering:

$$B_e = \frac{\hbar}{4e\mathcal{D}\tau_e}, \tag{8.31}$$

and connected to phase breaking:

$$B_\varphi = \frac{\hbar}{4eD\tau_\varphi},\tag{8.32}$$

respectively. The characteristic field B_φ is a measure for the width of a weak localization conductance dip, i. e. if the phase coherence length l_φ is large, B_φ is small and so
is the width of the weak localization minimum. The underlying reason is, that for large
l_φ the upper limit of the enclosed areas of phase coherent closed loops is large. Thus
a small magnetic field is already sufficient to lead to a dephasing between the partial
waves propagating in opposite directions. Using the definitions of the characteristics
fields equation (8.29) can be rewritten as

$$\Delta\sigma_{2D}(B) = \frac{e^2}{2\pi^2\hbar}\left[\Psi\left(\frac{1}{2} + \frac{B_\varphi}{B}\right) - \Psi\left(\frac{1}{2} + \frac{B_e}{B}\right) + \ln\left(\frac{B_e}{B_\varphi}\right)\right].\tag{8.33}$$

For a quasi one-dimensional system it is also possible to define a magnetic relaxation
time τ_B. However, in this case the enclosed area is limited in one dimension by the
width W of the wire, so that τ_B is given by [128]

$$\tau_B = \frac{3l_m^4}{W^2D}.\tag{8.34}$$

Under the condition that $l_e \ll W \ll l_m, l_\varphi$ the expression for the weak localization
correction of the conductivity is given by

$$\delta\sigma_{1D}(B) = -\frac{2\hbar}{m}\sigma_{2D}\int_0^\infty dt W^{-1}(4\pi Dt)^{-1/2}(1 - e^{-t/\tau_e})e^{-t/\tau_\varphi}e^{-t/\tau_B},\tag{8.35}$$

which results in

$$\delta\sigma_{1D}(B) = -\frac{e^2}{\pi\hbar}\frac{\sqrt{D}}{W}\left[\left(\frac{1}{\tau_\varphi} + \frac{1}{\tau_B}\right)^{-\frac{1}{2}} - \left(\frac{1}{\tau_\varphi} + \frac{1}{\tau_e} + \frac{1}{\tau_B}\right)^{-\frac{1}{2}}\right].\tag{8.36}$$

In Figure 8.8 the weak localization correction $\delta G_{1D} = \delta\sigma_{1D}(B)(W/L)$ to the classical
conductance calculated by equation (8.36) is shown for different wire widths. It can
be clearly seen that for decreasing the wire width W the dip becomes broader. This is
due to the fact that the area of the loops enclosing the magnetic flux becomes smaller
because it is limited in one direction by W.

Up to now have we considered electron interference effects where the spin orientation is conserved, and thus its contribution can be neglected. As a consequence of
the electron localization an increase of the sample resistance is observed. If spin-orbit
coupling is present, the opposite behavior is found, i. e. a decrease of the resistance.
Below, first the basics of the weak antilocalization are explained, and then in the second part spin-orbit coupling is included into the interference effects of ring structures.

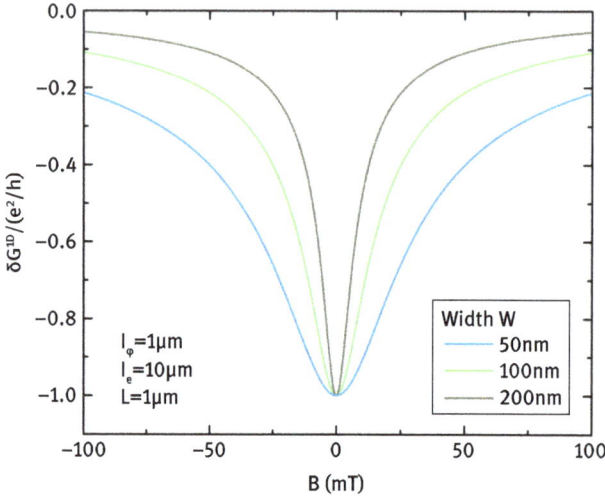

Figure 8.8: Weak localization correction of the conductance for a quasi one-dimensional system as a function of magnetic field for different wire width of 50, 100, and 200 nm. The calculations were performed according to equation (8.36), with $l_\varphi = 1\,\mu m$, $l_e = 10\,nm$, and $L = 1\,\mu m$.

8.3 Spin interference effects

The strength of spin-orbit coupling in semiconductor structures can be analyzed by means of the weak antilocalization effect, which is an extension of the weak localization effect by including spin precession or spin scattering. As we learned above, weak localization is an electron interference effect where a decrease of the sample conductance is observed due to electron localization. If spin-orbit coupling is present, the opposite behavior is found, i. e. a conductance increase. This is the reason why this effect is called weak antilocalization. Weak antilocalization can be utilized to gain information on the strength of spin-orbit coupling and is therefore an alternative method to the analysis of the beating pattern in the Shubnikov–de Haas oscillations discussed in the previous chapter. Very similar to the interference effects introduced above, spin interference can be observed in arrays of ring structures. These measurements can also be employed to analyze spin-orbit coupling.

8.3.1 Weak antilocalization

When we discussed the weak localization effects in Section 8.2.4, spin effects were neglected, i. e. it was assumed that the spin orientation was preserved while the electron partial waves propagate along a closed loop. Now we consider a situation where the spin orientation is altered while the electron moves through the sample. One possible mechanism which causes a change of the spin orientation is Rashba spin-orbit coupling, which results in spin precession during propagation [13]. Other effects leading to a change of the spin orientation are spin-orbit coupling due to crystal lattice asymmetry [105] or spin scattering at impurities [132]. In contrast to weak localization, for the weak antilocalization effect the presence of spin-orbit coupling or spin scattering results in a decrease of sample resistance around zero magnetic field.

Below we will give a simple explanation of the weak antilocalization effect, following the approach of Bergmann [133]. The situation is depicted in Figure 8.9 (a). When the spin orientation is changed while an electron propagates in a closed loop, the interference is not necessarily constructive, as for the spin-conserving weak localization effect. For the latter the interference is always constructive at zero magnetic field, no matter what shape the closed loops have. If we take into account a change of the spin orientation, the total spin rotation after the propagation through a closed loop depends on the geometry of this particular loop and/or the spin-orbit scattering at impurities. Thus each particular loop causes a different spin rotation of the electrons. In order to calculate the total interference amplitude, the spin rotation has to be considered in addition to the electron phase. It will be shown below that due to averaging effects of different interference amplitudes destructive interference dominates, leading to an enhanced conductance at zero field, compared to the classical case. Figure 8.9 (b) illustrates the enhanced conductivity at zero magnetic field.

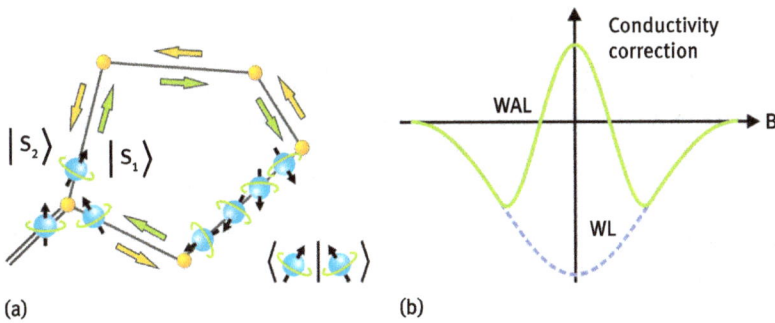

Figure 8.9: (a) Electron interference in closed loops leading to the weak antilocalization effect. Here spin precession due to spin-orbit coupling is included. (b) The weak antilocalization effect (WAL) leads to an enhanced conductivity at zero magnetic field, whereas the weak localization effect (WL) leads to a decrease of the conductivity.

Let us first analyze the interference in an arbitrarily chosen closed loop in the presence of spin-orbit coupling. The spin-related effects can be summarized by assigning a total spin rotation after propagation along the loop. The rotation of a spin-1/2 particle can be described by a rotation matrix $\hat{U}_r$, which is applied to the spinor representing the spin state. Any rotation can be represented by three characteristic rotation angles β, θ, and α (Euler's angles), where the rotation is decomposed in three rotations about certain axes. The first and third rotations (β, α) are chosen to be about the z-axis, while the second rotation (θ) is about the y-axis. Using this sequence the spin rotation matrix $\hat{U}_r$ can be expressed as

$$\hat{U}_r = \hat{R}_z(\alpha)\hat{R}_y(\theta)\hat{R}_z(\beta)$$

$$= \begin{pmatrix} e^{i\alpha/2} & 0 \\ 0 & e^{-i\alpha/2} \end{pmatrix} \begin{pmatrix} \cos\theta/2 & \sin\theta/2 \\ -\sin\theta/2 & \cos\theta/2 \end{pmatrix} \begin{pmatrix} e^{i\beta/2} & 0 \\ 0 & e^{-i\beta/2} \end{pmatrix}, \tag{8.37}$$

with $\hat{R}_z$ and $\hat{R}_y$ being the rotation matrices about the z- and y-axis, respectively. By explicitly multiplying the three matrices given above, $\hat{U}_r$ can be written as

$$\hat{U}_r = \begin{pmatrix} e^{i/2(\alpha+\beta)}\cos\theta/2 & e^{i/2(\alpha-\beta)}\sin\theta/2 \\ -e^{-i/2(\alpha-\beta)}\sin\theta/2 & e^{-i/2(\alpha+\beta)}\cos\theta/2 \end{pmatrix}. \tag{8.38}$$

The final state $|s'\rangle$, resulting from the rotation of the initial spin state $|s\rangle$ after propagation along a closed loop can be expressed by

$$|s'\rangle = \hat{U}_r|s\rangle. \tag{8.39}$$

For the propagation in the opposite direction along the time reversed path, the spin rotation is reversed. Therefore, the final spin state is given by

$$|s''\rangle = \hat{U}_r^{-1}|s\rangle. \tag{8.40}$$

Regarding the localization effects we are interested in the amplitude of the interference of the two states after propagation in opposite orientations along the loop:

$$\langle s''|s'\rangle = \langle(\hat{U}_r^{-1}s)|(\hat{U}_r s)\rangle = \langle s|(\hat{U}_r^{-1})^\dagger \hat{U}_r|s\rangle = \langle s|\hat{U}_r^2|s\rangle. \tag{8.41}$$

Here, we made use of the property that $\hat{U}_r$ is a unitary transformation, i.e. $\hat{U}_r^\dagger = \hat{U}_r^{-1}$, with $\hat{U}_r^\dagger$ the adjoint matrix of $\hat{U}_r$. The square of $\hat{U}_r$ is given by the matrix

$$\hat{U}_r^2 = \begin{pmatrix} e^{i(\alpha+\beta)}\cos^2\frac{\theta}{2} - \sin^2\frac{\theta}{2} & \frac{1}{2}(e^{i\alpha} + e^{-i\beta})\sin\theta \\ -\frac{1}{2}(e^{-i\alpha} + e^{i\beta})\sin\theta & -e^{i(\alpha+\beta)}\cos^2\frac{\theta}{2} - \sin^2\frac{\theta}{2} \end{pmatrix}. \tag{8.42}$$

If the spinor of the original spin state has the form $(a, b)^T$, with $a^2 + b^2 = 1$, one obtains for the expectation value of $\hat{U}_r^2$

$$\langle s''|s'\rangle = (|a|^2 e^{i(\alpha+\beta)} + |b|^2 e^{-i(\alpha+\beta)})\cos^2\frac{\theta}{2} - \sin^2\frac{\theta}{2}$$
$$+ \frac{1}{2}\sin\theta[a^*b(e^{i\alpha} + e^{-i\beta}) + ab^*(e^{-i\alpha} + e^{i\beta})]. \tag{8.43}$$

In order to understand under which condition weak antilocalization occurs, one has to analyze equation (8.43). For strong spin-orbit coupling the orientations of the final spin states are distributed statistically. The only term which does not vanish after averaging over all angles is $\sin^2(\theta/2)$, since it can be written as $(1 - \cos\theta)/2$, which finally yields a negative factor of $-1/2$. Owing to the negative sign destructive interference dominates for the case of strong spin scattering. In contrast, if there is no spin rotation $\alpha, \beta, \theta = 0$, only the $\cos^2$-term in equation (8.43) remains. Thus, in this case constructive interference occurs, leading to the weak localization effect discussed in the previous section. Please note that the correction to the conductivity due to the weak antilocalization effect is by a factor of $1/2$ smaller compared to the weak localization effect. This is also illustrated in Figure 8.9.

8.3.2 Spin relaxation mechanisms

There are a number of possible mechanisms which can result in a change of spin orientation. During diffusive transport the information on the spin orientation is lost after a characteristic time, i. e. the spin relaxation time. As the spin is a vector and the mechanism leading to a spin relaxation might also be anisotropic, the spin relaxation has to be described by the following general expression [134]

$$\frac{ds_i}{dt} = -\sum_j \frac{s_j}{\tau_{ij}}, \quad \text{with } i, j = x, y, z. \tag{8.44}$$

Here τ_{ij} is the spin relaxation time for a spin along the i-direction due to the presence of spins along the j-direction. In many cases it is sufficient to neglect cross relaxation, i. e. $i = j$. For a two-dimensional electron system confined in the z-direction with spin-orbit coupling one finds for the relaxation time [134]

$$\frac{1}{\tau_{so}} = \frac{1}{\tau_{zz}} = \frac{2}{\tau_{xx}} = \frac{2}{\tau_{yy}}. \tag{8.45}$$

Similarly, to the phase coherence length l_φ in a diffusive system the spin relaxation length can be defined as

$$l_{so} = \sqrt{D\tau_{so}}. \tag{8.46}$$

The spin relaxation length has to be distinguished from the spin precession length for Rashba spin-orbit coupling l_R. The latter is the length the spin precesses a full round on a straight nondiffusive path.

In addition to spin precession due to the Rashba or Dresselhaus spin-orbit coupling, the spin can also relax by other processes, i. e. by scattering at impurities. Below we will introduce the two most important spin relaxation mechanisms for semiconductor spintronic devices.

D'yakonov–Perel' spin relaxation

In semiconductor structures with the Rashba or Dresselhaus type of spin-orbit coupling the D'yakonov–Perel' spin relaxation mechanism is present [135]. As illustrated in Figure 8.10, on its path from one scattering center to the next the electron spin precesses about the internal spin-orbit field.

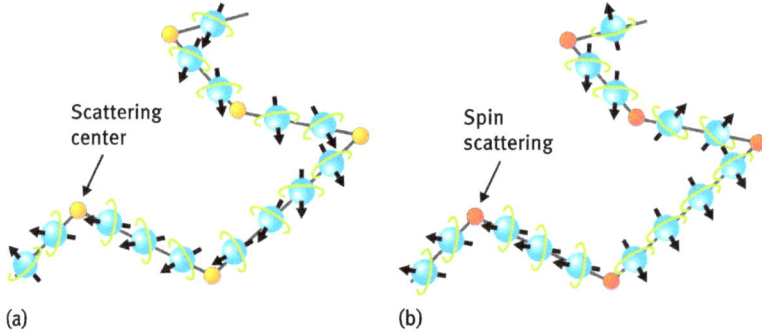

(a)   (b)

Figure 8.10: (a) D'yakonov–Perel' spin relaxation mechanisms: In between the scattering events the electron spin precesses about the internal magnetic field induced by spin-orbit coupling. (b) Elliott–Yafet spin relaxation: The spin relaxes by impurities or phonons. In between the scattering events the spin orientation is preserved.

This process can be described by defining a precession vector $\vec{\Omega}$ which depends on the different contributions of spin-orbit coupling. For the Rashba effect the corresponding Hamiltonian is given by

$$\hat{H}_{\text{so}} = \hat{\vec{\sigma}} \cdot \vec{\Omega}_{\text{R}}.\tag{8.47}$$

Very similarly to the classical case, where a magnetic moment precesses about a magnetic field, the form of the Hamiltonian implies that the electron spin precesses about $\vec{\Omega}_{\text{R}}$. For the Rashba effect with the Hamiltonian given by equation (7.5) the precession vector lies within the plane of the two-dimensional electron gas and has a length of

$$|\vec{\Omega}_{\text{R}}| = \alpha_{\text{R}}|\vec{k}|.\tag{8.48}$$

For a cylindrical coordinate system with the components of the k-vector given by $k_x = k \cos \theta$, $k_y = k \sin \theta$, and $k_z = 0$, the precession vector can be expressed as

$$\Omega_{\text{R},x} = \Omega_{\text{R}} \sin \theta, \quad \Omega_{\text{R},y} = -\Omega_{\text{R}} \cos \theta, \quad \text{and} \quad \Omega_{\text{R},z} = 0.\tag{8.49}$$

For a two-dimensional electron gas, where the electrons are confined along the z-direction the Dresselhaus or bulk inversion asymmetry contribution is given by equation (7.50). For this case the corresponding components of the precession vector $\vec{\Omega}_D$ are given by [134]

$$\Omega_{D,x} = -\Omega_{D,1} \cos\theta - \Omega_{D,3} \cos 3\theta, \qquad (8.50)$$

$$\Omega_{D,y} = \Omega_{D,1} \sin\theta - \Omega_{D,3} \sin 3\theta, \qquad (8.51)$$

$$\Omega_{D,z} = 0, \qquad (8.52)$$

with

$$\Omega_{D,1} = \eta_D k \left(\langle k_z^2 \rangle - \frac{1}{4}k^2 \right) \quad \text{and} \quad \Omega_{D,3} = \eta_D \frac{k^3}{4}, \qquad (8.53)$$

where $k^2 = k_x^2 + k_y^2$ and $\langle k_z^2 \rangle$ is the average squared wave vector along the z-direction. When Rashba and Dresselhaus spin-orbit coupling is present the total precession vector is given by the sum of both contributions:

$$\vec{\Omega} = \vec{\Omega}_R + \vec{\Omega}_D. \qquad (8.54)$$

Based on the precession vector defined above, we will now derive an expression for the spin relaxation rate $1/\tau_{so}$ in the presence of Rashba spin-orbit coupling. When the momentum direction is changed randomly during diffusive transport, the precession vector $\vec{\Omega}_R$ is changed as well. Assuming Rashba spin-orbit coupling, the spin precession angle between two scattering events scales with the elastic scattering time τ_e. The angle can be calculated using expression equation (7.15) for the spin precession length l_R, which corresponds to a full cycle of 2π. The average distance between two scattering centers is given by

$$x = |\vec{v}_F|\tau_e = \frac{\hbar|\vec{k}_F|}{m^*}\tau_e. \qquad (8.55)$$

Along that distance the spin precession angle is given by

$$\theta(\tau_e) = 2\pi\frac{x}{l_R} = 2\frac{|\vec{k}_F|\alpha_R}{\hbar}\tau_e. \qquad (8.56)$$

After a time t with $N = t/\tau_e$ scattering events, the spin orientation has changed on average by

$$\theta(t) = 2\frac{|\vec{k}_F|\alpha_R}{\hbar}\tau_e\sqrt{N} = 2\frac{|\vec{k}_F|\alpha_R}{\hbar}\sqrt{\tau_e t}. \qquad (8.57)$$

This relationship can also be expressed in terms of the precession vector $\vec{\Omega}_R$ given by equation (8.48)

$$\theta(t) = 2\frac{|\vec{\Omega}_R|}{\hbar}\sqrt{\tau_e t}. \qquad (8.58)$$

If we define the spin relaxation time τ_{so} as the time t by which the precession angle is assumed to be changed by 1, one arrives at the following expression for the spin relaxation rate

$$\frac{1}{\tau_{so}} = \tau_e \langle (2\vec{\Omega}_R/\hbar)^2 \rangle. \tag{8.59}$$

The angular brackets denote integration over all angles. One finds that a smaller electron mobility in a semiconductor, i. e. a smaller τ_e, results in a longer spin relaxation time τ_{so}. This is due to the fact that for smaller mobilities the mean free path is shorter, which means that along that path the spin precesses less. This tendency is called motional narrowing.

Elliott–Yafet spin relaxation

The Elliott–Yafet spin relaxation mechanism is illustrated in Figure 8.10 (b) [132, 136]. Here, the spin relaxation occurs because scattering at impurities or due to phonons is connected to a probability to flip the spin. In contrast to the D'yakonov–Perel' mechanism the spin orientation changes during the scattering event itself. As a consequence, the spin relaxes faster the more momentum scattering occurs. For degenerate III-V semiconductors the spin relaxation rate is found to be [137]

$$\frac{1}{\tau_{EY}} \sim \frac{\Delta_{so}^2}{(E_g + \Delta_{so})^2} \frac{E(\vec{k})^2}{E_g^2} \frac{1}{\tau_e}. \tag{8.60}$$

As can be seen here, for large band gap semiconductors the spin relaxation is negligible. The Elliott–Yafet spin relaxation rate is proportional to the elastic scattering rate $1/\tau_e$ in contrast to the D'yakonov–Perel' spin relaxation rate which is proportional to τ_e. Thus, by investigating the spin relaxation as a function of mobility the spin relaxation mechanism can be identified.

Beside the spin relaxation mechanism discussed above, other mechanisms exist which are not discussed here, e. g. Bir–Aronov–Pikus spin relaxation due to exchange interaction between electrons and holes [138].

8.3.3 Weak antilocalization in two-dimensional electron gases

The approach by Bergmann [133] gives us a general idea about the conductance increase when the spin orientation is changed while the electron partial waves propagate along time-reversed closed loops. The weak antilocalization effect can be employed to gain information on the spin relaxation length l_{so}. Usually, the experiments are performed by measuring the conductance as a function of an external magnetic field. A typical measurement on a two-dimensional electron gas in an $Al_{0.15}Ga_{0.85}N/GaN$ heterostructure is shown in Figure 8.11 [139].

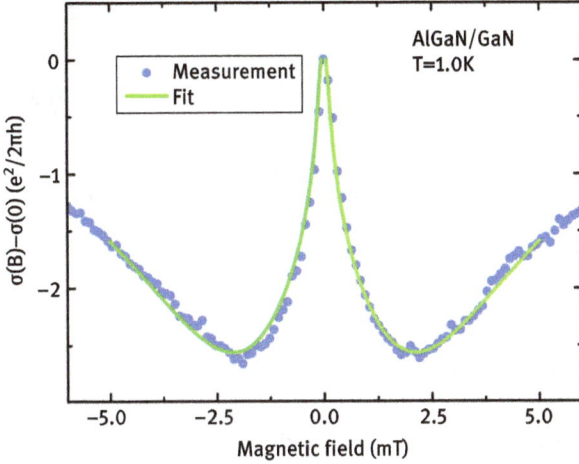

Figure 8.11: Measured weak antilocalization correction in an $Al_{0.15}Ga_{0.85}N/GaN$ two-dimensional electron gas at a temperature of 1.0 K. The solid line corresponds to a fit to a theoretical model of Iordanskii, Lyanda-Geller, and Pikus [140].

In order to preserve phase coherence, the measurements are conducted at low temperatures. As can be seen in Figure 8.11, a pronounced conductance peak is found at zero magnetic field. For two-dimensional systems this peak is usually restricted to relatively small magnetic fields in the mT-range.

In order to extract the spin relaxation time τ_{so}, the experimental curve needs to be fitted to a theoretical model. For a two-dimensional system the weak antilocalization effect is described by a model developed by Iordanskii, Lyanda-Geller, and Pikus (ILP) [140]. In this model the Rashba effect as well as contributions due to bulk inversion asymmetry (Dresselhaus term) are included in the conductivity correction. Considering only the Rashba effect, the relevant relaxation times are given by equation (8.59) with $|\vec{\Omega}_R| = |\vec{k}_F|\alpha_R$. In addition, the Dresselhaus term can also be included, which gives some additional k-linear and cubic contribution to the precession angle [140, 134]. Due to its complexity the exact derivation of the conductivity corrections is beyond the scope of this book. Therefore we only give the final expression for the conductivity correction:

$$
\begin{aligned}
\delta\sigma_{2D}(B) = -\frac{e^2}{4\pi^2\hbar}\Bigg\{ &\frac{1}{a_0} + \frac{2a_0 + 1 + \frac{B_{so}}{B}}{a_1(a_0 + \frac{B_{so}}{B}) - 2\frac{B_{so}}{B}} \\
&- \sum_{n=1}^{\infty}\left[\frac{3}{n} - \frac{3a_n^2 + 2a_n\frac{B_{so}}{B} - 1 - 2(2n+1)\frac{B_{so}}{B}}{[a_n + \frac{B_{so}}{B}]a_{n-1}a_{n+1} - 2\frac{B_{so}}{B}[2n+1]a_n - 1}\right] \\
&+ \ln\left(\frac{B_e}{B}\right) + \Psi\left(\frac{1}{2} + \frac{B_\varphi}{B}\right)\Bigg\},
\end{aligned}
\tag{8.61}
$$

with

$$a_n = n + \frac{1}{2} + \frac{B_\varphi}{B} + \frac{B_{so}}{B}. \tag{8.62}$$

Here we made use of the characteristic magnetic fields B_φ already defined in equation (8.32). In addition a corresponding quantity with respect to spin relaxation is defined by

$$B_{so} = \frac{\hbar}{4e\mathcal{D}\tau_{so}}, \tag{8.63}$$

with τ_{so} the spin relaxation time for Rashba spin-orbit coupling given by equation (8.59).

The experimental curve shown in Figure 8.11 was fitted to the model given above. As one can see here, a good agreement is achieved. From this fit a spin relaxation length of l_{so} = 290 nm and a phase coherence length of about 1 µm was extracted. The value of l_{so} corresponds to a Rashba coefficient of $\alpha_R = 8.5 \times 10^{-13}$ eVm.

In Figure 8.12 (a) the weak antilocalization effect in an InGaAs/InP two-dimensional electron gas is shown for different gate voltages [141]. Obviously the gate voltage and the corresponding change of the confinement potential have a large effect on the shape of the weak antilocalization curves. For smaller electron concentrations at more negative gate voltages the weak antilocalization feature is more pronounced. This is due to the fact that here the mobility and thus the elastic mean free path is smaller, thus the transport is more diffusive. By fitting the experimental weak antilocalization curves to the theoretical model of Golub [142], the Rashba coupling coefficient α_R can be extracted. The applied model is an extension of the ILP model, where in between the scattering centers the strong spin-orbit coupling results in a pronounced spin precession. This case can be quantified by the condition $\Omega\tau_e > 1$. In contrast, the ILP model is only valid for weak spin-orbit interaction, i. e. $\Omega\tau_e \ll 1$. In fact, for the two-dimensional electron gas employed for the measurements shown in Figure 8.12 (a), a relatively long elastic mean free path l_e between 0.5 and 2.5 µm was extracted [141]. As can be inferred from Figure 8.12 (b), the Rashba coefficient obtained from the fit to the Golub model decreases with increasing electron concentration n_{2D}. This is due to the fact that the $In_{0.53}Ga_{0.47}As/In_{0.77}Ga_{0.23}As/InP$ potential well becomes more symmetric with less negative gate voltages and increasing n_{2D}. Figure 8.12 (b) also shows the values of α_R obtained by analyzing the beating pattern in the Shubnikov–de Haas oscillations, as described in Section 7.5.1. The values obtained here show a good agreement with the values gained from the weak antilocalization measurements. However, at low electron concentrations the mobility is lower, which makes it difficult to resolve the beating pattern in the Shubnikov–de Haas oscillations. In this case it is easier to extract α_R from weak antilocalization, since this effect relies on diffusive transport.

The ILP model covers the situation that the spin precesses between two scattering events which corresponds to the D'yakonov–Perel' mechanism [135]. However, one

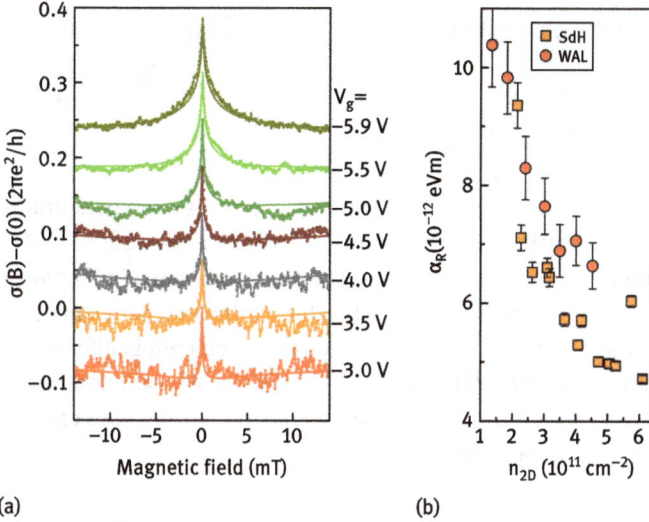

(a)

(b)

Figure 8.12: (a) Weak antilocalization in an InGaAs/InP two-dimensional electron gas for various gate voltages. The solid lines are fits to the model of Golub [142]. (b) Rashba coefficient α_R as a function of electron concentration n_{2D} determined from the beating pattern in the Shubnikov–de Haas oscillations (SdH) and from weak antilocalization (WAL) measurements.

can also imagine a scenario, where the scattering centers can result in spin-scattering due to lattice ions induced spin-orbit coupling in the electron wave function, i. e. the Elliott–Yafet mechanism [132]. Furthermore, one can imagine magnetic scattering by impurity spins. During the propagation between these scattering centers the spin orientation is preserved. For this case the following formula for a two-dimensional system was derived by Hikami–Larkin–Nagaoka [131]:

$$
\begin{aligned}
\Delta\sigma(B)_{2D} = \frac{e^2}{2\pi^2\hbar} \Bigg\{ &\Psi\left(\frac{1}{2} + \frac{B_\varphi}{B} + \frac{B_{so}}{B}\right) + \frac{1}{2}\Psi\left(\frac{1}{2} + \frac{B_\varphi}{B} + 2\frac{B_{so}}{B}\right) \\
&- \frac{1}{2}\Psi\left(\frac{1}{2} + \frac{B_\varphi}{B}\right) - \ln\left(\frac{B_\varphi + B_{so}}{B}\right) \\
&- \frac{1}{2}\ln\left(\frac{B_\varphi + 2B_{so}}{B}\right) + \frac{1}{2}\ln\left(\frac{B_\varphi}{B}\right) \Bigg\},
\end{aligned}
\tag{8.64}
$$

with $\Psi(x)$ being the digamma function.

8.3.4 Weak antilocalization in wire structures

In order to find the appropriate expression to describe the weak antilocalization effect for quasi one-dimensional structures, we can start with equation (8.20). Similar to the way we accounted for the phase shifts acquired due to a magnetic field, one has to include an average over the phase shifts owing to the spin-orbit induced effective magnetic field:

$$\langle\langle \mathcal{P}_{so}(t) \rangle\rangle = \frac{3}{2} e^{-4t/3\tau_{so}} - \frac{1}{2}. \tag{8.65}$$

This expression contains two interference contributions, i. e. the first exponentially decreasing part is the triplet contributions, while the second one is the singlet contribution. The total spin of the two counter-propagating partial waves can be subdivided into a singlet state, with zero total spin and a triplet state with the magnetic quantum numbers +1, 0, and −1. Only the triplet contribution with a finite spin is affected by spin-orbit coupling, while the singlet state with zero spin is not. An exponential decay as given above is found for a diffusive motion along the longitudinal direction of the wire. The factor $\mathcal{P}_{so}(t)$ is calculated by [127]

$$\mathcal{P}_{so} = \frac{1}{2} \mathrm{Tr} \left\{ \left[\mathcal{T} \exp\left(\frac{i}{\hbar} \int_0^t dt' \vec{\sigma} \cdot \vec{\Omega} \right) \right]^2 \right\}, \tag{8.66}$$

with $\mathcal{T}$ being the time-order operator, $\vec{\Omega}$ the precession vector, and $\mathrm{Tr}\,(\dots)$ the trace of the 2×2 matrix.

Including the exponential decay of $\langle\langle \mathcal{P}_{so}(t) \rangle\rangle$ into equation (8.20) and performing the integration over t leads to [143]

$$\delta G_{\mathrm{1D}}(B) = -\frac{e^2}{\pi\hbar}\frac{\sqrt{D}}{L}\left[\frac{3}{2}\left(\frac{1}{\tau_\varphi} + \frac{3}{4\tau_{so}} + \frac{1}{\tau_B} \right)^{-\frac{1}{2}} - \frac{1}{2}\left(\frac{1}{\tau_\varphi} + \frac{1}{\tau_B} \right)^{-\frac{1}{2}} \right.$$

$$\left. - \frac{3}{2}\left(\frac{1}{\tau_\varphi} + \frac{3}{4\tau_{so}} + \frac{1}{\tau_e} + \frac{1}{\tau_B} \right)^{-\frac{1}{2}} + \frac{1}{2}\left(\frac{1}{\tau_\varphi} + \frac{1}{\tau_e} + \frac{1}{\tau_B} \right)^{-\frac{1}{2}} \right]. \tag{8.67}$$

Here the three triplet contributions are represented by the terms with the prefactor $3/2$, while the terms with the prefactor $1/2$ are due to the singlet contribution. For strong spin-orbit coupling, i. e. small values of τ_{so} compared to τ_φ, the expressions with the prefactor $1/2$ dominate, resulting in a positive conductance correction. The triplet contribution is suppressed, since spin-orbit coupling only affects finite spin states, thus only the zero spin singlet contribution remains.

The magnitude of the conductance increase is at most half the value of the corresponding magnitude for the weak localization effect. This can be seen nicely in Figure 8.13, where the conductance correction $\delta G_{\mathrm{1D}}(B)$ as a function of a magnetic field is calculated for different spin-orbit coupling strengths. The spin relaxation length l_{so} was used as a parameter. As can be seen here, the amplitude of the conductance modulation for strong spin-orbit coupling, i. e. $l_{so} = 50\,\mathrm{nm}$, is about half as big as the one for negligible spin-orbit coupling ($l_{so} \to \infty$). The conductance peak decreases when the spin-orbit strength decreases and l_{so} increases.

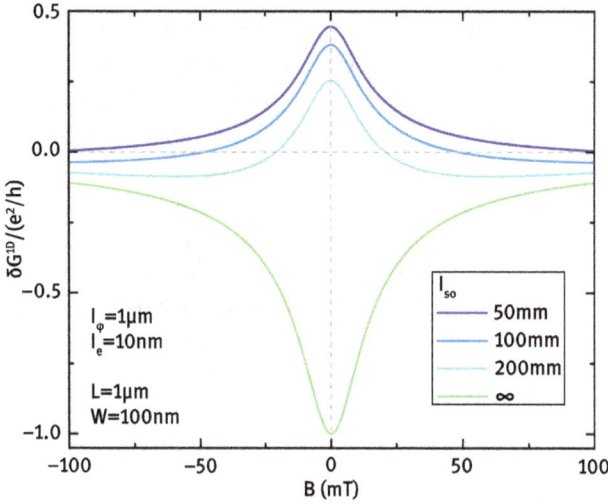

Figure 8.13: Conductance corrections calculated by applying equation (8.67). For $l_{so} \to \infty$ weak localization occurs. With decreasing l_{so}, thus increasing spin-orbit coupling, the conductance peak due to weak antilocalization increases.

As can be observed in Figure 8.14, the peak of $\delta G_{1D}(B)$ becomes broader for decreasing wire width.

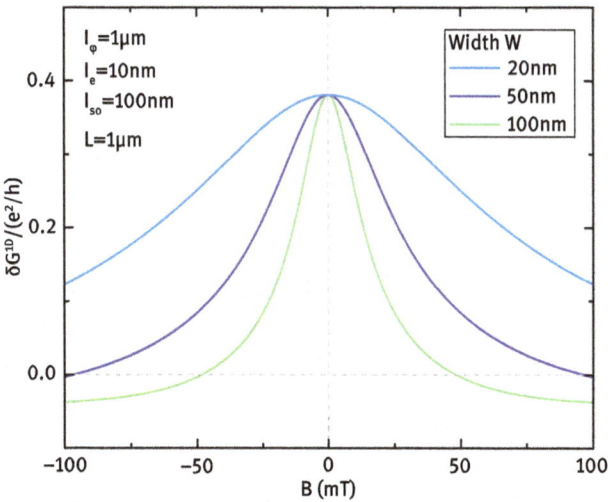

Figure 8.14: Conductance corrections calculated by applying equation (8.67). For l_{so} assumed to be fixed to 100 nm while the width of the wire is varied between 20 and 100 nm.

This once again has to do with the fact that the average flux trapped by the closed loops is smaller because these loops are more elongated with decreasing width.

Note that although the peak height is identical for all width W, the relative increase $\delta G_{1D}(B)/G_{1D}(0)$ is strongest for the smallest width, since $G_{1D}(0)$ scales with W.

In Figure 8.15 a scanning electron microscopy image of a wire array is shown, which is used to analyze the weak antilocalization effect in quasi one-dimensional structures [144]. The wires are defined by dry etching of an $In_{0.53}Ga_{0.47}As/In_{0.77}Ga_{0.23}As/InP$ heterostructure. Here many nanowires are contacted in parallel by ohmic contacts on both ends, in order to suppress the so-called universal conductance fluctuations [128]. Universal conductance fluctuations occur if the sample only contains a small finite number of scattering centers, i. e. in case of a single wire. The resulting conductance fluctuations are in the order of e^2/h. By measuring many wires in parallel these fluctuations can be averaged out, so that a smooth curve is obtained.

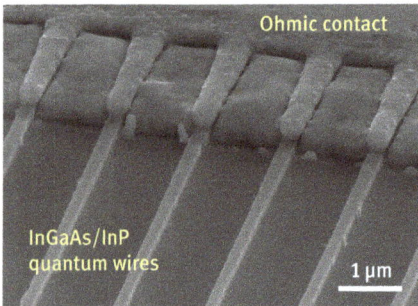

Figure 8.15: Scanning electron microscope image of a nanowire array. The $In_{0.53}Ga_{0.47}As/In_{0.77}Ga_{0.23}As/InP$ quantum wires defined by dry etching have a width of 230 nm and a lengths of 620 μm [144].

In Figure 8.16 conductivity measurements of wire arrays with different wire widths are shown. For wide wires a clear signature of weak antilocalization, i. e. a conductance peak at zero field, is observed due to the strong spin-orbit coupling in that material. However, with decreasing wire width the conductance peak height lowers while it totally vanishes for the set of wires with the smallest width of 250 nm. This behavior is not expected from the theoretical model expressed by equation (8.36).

Obviously there must be an addition mechanism, which results in a suppression of the weak antilocalization effect for narrow wire structures. As we learned above, the weak antilocalization results from interference of electron partial waves propagation in time reversed trajectories. In case of Rashba or Dresselhaus spin-orbit coupling the electrons are precessing about an effective magnetic field, while they are between two scattering centers. The orientation of the effective field is determined by the direction of propagation, i. e. in case of the Rashba effect it is perpendicular to the wave vector of the electron. The accumulated precession angle between two scattering events

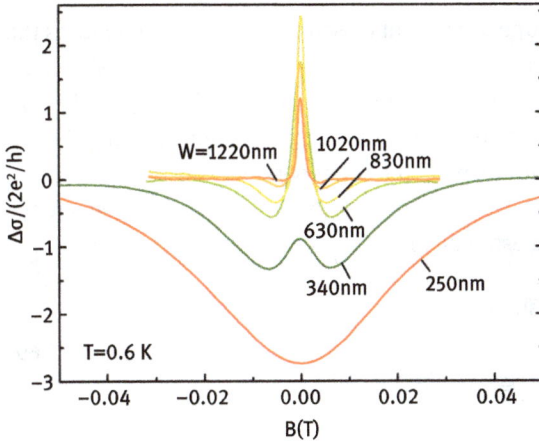

Figure 8.16: Magnetoconductivity corrections $\Delta\sigma$ in units of $2e^2/h$ for a set of wires defined in an $In_{0.53}Ga_{0.47}As/In_{0.77}Ga_{0.23}As/InP$ heterostructure at a temperature of 0.6 K. Arrays of 160 wires connected in parallel with a width W ranging from 250 nm to 1.22 µm were measured [144].

can be described by a corresponding rotation matrix for the electron spin. For a two-dimensional structure, one can imagine that there are many "round" closed loops, which implies that the direction of propagation is changed constantly. As in the discussion of the Bergmann model [133], the total spin rotation after propagation along a closed loop is given by multiplying the rotation matrices of all segments of the loop. For the time reversed path these rotation matrices have to be multiplied in reversed order. The matrices for spin rotations about different axes do not commute, e. g.

$$\hat{R}_z(\alpha)\hat{R}_y(\theta) \neq \hat{R}_y(\theta)\hat{R}_z(\alpha), \tag{8.68}$$

which implies that the spin orientation for the clockwise and counterclockwise paths is different. For wire structures the situation is different, since here the electrons propagate in one direction back and forth. As a consequence, the rotation axis is along one direction only. If the electron propagates and is back-scattered more or less along a single path along the same direction the accumulated total spin rotation is zero. This is the reason why in one-dimensional systems the weak antilocalization effect tends to be suppressed.

In order to interpret the suppression of weak antilocalization shown in Figure 8.16, we first have to assume that the Rashba coupling parameter α_R does not depend on the wire width [145]. Thus the suppression is not simply a consequence of a reduced spin-orbit coupling. The ballistic spin precession length $l_R = \pi\hbar^2/(m^*\alpha_R)$ is identical for all wires and given by $l_R = 200$ nm. It can be shown that in the two-dimensional limit, $l_{so} = l_R$ [140], thus the spin relaxation length corresponds to the spin precession length. This is obviously not the case for a wire structure, where l_{so} must have a strong width dependence. As discussed above we expect a suppression of spin relaxation in narrow wires, as the elongated shape of relevant closed paths effectively reduces the magnitude of accumulated random spin phases. A more quantitative analysis is possible using the concept of spin-orbit-induced effective magnetic fluxes [146, 147]. In complete analogy to equation (8.25), for an external magnetic field one can now insert the effective spin-orbit field and calculate the corresponding magnetic dephasing length [129]. Depending on the relation between the width W and the elastic mean free path l_e one arrives at the following estimate for the spin relaxation length [144]:

$$l_{so,1D} = \begin{cases} \frac{\sqrt{3}l_R}{W}^2 & \text{for } l_e < W \\ l_R\sqrt{\frac{C_1}{2}\frac{l_e l_R^2}{W^3} + \frac{C_2}{2}\frac{l_e^2}{W^2}} & \text{for } l_e > W. \end{cases} \tag{8.69}$$

The constants C_1 and C_2 depend on the type of boundary scattering, i. e. specular or diffusive, in the wire [129]. Using the Landauer–Büttiker formalism the correction of the conductivity can also be calculated numerically [144]. The resulting curves, shown in Figure 8.17 confirm that for decreasing wire width the peak due to weak antilocalization is suppressed and weak localization takes over.

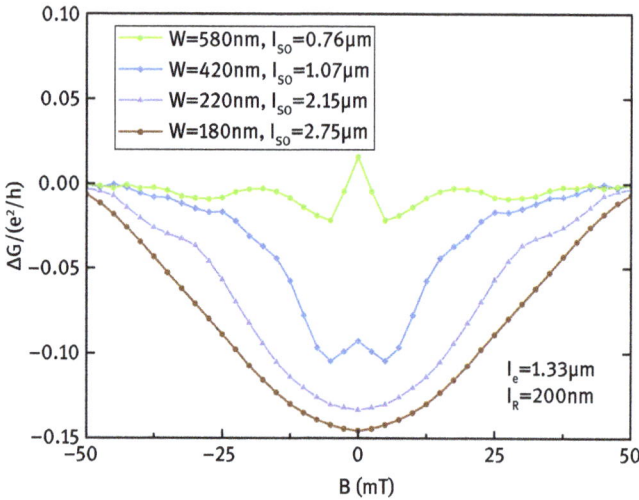

Figure 8.17: Magnetoconductance correction, calculated for disordered wires with length $L = 2\,\mu m$, Rashba coefficient $\alpha_R = 4.8 \times 10^{-12}$ eVm, and mean free path $l_e = 1.33\,\mu m$. From top to bottom, curves correspond to $W = 580$, 420, 220, and 180 nm. Indicated values for l_{so} were estimated using equation (8.69). Conductances were obtained by averaging over 50 different configurations of random bulk impurities and 50 slightly different energy values [144].

For the narrow wire limit $W\Delta k_R \ll 1$, with the Rashba wave vector Δk_R defined by equation (7.7), the following analytical formula has been derived by Kettemann

[148]:

$$\Delta\sigma = \frac{e^2}{h} \sqrt{B_W} \left(\frac{1}{\sqrt{B_\varphi + B^*(W)/4}} - \frac{1}{\sqrt{B_\varphi + B^*(W)/4 + B_s(W)}} \right.$$

$$\left. - 2\frac{1}{\sqrt{B_\varphi + B^*(W)/4 + B_s(W)/2}} \right). \tag{8.70}$$

Here, an effective external magnetic field is defined by

$$B^*(W) = \left[1 - \frac{1}{1 + W^2/3l_B^2} \right] B. \tag{8.71}$$

Furthermore, the spin relaxation field B_{so} given by equation (8.63) is modified to account for the reduced width. In the limit $W < l_{so}$ it is given by

$$B_s(W) = \frac{1}{12} \left(\frac{W}{l_R} \right)^2 B_{so}, \tag{8.72}$$

where l_R is the spin precession length defined by equation (7.15).

8.4 Spin interference effects in ring structures

In this section the spin interference effects in ring-like structures are treated. We will start with introducing the Berry phase, which gives an additional contribution to the phase the electron acquires in closed loops. In the second part the consequences for magnetotransport measurement on ring arrays will be discussed.

8.4.1 Berry phase

The Berry phase is an important parameter which needs to be included in the description of the spin transport in closed loops [149]. As explained in detail below a finite Berry phase appears if the electron spin propagates in a spatially varying magnetic field. This situation happens in the presence of spin-orbit coupling, where the orientation of the effective magnetic field depends on the direction of motion. Here we first introduce the general concept, and later on we will illustrate the Berry phase using a simple model system.

Let us consider a physical systems with a Hamiltonian that varies in time through a physical parameter $\vec{R}(t)$. Generally $\vec{R}(t)$ can be understood as a vector in parameter space, which can contain the wave vector, an external magnetic field or a magnetic flux. It should thus not be confused with the position vector. It is assumed that $\vec{R}(t)$

is varied slowly in time compared to the smallest energy scale in the system, i. e. we restrict ourself to the adiabatic limit. The Schrödinger equation for that system is

$$H(\vec{R(t)})|n(\vec{R}(t))\rangle = E_n(\vec{R}(t))|n(\vec{R}(t))\rangle, \tag{8.73}$$

where $|n(\vec{R}(t))\rangle$ is the instantaneous n^{th} eigenstate with eigen energy E_n. We assume that the eigenvalues are nondegenerate. Just to give an example, the nth eigenstate can be the nth band of a band structures. However, although the systems remains in the same band the wave vector as the parameter $\vec{R}(t)$ can vary resulting in a corresponding variation of the eigen energy E_n. We are interested in how the system evolves on a trajectory C in parameter space described by $\vec{R}(t)$ as t varies. It is important to realized that the phase is not really determined by the Schrödinger equation (8.73), i. e. if $|n(\vec{R}(t))\rangle$ is an eigenstate, then $\exp(i\varphi(\vec{R})(t))|n(\vec{R}(t))\rangle$ is also an eigenstate. Starting at the initial state at $t = 0$ given by $|n(\vec{R}(0))\rangle$, considering an adiabatic time evolution the system will remain in an instantaneous eigenstate of $H(\vec{R}(t))$ up to a phase. More explicitly the initial states evolves into

$$|n(\vec{R}(0))\rangle \rightarrow e^{i\varphi(t)}|n(\vec{R}(t))\rangle. \tag{8.74}$$

The value of $\varphi(t)$ can be determined by inserting the wave function $e^{i\varphi(t)}|n(\vec{R}(t))\rangle$ into the Schrödinger equation

$$H(\vec{R}(t))e^{i\varphi(t)}|n(\vec{R}(t))\rangle = i\hbar\frac{d}{dt}e^{i\varphi(t)}|n(\vec{R}(t))\rangle. \tag{8.75}$$

By making use of

$$H(\vec{R}(t))|n(\vec{R}(t))\rangle = E_n(\vec{R}(t))|n(\vec{R}(t))\rangle, \tag{8.76}$$

we can write

$$E_n(\vec{R}(t))|n(\vec{R}(t))\rangle = -\hbar\frac{d\varphi(t)}{dt}|n(\vec{R}(t))\rangle$$
$$+ i\hbar\frac{d}{dt}|n(\vec{R}(t))\rangle. \tag{8.77}$$

Calculating the scalar product with $\langle n(\vec{R}(t))|$ gives

$$E_n(\vec{R}(t)) - i\hbar\left\langle n(\vec{R}(t))\left|\frac{d}{dt}\right|n(\vec{R}(t))\right\rangle = -\hbar\frac{d\varphi(t)}{dt}. \tag{8.78}$$

By integrating up to t we can extract the solution for the phase $\varphi(t)$:

$$\varphi(t) = -\frac{1}{\hbar}\int_0^t E_n(\vec{R}(t'))\,dt' + i\int_0^t\left\langle n(\vec{R}(t'))\left|\frac{d}{dt'}\right|n(\vec{R}(t'))\right\rangle dt'.$$

Here the first integral is the conventional dynamical phase originating from the Hamilton evolution in time, while the second integral is the Berry phase:

$$\gamma_n = i \int_0^t \left\langle n(\vec{R}(t')) \left| \frac{d}{dt'} \right| n(\vec{R}(t')) \right\rangle dt'.\tag{8.79}$$

Since for a given time $R(t)$ is well associated with t, we can replace the integral over the time by an integral over the path C in the parameter space:

$$\gamma_n[C] = i \int_C \left\langle n(\vec{R}(t')) \left| \frac{d\vec{R}(t')}{dt'} \vec{\nabla}_{\vec{R}} \right| n(\vec{R}(t')) \right\rangle dt'$$

$$= \int_C i \langle n(\vec{R}) | \vec{\nabla}_{\vec{R}} | n(\vec{R}) \rangle \, d\vec{R}.\tag{8.80}$$

The formula for the Berry phase has the same structure as the one for the phase accumulation due to a vector potential $\vec{A}$ along a path C, as we discussed in connection with the Aharonov–Bohm effect in Section 8.2.2. Therefore, we define the integrand of equation (8.80) as the Berry vector potential or Berry connection:

$$\vec{\mathcal{A}}_n(\vec{R}) = i \langle n(\vec{R}) | \vec{\nabla}_{\vec{R}} | n(\vec{R}) \rangle.\tag{8.81}$$

We can even go a step further and in analogy to the magnetic field $\vec{B} = \vec{\nabla} \times \vec{A}$ define the Berry curvature as

$$\vec{\mathcal{B}}_n(\vec{R}) = \vec{\nabla}_{\vec{R}} \times \vec{\mathcal{A}}_n(\vec{R}).\tag{8.82}$$

We are particularly interested in closed-loop systems, where the starting and final points of path C are identical. The integral leading to the Berry phase is then written as

$$\gamma_n[C] = \oint_C i \langle n(\vec{R}) | \vec{\nabla}_{\vec{R}} | n(\vec{R}) \rangle d\vec{R}.\tag{8.83}$$

Applying Stokes' theorem, the integral along the closed loop can be expressed as a surface integral:

$$\gamma_n[C] = \int_F d\vec{f} \vec{\mathcal{B}}_n(\vec{R}),\tag{8.84}$$

with $\vec{f}$ being the area in parameter space. Obviously, this expression is in close analogy to the magnetic flux. The Berry connection $\vec{\mathcal{A}}_n(\vec{R})$ is not gauge invariant, similar to the case of the vector potential $\vec{A}$. Thus, if we transform

$$|n(\vec{R}(t))\rangle \rightarrow e^{i\chi(\vec{R})} |n(\vec{R}(t))\rangle,\tag{8.85}$$

for any $\chi(\vec{R})$, the Berry connection transforms as

$$\vec{A}_n(\vec{R}) \rightarrow \vec{A}_n(\vec{R}) - \frac{\partial \chi(\vec{R})}{\partial \vec{R}}. \qquad (8.86)$$

Apart from some exceptions it is usually not possible to perform a gauge transformation, so that finally the Berry phase cancels out. The latter would mean that the Berry phase is physically irrelevant. This can be most easily recognized by considering a closed path:

$$\gamma_n \rightarrow \gamma_n - \oint_C \frac{\partial \chi(\vec{R})}{d\vec{R}} d\vec{R}$$

$$\gamma_n - [\chi(\vec{R}(t)) - \chi(\vec{R}(0))], \qquad (8.87)$$

since for a closed loop $|n(\vec{R}(t))\rangle = |n(\vec{R}(0))\rangle$, owing to the condition that the eigenvalue is single valued. This implies that the phase χ in equation (8.85), accumulated after propagation along a closed loop, can only be a multiple of 2π. Therefore, we finally obtain

$$\gamma_n \rightarrow \gamma_n - 2\pi n, \qquad (8.88)$$

with n an integer. As long as γ_n is not a multiple of 2π, it cannot be cancelled. Thus, the Berry phase is physically relevant and needs to be considered to describe a system.

As a simple example we consider a spin-1/2 particle in a magnetic field, which rotates about the z-axis with a fixed polar angle θ, as depicted in Figure 8.18.

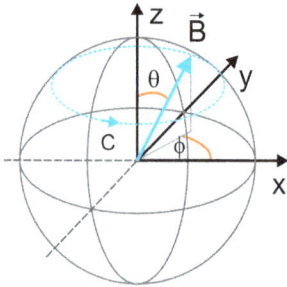

Figure 8.18: External magnetic field rotating about the z-axis.

In this case the parameter $\vec{R}$ acting on the spin is the magnetic field $\vec{B}$, which is described by

$$\vec{B} = B_0 \begin{pmatrix} \sin\theta\cos\phi \\ \sin\theta\sin\phi \\ \cos\theta \end{pmatrix}. \qquad (8.89)$$

As illustrated in Figure 8.18, the magnetic field rotates along a closed path C, by varying the azimuthal angle ϕ. The corresponding Hamiltonian for a spin-1/2 particle equation (3.29) was already introduced in Chapter 3, with the eigenfunctions

$$|n_+\rangle = \begin{pmatrix} \cos\frac{\theta}{2} \\ e^{i\phi}\sin\frac{\theta}{2} \end{pmatrix} \tag{8.90}$$

and

$$|n_-\rangle = \begin{pmatrix} -\sin\frac{\theta}{2} \\ e^{i\phi}\cos\frac{\theta}{2} \end{pmatrix}. \tag{8.91}$$

The corresponding eigenvalues are $E_\pm = \pm\frac{1}{2}g\mu_B B_0$. We have exactly a situation which was introduced in a more abstract way when the Berry phase was defined. By rotating the magnetic field about the z-axis, the spin orientation, i. e. the eigenstate, follows adiabatically the magnetic field orientation. With respect to the instantaneous external magnetic field corresponding to the parameter $\vec{R}$, the spin-up and -down state, is well defined. In order to get the Berry connection $\vec{A}_+$ we have to calculate

$$\vec{\nabla}|n_\pm\rangle = \left(\frac{\partial}{\partial r}\vec{e}_r + \frac{1}{r}\frac{\partial}{\partial\theta}\vec{e}_\theta + \frac{1}{r\sin\theta}\frac{\partial}{\partial\phi}\vec{e}_\phi \right)|n_\pm\rangle, \tag{8.92}$$

with the "radius" in parameter space given by the magnetic field $r = B_0$. This gives for the eigenstates $|n_\pm\rangle$

$$\vec{\nabla}|n_+\rangle = \frac{1}{B_0}\begin{pmatrix} -\frac{1}{2}\sin\frac{\theta}{2} \\ \frac{1}{2}e^{i\phi}\cos\frac{\theta}{2} \end{pmatrix}\vec{e}_\theta + \frac{1}{B_0\sin\theta}\begin{pmatrix} 0 \\ ie^{i\phi}\sin\frac{\theta}{2} \end{pmatrix}\vec{e}_\phi \tag{8.93}$$

and

$$\vec{\nabla}|n_-\rangle = \frac{1}{B_0}\begin{pmatrix} -\frac{1}{2}\cos\frac{\theta}{2} \\ -\frac{1}{2}e^{i\phi}\sin\frac{\theta}{2} \end{pmatrix}\vec{e}_\theta + \frac{1}{B_0\sin\theta}\begin{pmatrix} 0 \\ ie^{i\phi}\cos\frac{\theta}{2} \end{pmatrix}\vec{e}_\phi. \tag{8.94}$$

By calculating the scalar product with $\langle n_\pm|$ the Berry connections $\vec{A}_\pm$

$$\vec{A}_+ = i\langle n_+|\vec{\nabla}|n_+\rangle = -\frac{\sin^2(\frac{\theta}{2})}{B_0\sin\theta}\vec{e}_\phi \tag{8.95}$$

and

$$\vec{A}_- = i\langle n_-|\vec{\nabla}|n_-\rangle = -\frac{\cos^2(\frac{\theta}{2})}{B_0\sin\theta}\vec{e}_\phi \tag{8.96}$$

are obtained. Integration along curve C over ϕ from 0 to 2π, with $r = B_0$ and θ being constant gives for the Berry phase

$$\gamma_\pm = \oint_C \vec{A}_\pm B_0 \sin\theta d\phi \vec{e}_\phi$$

$$= \oint_C i\langle n_\pm|\vec{\nabla}|n_\pm\rangle B_0 \sin\theta d\phi \vec{e}_\phi$$

$$= -\pi(1 \mp \cos\theta), \tag{8.97}$$

where we made use of $\sin^2(\theta/2) = (1 - \cos\theta)/2$ and $\cos^2(\theta/2) = (1 + \cos\theta)/2$. The Berry phase can also be expressed in terms of the solid angle

$$\Omega = \int_0^{2\pi} (1 - \cos\theta)\, d\phi = 2\pi(1 - \cos\theta) \tag{8.98}$$

by

$$\gamma_+ = -\frac{1}{2}\Omega(C). \tag{8.99}$$

After introducing the general concept of the Berry phase, in the next section we will discuss how the Berry phase affects the interference in a ring-shaped spintronic device.

8.4.2 Spin-interference in a ring with Rashba spin-orbit coupling

From the weak antilocalization effect we learned that the presence of spin-orbit coupling modifies the electron interference. Instead of a resistance peak at zero field, in case of weak localization, a resistance dip evolves. However, no real control of the interference was obtained. This can be achieved by employing a ring structure, as depicted in Figure 8.19.

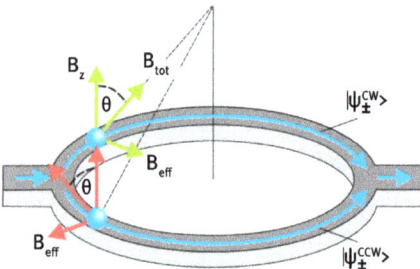

Figure 8.19: Clockwise (cw) and counter-clockwise (ccw) propagation of partial waves in a ring structure in the presence of Rashba spin-orbit coupling. The effective spin-orbit field B_{eff} lies within the plane of the ring. Its direction depends on the direction of motion. The external field B_z is applied along the z-axis. The total field B_{tot} is tilted by θ with respect to the normal.

Here, the electron partial waves propagating along both branches of the ring interfere. As for the Aharonov–Bohm effect (cf. Section 8.2.2), the interference has a direct impact on the conductance through the ring. The idea of the present device is that the interference is not controlled by the magnetic flux Φ penetrating the ring, but rather by the strength of Rashba spin-orbit coupling [150]. As we know from the previous chapter, the strength of the Rashba effect, quantified by the coupling parameter α_R, can be adjusted by means of a gate. Below we will first describe the system by setting up an appropriate Hamiltonian including the wave function propagation along the ring.

As depicted in Figure 8.19, the partial waves in the ring can propagate in clockwise (cw) and counter-clockwise (ccw) direction. The effective magnetic field B_{eff} due to the Rashba spin-orbit coupling lies within the plane of the ring and is perpendicular to the direction of motion and the internal electric field. We assume that the ring is formed in a two-dimensional electron gas in a semiconductor heterostructure so that the macroscopic electric field is normal to the ring. In addition, an external perpendicular magnetic field B_z is applied. The Hamiltonian representing this scenario can be expressed as [150, 151, 152]

$$\hat{H} = \frac{\hbar^2}{2mr_0^2}\left(-i\frac{\partial}{\partial\phi} + \frac{\Phi}{\Phi_0}\right)^2 + \frac{1}{2}g\mu_B B_z\hat{\sigma}_z$$

$$+ \frac{\alpha_R}{r_0}\hat{\sigma}_r\left(-i\frac{\partial}{\partial\phi} + \frac{\Phi}{\Phi_0}\right) - i\frac{\alpha_R}{2r_0}\hat{\sigma}_\phi. \tag{8.100}$$

The general structure has some similarities to the Hamiltonian of the tubular conductor discussed in Section 7.7.3. The first term is the kinetic term, with Φ the magnetic flux in the ring, induced by B_z. We assumed that the electrons are confined in a ring with a fixed radius of r_0. The second term in the Hamiltonian is the Zeeman contribution. The last two terms are due to the Rashba effect. Since the macroscopic electric field is oriented along the z-axis, the radial contribution $\hat{\sigma}_r$ (cf. equation (7.74)) is relevant in the third term. This is different from the tubular conductor, where the electric field points outwards. The last term is due to the fact that the electrons are confined within the ring [151].

The total magnetic field B_{tot} the electrons are exposed to is given by the sum of the external magnetic field B_z and the effective magnetic field due to the Rashba effect B_{eff}. As illustrated in Figure 8.19 the tilt angle θ of B_{tot} with respect to the z-axis is given by $\tan\theta = B_{eff}/B_z = \langle\hat{H}_R\rangle/\langle\hat{H}_Z\rangle$. As for the tubular conductor the total angular momentum is conserved, therefore, the eigenfunctions can be expressed in the form

$$\psi = e^{il\phi}\begin{pmatrix} \chi_1 \\ e^{i\phi}\chi_2 \end{pmatrix}. \tag{8.101}$$

Here, l is the orbital angular momentum quantum number. The $\exp(i\phi)$ factor in the second component of the spinor lifts the angular momentum by one, which ensures that for the spin-down component the total angular momentum is the same as for the spin-up component. We are mostly interested to operate the ring-interference device at zero external magnetic field $B_z = 0$, which implies $\Phi/\Phi_0 = 0$ and $\langle \hat{H}_z \rangle = 0$. In this case the magnetic field lies in the plane of the ring $\theta = \pi/2$. Applying the Hamiltonian equation (8.100) to the wavefunction given by equation (8.101) and performing the derivatives with respect to ϕ results in the following matrix equation for the spinor [152]:

$$
\begin{pmatrix} \frac{\hbar^2}{2mr_0^2}l^2 & \frac{\alpha_R}{r_0}(l+\frac{1}{2}) \\ \frac{\alpha_R}{r_0}(l+\frac{1}{2}) & \frac{\hbar^2}{2mr_0^2}(l+1)^2 \end{pmatrix} \begin{pmatrix} \chi_1 \\ \chi_2 \end{pmatrix} = E_{\pm}^\lambda \begin{pmatrix} \chi_1 \\ \chi_2 \end{pmatrix}, \tag{8.102}
$$

with $\lambda = $ ccw and cw for counter-clockwise ($l > 0$) and clockwise ($l < 0$) propagation, respectively. For the above equation the phase factors $\exp(\mp i\phi)$ in the off-diagonal elements of the Hamiltionian were removed by replacing the spinor $(\chi_1, e^{i\phi}\chi_2)^T$ by $(\chi_1, \chi_2)^T$. The normalized eigenfunctions are given by

$$
\begin{pmatrix} \chi_1 \\ \chi_2 \end{pmatrix} = \frac{1}{\sqrt{1 + \Gamma_\pm^\lambda}} \begin{pmatrix} 1 \\ \Gamma_\pm^\lambda \end{pmatrix}, \tag{8.103}
$$

with

$$
\Gamma_\pm^\lambda = \frac{E_\pm^\lambda - \frac{\hbar^2}{2m^*r_0^2}l^2}{\frac{\alpha_R}{r_0}(l+1/2)}. \tag{8.104}
$$

The energy eigenvalues are

$$
E_\pm^\lambda = \frac{\hbar^2}{2mr_0^2}\left\{ \left[\left(l+\frac{1}{2}\right)^2 + \frac{1}{4} \right] \pm \sqrt{\left(l+\frac{1}{2}\right)^2 \left[1 + \left(\frac{2m^*\alpha_R r_0}{\hbar^2}\right)^2 \right]} \right\}. \tag{8.105}
$$

As an example we consider a counter-clockwise spin-up state (+). If we abbreviate $Q_R = 2m^*\alpha_R r_0/\hbar^2$ the component of the spinor can be written as

$$
\chi_1 = \frac{1}{\sqrt{2}} \frac{Q_R}{(1 + \sqrt{1 + Q_R^2} + Q_R^2)^{1/2}},
$$

$$
\chi_2 = \frac{1}{\sqrt{2}} \frac{1 + \sqrt{1 + Q_R^2}}{(1 + \sqrt{1 + Q_R^2} + Q_R^2)^{1/2}}. \tag{8.106}
$$

We can transfer χ_1 and χ_2 to a more simple and intuitive form. For that we define the tilt angle ξ by $\tan\xi = Q_R$, so that the first component of the spinor can be written as

$$
\begin{aligned}
\chi_1 &= \frac{1}{\sqrt{2}} \frac{\tan\xi}{(1 + \sqrt{1 + \tan^2\xi} + \tan^2\xi)^{1/2}} \\
&= \frac{1}{\sqrt{2}} \frac{\tan\xi}{(1 + \frac{1}{\cos\xi} + \frac{\sin^2\xi}{\cos^2\xi})^{1/2}} \\
&= \frac{1}{\sqrt{2}} \frac{\sin\xi}{\sqrt{1 + \cos\xi}} \\
&= \sin\frac{\xi}{2}.
\end{aligned}
\tag{8.107}
$$

By calculating the second component in the same fashion, we finally arrive at

$$
|\psi_+^{\text{ccw}}\rangle = e^{il\phi} \begin{pmatrix} \sin\frac{\xi}{2} \\ e^{i\phi}\cos\frac{\xi}{2} \end{pmatrix}
\tag{8.108}
$$

for the spin-up eigenstate propagating in counter-clockwise direction. Here we rein-serted the phase factor $\exp(i\phi)$ in the second component of the spinor. The state $|\psi_+^{\text{ccw}}\rangle$ is an eigenvalue of the spin-operator $\sin\xi\hat{\sigma}_r - \cos\xi\hat{\sigma}_z$. Similarly, the remaining three eigenfunctions of the above Hamiltonian for clockwise (cw) and counter-clockwise (ccw) propagation for spin-up (+) and spin-down (−) are given by

$$
|\psi_+^{\text{cw}}\rangle = e^{-il\phi} \begin{pmatrix} \cos\frac{\xi}{2} \\ -e^{i\phi}\sin\frac{\xi}{2} \end{pmatrix},
\tag{8.109}
$$

$$
|\psi_-^{\text{ccw}}\rangle = e^{+il\phi} \begin{pmatrix} \cos\frac{\xi}{2} \\ -e^{i\phi}\sin\frac{\xi}{2} \end{pmatrix},
\tag{8.110}
$$

$$
|\psi_-^{\text{cw}}\rangle = e^{-il\phi} \begin{pmatrix} \sin\frac{\xi}{2} \\ e^{i\phi}\cos\frac{\xi}{2} \end{pmatrix},
\tag{8.111}
$$

with $l \geq 0$ being the angular momentum quantum number. One finds that counter-clockwise and clockwise states of equal spin are orthogonal.

As a next step, the conductance through the ring will be determined on the basis of the four eigenfunctions given above. According to equation (8.6) the probability for an electron transported through the ring is given by the square of the total amplitude. Thus, in case of a single non-spin-degenerate channel transport the conductance can be expressed as

$$
G = \frac{e^2}{h} |\psi_+^{\text{ccw}} + \psi_+^{\text{cw}} + \psi_-^{\text{ccw}} + \psi_-^{\text{cw}}|^2.
\tag{8.112}
$$

As mentioned above, spinors of equal spin states are orthogonal, and thus their interference is zero, whereas spinors of opposite spin and opposite propagation are parallel and thus contribute to the interference. Taking all interference contributions into account, one finally obtains for the conductance [152]

$$G = \frac{e^2}{h}\left\{1 + \cos\left[\pi\frac{2m^*\alpha_R r_0}{\hbar^2}\sin\xi - \pi(1 - \cos\xi)\right]\right\},\tag{8.113}$$

with $\tan\xi = 2m^*\alpha_R r_0/\hbar^2$. As a matter of fact, one can deduce from the expression for the conductance that it can be controlled very efficiently by controlling the Rashba coupling parameter. This can be realized by covering the complete ring structure by a gate electrode. By applying a gate voltage, the electric field and thus α_R can be controlled. In Figure 8.20 the conductance G is plotted as a function of the normalized Rashba parameter α_R. As can be seen here, G oscillates with increasing α_R. At larger spin-orbit coupling the tilt angle ξ approaches $\pi/2$. In that case equation (8.113) simplifies to

$$G \approx \frac{e^2}{h}\left[1 - \cos\left(\frac{2\pi m^*\alpha_R r_0}{\hbar^2}\right)\right],\tag{8.114}$$

whereas at smaller values of α_R the period shifts slightly towards larger values.

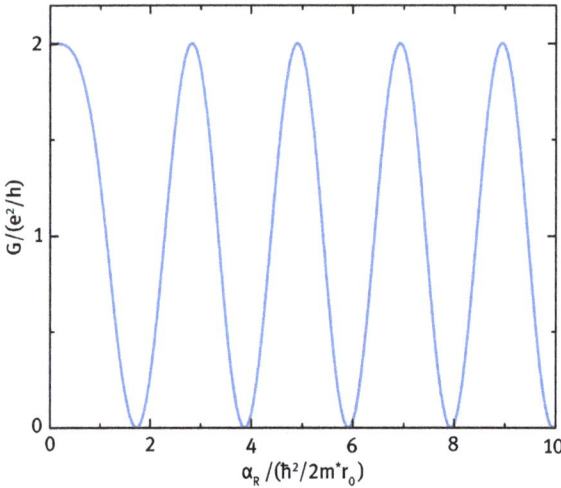

Figure 8.20: Calculated conductance G according to equation (8.113) as a function of $\alpha_R/(\hbar^2/2m^*r_0)$.

For the sake of simplicity we assumed the external magnetic field to be zero. However, if a magnetic field is applied, one obtains an additional phase shift between both arms of the ring due to the Aharonov–Bohm effect. The angle ξ is the tilt angle of spin state propagating along the ring structures. However, while changing ϕ the spin orientation changes. As we know from Section 8.4.1 where we discussed the Berry phase, this will lead to a phase accumulation. Indeed, the phase factor $-\pi(1-\cos\xi)$ in equation (8.113), the so-called the Aharonov–Anandan phase [153], originates from this phase accumulation. In the adiabatic limit, when the spin orientation is aligned to the direction of the effective magnetic field $\theta = \pi/2$, it corresponds to the Berry phase.

In Figure 8.21 a scanning electron beam image of a typical device structure is shown [154]. The pattern is defined by plasma etching an $In_{0.52}Al_{0.48}As/In_{0.53}Ga_{0.47}As/In_{0.52}Al_{0.48}As$ heterostructure. The structure consists of interconnected rings, in order to suppress random phase shifts of single rings. Thus, the obtained interference is of Altshuler–Aronov–Spivak type, i. e. the interference occurs between partial waves on time-reversed paths along the complete ring (cf. Section 8.2.3).

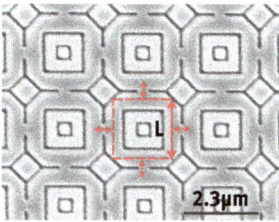

Figure 8.21: Scanning electron beam micrograph of a square-loop lattice. The structure was patterned into an $In_{0.52}Al_{0.48}As/In_{0.53}Ga_{0.47}As/In_{0.52}Al_{0.48}As$ heterostructure by plasma etching. The electron path of a closed loop is indicated by the red line, with L the side length. Figure provided by T. Koga, Hokkaido University, and J. Nitta, Tohoku University [154].

The corresponding magneto-transport measurements are shown in Figure 8.22 [154].

One finds that the conductivity oscillates with a period of $h/2e$ corresponding to the above-mentioned Alshuler–Aronov–Spivak oscillations. By changing the gate voltage the $h/2e$ period remains, however, the phase of the oscillations changes, i. e. from a conductivity minimum at zero field to a maximum and vice versa. Thus, σ_{2D} at $B = 0$ oscillates with increasing the gate voltage V_g. As is known for this kind of heterostructure, by varying V_g the strength of spin-orbit coupling and thus α_R is changed. This suggests that indeed G is controlled by interference effects due to adjusting the strength of spin-orbit coupling. In contrast to the case discussed theoretically, here the interference occurs between two time-reversed paths, therefore the formula given by equation (8.113) is not directly applicable here.

8.5 Summary

- At low temperatures electron interference effects contribute to conductance. The phase of the electron waves is affected by the vector potential.
- In ring structures regular magnetoconductance oscillations can be observed, e. g. due to the Aharonov–Bohm effect.

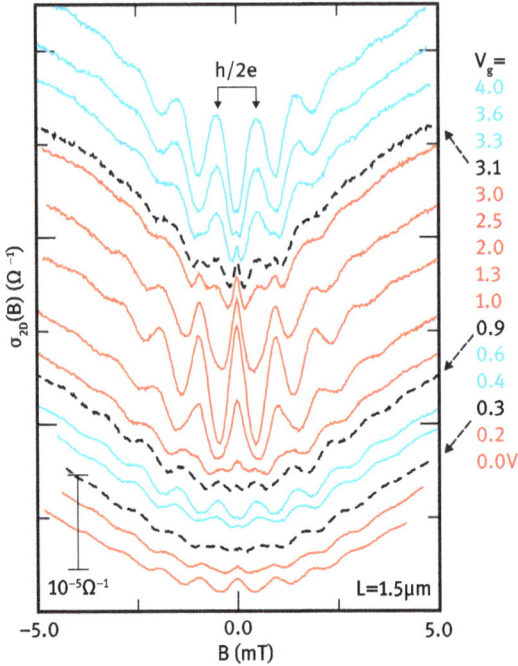

Figure 8.22: Sheet conductivities σ_{2D} as a function of the magnetic field B at different gate voltages for a square loop array sample with $L = 1.5$ μm. The plotted curves are offset for clarity. The range corresponding to a magnetic flux half quanta $h/2e$ piercing the square lattice is indicated. Red and blue curves correspond to measurements with a conductance maximum and minimum at $B = 0$, respectively. Figure provided by T. Koga, Hokkaido University, and J. Nitta, Tohoku University [154].

- In diffusive samples without spin-orbit coupling or spin scattering weak localization occurs at low temperatures, which results in a decrease of conductance at zero magnetic field.
- In the presence of spin-orbit coupling or spin scattering weak antilocalization is found, which gives a conductance peak at zero magnetic field. For this effect the change of the spin orientation during diffusive motion is included.
- In ring-shaped structures the interference pattern can be controlled by means of a gate electrode via the Rashba effect.

Exercises

Problem 8.1. Show that if $\chi = (\chi_1, e^{i\phi}\chi_2)^T$ is an eigenstate of

$$\begin{pmatrix} \alpha & e^{-i\phi}\beta \\ e^{i\phi}\beta & \alpha \end{pmatrix},$$

then $\tilde{\chi} = (\chi_1, \chi_2)^T$ is an eigenstate of

$$\begin{pmatrix} \alpha & \beta \\ \beta & \alpha \end{pmatrix}.$$

This relation was used to set-up the Schrödinger equation (8.102) for the spinor in a ring-shaped conductor.

Problem 8.2. Calculate the spin expectation value of the state

$$|\psi_+^{ccw}\rangle = e^{il\phi} \begin{pmatrix} \sin\frac{\xi}{2} \\ e^{i\phi} \cos\frac{\xi}{2} \end{pmatrix}$$

in radial, tangential, and z-direction, i. e.

$$\langle\psi_+^{ccw}|\hat{\sigma}_r|\psi_+^{ccw}\rangle, \quad \langle\psi_+^{ccw}|\hat{\sigma}_\phi|\psi_+^{ccw}\rangle, \quad \text{and} \quad \langle\psi_+^{ccw}|\hat{\sigma}_z|\psi_+^{ccw}\rangle,$$

respectively.

Problem 8.3. Show explicitly that the spin rotation matrices $\hat{R}_z(\alpha)$ and $\hat{R}_y(\theta)$ do not commute. The corresponding matrices are given by

$$\begin{pmatrix} e^{i\alpha/2} & 0 \\ 0 & e^{-i\alpha/2} \end{pmatrix} \quad \text{and} \quad \begin{pmatrix} \cos\theta/2 & \sin\theta/2 \\ -\sin\theta/2 & \cos\theta/2 \end{pmatrix}, \tag{8.115}$$

respectively.

Problem 8.4. We assume a ring-shaped conductor formed in a semiconductor heterostructure. In the presence of Rasbha effect the effective magnetic field B_{eff} will point outwards of the ring within the plane of the two-dimensional electron gas. The magnetic field is perpendicular to the direction of motion. We assume the adiabatic limit. Calculate the Berry phase the electron acquires along a full round in counterclockwise direction.

9 Spin Hall effect

9.1 Introductory remarks

As we learned in the previous chapters, for many spintronic device applications injection of spin-polarized carriers is required. There are various methods to achieve this. In Chapter 6 the electrical spin injection from a magnetic electrode was discussed. Alternatively, spin-polarized carriers can be generated by utilizing circularly polarized light. In this chapter we will discuss another very elegant method, where the so-called spin Hall effect is used. The basic phenomenon is shown in Figure 9.1 (a). At zero magnetic field a current is driven through a conducting bar. It is observed that on one side perpendicularly to the current flow the spins are polarized in up direction, while on the opposite side the spins are polarized in down direction.

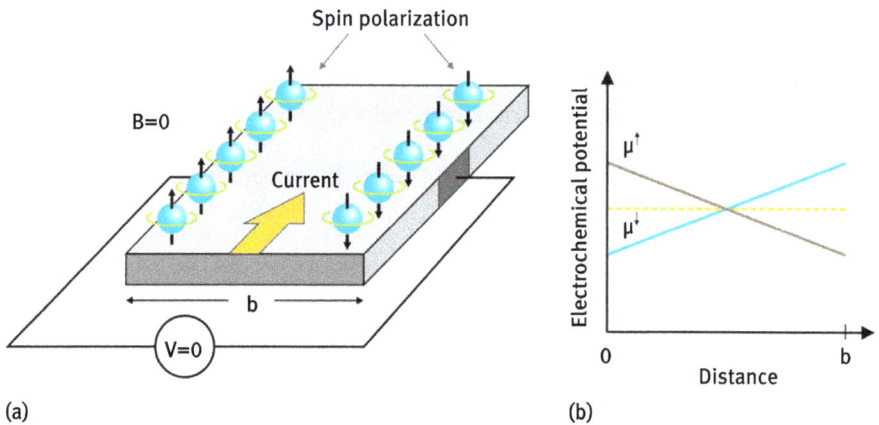

Figure 9.1: (a) Schematic illustration of the spin Hall effect. At zero magnetic field an electrical current flows through a conductor. On one side of the conductor spin-up electrons are accumulated, while on the opposite side spin-down electrons are found. (b) Electrochemical potential $\mu^\uparrow$ and $\mu^\downarrow$ across the sample for spin-up and spin-down electrons, respectively.

The set-up of the spin Hall effect measurement is very similar to the well-known set-up of the Hall effect, which we discussed before in Section 4.4. The only difference is that in case of the Hall effect a magnetic field is applied so that the carriers are deflected due to the Lorentz force and accumulated at the boundary of the sample. In case of the spin Hall effect, the carriers experience a spin dependent scattering, which results in an accumulation of spin-polarized carriers at the boundary.

https://doi.org/10.1515/9783110639001-009

9.2 Basic phenomena

The spin Hall effect was predicted in 1971 by the Russian physicists D'yakonov und Perel' [155]. In 1999 it was re-discovered by Hirsch [156]. As shown in Figure 9.1 (a), a spin polarization is observed on each side of the conductor but no voltage drop occurs. As long as no external magnetic field is applied, the average electro-chemical potential remains constant across the conductor. Thus, no electrical Hall voltage builds up. The fact that the net electrochemical potential is constant does not mean that the corresponding potentials for each spin orientation are constant. In fact, as shown in Figure 9.1 (b) the accumulation of a certain spin orientation on one side implies that the electrochemical potential is raised. However, since the spins of the opposite orientation are depleted on this side, the electrochemical potential is lowered correspondingly. Thus, in total the electrochemical potential remains constant.

Similarly to the Hall effect discussed in Section 4.4, the spin Hall effect originates from the deflection of propagating carriers; however in this case spin-dependent. In principle one can distinguish between two mechanisms, both of which originate from spin-orbit coupling.

- *Extrinsic spin Hall effect.* The scattering at impurities leads to a spin-dependent scattering [157]. The corresponding Hamiltonian for the extrinsic spin Hall effect can be expressed as

$$\hat{H}_{\text{ext}} = \lambda \hat{\vec{\sigma}} \cdot (\vec{k} \times \vec{\nabla}V), \tag{9.1}$$

with V being the impurity potential, $\vec{k}$ the wave vector of the propagation charge carrier and $\hat{\vec{\sigma}}$ the spin operator. The prefactor λ quantifies the strength of the scattering. The name extrinsic spin Hall effect originates from the fact that gradients of impurity potentials are responsible for the spin polarization rather than intrinsic spin-orbit coupling effects present in pure materials.

- *Intrinsic spin Hall effect.* In this case the intrinsic properties of the material, e. g. Rashba spin-orbit coupling, lead to spin-dependent scattering of the carriers [158]. The Hamiltonian describing this situation can be written as

$$\hat{H}_{\text{int}} = -\frac{1}{2}\mu_{\text{B}}\vec{B}_{\text{eff}}(\vec{k}) \cdot \hat{\vec{\sigma}}. \tag{9.2}$$

Here, $\vec{B}_{\text{eff}}$ is the effective magnetic field due to spin-orbit coupling.

As illustrated in Figure 9.2, there are two mechanisms connected to the deflection of electrons in the extrinsic spin Hall effect. The first one is skew scattering [155, 156]. When a carrier scatters at an impurity potential V, the scattering cross section depends on the spin state [159].

This mechanism, known as Mott skew scattering [160], can be described by the Hamiltonian given in equation (9.1). The skew scattering mechanism cannot be easily derived in a simple straightforward way; it only appears in the third-order Born ap-

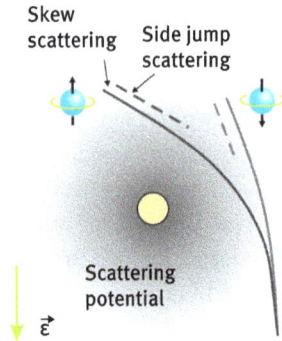

Figure 9.2: Spin-dependent scattering at an impurity. There are two mechanisms: First, the skew scattering and second, the side jump scattering. Figure adapted from Engel et al. [157].

proximation in scattering theory. The second mechanism is the side jump scattering [161], which describes the lateral displacement of the wave function during the scattering event (cf. Figure 9.2).

9.3 Boltzmann equation and skew scattering

The spin Hall effect takes place in the diffusive transport regime, which can be described within the framework of the Boltzmann equation. The Boltzmann equation covers the situation that the carriers are driven by an electric field $\vec{\mathcal{E}}$ and being scattered by different mechanisms, e. g. ionized impurity scattering or phonon scattering. By means of the Boltzmann equation the corresponding nonequilibrium carrier distribution can be determined. Once this is known, other transport parameters, e. g. the current density, can be calculated. The form of the nonequilibrium distribution function is determined by the relevant scattering mechanisms. In most cases the scattering processes are independent of the spin orientation of the electrons. However, as we know from the previous section, spin-dependent scattering is required in order to observe the spin Hall effect. Here we first introduce the basic concept of the Boltzmann equation. After that, we shall have a closer look on the spin-dependent scattering processes.

9.3.1 Boltzmann equation

The Boltzmann equation describes the situation where the carrier distribution is modified by an external field and by scattering processes. Let us consider a number of particles dN in a phase space volume $d\vec{r}d\vec{k}$. At time t this number is given by

$$dN = f(\vec{r}, \vec{k}, t)\, d\vec{r}\, d\vec{k}. \tag{9.3}$$

In the absence of scattering, the position and momentum at time $t + dt$ will be given by $\vec{r} + \vec{v}dt$ and $\vec{k} - e\vec{\mathscr{E}}dt/\hbar$, with $\vec{v}$ being the velocity of the particles and $\vec{\mathscr{E}}$ an electric field. Without scattering, the number of particles in a volume is conserved:

$$f(\vec{r} + \vec{v}\,dt, \vec{k} - e\vec{\mathscr{E}}\,dt/\hbar, t + dt)\,d\vec{r}\,d\vec{k} = f(\vec{r}, \vec{k}, t)\,d\vec{r}\,d\vec{k}. \tag{9.4}$$

If scattering occurs, the number of particles in a phase space volume $d\vec{r}d\vec{k}$ changes according to

$$dN_{sc} = \left(\frac{\partial f}{\partial t}\right)_{sc} dt\,d\vec{r}\,d\vec{k} = f(\vec{r} + \vec{v}\,dt, \vec{k} - e\vec{\mathscr{E}}\,dt/\hbar, t + dt)\,d\vec{r}\,d\vec{k} - f(\vec{r}, \vec{k}, t)\,d\vec{r}\,d\vec{k}, \tag{9.5}$$

which results in

$$f(\vec{r} + \vec{v}\,dt, \vec{k} - e\vec{\mathscr{E}}\,dt/\hbar, t + dt) - f(\vec{r}, \vec{k}, t) = \left(\frac{\partial f}{\partial t}\right)_{sc} dt. \tag{9.6}$$

After linear expansion in dt, the Boltzmann equation

$$\frac{\partial f}{\partial t} + \vec{v}\,\vec{\nabla}_r f - \frac{e}{\hbar}\vec{\mathscr{E}}\vec{\nabla}_k f = \left(\frac{\partial f}{\partial t}\right)_{sc} \tag{9.7}$$

is obtained. We are interested in the case of a spatially homogeneous distribution $(\vec{\nabla}_r f = 0)$ in the steady state $(\partial f/\partial t = 0)$:

$$-\frac{e}{\hbar}\vec{\mathscr{E}}\vec{\nabla}_k f(\vec{k}) = \left(\frac{\partial f(\vec{k})}{\partial t}\right)_{sc}. \tag{9.8}$$

The main task for solving the Boltzmann equation is to include the relevant scattering processes on the right-hand side. By making use of the quantum mechanical transition probabilities

$$W_{\vec{k}'\vec{k}} \sim |\langle\vec{k}'|\hat{H}_{sc}|\vec{k}\rangle|^2, \tag{9.9}$$

with $\hat{H}_{sc}$ being the scattering Hamiltonian, this contribution can be written as

$$\left(\frac{\partial f(\vec{k})}{\partial t}\right)_{sc} = \sum_{\vec{k}'} W_{\vec{k}\vec{k}'}[1 - f(\vec{k})]f(\vec{k}') - W_{\vec{k}'\vec{k}}[1 - f(\vec{k}')]f(\vec{k}). \tag{9.10}$$

The second part of the sum considers the scattering from an occupied $\vec{k}$ state into an unoccupied $\vec{k}'$ state, while the first part describes the reverse process. Assuming detailed balance $W_{\vec{k}\vec{k}'} = W_{\vec{k}'\vec{k}}$ one can simplify the scattering contribution to

$$\left(\frac{\partial f(\vec{k})}{\partial t}\right)_{sc} = -\sum_{\vec{k}'} W_{\vec{k}\vec{k}'}[f(\vec{k}) - f(\vec{k}')]. \tag{9.11}$$

For elastic scattering, the energy is conserved during the scattering event, i. e. $E_{\vec{k}} = E_{\vec{k}'}$. Furthermore, it is more convenient to only consider the deviation from the equilibrium distribution f_0, i. e. the Fermi distribution:

$$f(\vec{k}) = f_0 + \delta f(\vec{k}), \tag{9.12}$$

so that one can write

$$\left(\frac{\partial f(\vec{k})}{\partial t}\right)_{sc} = -\sum_{\vec{k}'} W_{\vec{k}\vec{k}'}[\delta f(\vec{k}) - \delta f(\vec{k}')]\delta(E_{\vec{k}} - E_{\vec{k}'}). \tag{9.13}$$

Since the summation is only taken over $\vec{k}'$, one can write

$$\left(\frac{\partial f(\vec{k})}{\partial t}\right)_{sc} = -\delta f(\vec{k})\sum_{\vec{k}'} W_{\vec{k}\vec{k}'}\left[1 - \frac{\delta f(\vec{k}')}{\delta f(\vec{k})}\right]\delta(E_{\vec{k}} - E_{\vec{k}'}). \tag{9.14}$$

By assuming isotropic energy bands, where the energy only depends on the absolute value of $\vec{k}$, and assuming an isotropic scattering probability, i. e. $W_{\vec{k}\vec{k}'}$ depends only on the angle θ between $\vec{k}$ and $\vec{k}'$, one can express the sum in the expression given above by $1/\tau$, with τ being the relaxation time. Thus, in this approximation the scattering contribution is given by

$$\left(\frac{\partial f}{\partial t}\right)_{sc} = -\frac{\delta f(\vec{k})}{\tau}. \tag{9.15}$$

We can insert this relation in the Boltzmann equation for a spatially homogeneous distribution in the steady state equation (9.8):

$$f(\vec{k}) = f_0(\vec{k}) + \frac{e}{\hbar}\tau\vec{\mathcal{E}}\vec{\nabla}_{\vec{k}}f(\vec{k}). \tag{9.16}$$

As a next step we assume for the distribution in $\vec{\nabla}_{\vec{k}}f(\vec{k})$ only the equilibrium distribution f_0, which gives

$$f(\vec{k}) = f_0(\vec{k}) + \frac{e}{\hbar}\tau\vec{\mathcal{E}}\vec{\nabla}_{\vec{k}}f_0(\vec{k}). \tag{9.17}$$

We can interpret the right side of the equation as an expansion around $\vec{k}$, so that one can finally write

$$f(\vec{k}) = f_0\left(\vec{k} + \frac{e}{\hbar}\tau\vec{\mathcal{E}}\right). \tag{9.18}$$

As illustrated in Figure 9.3, the steady state distribution thus corresponds to an equilibrium distribution, which is shifted by $-(e/\hbar)\tau\vec{\mathcal{E}}$, due the effect of the external electric field $\vec{\mathcal{E}}$ and the scattering events, expressed by τ. This distribution can then for example be used to determine the current density in a conductor [24].

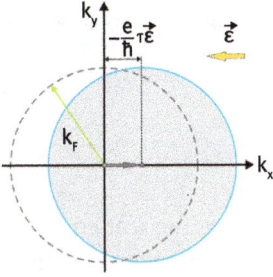

Figure 9.3: Shift of the electron distribution in k-space after applying an electric field $\vec{\mathcal{E}}$.

9.3.2 Intrinsic spin Hall effect

We can take the shift of the distribution function as depicted in Figure 9.3 as a basis to discuss the intrinsic spin Hall effect. According to equation (9.2), this effect relies on an effective magnetic field $\vec{B}_{\mathrm{eff}}$, which results from the spin-orbit coupling effect in the crystal, e. g. the Rashba effect. As we know from Section 7.3 the magnitude and orientation of $\vec{B}_{\mathrm{eff}}(\vec{k})$ depends on the k-vector. In case of the Rashba effect, the effective field is perpendicular to the direction of motion, as shown in Figure 9.4.

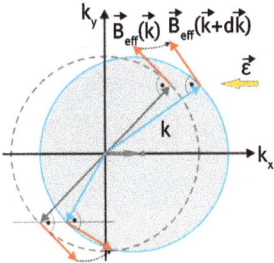

Figure 9.4: Schematic illustration of the intrinsic spin Hall effect. The application of an electric field induces a change of the k vectors. This results in a corresponding change of the effective magnetic field from $\vec{B}_{\mathrm{eff}}(\vec{k})$ to $\vec{B}_{\mathrm{eff}}(\vec{k} + d\vec{k})$. The change of the effective field forces the spins to adjust their orientation. On the opposite sides of the Fermi surface the change of spin orientation is in different directions. As a result a spin current develops. Figure adapted from Vignale [162].

The corresponding spin precesses about $\vec{B}_{\mathrm{eff}}(\vec{k})$ and will finally be aligned. By applying an external electric field $\vec{\mathcal{E}}$ the distribution function is shifted. The electrons are accelerated and change their k-vector from $\vec{k}$ to $\vec{k} + d\vec{k}$. Directly connected to this, the direction of the effective magnetic field is also changed, as illustrated in Figure 9.4. Thus, the electron spin is no longer aligned. As a consequence, the electron spins tilt away from their original orientation to align to the altered effective field $\vec{B}_{\mathrm{eff}}(\vec{k} + d\vec{k})$.

As one can infer from Figure 9.4, the tilting is in opposite directions on opposite sides of the Fermi surface.

9.3.3 Extrinsic spin Hall effect: skew scattering contribution

As long as spin-orbit coupling is neglected for electron scattering, the scattering rate from $\vec{k}$ to $\vec{k}'$ given by $W_{\vec{k}\vec{k}'}$ is symmetric with respect to the scattering angle θ. However, in case of spin-orbit coupling the scattering probability contains an asymmetric contribution [162, 163]

$$W_{\vec{k}\vec{k}',\sigma} = [W^{s}_{\check{k}\check{k}'} + \sigma(\check{k} \times \check{k}')_{z} W^{a}_{\check{k}\check{k}'}]\delta(E_{\vec{k}} - E_{\vec{k}'}). \tag{9.19}$$

Here we assume a two-dimensional system for simplicity, where $\check{k}$ and $\check{k}'$ are unit vectors on the direction of $\vec{k}$ to $\vec{k}'$, and $\sigma = \pm 1$ represents the two spin orientations perpendicular to the plane. The second part in the brackets reflects the contribution of the spin-orbit coupling. Note that $W^{s}_{\check{k}\check{k}'}$ and $W^{a}_{\check{k}\check{k}'}$ only depend on the magnitude of the vectors $\check{k}$ to $\check{k}'$, and that they are both symmetric under exchange of $\check{k}$ to $\check{k}'$. The antisymmetry is explicitly included in equation (9.19) by

$$(\check{k} \times \check{k}')_{z} = \sin\theta. \tag{9.20}$$

The symmetry relation between the spin and the scattering angle is illustrated in Figure 9.5. According to equation (9.19) the same scattering probability is expected for inversed spin if the scattering angle θ is inversed to $-\theta$.

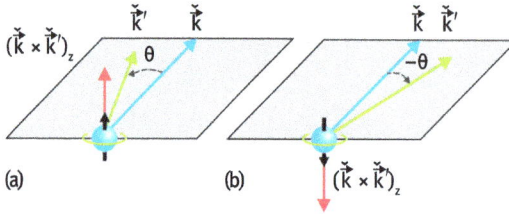

Figure 9.5: Symmetry relation with regard to spin and scattering angle θ of a skew scattering process. (a) and (b) for spin-up and spin-down electrons, respectively.

Based on the scattering probabilities the relaxation times can be determined in a similar way as expressed by equation (9.14), i. e. the sum over $\vec{k}'$. The sum can be transformed to an integral over the scattering angle θ, with the scattering probability $W(k, \theta)$ being only dependent on the scattering angle θ and the magnitude of the wave vector

$$W(k, \theta) = W^{s}(k, \theta) + \sigma W^{a}(k, \theta) \sin\theta. \tag{9.21}$$

From the symmetric contribution $W^s(k, \theta)$ of the scattering probability the elastic scattering time τ_e can be determined:

$$\frac{1}{\tau_e} = \frac{m^* \mathcal{A}}{4\pi^2 \hbar^2} \int_0^{2\pi} d\theta W^s(k_F, \theta)(1 - \cos \theta), \tag{9.22}$$

with $\mathcal{A}$ being the area of the two-dimensional electron gas. We assumed zero temperature so that one can replace k by the Fermi wave number k_F. From the asymmetric contribution $W^a(k, \theta)$ we obtain the antisymmetric scattering time:

$$\frac{1}{\tau_a} = \sigma \frac{m^* \mathcal{A}}{4\pi^2 \hbar^2} \int_0^{2\pi} d\theta W^a(k_F, \theta) \sin^2 \theta. \tag{9.23}$$

Taking a closer look to the integral one finds that, owing to the sine function, scattering to a perpendicular direction is contributing dominantly. Thus $1/\tau_a$ expresses the scattering processes perpendicular to the direction of the current, i. e. electric field. This pumping can be described by the action of a spin electric field:

$$\mathcal{E}_{s,y}^z = \rho_s j_{c,x}, \tag{9.24}$$

which is related to the electrical charge current $j_{c,x}$ along the x-direction by the resistivity

$$\rho_s = \frac{m^*}{n_{2D} e^2 \tau_a}. \tag{9.25}$$

Here, n_{2D} is the electron concentration of the two-dimensional electron gas. The spin electric field $\mathcal{E}_{s,y}^z$ is oriented along the y-direction. The spins are aligned along the z-direction, thus perpendicular to the current and to the spin electric field. The scattering time τ_a can be positive or negative, depending of the spin orientation σ. Thus, as illustrated in Figure 9.6, by reversing the spin orientation, the spin electric field is reversed as well, and the spins are scattered in opposite directions.

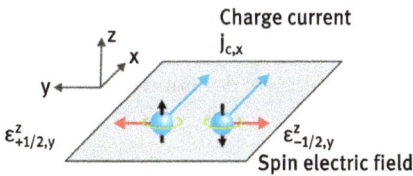

Figure 9.6: Relation between the charge current and the spin electric field for spins aligned along the z-axis.

Similar to the classical Hall effect, the spin of opposite orientations are accumulated at the left and right edge of the sample. Thus, the essence of the spin Hall effect is that the scattering probability has an asymmetric component, which leads to a spin-dependent scattering.

9.3.4 Skew scattering in a two-dimensional system

In order to be more specific, we address a two-dimensional system with scattering centers [163]. As depicted in Figure 9.7, a circular well potential of height V_0 is assumed, which can be described by

$$V(r) = V_0\theta(a - r) + \bar{a}aL_zS_z\delta(r - a)V_0, \tag{9.26}$$

with $\theta(x)$, the Heaviside function, being 0 for $x < 0$ and 1 for $x \geq 0$. The radius of the scattering center is a and $\bar{a}$ corresponds to $a\hbar/a^2$, with a being the spin-orbit coupling parameter. The orbital angular momentum and spin along the z-direction are $L_z = l$ and $S_z = \sigma$, respectively. Both are conserved. The δ-function in the second term of equation (9.26) is used to model the strong increase of the potential, corresponding to an effective electric field. The latter is responsible for the spin-orbit coupling.

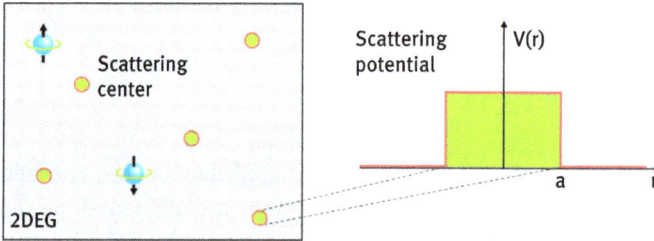

Figure 9.7: Two-dimensional system with circular potential barriers.

The wave function can be separated into a radial and orbital contribution:

$$\Psi_{kl\sigma}(r, \theta) = R_{kl\sigma}(r)e^{il\theta}. \tag{9.27}$$

By apply the Hamiltonian to $e^{il\theta}$ one arrives at the Schrödinger equation for the radial contribution

$$R''_{kl\sigma} + \frac{1}{r}R'_{kl\sigma} + \left(k^2 - v_0 - \frac{l^2}{r^2}\right)R_{kl\sigma} = 0, \quad r < 1 \tag{9.28}$$

and

$$R''_{kl\sigma} + \frac{1}{r}R'_{kl\sigma} + \left(k^2 - \frac{l^2}{r^2}\right)R_{kl\sigma} = 0, \quad r > 1. \tag{9.29}$$

In the above equations dimensionless parameters were used, i. e. r corresponds to r/a and k to ka. Furthermore, the parameter $v_0 = 2mV_0a^2/\hbar^2$ measures the barrier height. The corresponding solutions of the Schrödinger equation are

$$R_{kl\sigma}(r) = J_{|l|}(vr), \quad r < 1 \tag{9.30}$$

and

$$R_{kl\sigma}(r) = e^{i\delta_{l\sigma}}\left[\cos\delta_{l\sigma}J_{|l|}(kr) - \sin\delta_{l\sigma}Y_{|l|}(kr)\right] \quad r > 1, \tag{9.31}$$

with $J_{|l|}(kr)$ and $Y_{|l|}(kr)$ the Bessel function of the first and second kinds, $v = \sqrt{k^2 - v_0}$. The matching conditions lead to the following expression for the phase factor [163]:

$$\cot\delta_{l\sigma} = \frac{kY'_{|l|}(k) - \beta_{l\sigma}Y_{|l|}(k)}{kJ'_{|l|}(k) - \beta_{l\sigma}J_{|l|}(k)}, \tag{9.32}$$

with $\beta_{l\sigma} = vJ'_{|l|}(v)/J_{|l|}(v) + \bar{a}l\sigma v_0$. At large distances $r \to \infty$ the wave function can be written as

$$\Psi_{kl\sigma}(r,\theta) \sim \Psi^0_{kl\sigma}(r,\theta) + \frac{e^{2i\delta_{l\sigma}} - 1}{\sqrt{2\pi kr}}e^{i(kr - |l|\pi/2 - \pi/4)}e^{il\theta}, \tag{9.33}$$

with

$$\Psi^0_{kl\sigma}(r,\theta) = \sqrt{\frac{2}{\pi kr}}\cos(kr - |l|\pi/2 - \pi/4)e^{il\theta} \tag{9.34}$$

being the free wave function in the channel of angular momentum l. The scattering amplitude $f_\sigma(k,\theta)$ is the factor by which the outgoing wave $e^{ikr}/\sqrt{r}$ is multiplied in the above equation:

$$f_\sigma(k,\theta) = \sum_{l=-\infty}^{\infty}\frac{e^{2i\delta_{l\sigma}} - 1}{\sqrt{2\pi k}}e^{-i(|l|\pi/2+\pi/4)}e^{il\theta}. \tag{9.35}$$

The differential cross section is given by

$$\left(\frac{d\sigma_c}{d\theta}\right)_\sigma = |f_\sigma(k,\theta)|^2$$

$$= \frac{1}{2\pi k}\sum_{l,l'}(e^{2i\delta_{l\sigma}} - 1)(e^{-2i\delta_{l'\sigma}} - 1)e^{-i\pi/2(|l|-|l'|)}e^{i(l-l')\theta}. \tag{9.36}$$

The total scattering rate is related to the differential scattering cross section for a single impurity by

$$W(k,\theta) = W^s(k,\theta) + \sigma W^a(k,\theta)\sin\theta = n_i\frac{4\pi^2\hbar^3k}{m^2\mathcal{A}}\frac{d\sigma_c}{d\theta}, \tag{9.37}$$

with $n_i = N_i/\mathcal{A}$ being the areal density of impurities. Combining this with equation (9.36) one gets

$$W(k,\theta) = n_i \frac{2\pi^2\hbar^3 k}{m^2 \mathcal{A}} \sum_{l,l'} (e^{2i\delta_{l\sigma}} - 1)(e^{2i\delta_{l'\sigma}} - 1) i^{(|l'|-|l|)} e^{i(l-l')\theta}. \tag{9.38}$$

By using $e^{i(l-l')\theta} = \cos[(l-l')\theta] + i\sin[(l-l')\theta]$ and $e^{\pm 2i\delta_{l\sigma}} - 1 = \pm 2i/(\cot\delta_{l\sigma} \mp i)$ one can separate the symmetric and antisymmetric contributions of $W(k,\theta)$, so that one finally obtains for the scattering rates

$$W^s(k,\theta) = n_i \frac{8\pi^2\hbar^3}{m^2 \mathcal{A}} \sum_{l,l'} \frac{i^{|l'|-|l|} \cos[(l-l')\theta]}{(\cot\delta_{l\sigma} - i)(\cot\delta_{l'\sigma} + i)} \tag{9.39}$$

and

$$W^a(k,\theta) = \sigma n_i \frac{8\pi^2\hbar^3}{m^2 \mathcal{A}} \sum_{l,l'} \frac{i^{|l'|-|l|+1} \sin[(l-l')\theta]}{(\cot\delta_{l\sigma} - i)(\cot\delta_{l'\sigma} + i)}. \tag{9.40}$$

Let us now have a closer look at the properties of $W^s(k,\theta)$ and $W^a(k,\theta)$. The phase shifts in the above formulas have the symmetries

$$\delta_{-l,-\sigma}(\alpha) = \delta_{l,\sigma}(\alpha), \tag{9.41}$$

$$\delta_{-l,\sigma}(-\alpha) = \delta_{l,\sigma}(\alpha). \tag{9.42}$$

Since the sum over l and l' in equations (9.39) and (9.40) runs from $-\infty$ to ∞ one finds because of equation (9.41) that $W^s(k,\theta)$ and $W^a(k,\theta)$ are invariant under spin reversal $\sigma \to -\sigma$. The sign change in the sine function in equation (9.40) with changing the sign of θ is compensated by the σ-prefactor. Thus the asymmetry regarding the spin is solely given by the σ in equation (9.37). Furthermore, because of equation (9.42) the asymmetric contribution $W^a(k,\theta)$ changes sign with a change of sign of α. Thus, if the spin-orbit contribution is inversed, the spins are deflected to the opposite side. In the formula the reason for the sign change is that by reversing α the sign of l and l' in the sine-function changes.

9.4 Experiments on spin Hall effect in semiconductor layers

The presence of the spin Hall effect cannot be detected directly by electrical means, since no voltage drop is expected. However, by using the magneto-optical Kerr effect (MOKE) it is possible to obtain information on the spin polarization and thus the magnetization in a material, in case of the spin Hall effect on the spin accumulation at the sides of the conductor. As already explained in Section 6.9.3, for this method a linearly polarized laser beam is focused on the surface of the magnetized semiconductor layer.

The setup for measuring the spin Hall effect is shown in Figure 9.8 (a). The rotation of the polarization of the reflected beam is a measure of the out-of-plane magnetization. Thus, in case of the spin Hall effect it is expected that the polarization is rotated in opposite directions on both sides, as illustrated in Figure 9.8 (a).

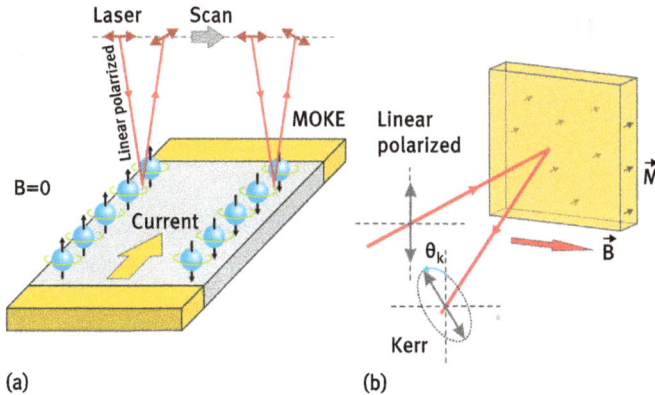

Figure 9.8: (a) Measurements of the spin accumulation at the edges of a semiconductor bar by means of the magneto-optical Kerr effect (MOKE). (b) Detail of the magneto-optical Kerr effect measurement. A linear polarized laser beam is focused on the sample. The magnetization $\vec{M}$ of the sample is aligned along the incoming linearly polarized beam. For the Kerr effect measurement the polarization angle θ_K of the reflected beam is measured.

A more detailed illustration of a magneto-optical Kerr effect measurement is given in Figure 9.8 (b). It can be demanding to precisely measure the Kerr angle, because it is relatively small and difficult to distinguish from spurious effects. In order to clearly single out the Kerr effect, an external magnetic field can be applied, which leads to a rotation of the spin orientation by precession. Thus, the Kerr angle is measured as a function of a magnetic field. The underlying mechanism is called the Hanle effect. The expected spin signal as a function of an external magnetic field $\vec{B}$ is shown in Figure 9.9 for different orientations of the initial magnetization with respect to the incoming laser beam.

Let us begin with the situation where the magnetization $\vec{M}$ is initially aligned along the laser beam (cf. situation (1) in Figure 9.9). At $\vec{B} = 0$ the net magnetization remains in its initial direction. In our case of the spin Hall effect spins are continuously generated in a certain direction, depending on the current direction. However, after a certain time, these spins also dephase, due to spin scattering. This leads to successively smaller contribution to the local net spin polarization $\langle S_z \rangle$ and thus to the magnetization, as illustrated by the series of red arrows in Figure 9.9. If an external magnetic field is applied, the spins generated by the spin Hall effect also start to precess. The longer the time after generation, the larger is the precession angle. At the

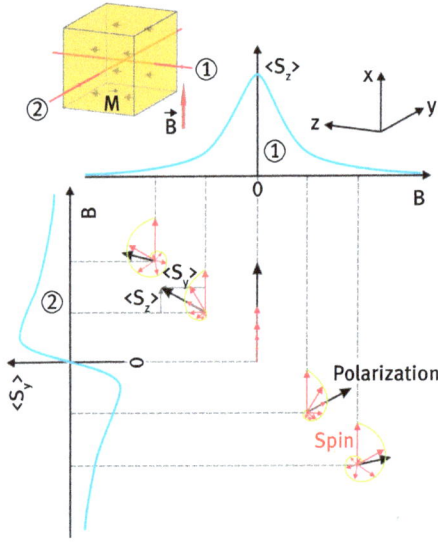

Figure 9.9: Illustration of the Hanle effect resulting in a depolarization of a spin ensemble with increasing external magnetic field. (1) For $\langle S_z \rangle$ measured by means of the magneto-optical Kerr effect the laser beam for spin detection is aligned to the magnetization $\vec{M}$ at $\vec{B} = 0$. (2) For $\langle S_y \rangle$ the laser beam is perpendicular to $\vec{M}$.

same time the spins dephase. As depicted in Figure 9.9, in total this leads to a spin distribution with successively smaller contributions the larger the precession angle is. As a result the magnetization connected to the net spin polarization $\langle S_z \rangle$ is rotated and reduced, compared to the case at $B = 0$. The larger the external magnetic field is the smaller is the net spin polarization. Note that only the component along the laser beam is detected. In total one expects a symmetric curve of $\langle S_z \rangle$ as a function of external magnetic field. Being more specific, the dependence of $\langle S_z \rangle$ on B can be expressed by a Lorentz curve

$$\langle S_z \rangle = \frac{1}{T_2^*} \int_0^\infty S_0 e^{-t/T_2^*} \cos(\omega_L t)\, dt = S_0 \frac{1}{1 + \frac{B^2}{B_{1/2}^2}}, \tag{9.43}$$

with T_2^* being the spin lifetime of the spin ensemble, $\omega_L = g\mu_B B/\hbar$ the Larmor frequency, and $B_{1/2}$ the half-width of the Lorentz curve. The latter is given by

$$B_{1/2} = \frac{\hbar}{g\mu_B T_2^*}. \tag{9.44}$$

Let us now address the case where the laser beam is oriented perpendicularly to the initial magnetization along the z-axis, as illustrated by situation (2) in Figure 9.9. Here, the spin distribution $\langle S_y \rangle$ is measured as a function of an external magnetic field. At

$B = 0$ no Kerr effect is detected, since the magnetization direction is perpendicular to the laser beam. As soon as a finite magnetic field is applied along the x-direction, the spins precess and a finite value of $\langle S_y \rangle$ is found. With increasing or decreasing magnetic field first a maximum or minimum is observed for $\langle S_y \rangle$, while for even larger magnetic field strength a continuous decrease of the spin polarization amplitude along the y-direction is expected. In contrast to the previous case, $\langle S_y \rangle$ vs. B is an antisymmetric Lorentz curve given by

$$
\langle S_y \rangle = \frac{1}{T_2^*} \int_0^\infty S_0 e^{t/T_2^*} \sin(\omega_L t)\, dt
$$

$$
= S_0 \frac{\dfrac{B}{B_{1/2}}}{1 + \dfrac{B^2}{B_{1/2}^2}}. \tag{9.45}
$$

After explaining the general concepts of magneto-optical measurements, we now return to the local detection of spin-polarized carriers generated by the spin Hall effect. In Figure 9.10 a schematic of a MOKE measurement of the spin Hall effect in a semiconductor layer, e. g. GaAs, is shown [164]. The MOKE signal is measured on two spots on opposite sides of the conductor. At zero external magnetic field a positive Kerr rotation is found on the left-side spot, while a negative value is observed on the right side. Thus, an out-of-plane spin polarization is detected in opposite orientations on each side of the GaAs bar, as expected for the spin Hall effect. When an increasing external in-plane magnetic field is applied, the Kerr rotation angle monotonously drops to zero. The reason for this behavior is that in the presence of an external magnetic field the spins are rotated from an out-of-plane to an in-plane direction as explained above.

The fact that the spins are only accumulated at the sidewalls of a GaAs bar was confirmed by measurement, where the laser beam was scanned across the sample from one sidewall to the opposite one [164]. At zero external field, a Kerr signal was observed at the edges. In the area in between the edges the Kerr rotation angle was zero. A complete scan of Kerr rotation angle on the surface of the GaAs bar confirmed that an out-of-plane spin polarization is only located at the edges, while no spin polarization is observed in between. The Kerr rotation observed in the measurements discussed above is assigned to the extrinsic spin Hall effect. The GaAs film is n-type doped and thus contains a large number of scattering centers. On these impurities the spin-dependent scattering occurs leading to a spin accumulation on each side of the bars.

9.5 Detection of the spin Hall effect by electroluminescence

Spin polarized carriers generated by the spin Hall effect can also be detected directly by means of circularly polarized light emitted from a light emitting diode (LED) structure [165]. The layer system used for that experiment is shown in Figure 9.11 (a).

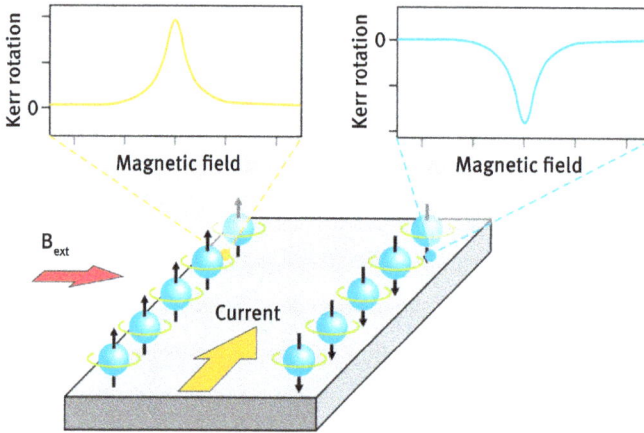

Figure 9.10: Illustration of the measurement of the spin Hall effect in a semiconductor layer. Schematic result of a measurement of the Kerr rotation angle as a function of an external in-plane magnetic field for a spot on the left and right edges, respectively.

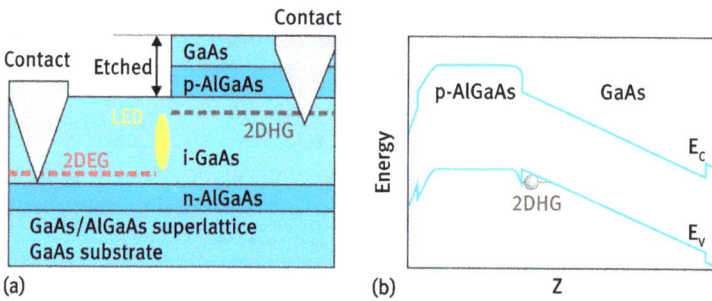

(a)   (b)

Figure 9.11: (a) Layer system with a two-dimensional hole gas (2DHG) formed in the nonetched area and a two-dimensional electron gas (2DEG) in the etched area. The light-emitting diode (LED) is formed where the two-dimensional hole gas and two-dimensional electron gas face each other. (b) Conduction and valence band profile of the nonetched layer system. At the p-AlGaAs/GaAs interface a two-dimensional hole gas is present. Figure adapted from Wunderlich et al. [165].

It consists of an AlGaAs/GaAs heterostructure with a bottom n-type doped AlGaAs layer and a top p-type AlGaAs layer. In the as-grown layer system the doping concentrations are adjusted in such a fashion that a two-dimensional hole gas (DHG) is formed below the p-type AlGaAs layer at the AlGaAs/GaAs interface. Although the bottom AlGaAs is n-type, no two-dimensional electron gas (DEG) is formed at the bottom GaAs/AlGaAs interface (cf. Figure 9.11 (b)). However, this situation changes if the top GaAs and AlGaAs layers are removed by wet chemical etching. Since no holes are supplied anymore, a two-dimensional electron gas is formed at the bottom GaAs/AlGaAs interface. Thus the p- and n-regions defining the light emitting diode are realized by adjacent two-dimensional hole and electron gases, respectively.

The device structure employed to detect carriers, which are polarized due to the spin Hall effect, is shown in Figure 9.12. Here, a current I_p flows through a stripe formed by a two-dimensional hole gas. It is expected that at the edges of this p-type conductor the holes are spin-polarized due to the spin Hall effect. As illustrated in Figure 9.12, the spin polarization is detected by two light emitting diodes, one on each edge of the stripe. If the spins are polarized out of plane, circularly polarized light is expected to be emitted, with a different orientation on each side of stripe. Details on the operation principle of a spin-LED can be found in Section 6.4.

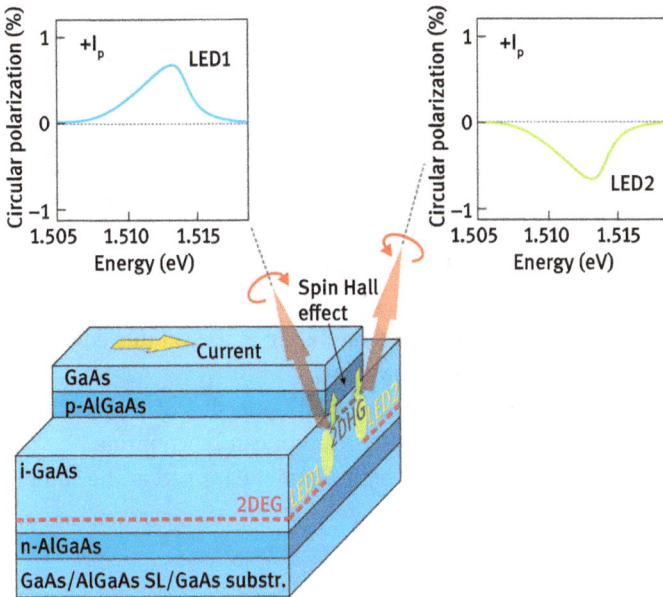

Figure 9.12: Schematics of the structure comprising a conductive p-type channel with two light-emitting diode (LED) structures on the edges. A hole current flows through the upper ridge and is spin-polarized on each edge due to the spin Hall effect. One each edge one finds an adjacent two-dimensional electron gas to form the light-emitting diode structures. The two-dimensional electron gas is formed by removing the top p-doped AlGaAs layer. The emission of circularly polarized light gives information on the spin polarization. The graphs show the degree of circularly polarized light as a function of photon energy for a fixed current direction emitted from LED 1 and LED 2, respectively. Graphs adapted from Wunderlich et al. [165].

A schematic of the corresponding measurement of the circularly polarized light emission for a fixed current I_p in the p-type stripe is shown in Figure 9.12. As can be seen in the upper left panel, a peak in the emission of circularly polarized light from LED 1 is found at an energy expected for optical transitions between the two-dimensional hole and electron gases. This circularly polarized light can be attributed to the recombination of spin-polarized holes from the edge of the two-dimensional hole gas with unpolarized electrons from the two-dimensional electron gas. On the other side, the measurement of the emission from LED 2 shows a circular polarization in the opposite direction (upper right panel). This is due to the fact, that on this side the holes are polarized in opposite direction.

A detailed analysis of the underlying mechanism shows, that the spin Hall effect for holes can be attributed to the intrinsic spin Hall effect [165]. One reason is that the mobility of the two-dimensional hole gas is relatively large owing to the modulation doping. Thus, only a few scattering centers are present which could in principle be responsible for the extrinsic spin Hall effect. Furthermore, the spin-orbit coupling in the valence band is much stronger than in the conduction band.

9.6 Summary

- In the spin Hall effect spins of opposite orientation are accumulated at the side of a conductor when a current flows through the structure.
- Two cases are distinguished: the extrinsic spin Hall effect owing to spin dependent scattering at impurities and the intrinsic spin Hall effect due to Rasbha or Dresselhaus spin-orbit coupling.
- In the diffusive transport regime the spin accumulation at the sample edges can be described by means of the Boltzmann equations.
- The spin polarization at the edges can be detected by means of the magneto-optical Kerr effect. Alternatively, the spin-polarization can be confirmed by circularly polarized light emitted from a light emitting diode formed at the sample edges.

Exercises

Problem 9.1. The time-dependence of a spin ensemble in the presence of an external magnetic field along the z-direction $\vec{B} = (0, 0, B)$ is given by

$$\frac{d}{dt}\langle \vec{S} \rangle = -\frac{\omega_\mathrm{L}}{B}(\langle \vec{S} \rangle \times \vec{B}),$$

with $\omega_\mathrm{L} = (g\mu_\mathrm{B}/\hbar)B$, the Larmor frequency. Find the solution of the equation given above.

Problem 9.2. For the Hanle effect the generated spin precesses while it also dephases. The corresponding value of $\langle S_z \rangle$ is given by

$$\langle S_z \rangle = \frac{1}{T_2^*} \int_0^\infty S_0 e^{-t/T_2^*} \cos(\omega_\mathrm{L} t)\, dt,$$

with S_0 being the value at $B = 0$, T_2^* the spin lifetime, and ω_L the Larmor frequency. Show explicitly that the integration gives

$$\langle S_z \rangle = S_0 \frac{1}{1 + \frac{B^2}{B_{1/2}^2}},$$

where the half-width is given by $B_{1/2} = \hbar/(g\mu_B T_2^*)$.

Problem 9.3. Let us assume a semiconductor material, where the spin Hall effect is observed. Let us now assume that a spin polarized current is injected. Discuss the effect on two voltage probes across the current path.

10 Quantum spin Hall effect

10.1 Introductory remarks

The quantum spin Hall effect is a phenomenon which was only recently discovered. Similar to the spin Hall effect, the spin of the propagating carriers is spatially separated at zero magnetic field. However, the origin is of different nature. The quantum spin Hall effect is observed in a two-dimensional structure in a quantum well, which is formed in a II-VI semiconductor heterostructure. In contrast to well-known quantum wells formed in III-V semiconductors, here the spin-orbit coupling is very strong due to the heavy element Hg used in the HgTe/CdTe heterostructure. The strong spin-orbit coupling leads to an inversion of the band structure. As a consequence edge channels are formed at the border of the sample, as illustrated in Figure 10.1. In this respect the transport phenomenon is very similar to the integer quantum Hall effect described in Section 2.8.3, where carrier transport in edge channels can explain the quantized steps in the Hall resistance. Let us assume a situation, as depicted in Figure 10.1 (a). At very large magnetic fields $B > 0$ applied perpendicularly to a two-dimensional electron gas in a semiconductor heterostructure the electron transport is carried by a single edge channel. Thus, the electrons are flowing along the edge of the sample, i. e. for the given magnetic field clockwise. Within this edge channel the spins are polarized, owing to the Zeeman effect. If the magnetic field is inversed ($B < 0$), the direction of the edge channel transport as well as the spin orientation are inverted. If we combine these two copies of the quantum Hall edge channels, we end up with a scenario depicted in the bottom layer shown in Figure 10.1 (a). Here two counter-flowing edge channels are present, with a net magnetic field being zero. The carrier transport in the quantum spin Hall effect corresponds to this situation. The external magnetic field is indeed zero, but the spin polarizing field is provided by a very strong internal spin-orbit field. Since the orientation of the spin-orbit field is coupled to the direction of motion, the spin-polarization is locked to the direction as well. This situation is depicted in Figure 10.1 (b). Because of the rigid connection of the direction of motion with the spin orientation, these edge channel states are called helical. Based on the edge channel transport already introduced in Section 2.8.3, in connection with the integer quantum Hall effect, this scheme will be modified for the special case of the quantum spin Hall effect. We will see that the quantization effects observed in the experiments can be consistently described using this framework.

In a more general concept, the quantum spin Hall system can be regarded as a topological insulator, or more precisely as a two-dimensional topological insulator. These materials belong to a novel state of matter between classical insulators and metallic conductors. In a two-dimensional topological insulator the two-dimensional plane is insulating while the edges are conductive, whereas for three-dimensional systems the bulk is insulating while the surface is conductive. The edge channels or surface states are topologically protected, which means that they are robust against dis-

https://doi.org/10.1515/9783110639001-010

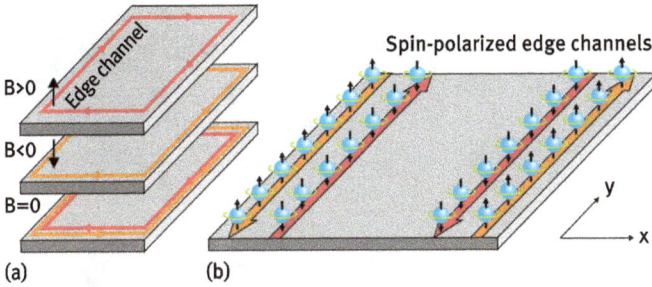

Figure 10.1: (a) The quantum spin Hall effect can be envisioned as a combination of two copies of edge channels in the integer quantum Hall effect. The upper two layers show a single edge channel for opposite magnetic field orientations, respectively. The lower layer combines both cases, with a net magnetic field being zero. (b) In the quantum spin Hall effect the carrier transport takes place in edge channels, which are formed at the border of a two-dimensional system. Carriers with opposite spin propagate in opposite direction. Figure adapted from König et al. [167].

tortions. The underlying reason is the peculiar bandstructure, which is inversed in its order compared to the bandstructure of classical insulators or conductors. In this chapter we will restrict ourself to the two-dimensional case. The properties of three-dimensional topological insulators as well as some more fundamental aspects of topo-logical protected states are covered in the next chapter. Overviews on the quantum spin Hall effect can be found in [14, 15, 166].

10.2 Inverted quantum well in HgTe/CdTe

Before we discuss the peculiar properties of a HgTe/CdTe quantum well, we will first recall the more common quantum well in III-V semiconductor heterostructures. A typ-ical example is shown in Figure 10.2. Here, the quantum well is formed by a GaAs layer sandwiched between two AlGaAs barrier layers. A schematic of the band structure of GaAs is shown in Figure 10.2 (a). The Γ_6-conduction band is separated by the band gap E_g from the Γ_8-light and heavy hole valence bands. The Γ_7-split-off band is separated by the energy Δ_{so} from the upper valence bands, due to the presence of spin-orbit cou-pling. The GaAs/AlGaAs heterostructure shown in Figure 10.2 (b) is a normal type het-erostructure, where the conduction band minimum and the valence band maximum of GaAs are located within the band gap of AlGaAs. The electrons in the conduction band are confined in a quantum well. The potential profile of the quantum well is con-stituted by joining bands of the same symmetry, e. g. in case of the conduction band by the Γ_6-bands. The barrier height corresponds to the conduction band offset. How-ever, for the sake of simplicity we neglect the finite barrier height and only consider the dependence of the energy levels on the effective mass and quantum well width d.

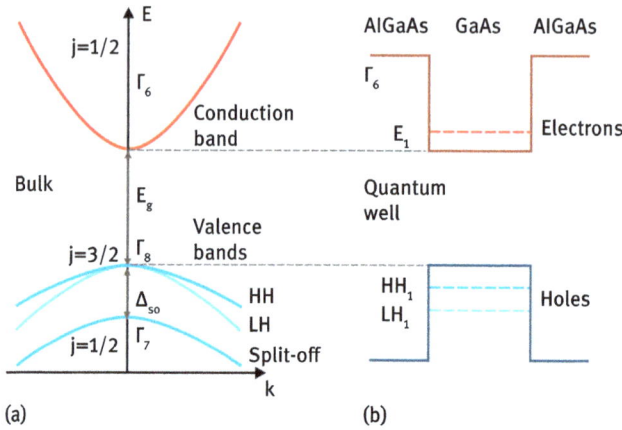

Figure 10.2: (a) Schematics of the band structure of GaAs. The upper Γ_6-band is the conduction band. The valence bands consist of the light hole (LH) and heavy hole (HH) Γ_8-bands and the Γ_7-split-off band. E_g is the band gap and Δ_{so} is the split-off energy due to spin-orbit coupling. (b) Al-GaAs/GaAs/AlGaAs quantum well. Confined states are formed in the conduction and valence band. Owing to the different effective masses, the light hole (LH) and heavy hole (HH) states are different in energy.

The confinement energy is then given by

$$E_n = \frac{\hbar^2}{2m^*} \frac{\pi^2}{d^2} n^2, \tag{10.1}$$

with $n = 1, 2, \ldots$ and m^* being the effective electron mass. The hole states are also confined in a quantum well formed in the valence band. Here the confinement energy is measured from the valence band maximum. However, according to equation (10.1) the different effective masses of light and heavy holes result in different confinement energies, i. e. the confinement energy is larger for the light holes (cf. Figure 10.2 (b)).

The quantum well structure used for the measurement of the quantum spin Hall effect is shown in Figure 10.3 [167]. Here the carriers are confined in the HgTe layer, while the CdTe layers serve as barriers. The quantum well is n-type modulation doped by using I-doping on both sides. The Fermi level can be adjusted by applying a voltage to the gate electrode.

A schematic of the band structure of HgTe around the Γ-point is shown in Figure 10.4 (a). Here, Hg, as a heavy element, leads to a very strong spin-orbit splitting between the Γ_7-valence band (split-off band) and the Γ_8-valence band (heavy and light holes). The splitting is even that strong that the Γ_8-band is located above the s-type Γ_6-conduction band, thus the usual order of bands is inverted. A zero-gap semiconductor is formed because of the degeneracy of the heavy and light hole band at the Γ-point. If Hg is replaced by the lighter element Cd the spin-orbit coupling is smaller, thus the usual order of bands is preserved, as shown in Figure 10.4 (b).

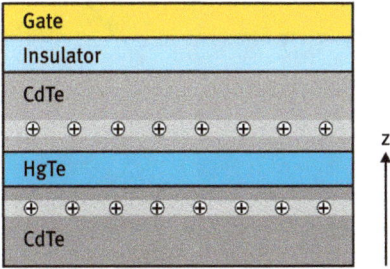

Figure 10.3: HgTe/CdTe quantum well structure. The carriers are confined in the HgTe layer. The barriers are formed by the modulation doped CdTe layers. The Fermi level in the quantum well can be adjusted by means of the gate electrode on top.

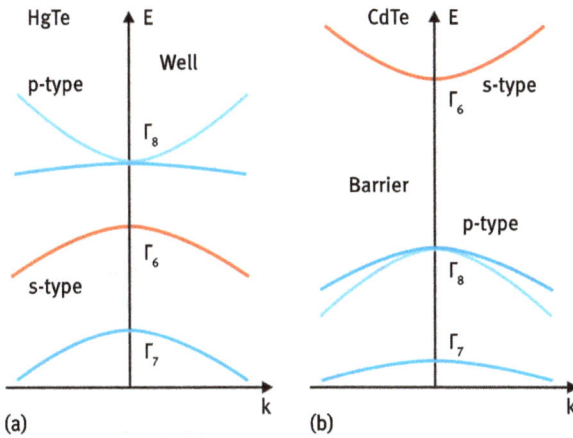

(a)

(b)

Figure 10.4: (a) Schematics of the band structure of the HgTe well material. Owing to the strong spin-orbit coupling the band structure is inversed, i. e. the Γ_8 band is located above the Γ_6-bands. (b) Normal band structure of CdTe used as barrier material.

For a HgTe quantum well with CdTe barriers, two different situations can be realized, depending on the quantum well width. In case of a wider quantum well, comprising a width d above a critical width d_c, the order given by the HgTe material is preserved, i. e. the E_1-level belonging to the Γ_6-band is located below the H_1-level belonging to the heavy hole band (cf. Figure 10.5 (a)). Consequently, this configuration is called an inverted quantum well. Owing to the lower effective mass, the energy level originating from the light hole is below the E_1-level, because of the larger quantization energy according to equation (10.1). Therefore this level is disregarded here. For $d < d_c$ the larger confinement energies result in a reversal of the levels, so that the E_1-level is located above the H_1-level, as it is known for conventional GaAs/AlGaAs quantum wells. This configuration, depicted in Figure 10.5 (b), is therefore called normal quantum well. As we will see later, the energetic order of the E_1 and H_1 levels has a large impact on the

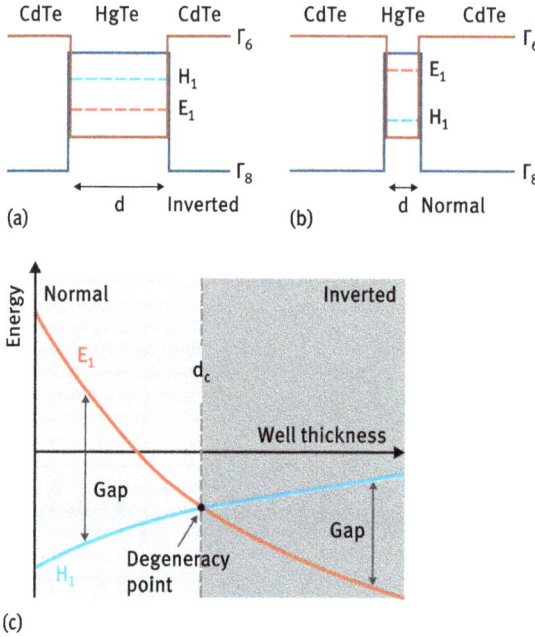

CdTe HgTe CdTe   CdTe HgTe CdTe

(a) d Inverted   (b) d Normal

(c)

Figure 10.5: (a) Band profile of an inverted HgTe/CdTe quantum well ($d > d_c$), where the E_1 level is located below the H_1 level. (b) Corresponding band profile for a normal quantum well ($d < d_c$), with H_1 below E_1. (c) Dependence of the energy levels E_1 and H_1 on the quantum well thickness d. Schematics adapted from [106].

electronic structure. In Figure 10.5 (c), the dependence of the energy levels E_1 and H_1 is plotted as a function of the well thickness. Below a critical thickness d_c the level E_1 is located above the H_1 level (normal structure) and a band gap is formed. At the degeneracy point $d = d_c$ the band gap is closed, and the energy depends linearly on the momentum (Dirac cone). For $d > d_c$ the band gap reappears, but now with H_1 located above E_1.

We have seen that in case of an inverted structure an energy gap is formed. However, the upper band originates from p-type orbitals, while the lower band is formed by s-type orbitals. As a next step, we assume that the material does not extend to infinity but has a limited size and is connected to a conventional material, i. e. a material with an s-type conduction band located above a p-type valence band. At the interface the bands of a certain type are continuously transfered to the other side. This is illustrated in Figure 10.6. As one can see, so-called topologically protected gap states are formed at the interface, while both materials possess a gap in the bulk. In contrast to interface states, which exist due to the chemical binding at the interface [33], these topologically protected interface states are robust and exist even under distortions. In a two-dimensional system as discussed here, edge channels are formed along the in-

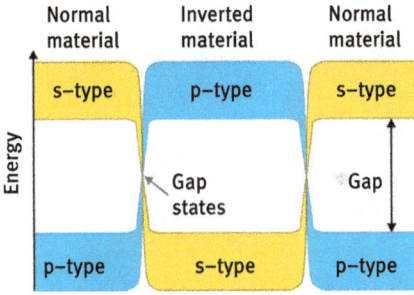

Figure 10.6: Illustration on the formation of gap states at the interface of an inverted quantum well material to a material with a normal order of bands.

terface, very similar to the edge channels in quantum Hall systems. The only difference is that no magnetic field is required to form edge channels.

The band structure, including the interface can also be calculated directly, either by exact diagonalization or by analytic methods. The corresponding graph of the analytically calculated states are depicted in Figure 10.7 [106].

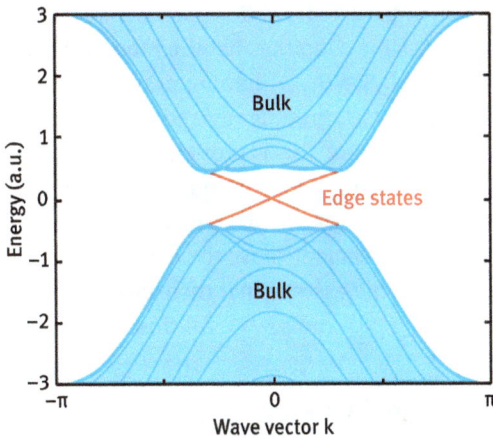

Figure 10.7: Schematic band structure of a HgTe/CdTe quantum well, including bulk and interface states. Figure adapted from König et al. [14].

10.3 Band structure

In order to get a better idea on the transport in inverted quantum wells, we will discuss the band structure in more detail [106]. It is sufficient to consider only states in the s-type band, i. e. E_1 states and the heavy hole states H_1 in the p-type band (cf. Figure 10.8).

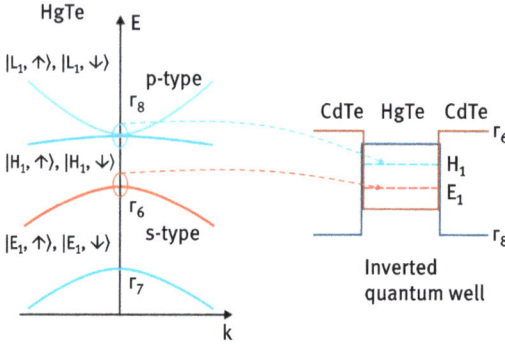

Figure 10.8: Relevant states in a material with an inverted band structure. In the model described, here the light hole states $|L_1\pm\rangle$ are neglected and only the $|H_1\pm\rangle$ and $|E_1\pm\rangle$ states are considered.

If inversion symmetry is assumed at the beginning, the subbands E_1 and H_1 must be double degenerate with opposite spins. Thus, the basis can be expressed as

$$\{|E_1,\uparrow\rangle, |H_1,\uparrow\rangle, |E_1,\downarrow\rangle, |H_1,\downarrow\rangle\}, \tag{10.2}$$

with $\{|E_1,\uparrow\rangle, |E_1,\downarrow\rangle\}$ and $\{|H_1,\uparrow\rangle, |H_1,\downarrow\rangle\}$ being two sets of Kramers partners. As a next step, the appropriate Hamiltonian has to be constructed. For this purpose, we have to consider the symmetry of the system and possible couplings between the four basis states [95]. For example, $\{|E_1,\uparrow\rangle, |E_1,\downarrow\rangle\}$ and $\{|H_1,\uparrow\rangle, |H_1,\downarrow\rangle\}$ have opposite parity, because they originate from different types of orbitals. Consequently, the matrix element connecting must both be odd, in order to be nonzero. This is provided by a coupling linear in k. The heavy hole state $|H_1,\uparrow\rangle$ is formed from the spin-orbit coupled p-type orbitals $|(p_x+ip_y),\uparrow\rangle$, while $|H_1,\downarrow\rangle$ is composed by $|-(p_x-ip_y),\downarrow\rangle$. In order to preserve rotational symmetry about the z-axis the matrix elements have to be proportional to $k_\pm = k_x \pm ik_y$. As an additional prerequisite, we can assume that there is no coupling between $|E_1,\uparrow\rangle$ and $|H_1,\downarrow\rangle$ as well as between $|E_1,\downarrow\rangle$ and $|H_1,\uparrow\rangle$. All these arguments lead to the following model Hamiltonian:

$$\hat{H} = \begin{pmatrix} \hat{h}(\vec{k}) & 0 \\ 0 & \hat{h}^*(-\vec{k}) \end{pmatrix}, \tag{10.3}$$

where the 2×2 matrix components are given by

$$\hat{h}(\vec{k}) = \epsilon(\vec{k})\hat{\tau}_0 + Ak_x\hat{\tau}_x - Ak_y\hat{\tau}_y + M(\vec{k})\hat{\tau}_z, \tag{10.4}$$

with τ_x,τ_y,τ_z being the Pauli matrices in orbital space, τ_0 the 2×2 unit matrix,

$$\epsilon(\vec{k}) = C + D(k_x^2 + k_y^2), \tag{10.5}$$

and

$$M(\vec{k}) = M - B(k_x^2 + k_y^2). \tag{10.6}$$

Here, A, B, C, D, and M are material parameters which depend on the geometry of the quantum well. The special form of the Hamiltonian is governed by the symmetry of the system, as outlined above. The coupling being linear in k between $|E_1, \uparrow\rangle$ and $|H_1, \uparrow\rangle$ or $|E_1, \downarrow\rangle$ and $|H_1, \downarrow\rangle$ is quantified by A. The mass parameter M together with the τ_z matrix induces a gap between the $\{|E_1, \uparrow\rangle, |E_1, \downarrow\rangle\}$ and $\{|H_1, \uparrow\rangle, |H_1, \downarrow\rangle\}$ bands. Even more important, at the critical thickness d_c the parameter M changes sign, and thus the energetic order between $\{|E_1, \uparrow\rangle, |E_1, \downarrow\rangle\}$ and $\{|H_1, \uparrow\rangle, |H_1, \downarrow\rangle\}$ is reversed. For $d > d_c$ the E_1-level drops below the H_1-level at the Γ-point, i. e. the mass M becomes negative. By means of parameter B the dispersion of the $\{|E_1, \uparrow\rangle, |E_1, \downarrow\rangle\}$ and $\{|H_1, \uparrow\rangle, |H_1, \downarrow\rangle\}$ bands can be set to different bendings. In the next section it will be shown that the signs of M and D are important to determine whether topologically protected edge channels form. With all parameters introduced the complete Hamiltonian is given by

$$\hat{H} = \epsilon(\vec{k})I_{4\times4} + \begin{pmatrix} M(\vec{k}) & Ak_+ & 0 & 0 \\ Ak_- & -M(\vec{k}) & 0 & 0 \\ 0 & 0 & M(\vec{k}) & -Ak_+ \\ 0 & 0 & -Ak_- & -M(\vec{k}) \end{pmatrix}, \tag{10.7}$$

with $k_\pm = k_x \pm i k_y$. Using this Hamiltonian the bulk energy spectrum can be calculated:

$$E_\pm(\vec{k}) = \epsilon(\vec{k}) \pm \sqrt{A^2(k_x^2 + k_y^2) + M^2(\vec{k})}. \tag{10.8}$$

As mentioned above, the mass M corresponds to the energy difference between the E_1- and H_1-levels. At the critical point, i. e. $M = 0$, the linear contribution expressed by the parameter A dominates close the Γ-point. In this case one obtains two copies of massless Dirac fermions, one for each spin with a linear $E - k$ dispersion. In Table 10.1 the parameters A, B, D, and M of a HgTe/CdTe quantum well are listed for different quantum well thicknesses [15].

Table 10.1: Material parameters of HgTe/CdTe quantum wells with different thicknesses d [15].

Type	d (Å)	A (eV Å)	B (eV Å^2)	D (eV Å^2)	M (eV)
normal	55	3.87	−48.0	30.6	0.009
massless	61	3.78	−55.3	37.8	−0.00015
inverted	70	3.65	−68.6	51.2	−0.010

10.4 Helical edge states

Based on the band structure discussed in the previous section, we will deduce the existence of topologically protected edge channels leading to edge channel transport

similar to the quantum Hall effect [15]. These states are helical, since the spin orientation is directly connected to the direction of motion. For the formation of edge states
we need a spatially restricted system. For our purposes we consider a half-space $x > 0$
in the x-y plane, as depicted in Figure 10.9 (a).

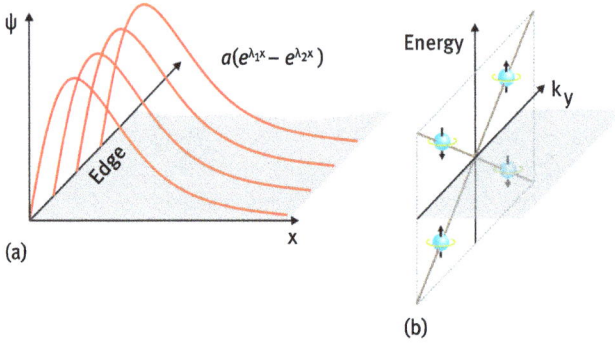

(a)

(b)

Figure 10.9: (a) Edge state in a two-dimensional system in the half-space $x > 0$ with a border along
the y-axis. (b) Linear energy-momentum dispersion of the edge channels. The spin is oriented along
the z-axis.

For the description of the edge channels at the interface it is convenient to subdivide
the Hamiltonian given by equation (10.3) in two parts:

$$\hat{H} = \hat{H}_0 + \hat{H}_1, \tag{10.9}$$

with

$$\hat{H}_0 = \tilde{\epsilon}(k_x)I_{4\times4} + \begin{pmatrix} \widetilde{M}(k_x) & Ak_x & 0 & 0 \\ Ak_x & -\widetilde{M}(k_x) & 0 & 0 \\ 0 & 0 & \widetilde{M}(k_x) & -Ak_x \\ 0 & 0 & -Ak_x & -\widetilde{M}(k_x) \end{pmatrix} \tag{10.10}$$

and

$$\hat{H}_1 = Dk_y^2 I_{4\times4} + \begin{pmatrix} -Bk_y^2 & iAk_y & 0 & 0 \\ -iAk_y & Bk_y^2 & 0 & 0 \\ 0 & 0 & -Bk_y^2 & -iAk_y \\ 0 & 0 & +iAk_y & Bk_y^2 \end{pmatrix}. \tag{10.11}$$

Here we used the definitions $\tilde{\epsilon}(k_x) = C + Dk_x^2$ and $\widetilde{M}(k_x) = M - Bk_x^2$. All k_x-dependent
terms were put into $\hat{H}_0$. However, for a semiinfinite system, k_x has to be replaced
by the operator $-i\partial_x$. On the contrary, translation symmetry is preserved along the

y-direction, and thus k_y is still a good quantum number. We assume the most simple case with $k_y = 0$ so that $\hat{H}_1 = 0$. The wave equation is then reduced to

$$\hat{H}_0(k_x \rightarrow -i\partial_x)\Psi(x) = E\Psi(x). \tag{10.12}$$

The corresponding eigenstates have the form

$$\Psi_\uparrow = \begin{pmatrix} \psi_0 \\ 0 \end{pmatrix}, \quad \Psi_\downarrow = \begin{pmatrix} 0 \\ \psi_0 \end{pmatrix}. \tag{10.13}$$

Both are related by time-reversal symmetry. $\Psi_\uparrow$ and $\Psi_\downarrow$ are four-component vectors, with ψ_0 being a two-component spinor. For the edges states the wave function has to be localized at the edge. They satisfy the wave equation

$$\left[\tilde{\epsilon}(-i\partial_x) + \begin{pmatrix} \widetilde{M}(-i\partial_x) & -iA\partial_x \\ -iA\partial_x & -\widetilde{M}(-i\partial_x) \end{pmatrix} \right]\psi_0(x) = E\psi_0(x). \tag{10.14}$$

For the $\Psi_\downarrow$ state the sign of A is reversed compared to $\Psi_\uparrow$. In order to further simplify the system, we neglect $\tilde{\epsilon}$. In that case we get particle-hole symmetry, implying that a special edge state with $E = 0$ exists. As a reasonable ansatz for the wave function one can choose

$$\psi_0 = e^{\lambda x}\phi, \tag{10.15}$$

with ϕ being a two-component spinor. With this, equation (10.14) can be written as

$$\begin{pmatrix} M + B\lambda^2 & -iA\lambda \\ -iA\lambda & -(M + B\lambda^2) \end{pmatrix}\phi = 0. \tag{10.16}$$

By using the Pauli spin matrices the equation has the following form:

$$(M + B\lambda^2)\hat{\tau}_z\phi = iA\lambda\hat{\tau}_x\phi. \tag{10.17}$$

This equation can be equivalently written as

$$(M + B\lambda^2)\hat{\tau}_y\phi = -A\lambda\phi. \tag{10.18}$$

Here we used the following relations for the Pauli spin matrices:

$$\tau_x\tau_z = -i\tau_y, \quad \tau_x^2 = \tau_0 = I_{2\times2}. \tag{10.19}$$

Being more specific about ϕ one can define a two-component spinor $\phi_\pm$ by

$$\hat{\tau}_y\phi_\pm = \pm\phi_\pm, \tag{10.20}$$

being the eigenvectors of the τ_y Pauli spin matrix

$$\phi_+ = \frac{1}{\sqrt{2}}\begin{pmatrix}1\\i\end{pmatrix}, \quad \phi_- = \frac{1}{\sqrt{2}}\begin{pmatrix}1\\-i\end{pmatrix}. \tag{10.21}$$

This directly shows that the interface states are a superposition of the $|E_1, \uparrow\rangle$ and $|H_1, \uparrow\rangle$ states. Thus, equation (10.18) simplifies to two quadratic equations for λ. If λ is a solution for ϕ_+ then $-\lambda$ is a solution for ϕ_-. The general solution is therefore given by

$$\psi_0 = (ae^{\lambda_1 x} + be^{\lambda_2 x})\phi_- + (ce^{-\lambda_1 x} + de^{-\lambda_2 x})\phi_+, \tag{10.22}$$

where $\lambda_{1,2}$ satisfies

$$\lambda_{1,2} = \frac{1}{2B}(-A \pm \sqrt{A^2 - 4MB}) = \frac{1}{2}\left(-\frac{A}{B} \pm \sqrt{\frac{A^2}{B^2} - 4\frac{M}{B}}\right). \tag{10.23}$$

The existence conditions for an edge state with an amplitude decreasing towards zero far from the interface are either

$$\operatorname{Re}\lambda_{1,2} < 0 \quad (c = d = 0) \quad A/B > 0 \tag{10.24}$$

or

$$\operatorname{Re}\lambda_{1,2} > 0 \quad (a = b = 0) \quad A/B < 0. \tag{10.25}$$

These conditions can only be satisfied in the inverted regime $M/B > 0$! The latter condition makes sure that $|\sqrt{\ldots}|$ is smaller than $|A/B|$, so that $\lambda_{1,2}$ have the same sign. The same sign of $\lambda_{1,2}$ is required to make sure that the wave function diminishes far from the interface. Thus, the wave functions of the edge states at the Γ-point are finally given by

$$\psi_0(x) = \begin{cases} a(e^{\lambda_1 x} - e^{\lambda_2 x})\phi_- & A/B > 0 \\ c(e^{-\lambda_1 x} - e^{-\lambda_2 x})\phi_+ & A/B < 0. \end{cases} \tag{10.26}$$

A schematic wave function can be found in Figure 10.9 (a). Under the conditions given above $a = -b$ and $c = -d$ to satisfy the condition that the wave function is zero at the interface. For a given material either a and b or c and d are zero, since A/B is determined by the material properties only (cf. Table 10.1). Please keep in mind that nevertheless according to equation (10.13) two interface states with opposite spins exist. In order to calculate the state for the reversed spin, the parameter A has to be replaced by $-A$.

Up to now, we have restricted ourselves to a zero energy state. For finite k_y values, and thus propagating states along the interface, the Hamiltonian for the helical edge states is given by [14]

$$\hat{H}_{\text{edge}} = Ak_y\hat{\sigma}_z, \tag{10.27}$$

with σ_z being the Pauli matrix in spin space. This corresponds to an one-dimensional Dirac Hamiltonian for massless particles, with a Dirac velocity given by $v = A/\hbar$. The energy-momentum dispersion is depicted in Figure 10.9 (b). The spin is oriented along the z-axis, thus being perpendicular to the plane of the two-dimensional system. Inversing the wave vector inverses the spin, as requested by time-reversal symmetry.

10.5 Conductance in a normal and inverted HgTe/CdTe quantum well

In the previous section we have convinced ourself that, depending on the band parameters, edge states may form at the interface. These edge states should only exist in case of an inverted band structure with $M < 0$. In the following we will compare the transport properties of two HgTe quantum wells with quantum well thicknesses above and below the critical thickness d_c.

In Figure 10.10 (a) a quantum well with $d < d_c$ is depicted. The E_1-level is located above the H_1-level with a band gap in between. During the transport measurements the Fermi level is shifted by using a gate electrode on top of the quantum well (cf. Figure 10.3). The expected conductance is shown in Figure 10.10 (b). For large Fermi energies $E_F > E_1$ the quantum well is filled with electrons with a finite kinetic energy so that electron transport takes place, i. e. the conductance is finite. When the Fermi level is pulled downwards and shifted into the band gap the conductance is zero, since no states are available for electron transport. Only when the Fermi level is moved far enough to pass the H_1-level does hole transport take place, and thus the conductance increases again.

In Figure 10.11, it is shown that the transport properties, as described above, are indeed observed. When the gate voltage V_g is varied from 0 to -0.75 V the Fermi level is shifted downwards. The overall slope is positive with electrons contributing to the transport. Since it is a high mobility two-dimensional system, quantum Hall steps are observed. The more negative the gate voltage gets the larger the slope is. Thus the electron concentration is lowered. For an even larger negative voltage ($V_g = -1.75$ V) the slope is inversed, i. e. hole transport takes over. Between -0.75 V and -1.75 V the Fermi level passes the band gap. No Hall effect is measured because of the lack of free carriers.

A completely different behavior is expected for inverted quantum wells with $d > d_c$ (cf. Figure 10.10 (c)). As discussed in the previous section, edge states form within the band gap, which contribute to the transport. As a consequence, when the Fermi level is shifted into the band gap a finite conductance is expected. Ideally the conductance G should be quantized at a value of $2e^2/h$. The dependence of the conductance as a function of the Fermi energy is illustrated in Figure 10.10 (d). Outside the gap the transport is carried by bulk states of the two-dimensional system, similar to the situation in the normal quantum well [167].

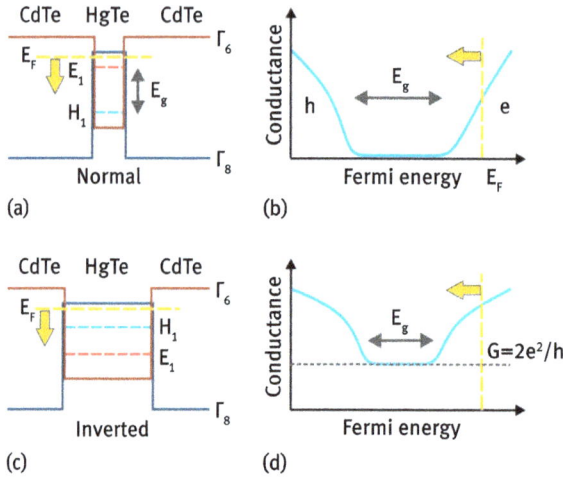

(a)

(b)

(c)

(d)

Figure 10.10: (a) HgTe/CdTe quantum well with a thickness smaller than the critical thickness ($d < d_c$). The Fermi level is shifted downwards by using a gate electrode. (b) Corresponding conductance as a function of Fermi energy E_F for a quantum well with $d < d_c$. The Fermi energy is shifted by varying the gate voltage. (c) Band profile of a HgTe/CdTe quantum well with $d > d_c$. (d) Corresponding conductance as a function of E_F. Figure adapted from Bernevig et al. [106].

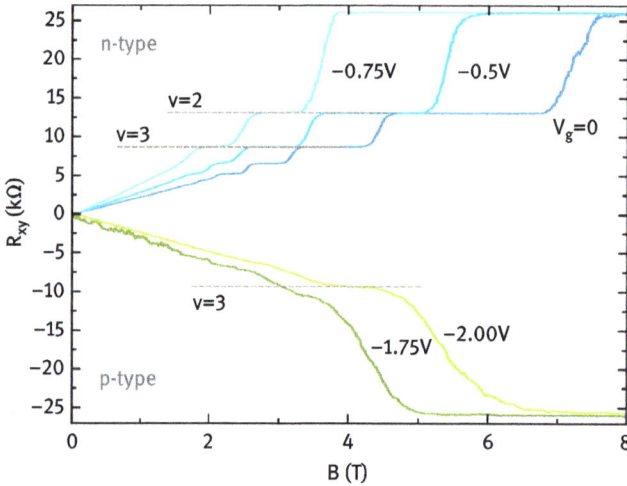

Figure 10.11: Quantum Hall effect measured on a normal HgTe/CdTe quantum well ($d < d_c$). The filling factors $v = 2$ and 3 for the electron and $v = 3$ hole systems are indicated, respectively [14]. Figure provided by H. Buhmann, Würzburg University.

In Figure 10.12 transport measurements are shown for HgTe/CdTe heterostructures with a quantum well width larger than the critical width d_c. The 4-terminal resistance $R_{14,23}$ is plotted as a function of gate voltage, i. e. Fermi energy. A finite resistance is

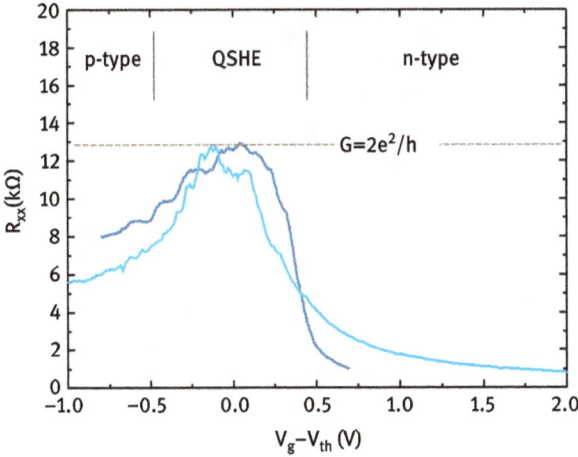

Figure 10.12: Measured 4-terminal resistance a function of the normalized gate voltage $V_g - V_{th}$ for two HgTe/CdTe quantum wells with thicknesses larger than the critical width d_c. Here, V_{th} is the threshold voltage. The ranges of different type of conductance, i. e. p-type, QSHE, or n-type are indicated. Figure provided by H. Buhmann, Würzburg University.

observed around zero gate voltage, with $V_g - V_{th}$ the normalized gate voltage. The threshold voltage V_{th} corresponds to the voltage required to put the Fermi level in midgap position. At zero voltage the Fermi energy lies within the band gap. The finite resistance is attributed to the existence of edge states located in the band gap. As long as the sample length is short, scattering can be neglected. In that case a conductance close to $2e^2/h$ is found, being consistent with two edge channels at each interface (cf. Figure 10.1). On finds that the conductance in the band gap region does not depend on the width of the sample, supporting the assumption that the transport takes place at the interface only [167]. For longer samples scattering occurs. This increases the resistance in the band gap region.

10.6 Edge channel transport

The finite quantized conductance observed in the band gap region for inverted quantum wells can be consistently explained within the Landauer–Büttiker model [39]. The corresponding scheme of the sample is shown in Figure 10.13.

The geometry is similar to the geometry used for the quantum Hall effect. Contacts 1 and 4 are used for feeding the current. Contacts 2, 3, 5, and 6 are voltage probes, and thus no current flows from the leads into or out of the contacts ($I = 0$). In an inverted quantum well edge channels are formed. On each side there are two counter propagating helical edge channels, i. e. one for each spin orientation. The carrier transport is in opposite directions.

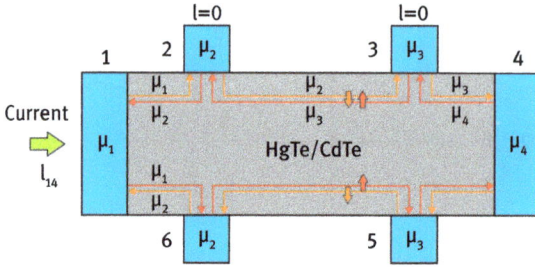

Figure 10.13: Schematics of a sample for measuring the quantum spin Hall effect in inverted quantum wells. Edge states with carriers moving in opposite directions for opposite spins are formed. The bias current flows between contact 1 and 4. The other contacts are voltage probes.

As discussed in conjunction with the quantum Hall effect, the current in a spin-polarized edge channel is given by

$$I_i = \int_0^{\mu_i} v(E) D_{1D}(E)\, dE = \frac{e}{h}\mu_i, \tag{10.28}$$

with $D_{1D}(E) \propto 1/\sqrt{E} \propto 1/v(E)$ being the one-dimensional density of states. Since D_{1D} is proportional to the inverse of the velocity, the current only depends on the electrochemical potential μ_i of the supplying contact i. The four-terminal resistance $R_{14,23}$ is defined as

$$R_{14,23} = \frac{V_{23}}{I_{14}} = \frac{(\mu_2 - \mu_3)/e}{I_{14}}. \tag{10.29}$$

Here the current is driven from contact 1 to 4, while the voltage is measured between two different contacts 2 and 3. The current from contact 1 to 4 is given by

$$I_{14} = \frac{e}{h}(2\mu_1 - 2\mu_2), \tag{10.30}$$

since for contact 1 the carriers occupied up to μ_1 are leaving the contact via two edge channels, while carriers are entering from contacts 2 and 6 with electro/chemical potential μ_2. Contacts 2 and 6 must be on the same potential, since no magnetic field is applied, and thus no Hall effect occurs. As the net current in the voltage probes 2 and 3 must be zero, the following relations hold:

$$(e/h)(\mu_1 - 2\mu_2 + \mu_3) = 0, \tag{10.31}$$
$$(e/h)(\mu_4 - 2\mu_3 + \mu_2) = 0. \tag{10.32}$$

Using equation (10.31) one can replace μ_3 in V_{23} by $2\mu_2 - \mu_1$, so that finally the 4/terminal resistance is given by

$$R_{14,23} = \frac{V_{23}}{I_{14}} = \frac{h}{2e^2} = 12.6\text{k}\Omega. \tag{10.33}$$

A comparison with the results shown in Figure 10.12 confirms that indeed the 4-terminal resistance settles to a value of $h/2e^2$, as expected for two contributing spin-separated channels.

Using the same approach as explained above, the 2-terminal resistance $R_{14,14}$ can also be calculated, which is given by

$$R_{14,14} = \frac{V_{14}}{I_{14}} = \frac{3h}{2e^2}. \tag{10.34}$$

This value differs from the value of $h/2e^2$, which is expected for the two-terminal resistance in the quantum Hall effect regime. The difference is due to the fact that in the quantum spin Hall effect the carriers in the edge channels on each side propagate in opposite directions, while in the quantum Hall regime the carriers move in the same direction.

10.7 Spin-polarized transport

So far it has been only shown that in the case of an inverted quantum well the carrier transport occurs in one-dimensional edge channels as long as the Fermi level lies in the band gap located between the H_1- and E_1-levels (cf. Figure 10.10). What still needs to be proven is that the transport within these edge channels is indeed spin-polarized. In order to show this feature, the H-shaped device structure shown in Figure 10.14 can be employed [168]. For this sample an inverted quantum well in a HgTe/CdTe heterostructure is used. The type of transport, i. e. quantum spin Hall transport or metallic, in the upper and lower section can be controlled by means of gates A and B, respectively. In the configuration shown here, gate B is adjusted to the voltage so that the lower section of the H-shaped structure is in the metallic regime. By driving a current from contact 3 to contact 4 spins of opposite orientation are accumulated at the bottom and top side due to the strong spin Hall effect (SHE) in this material (cf. Chapter 9). For the case shown in Figure 10.14, spin-up electrons are accumulated at the bottom, while spin-down electrons are accumulated at the top.

By adjusting the voltage at gate A to a value so that the Fermi level lies in between the top H_1 and the bottom E_1 quantum well levels the top section of the H-shaped structure is put into the quantum spin Hall effect regime (QSHE). Here we expect edge channel transport. More precisely, helical transport is expected with the spin orientation locked to the direction of motion in the edge channel. For the situation shown in Figure 10.14 only the right-moving spin-polarized edge channel at the bottom picks up spin-down electrons provided by the spin Hall effect. For the remaining edge channels, either the spin orientation does not fit the orientation given by the accumulated carriers, i. e. spin-down, or the transport imposed by the edge channel is in the wrong direction, i. e. the left-bottom spin-down edge channel emerging from contact 1. Due to the exclusive transport of spin-down carriers across the SHE-QSHE interface by a

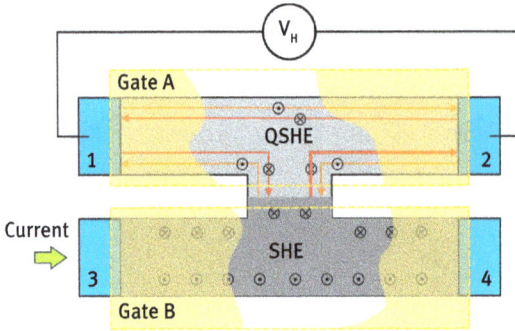

Figure 10.14: Spin injection into a spin-polarized edge channel: The lower part of the H-shape structure is in a normal conducting metal state by applying an appropriate gate voltage to gate B. By driving a current through this section, the spins of opposite orientations are accumulated at the top and bottom due to the spin Hall effect (SHE). The top section of the H-shaped structure is put into the quantum spin Hall effect state (QSHE) by adjusting gate A. Due to the helical transport in the edge channels only the right-moving edge channel with down spin orientation ⊗ can pick up spin-polarized carriers generated by the spin Hall effect from the bottom part. The charge accumulation at the top contact on the right side (2) leads to a build-up of a voltage drop V_H between contacts 1 and 2. Figure adapted from Brüne et al. [168].

spin-polarized edge channel into contact 2, carriers are accumulated in this contact. Because of this imbalanced situation a voltage V_H builds up between contact 1 and 2. If we would assume a non-spin-locked transport in the edge channels the carriers would be symmetrically distributed while being transferred across the SHE-QSHE interface. Thus no voltage drop is expected in this case, i. e. $V_H = 0$.

A typical outcome of a 4-terminal resistance measurement as a function of the voltage at the bottom gate (gate B) is shown in Figure 10.15 [168]. Here the top gate (gate A) was adjusted to zero bias voltage, so that the quantum well underneath is in the quantum spin Hall regime. By changing voltage at the gate at the bottom (gate B) from negative to positive values, the quantum well underneath is transferred from a metallic hole conductor via the quantum spin Hall regime around zero gate voltage to a metallic electron conductor. The largest signal of the 4-terminal resistance $R_{34,12}$ is measured when the bottom section is in the quantum spin Hall effect regime similar to the top section. For $R_{34,12} = V_{12}/I_{34}$ the current is driven between contacts 3 and 4 while the voltage is measured between contacts 1 and 2. In this case the transport is totally governed by the edge channel transport between all four contacts. The resistance can be calculated within the framework of the Landauer–Büttiker model, as discussed above. For this particular case a value of $h/4e^2$ is expected. The experimental value is slightly below that value.

The proof that spin polarized transport takes place is given by the finite resistance values when the bottom section is either in the metallic hole or metallic electron regime. In these cases the spin Hall effect, which occurs due to the current flowing between contacts 3 and 4, results in a spin accumulation. Owing to the spin selective

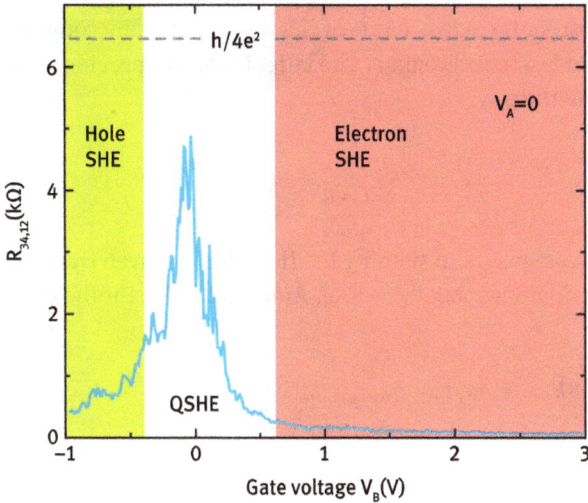

Figure 10.15: Plot of the 4-terminal resistance $R_{34,12}$ as a function of gate voltage V_B. The voltage at the top gate (gate A) was kept at zero, and thus the quantum well underneath is in the quantum spin Hall regime. At negative voltages applied to gate B holes are induced (green range) in the lower section of the H-shaped sample shown in Figure 10.14, while at positive gate voltages electrons are induced (red range). By driving a current through this section spin polarized carriers are induced due to the spin Hall effect (SHE). Around $V_B = 0$ the lower section is in the quantum spin Hall regime [168]. Figure provided by H. Buhmann, Würzburg University.

pick-up by the edge channels in the quantum spin Hall conductor in the upper section a finite voltage V_{12} is built up between contacts 1 and 2, and thus a finite nonlocal resistance $R_{34,12} = V_{34}/I_{12}$ is measured. The experiment showed a slightly larger value of $R_{34,12}$ for hole spin Hall effect in the bottom section.

10.8 Summary

- In HgTe an inverted band structure is found, i. e. the p-type band is located above the s-type band. The band inversion is caused by the very strong spin-orbit coupling.
- In a CdTe/HgTe/CdTe quantum well comprising a quantum well thickness above a critical thickness the lowest quantized level originating from the Γ_6-band is located below the corresponding level from the Γ_8-band. This type of quantum well is called inverted.
- At the surface or at the interface to a normal insulator topologically protected states are formed. These states are helical, i. e. the spin orientation is locked to the direction of motion.
- The transport through the topologically protected states can be describe in the edge channel picture. The 4-terminal resistance is quantized at a value of $h/2e^2$.

– The spin-polarized transport in the edge channels can be confirmed by coupling to an area being in the spin Hall effect regime. The latter is used to provide spin-polarized carriers at the interface.

Exercises

Problem 10.1. Sometimes it is convenient to simplify the Hamiltonian given by equation (10.4) by using a simplified lattice model [14, 106]. Around the Γ-point the components of the Hamiltonian

$$\hat{h}(\vec{k}) = \epsilon(\vec{k})\hat{\sigma}_0 + Ak_x\hat{\tau}_x - Ak_y\hat{\tau}_y + M(\vec{k})\hat{\tau}_z$$

can be approximated by

$$\epsilon(\vec{k}) \rightarrow C + 2Da^{-2}(2 - \cos k_x a - \cos k_y a)$$
$$Ak_x\sigma_x \rightarrow Aa^{-1}\sin k_x a$$
$$Ak_y\sigma_y \rightarrow -Aa^{-1}\sin k_y a$$
$$M(\vec{k}) \rightarrow M - 2Ba^{-2}(2 - \cos k_x a - \cos k_y a),$$

with a being the lattice constant. Calculate the energy eigenvalues for this Hamiltonian. Discuss the situation, when $M \rightarrow 0$.

Problem 10.2. Let us assume a Hall bar as shown in Figure 10.13. Calculate 2-terminal resistance $R_{14,14}$ for a system in the quantum spin Hall effect regime using the Landauer–Büttiker formalism.

Problem 10.3. The states in a quantum spin Hall system can be described by the effective Hamiltonian given by equation (10.7). For the edge channel state at $x > 0$ we get for $A/B > 0$ the solution equation (10.26)

$$\psi_0(x) = a(e^{\lambda_1 x} - e^{\lambda_2 x})\phi_- = \xi(x)\phi_-,$$

with

$$\phi_- = \frac{1}{\sqrt{2}}\begin{pmatrix} 1 \\ -i \end{pmatrix}.$$

Show that by projecting the bulk Hamiltonian onto the edge state the following Hamiltonian is obtained

$$\hat{H}_{edge} = Ak_y\hat{\sigma}_z.$$

Problem 10.4. Let us consider quantum wells of widths d for electrons in the conduction band and heavy holes in the valence band. For both quantum wells the barrier height is assumed to be infinitive. The bottom of the conduction band is located 0.4 eV below the top of the valence band, i. e. the band gap E_g is –0.4 eV. The effective electron mass is $m_e^* = 0.4m_0$, while the heavy hole mass is assumed to be $m_h^* = 0.8\,m_0$, with m_0 being the free electron mass. Calculated the energy difference ΔE between the lowest electron level $E_{1,e}$ and the highest hole band $E_{1,h}$. Determine the critical width d_c, where both levels match.

11 Topological insulators

11.1 Introductory remarks

In the previous chapter, the quantum spin Hall effect was discussed. The corresponding material system already belongs to the class of topological insulators. In that case the two-dimensional plane is insulating, while the edges are conductive. Moreover, these edge channels are topologically protected, which means that they cannot be destroyed by distortions, e. g. potential fluctuations. In this chapter we will extend the concept of topological insulators to three-dimensional structures, i. e. the bulk is insulating while the surface is conductive. To illustrate this special property, the normal insulator is compared to a topological insulator in Figure 11.1 (a) and (b), respectively.

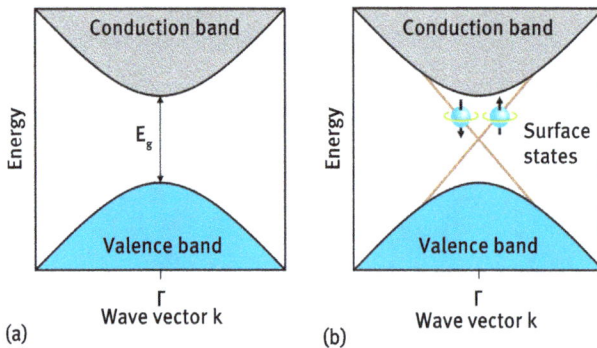

Figure 11.1: (a) Schematic illustration of a band structure of an insulator. The empty conduction band is separated from the fully occupied valence band by a band gap E_g. (b) In a three-dimensional topological insulator topologically protected surface states exist within the bulk band gap. The spin orientation is locked to the direction of motion.

In a normal insulator, the empty conduction band is separated by the band gap E_g from the fully occupied valence band. In an insulator E_g is so large that at room temperature thermal excitations from the valence band into the conduction band are negligible. Since completely filled bands cannot contribute to the carrier transport, the material is insulating. Conceptually, an intrinsic semiconductor is an insulator as well, although due to the smaller band gap some carriers might get thermally excited from the valence into the conduction band. In contrast, as shown in Figure 11.1 (b), in three-dimensional topological insulators surface states exist within the band gap. These surface states have a linear dispersion and cross each other at the so-called Dirac point. Due to the linear dispersion, the electrons behave like massless relativistic particles. As in the quantum spin Hall system, the surface states are helical, i. e. the spin orientation is locked to the direction of motion. As we will see later, this has important consequences for the carrier transport, since direct backscattering is suppressed. The

https://doi.org/10.1515/9783110639001-011

reason is that, apart from inversion of the direction of propagation, the spin orientation also needs to be inverted. This is very unlikely. Such property makes this material
particularly interesting for spintronic applications, because high mobility devices are
expected, due to the low scattering rate. In order to obtain a three-dimensional topological insulator, strong spin-orbit coupling is required. This can be realized by using
heavy elements constituting the crystal. Typical representatives are Bi_2Se_3, Bi_2Te_3, or
Sb_2Te_3 [169]. For these materials the bulk band gap is large, i. e. in the order of some
tenths of eV, so that room temperature spintronics might be feasible.

In this chapter, first the material system is introduced. Subsequently, the bandstructure is discussed theoretically and compared to the experimentally obtained one.
Finally, transport phenomena are explained, which are a direct consequence of the existence of the topologically protected surface states.

11.2 Material system

The typical three-dimensional topological insulators, such as Bi_2Se_3, Bi_2Te_3, or
Sb_2Te_3, crystallize in a tetradymite structure, with a threefold rotation symmetry
about the z-axis. The unit cell is depicted in Figure 11.2 (a). It consists of a stack of
three quintuple layers, each having a different stacking sequence. The quintuple
layer is formed by five layers, which are covalently bonded. The atomic layers in the
quintuple layer alternately consist of one of the constituting elements, e. g. of a Bi
and a Te layer for Bi_2Te_3. The quintuple layers are weakly connected by van der Waals
forces. The lattice parameters for Bi_2Te_3 and Sb_2Te_3 are practically identical and have
a length of $c = 30.49$ Å along the crystalline z-axis, while the atomic distance within
the hexagonal layers is about $a = 4.4$ Å. In Figure 11.2 (b) a transmission electron
microscopy image of a Bi_2Te_3 layer is shown. The quintuple layers are separated by
the dark stripes.

Epitaxial layers of three-dimensional topological insulators can be grown by
means of molecular beam epitaxy (cf. Section 2.5.1) [170]. Although the lattice constant is off by about 14 %, mostly a Si(111) wafer is used as a substrate. The surface
of Si(111) has a hexagonal symmetry. In Figure 11.3 (a) an atomic force microscopy
image of an epitaxial Bi_2Te_3 layer is shown. Extended areas of same height are found.
The height profile shown in Figure 11.3 (b) reveals that the step height between the
different areas corresponds to the thickness of a quintuple layer.

11.3 Bulk band structure

In this section we will go deeper into the peculiarities of the band structure of topological insulators, as schematically illustrated in Figure 11.1 (b). We have to find out
what the underlying reason is for forming the surface states. One important aspect

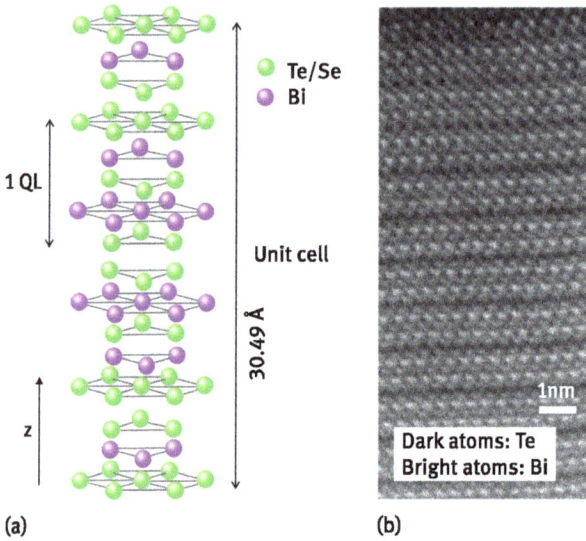

Figure 11.2: (a) Crystal structure of a typical topological insulator, i. e. Bi_2Te_3 or Sb_2Te_3. The unit cell has a height of 30.49 Å and consist of three stacked quintuple layers (QL). The layers within the quintuple layers are covalently bonded, while the quintuple layers are bonded by van der Waals force. Figure provided by G. Mussler, Forschungszentrum Jülich. (b) Transmission electron microscope image of a Bi_2Te_3 layer. The quintuple layers can be easily identified. Image provided by M. Luysberg, Forschungszentrum Jülich.

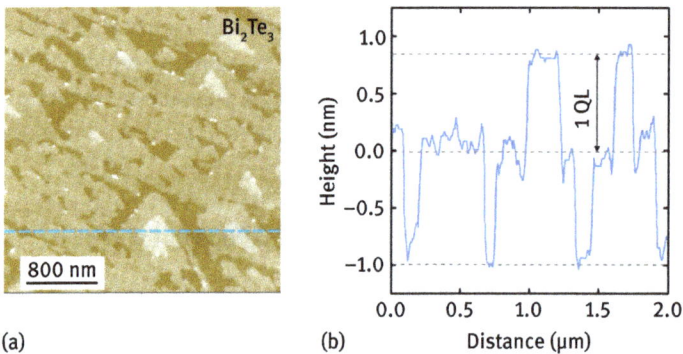

Figure 11.3: (a) Atomic force microscope image of a Bi_2Te_3 layer grown by molecular beam epitaxy. The surface consists of extended areas of continuous single quintuple layers. (b) The height profile of the surface topography along the dashed line in the atomic force microscope image in (a) shows the quintuple steps of about 1 nm. Figure provided by G. Mussler, Forschungszentrum Jülich.

is the symmetry of the crystal lattice. In Figure 11.4 (a), a single Bi_3Se_2 quintuple layer is shown. The layers within the quintuple layer have the stacking sequence A–B–C–A–B.

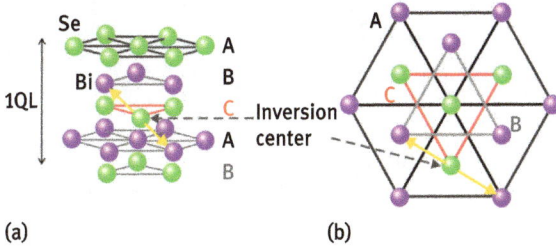

(a)   (b)

Figure 11.4: (a) Quintuple layer of Bi_2Se_3 with the stacking sequence A–B–C–A–B. The yellow arrows illustrate the inversion around the inversion center. (b) Top view along the z-direction. The triangle lattice in one quintuple layer has three different positions, corresponding to the layers A, B, and C, shown in (a).

The Se atoms at the center layer serve as an inversion center, i. e. for the atoms of the first B and second A layer, as illustrated in Figure 11.4 (a). The top view of the quintuple layers is depicted in Figure 11.4 (b), also including the inversion between atom positions in layer B and A with the inversion center in layer C. The fact that inversion symmetry exists allows us to construct eigenfunctions with a defined parity, i. e. even or odd. As will be outlined below, an important criteria for whether or not a material is a topological insulator is the change of parity of the occupied bands.

11.3.1 Level evolution

In Figure 11.5 the evolution of the atomic states of Bi_2Se_3 into the conduction and valence band is schematically illustrated [169].

On the first stage we consider the chemical bonds within the layers of the quintuple layer. This leads to a level splitting similar to the symmetric bonding state and antisymmetric antibonding states in a hydrogen molecule, i. e. states with even and odd parity, respectively (cf. Section 3.4.1). Accordingly, the states in a unit cell of a Bi_2Se_3 crystal can be subdivided into states with even (+) and odd (−) parity. Next, the effect of the crystal-field splitting between different p-orbitals is considered. Since the symmetry of the crystal is different along the z-direction and in the xy-plane, the p_z-orbital is split-off from the p_x- and p_y-orbital. The latter two remain degenerate. The states closest to the Fermi energy are the $|P2_z^-\rangle$ state from Se and the $|P1_z^+\rangle$ state from Bi, as can be seen in Figure 11.5. Finally, spin-orbit coupling is included and the spin orientation matters. The energy is then determined by the total angular momentum combining the orbital and spin angular momentum. This leads to the crossing of two states mentioned before. Hence, the states $|P1_z^+, \uparrow\rangle$ and $|P1_z^+, \downarrow\rangle$ with even parity are below the $|P2_z^-, \uparrow\rangle$ and $|P2_z^-, \downarrow\rangle$ states with odd parity.

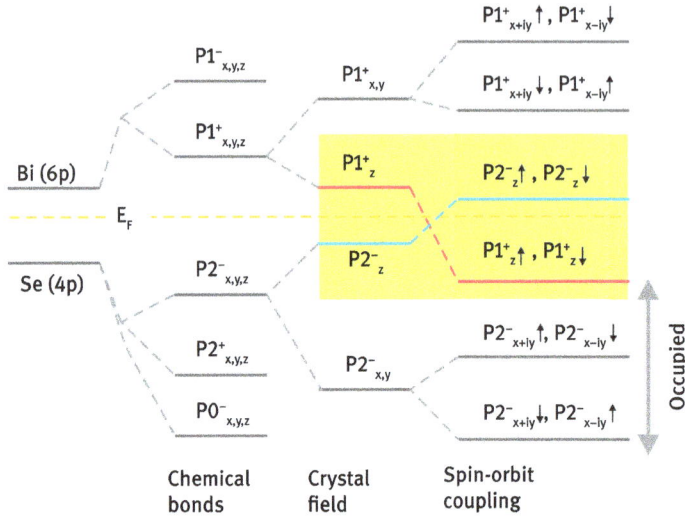

Figure 11.5: Evolution from the atomic p_x-, p_y-, and p_z-orbitals of Bi and Se into the conduction and valence bands of Bi_2Se_3 at the Γ-point by consecutively including chemical bonds, the crystal field, and spin-orbit coupling. The dashed line corresponds to the Fermi energy. The level crossing due to spin-orbit coupling are highlighted. Scheme adapted from Zhang et al. [169].

11.3.2 Effective Hamiltonian

The four states given above can be used as a basis to construct an effective Hamiltonian around the Γ-point. This is reasonable, since most measured phenomena originate from this region. This approach is similar to the $k \cdot p$ approximation discussed in Chapter 7, where we were also interested in a range around the Γ-point. The procedure followed here is comparable to the one employed to obtain the Hamiltonian equation (10.7) describing the quantum spin Hall effect. In the present case, the basis states are the four states located closest to the Fermi level:

$$\{|P1_z^+, \uparrow\rangle, |P2_z^-, \uparrow\rangle, |P1_z^+, \downarrow\rangle, |P2_z^-, \downarrow\rangle\}. \tag{11.1}$$

The Hamiltonian can be built by accounting for the symmetry of the system, i. e. time-reversal symmetry demands that after applying the time reversal operator given by $\hat{\mathcal{T}} = i\hat{\sigma}_y\hat{\mathcal{K}} \otimes I_{2\times2}$ the Hamiltonian should be the same. Here, $\hat{\mathcal{K}}$ is the complex conjugate operator. The same holds for the inversion symmetry operator $\hat{\mathcal{I}} = I_{2\times2} \otimes \hat{\tau}_z$, with $\hat{\tau}_z$ the Pauli matrix in orbital space. Furthermore, as can be inferred from Figure 11.4, the crystal structure has a threefold rotation symmetry about the z-axis. Thus, by applying the corresponding rotation operator $\hat{\mathcal{C}}_3 = \exp\left[i(\pi/3)\right]\hat{\sigma}_y \otimes I_{2\times2}$, the Hamiltonian should be the same again. Obeying all these symmetry relations demands a special form of the Hamiltonian, which can be written as

$$\hat{H} = \epsilon(\vec{k})I_{4\times4} + \begin{pmatrix} M(\vec{k}) & A_1 k_z & 0 & A_2 k_- \\ A_1 k_z & -M(\vec{k}) & A_2 k_- & 0 \\ 0 & A_2 k_+ & M(\vec{k}) & -A_1 k_z \\ A_2 k_+ & 0 & -A_1 k_z & -M(\vec{k}) \end{pmatrix}, \tag{11.2}$$

with $k_\pm = k_x \pm i k_y$ and $\epsilon(\vec{k}) = C + D_1 k_z^2 + D_2 k_\perp^2$. The contribution $M(\vec{k}) = M - B_1 k_z^2 - B_2 k_\perp^2$ is the mass term, as already introduced for quantum spin Hall systems. The parameters $A_1, A_2, C, \ldots$ are obtained from fitting to ab initio band structure calculations [169]. A negative sign of the mass term $M(\vec{k})$ accounts for the band inversion. For the topological insulators Bi_2Te_3, Sb_2Te_3, and Bi_2Se_3 the mass parameter at the Γ-point has a negative value, which reflects the inversed order, i. e. the $|P1_z^+, \uparrow\rangle$ and $|P1_z^+, \downarrow\rangle$ states have a lower energy than the corresponding $|P2_z^-, \uparrow\rangle$ and $|P2_z^-, \downarrow\rangle$ states [171]. The advantage of the effective Hamiltonian is that it can be used to obtain the wave function of the surface states in an easy way.

11.3.3 Ab-initio band structure calculations

As mentioned in the previous section, the parameters of the effective Hamiltonian given by equation (11.2) are obtained by fitting to band structures gained by ab initio methods. Here, often the local density approximation (LDA) of the density functional theory or generalized gradient (GGA) approximations are employed [169, 172]. These calculations show good agreement with the experimental results. More subtle features of the band structure can be reproduced by employing GW calculations. Here, G stands for Greens function and W for screened Coulomb interaction. In this approximation many-body effects are included [173, 174]. Using the GW approximation the nature of the band gap, i. e. indirect or direct, its magnitude, and the effective masses are described correctly [174]. In Figure 11.6 the bulk band structure of Bi_2Te_3 is shown. At the Γ-point, the gap is indirect, indicated by the maxima of the valence bands at finite k values. Since the sample boundary is not considered here, no surface states appear in the band structure.

In order to decide whether a material is a topological insulator or not, it is sufficient to know that a band inversion occurs at the Γ-point. The Γ-point is one of the special points in the Brillouin zone. It belongs to the time-reversed invariant momenta (TRIM), where a two-fold degeneracy occurs owing to Kramer's theorem. When moving from the Γ-point to other points in the Brillouin zone, the order of the energy levels might change, i. e. the level crossing due to spin-orbit coupling is lifted. Of special interest here are other points of time-reversal invariant momenta, e. g. F, L, and Z. Their location in the Brillouin zone is depicted in Figure 11.7.

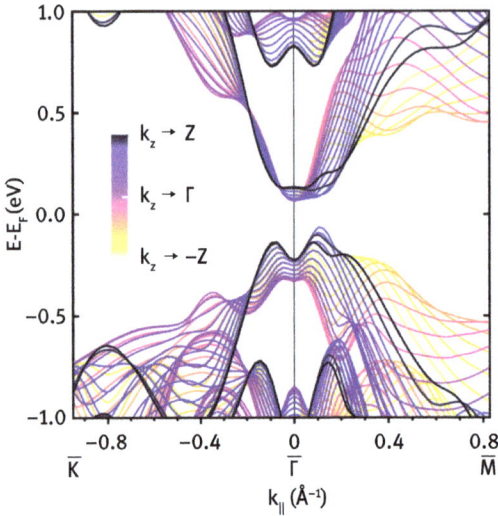

Figure 11.6: Calculated bulk band structure of Bi_2Te_3 employing the *GW* approximation. The bulk bands are projected onto the (111) surface along the $\bar{\Gamma} - \bar{K}$ and $\bar{\Gamma} - \bar{M}$ directions. Figure provided by I. Aguilera, Forschungszentrum Jülich.

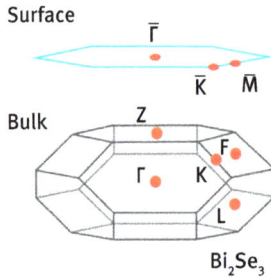

Figure 11.7: Brillouin zone of Bi_2Te_3. The upper hexagon shows the two-dimensional Brillouin zone of the projected (1, 1, 1) surface, with the high symmetry points $\bar{\Gamma}$, $\bar{K}$ and $\bar{M}$. The three-dimensional Brillouin zone is shown below. The time-reversal invariant points Γ, Z, F, and L are indicated.

In total there are eight TRIMs for a three-dimensional Brillouin zone. Using the method of Fu and Kane [175] it can be determined whether a material is a topological or trivial insulator, i. e. for a so-called strong topological insulator the product of the parity of the occupied states at the Γ-point is changed upon including spin-orbit coupling, while it remains unchanged on the other seven time-inversal invariant momenta. Using that criteria it could be confirmed that Bi_2Se_3, Bi_2Te_3, and Sb_2Te_3 are all topological insulators, while Sb_2Se_3 is a trivial insulator [169].

11.4 Surface states

For the quantum spin Hall systems we found that at the boundary one-dimensional edge channels are formed which have a linear energy-momentum relation. For three-dimensional topological insulators the equivalent scenario is the formation of two-dimensional topologically protected surface states.

11.4.1 Surface states deduced from the effective Hamiltonian

As we shall show below, the surface states directly emerge from the effective Hamiltonian given by equation (11.2) [15]. The procedure follows the one used to demonstrate the existence of edge channels in a quantum spin Hall system. Here we consider the half space with $z > 0$ and the model Hamiltonian

$$\hat{H} = \hat{H}_0 + \hat{H}_1, \tag{11.3}$$

with

$$\hat{H}_0 = \tilde{\epsilon}(k_z)I_{4\times4} + \begin{pmatrix} \widetilde{M}(k_z) & A_1 k_z & 0 & 0 \\ A_1 k_x & -\widetilde{M}(k_z) & 0 & 0 \\ 0 & 0 & \widetilde{M}(k_z) & -A_1 k_z \\ 0 & 0 & -A_1 k_z & -\widetilde{M}(k_z) \end{pmatrix} \tag{11.4}$$

and

$$\hat{H}_1 = D_2 k_\perp^2 I_{4\times4} + \begin{pmatrix} -B_2 k_\perp^2 & 0 & 0 & A_2 k_- \\ 0 & B_2 k_\perp^2 & A_2 k_- & 0 \\ 0 & A_2 k_+ & -B_2 k_\perp^2 & 0 \\ A_2 k_+ & 0 & 0 & B_2 k_\perp^2 \end{pmatrix}, \tag{11.5}$$

where,we used the definitions $\tilde{\epsilon}(k_x) = C + D_1 k_z^2$ and $\widetilde{M}(k_z) = M - B_1 k_z^2$. The first part contains all components of equation (11.2) with momenta k_z in z-direction, while the second part contains the momenta $k_\perp$ within the xy-plane. By setting $k_x = k_y = 0$ and replacing k_z by $-i\partial_z$, the surface state wave functions can be deduced in a similar fashion outlined in Section 10.4 [15]. A surface state exists for $M/B_1 > 0$. For finite k_x and k_y the Hamiltonian for the surface states is given by [169, 171]

$$\hat{H}_{surf}(k_x, k_y) = C + A_2(\hat{\sigma}_x k_y - \hat{\sigma}_y k_x). \tag{11.6}$$

This Hamiltonian corresponds to a 2×2 Dirac Hamiltonian for massless particles. The energy depends linearly on the in plane wave vector and corresponds to a single Dirac cone. The dispersion is depicted in Figure 11.8. As can be deduced from equation (11.6), the upper Dirac cone has a left-handed helicity.

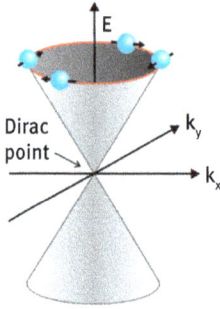

Figure 11.8: Dirac-cone of the surface states with a momentum in the $k_x k_y$-plane. The crossing point is the Dirac point. The spin orientation is perpendicular to the direction of propagation.

11.4.2 Surface states obtained from ab initio calculations

The existence of surface states is also confirmed by ab initio band structure calculations. In Figure 11.9 the band structure of Bi_2Te_3 is shown [172].

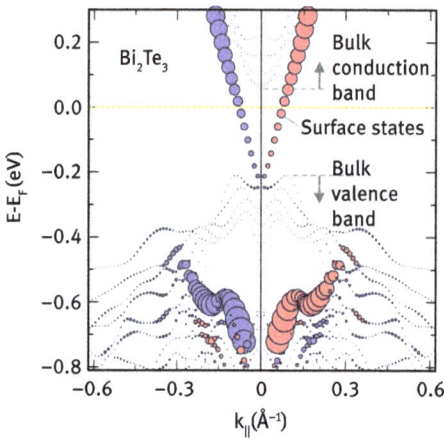

Figure 11.9: Simulated band dispersion of a Bi_2Te_3 slab along the $\bar{\Gamma} - \bar{K}$ direction. The spin-polarized spectral weight is expressed by the circles. The size of the circles gives the absolute spin polarization of the states for the in-plane p_x-component. The color corresponds to the sign. Figure provided by G. Bihlmayer, Forschungszentrum Jülich.

Around the Γ-point at $k = 0$ surface states are present which connect the valence band with the conduction band. Their dispersion is almost linear, thus corresponding to relativistic states. For this particular material the Dirac point is hidden in the valence band region. The surface states are spin-polarized as expressed by the color and size of the circles in Figure 11.9. The spin polarization is in-plane. The states with linear dispersion in the band gap only exist at the surface or at the interface to a trivial insulator. Within the bulk of the topological insulator, the presence of the band gap implies an insulating behavior, while the continuous evolution of the surface state in the band gap results in a finite conductance. Since the surface states are continuously connected within the gap the surface is conductive under all circumstances.

11.4.3 Topological protection of surface states

The deeper reason why topologically protected surface states exist within the bulk band gap is the band inversion due to spin-orbit coupling. This can most easily be understood by assuming a situation where the topological insulator is connected to a trivial insulator, where no band inversion due to strong spin-orbit coupling occurred. In the bulk both materials are insulating, because the Fermi energy is assumed to be located within the corresponding band gaps. However, at the interface the bands of both materials have to be matched by a continuous transfer to each other. For example the upper band in the topological insulator has to be matched to the lower, non-inverted band of the trivial insulator, since they are of the same type. The same is true for the lower band in the topological insulator, which transfers to the upper band in the trivial insulator. In any case, these states at the interface are located within the band gaps of both materials and thus allow electrical conductance.

We can generalize the case discussed above and make up our mind about under which condition the surface states are topologically protected. Let us begin with the topologically trivial situation depicted in Figure 11.10 (a).

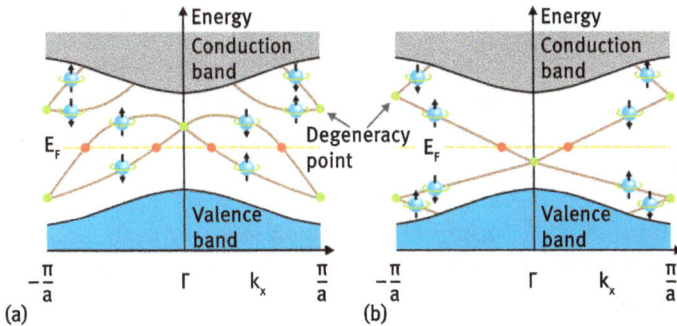

Figure 11.10: Electronic dispersion between two boundary Kramers degenerate points, i. e. the Γ point and some other degeneracy point at the boundary of the Brillouin zone. (a) In a normal insulator an even number of surface states cross the Fermi level E_F for k_x in the range between Γ and π/a, with a the lattice constant. (b) Topologically protected surface states with an odd number of crossing points between Γ and π/a.

Here, the band gap region between the bulk valence and conduction bands is shown, including the surface states within the gap. The wave vector k_x spans the complete Brillouin zone from $-\pi/a$ to π/a with the Γ-point at $k_x = 0$. Here, a is the lattice constant. We know from Section 7.6.3 that the presence of spin-orbit coupling can lift the spin degeneracy of states, except for special time-invariant symmetry points, e. g. at the Γ point or at $k = \pm\pi/a$, where a two-fold degeneracy is required, owing to Kramer's theorem. At zero magnetic field time-inversion symmetry holds, so that at finite k vectors, the states $|k, \uparrow\rangle$ and $|-k, \downarrow\rangle$ are time-inversed pairs and are energetically degenerate. However, due to the presence of spin-orbit coupling states with the same k vector but opposite spins are usually nondegenerate. This situation is depicted in Figure 11.10 (a). The surface states split-up at the Γ point and merge pairwise at a Kramers degeneracy points at the border of the Brillouin zone, i. e. at $-\pi/a$ and π/a. If we have a look at the crossings of the surface states with the Fermi level in the range between Γ and π/a we find an even number. Due to the crossing with E_F the surface is conductive. However, by changing the outer potential slightly, one can imagine a situation where the surface states are pushed below E_F, so that there are no crossing points left, and thus no surface conductance occurs. There is no topological protection, since a simple shift of these states can suppress the surface conductance. As shown in Figure 11.10 (a), in case of a topological insulator the bands intersect E_F an odd number of times. The spin-split surface states are not pairwise connected between the symmetry points. Even in case that the surface states are moved by some outer potentials, it is never possible to prevent a crossing of at least one state with E_F. The surface conductance is thus topologically protected.

11.5 Angle-resolved photo-emission spectroscopy

The band structure of a topological insulator can be determined experimentally by using angle-resolved photo-emission spectroscopy (ARPES), which is based on the photoelectric effect [33]. A scheme of the setup is shown in Figure 11.11 (a).

Here, the sample is irradiated by a photon source emitting photons of energy $h\nu$, with ν the frequency. As a source one can employ a gas discharging lamp, e. g. using He gas, which has two main photon lines at 21.2 eV (He I) and 40.8 eV (He II), respectively. Alternatively, synchrotron radiation can be utilized to supply monochromatic photons in a broad energy range. When light from the photon source is incident on the surface of the sample an electron can absorb a photon and escape from the material with the kinetic energy E_{kin}. The corresponding energy scheme is shown in Figure 11.11 (b). The kinetic energy of the escaped electron is determined by

$$E_{kin} = h\nu - \phi - |E_B|. \tag{11.7}$$

In order to be released from the surface, the electron has to overcome the binding energy E_B and the material work function ϕ. The latter is typically around 4–5 eV for metals. Thus with a given photon energy $h\nu$ and a known value for ϕ, the binding energy E_B can be determined by measuring the kinetic energy. This can be done by using

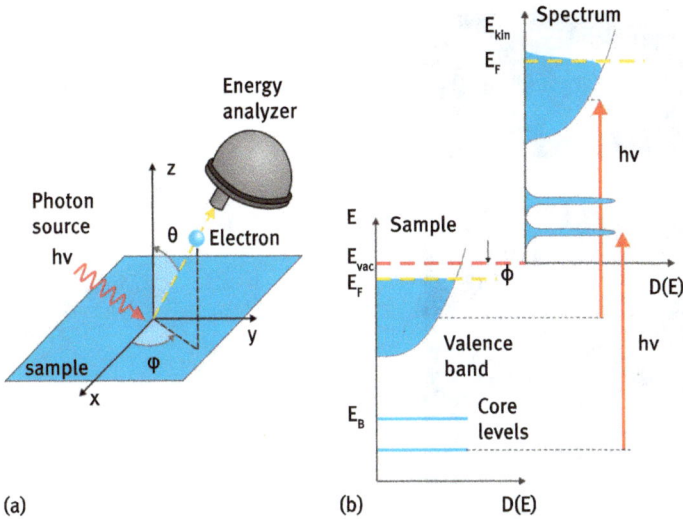

Figure 11.11: (a) Setup for angle-resolved photo-emission spectroscopy. The sample is irradiated by a photon source with a photon energy of hv. The electron emitted under a polar θ and azimuthal φ emission angle are detected by an electrostatic hemispherical energy analyzer. (b) Simplified scheme illustrating the excitation of a photo electron. On the left the density of states of a metal is shown consisting of core levels with a large binding energy and a metallic band crossing the Fermi level E_F. The vacuum level is defined to be zero. The work function ϕ quantifies the energy an electron at E_F needs to escape to the vacuum. On the right side the measured energy spectrum is shown. The stronger the electrons are bound, the lower is the kinetic energy of the escaped electrons.

an energy analyzer as depicted in Figure 11.11 (a). The entrance opening of the energy analyzer can be set to different orientations so that only electrons emitted along a direction characterized by the polar θ and azimuthal φ emission angles are measured. By this, the momentum of the electron $\hbar\vec{k}$ is completely determined. For the component perpendicular to the surface $\hbar k_\perp$ one has to put in the information of the work function, since in this direction the momentum of the electron is changed when it escapes the material. However, for the component along the surface $\hbar k_\parallel$ translation symmetry is conserved so that one directly finds

$$\hbar k_\parallel = \sqrt{2mE_{kin}} \sin \theta. \tag{11.8}$$

Thus, with ARPES the band structure of a material can be determined. One has to keep in mind that the electrons are only emitted from the surface layer with a thickness of a few nanometers [33]. This means that band bending effects at the surface are basically not detected. Therefore, it might be possible that the occupation of the bands further inside the material is different.

A typical ARPES $|E_B|$ vs. k_x map of a Bi_2Te_3 layer is shown in Figure 11.12 (b) [172, 176, 177]. The binding energy is defined with respect to the Fermi energy E_F. In

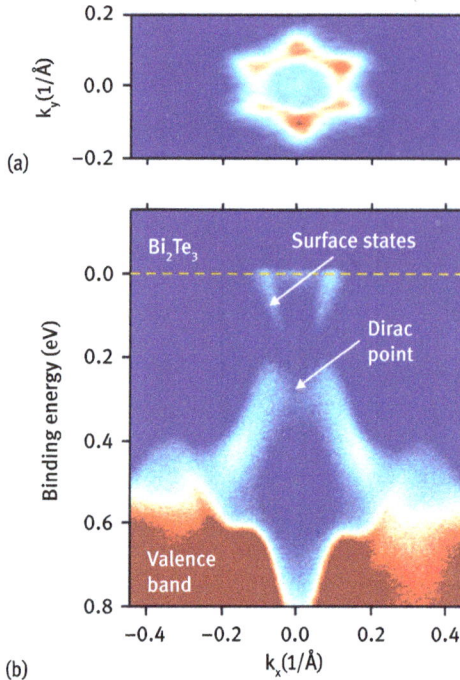

Figure 11.12: Angle-resolved photo/emission spectroscopy (ARPES) map of Bi_2Te_3. (a) Fermi surface with the wave vectors k_x and k_y lying within the sample surface. The Fermi surface has a hexagonal symmetry. (b) Binding energy as a function of k_x. The Fermi energy is defined to be zero. Right below the Fermi energy down to 0.2 eV the surface states are visible. The Dirac point is located in a pocket of the bulk valence band. Figure provided by L. Plucinski and M. Eschbach, Forschungszentrum Jülich.

the range between the Fermi energy and about 0.2 eV the surface states with a linear energy-momentum dispersion are located. The Dirac point is hidden in the bulk valence band. This is in accordance to the band structure calculations shown in Figure 11.9. In Figure 11.12 (a) the corresponding Fermi surface is shown. It can be seen that it has a hexagonal symmetry, in contrast to the ideal case depicted in Figure 11.8. The more complex structure can be attributed to the Dresselhaus spin-orbit coupling contribution together with the hexagonal symmetry of the surface.

The ARPES energy vs. momentum map of an epitaxial Bi_2Se_3 layer is shown in Figure 11.13.

Here, one finds that the Fermi level is located within the conduction band, i. e. the material is n-type. This is a common problem of three-dimensional topological insulators. Due to vacancies or antisite defects the material is doped. In case of Bi_2Se_3 usually degenerately n-type so that metallic bulk conductance dominates the transport. This makes it difficult to obtain any information on surface state conductance in transport experiments. Nevertheless, the surface states within the band gap are re-

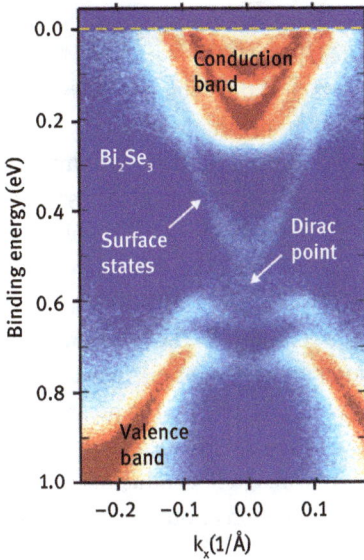

Figure 11.13: Angle-resolved photo-emission spectroscopy map of Bi_2Se_3. The Fermi level is located within the conduction band. The surface states within the band gap have a linear dispersion. The Dirac point is located close to the valence band edge. Figure provided by L. Plucinski and M. Eschbach, Forschungszentrum Jülich.

solved by ARPES measurements. As can be seen in Figure 11.13, the dispersion is linear with the Dirac cone slightly above the valence band maximum.

11.6 Transport experiments

The existence of topologically protected surface states could be proven unambiguously by means of angle-resolved photo-emission spectroscopy measurements. However, for the realization of spin electronic devices it is also necessary to reveal the presence of surface states in electrical transport measurements. Here, one often meets with difficulties which originate from the large unintentional background doping of the material due to antisite defects and vacancies. The Fermi level lies either in the conduction or valence band and not within the bulk band gap, i. e. Bi_2Te_3 is found to be n-type, while for Sb_2Te_3 a p-type conductance is usually observed. As a consequence there is a significant contribution from bulk carriers to the conductance in addition to the surface channel. In order to eliminate the bulk contribution, the material can be doped to compensate the background doping, i. e. the conductance of Bi_2Te_3 can be controlled by Sn doping [178]. Furthermore, the conductance of Bi_2Se_3 can be tuned from n- to p-type by doping with Ca [179]. Another possible solution to reduce the bulk contribution is the growth of ternary or even quaternary material. $(Bi_{1-x}Sb_x)_2Te_3$ is formed by alloying n-type Bi_2Te_3 with p-type Sb_2Te_3. Here the different types of doping for each component result in a compensation effect [180, 181].

In order to resolve the topological protected surface states, one has to look for magnetotransport phenomena, which are solely found in two-dimensional systems. In Section 2.8 we discussed Shubnikov–de Haas oscillations and the quantum Hall effect, which rely on the two-dimensional nature of the carrier system. These phenomena were observed in transport measurements on three-dimensional topological insulators and indicate the presence of two-dimensional topological protected surface states [182, 183, 184, 185]. Another possibility to reveal surface states in transport experiments is to search for Aharonov–Bohm type oscillations in the magnetoresistance of nanowires or nanoribbons [186, 187]. The latter term is sometimes used for flat ribbon-like structures. The magnetoconductance oscillations rely on the presence of coherent surface states on the circumference, which are penetrated by a magnetic flux. Since this effect is in close connection to the flux-periodic magnetoconductance oscillation of semiconductor nanowires, we will discuss this phenomenon in detail below.

11.6.1 Topologically protected surface states in nanowires

A signature of transport in topologically protected surface states are flux-periodic magnetoconductance oscillations in topological insulator nanowires [188, 189, 190]. A schematic illustration of a nanowire with circular cross section is depicted in Figure 11.14. The magnetic field is oriented axially. The approach is conceptually very similar to the one previously discussed in connection to semiconductor nanowires, where the presence of surface states due to an accumulation layer resulted in magnetoconductance oscillations with a period of a magnetic flux quantum h/e (cf. Section 7.7.3).

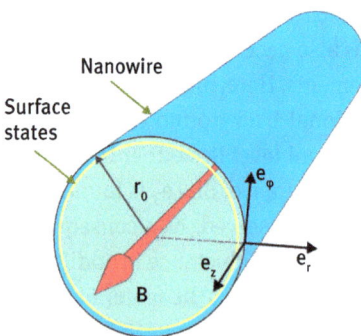

Figure 11.14: Schematic illustration of a topological insulator nanowire with circular cross section. The magnetic field $\vec{B}$ is oriented along the nanowire axis. The surface states circumvent the magnetic flux given by the magnetic field times the nanowire cross section πr_0^2, with r_0 the radius.

The main difference between both systems is the linear energy-momentum relation when it comes to topological insulators. We will first derive an expression for the energy levels of angular momentum states as a function of a magnetic flux penetrating the cross section of the wire. Magnetoconductance measurements on Sb_2Te_3 nanowires will be discussed later

In order to find an expression for the energy levels of bound states at the surface of a topological insulator nanowire, we start with an idealized situation. Instead of referring to the effective Hamiltonian, as given by equation (11.2), which reflects the special symmetry of the band structure, we begin with the three-dimensional massive Dirac Hamiltonian

$$\hat{H} = v \begin{pmatrix} Mv & \vec{\sigma}\vec{p} \\ \vec{\sigma}\vec{p} & -Mv \end{pmatrix}, \tag{11.9}$$

with v being the Dirac velocity and M the mass. We assume a flat interface at $z = 0$, with the topological insulator extending into $z < 0$. For this case the surface state can be expressed as

$$\Psi(\vec{r}) = e^{-\frac{Mv}{\hbar}z} \begin{pmatrix} \chi(x,y) \\ \psi(x,y) \end{pmatrix}. \tag{11.10}$$

Similar to the case of the quantum spin Hall effect, we assumed a negative mass ($M < 0$) in the topological insulator, which reflects the band inversion. For $z < 0$ the surface state exponentially decays into the bulk. For vacuum, i. e. $z > 0$, we set $M \to \infty$, and thus $\Psi(\vec{r})$ is zero. At the boundary between a topological insulator and a trivial insulator or to vacuum the mass has to change sign while crossing the barrier.

Inserting the ansatz equation (11.10) into the Dirac equation for energy $E = 0$ and assuming zero momentum in x- and y-direction leads to

$$Mv^2\chi + i\hat{\sigma}_z v^2 M\psi = 0, \tag{11.11}$$

which results in

$$\chi = -i\hat{\sigma}_z\psi. \tag{11.12}$$

With $\vec{e}_z$ being the normal vector on the surface, one can also write

$$\chi = -i\hat{\vec{\sigma}}\vec{e}_z\psi, \tag{11.13}$$

which will help us below to express the appropriate relation for a curved surface. Using the relation between χ and ψ for the Dirac Hamiltonian given by equation (11.9), one finds for the surface state Hamiltonian

$$\hat{H}_{surf} = -iv(\hat{\sigma}_x\hat{p}_x + \hat{\sigma}_y\hat{p}_y)\hat{\sigma}_z, \tag{11.14}$$

which ends up as

$$\hat{H}_{surf} = v(\hat{p}_y\hat{\sigma}_x - \hat{p}_x\hat{\sigma}_y) \tag{11.15}$$

when we use the relations $\hat{\sigma}_x\hat{\sigma}_z = -i\hat{\sigma}_y$ and $\hat{\sigma}_y\hat{\sigma}_z = i\hat{\sigma}_x$. This Hamiltonian corresponds to equation (11.6), which was derived from the effective Hamiltonian for three-dimensional topological insulators. Assuming plane wave solutions along the x- and y-direction the energy momentum relation is linear:

$$E = \pm v\hbar|k|, \tag{11.16}$$

as illustrated in Figure 11.8.

Up to now we did not include any curvature of the boundary. The relation between χ and ψ can be generalized for any normal vector on a curved surface of a topological insulator. Since we assumed a cylindrical nanowire with a circular cross section and a radius r_0, as shown in Figure 11.14, the radial unit vector $\vec{e}_r$ in cylindrical coordinates corresponds to the normal vector of the outer surface. Analog to equation (11.13) one obtains for the relation between χ and ψ

$$\chi = -i\hat{\sigma}\vec{e}_r\psi. \tag{11.17}$$

Regarding the corresponding system for the cylindrical surface one has to make sure that the Hamiltonian is Hermitian. This is achieved by translating the Hamiltonian given by equation (11.14) to

$$\hat{H}_{surf} = -\frac{i}{2}v[\hat{\sigma}\hat{p}, \hat{\sigma}\vec{e}_r] = -\frac{1}{2}v[\hbar\vec{\nabla}\vec{e}_r + \vec{e}_r(\hat{p} \times \hat{\sigma}) + (\hat{p} \times \hat{\sigma})\vec{e}_r]. \tag{11.18}$$

The commutator ensures that $\hat{H}_{surf}$ is Hermitian. In order to write the Hamiltonian more explicitly, we use a cylindrical coordinate system. According to Section 7.7.3, the Pauli spin matrices σ_r and σ_ϕ for a cylindrical system are given by equations (7.74) and (7.75), respectively. For a magnetic field along the nanowire axis the corresponding vector potential can be expressed by

$$\vec{A} = \frac{1}{2}Br_0\vec{e}_\phi. \tag{11.19}$$

This vector potential contributes to the kinetic momentum $\vec{\pi} = \vec{p}+e\vec{A}$. The Hamiltonian resulting from equation (11.18) for the surface states is expressed by

$$\hat{H}_{surf} = -\frac{v\hbar}{2r_0}I_{2\times2} - v\begin{pmatrix} \frac{1}{r_0}(\frac{\hbar}{i}\frac{\partial}{\partial\phi} + \frac{1}{2}eBr_0^2) & \hbar\frac{\partial}{\partial z}e^{-i\phi} \\ -\hbar\frac{\partial}{\partial z}e^{i\phi} & -\frac{1}{r_0}(\frac{\hbar}{i}\frac{\partial}{\partial\phi} + \frac{1}{2}eBr_0^2) \end{pmatrix}. \tag{11.20}$$

Motion is allowed along the nanowire axis free carrier, so that one can use a plane wave solution for the propagation along the z-axis. Furthermore, the total angular momentum of the system is conserved. This brings us to the following ansatz for the wave function:

$$\Psi = e^{ikz} e^{i\phi l} \begin{pmatrix} \chi_1 \\ e^{i\phi} \chi_2 \end{pmatrix}, \tag{11.21}$$

with $l = 0, \pm1, \pm2, \ldots$ being the orbital angular momentum quantum number, while k is the wave vector along the z-direction. Inserting this wave function into the surface Hamiltonian results in

$$\hat{H}_{surf} = -\frac{v\hbar}{2r_0} I_{2\times2} - v \begin{pmatrix} \frac{\hbar}{r_0}(l + \frac{\Phi}{\Phi_0}) & i\hbar k e^{-i\phi} \\ -i\hbar k e^{i\phi} & -\frac{\hbar}{r_0}(l+1+\frac{\Phi}{\Phi_0}) \end{pmatrix}. \tag{11.22}$$

Joining both matrices finally gives

$$\hat{H}_{surf} = -v \begin{pmatrix} \frac{\hbar}{r_0}(l + \frac{1}{2} + \frac{\Phi}{\Phi_0}) & i\hbar k e^{-i\phi} \\ -i\hbar k e^{i\phi} & -\frac{\hbar}{r_0}(l + \frac{1}{2} + \frac{\Phi}{\Phi_0}) \end{pmatrix}. \tag{11.23}$$

The energy eigenvalues obtained from this Hamiltonian can be straightforwardly calculated. They are given by

$$E_{lk} = \pm\hbar v \sqrt{k^2 + \frac{(l + \frac{1}{2} + \frac{\Phi}{\Phi_0})^2}{r_0^2}}. \tag{11.24}$$

The corresponding energy spectrum for zero magnetic field ($\Phi = 0$) is shown in Figure 11.15 (a).

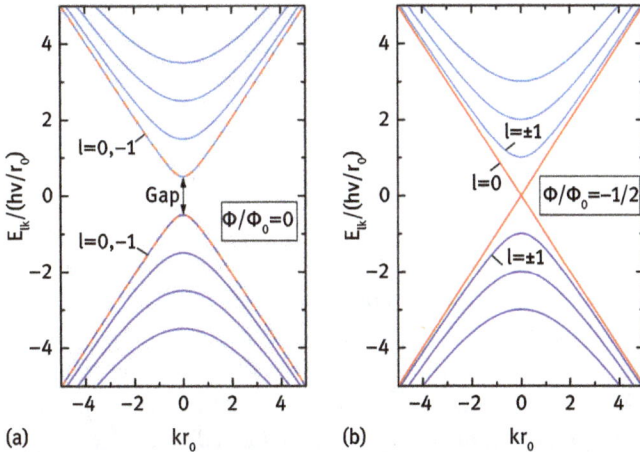

Figure 11.15: (a) Normalized band structure of a topological insulator nanowire at zero magnetic field, i. e. $\Phi/\Phi_0 = 0$. The angular momentum quantum numbers of the states split by the gap are $l = 0$ and -1. (b) Corresponding band structure for $\Phi/\Phi_0 = -1/2$.

At $k = 0$ an energy gap is formed between the $l = 0$ and $l = -1$ states. This is different from an extended flat surface, where a linear dispersion and no gap are found (cf. equation (11.16)). The gap scales with $1/r_0$, i. e. the smaller the radius the larger the gap is. The gap is a result of the Berry phase of π accumulated while circulating about the outer surface.

As can be seen in Figure 11.15 (b), the gap is closed for the states with $l = 0$ if the flux corresponds to a half flux quantum $\Phi = -\Phi_0/2$. More generally the gap vanishes at $\Phi = \Phi_0(n + 1/2)$, with $n = 0, \pm1, \pm2, \ldots$, as one can infer from equation (11.24). The flux periodicity of the energy spectrum can be seen in Figure 11.16. In contrast to the cylindrical semiconductor nanowire, as discussed in Section 7.7.3, here the energy depends linearly on the flux for $k = 0$. Around $E = 0$ the energy gap periodically opens and closes having a diamond shape.

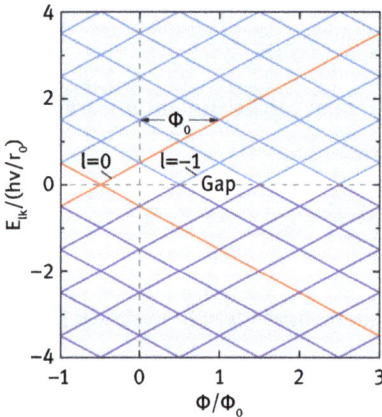

Figure 11.16: (a) Normalized band structure of a topological insulator nanowire as a function of Φ/Φ_0 at $k = 0$ and various values of the orbital angular momentum quantum number l.

11.6.2 Flux-periodic oscillations in nanowire structures

The energy spectrum discussed above has a direct impact on the transport properties. Very similar to the transport in semiconductor nanowires, it can be expected that the flux-periodicity in the energy spectrum will affect the magnetoconductance of a topological insulator nanowire or nanoribbon. Indeed, magnetoconductance oscillations were observed in measurements on Bi_2Se_3 [186] and Bi_2Te_3 nanoribbons [187]. Below we will discuss the transport properties of an Sb_2Te_3 nanowire in detail.

A schematic illustration of a topological insulator nanowire is shown in Figure 11.17 (a). Owing to the layered crystal structure, the cross section is often rectangular.

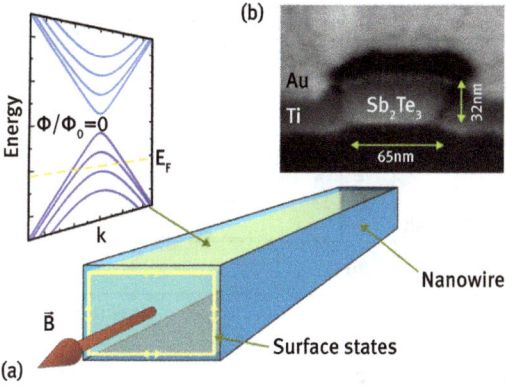

Figure 11.17: (a) Schematic illustration of a topological insulator nanowire. The magnetic field is oriented along the nanowire axis. The surface states circumvent the magnetic flux given by the magnetic field times the nanowire cross section. A schematic of the energy spectrum is shown for $\Phi/\Phi_0 = 0$. (b) Scanning electron microscope image of an Sb_2Te_3 nanowire cross section (65×32 nm^2). The nanowire is covered by a Ti/Au layer. Image provide by Y. Arango, Forschungs-zentrum Jülich, and J. G. Lu, University of Southern California, Los Angeles.

Nevertheless, one can assume that topologically protected surface states are formed continuously around the circumference of the nanowire, similar to the previously discussed case of the nanowire comprising a circular cross section. Therefore, the energy spectrum will have the same features, i. e. a gap at zero magnetic field. A scanning electron microscopy image of the cross section of an Sb_2Te_3 nanowire is shown in Figure 11.17 (b). The nanowire has an almost rectangular cross section. For transport measurements it is covered by a Ti/Au contact.

The corresponding magnetoresistance at 1.8 K is shown in Figure 11.18 (a).

Here, ΔR is plotted, where the slowly varying background resistance was subtracted. The magnetic field is aligned along the nanowire axis. The material Sb_2Te_3 is usually p-type, and thus the Fermi level is located within the bands below the energy gap, as illustrated in Figure 11.17 (a). The magnetoresistance shows a clear oscillating behavior with a period of 2.2 T. In the Fourier transform depicted in Figure 11.18 (b) the periodic oscillation results in a peak at about $0.45\,T^{-1}$. The cross section of the nanowire is approximately 2.1×10^{-15} m^{-2}, which results in a flux period of 4.6×10^{-15} Tm^2. The flux period is close to a magnetic flux quantum $h/e = 4.14 \times 10^{-15}$ Tm^2. Figure 11.18 (c) shows the magnetoresistance oscillations for an Sb_2Te_3 nanowire with a larger cross section area of 5.3×10^{-15} m^2. Consequently, the observed oscillation period is smaller, i. e. $\Delta B = 0.75$ T. As can be seen in Figure 11.18 (d), a peak is found at around $1.34\,T^{-1}$ in the Fourier spectrum. For this sample the corresponding flux period is 3.97×10^{-15} Tm^2, which is once again close to the magnetic flux quantum. Both samples show a behavior, which is consistent with the theoretical model introduced in the previous section. As shown in Figure 11.16, the energy spectrum is periodic with the magnetic flux. Assuming a fixed Fermi energy, the number of

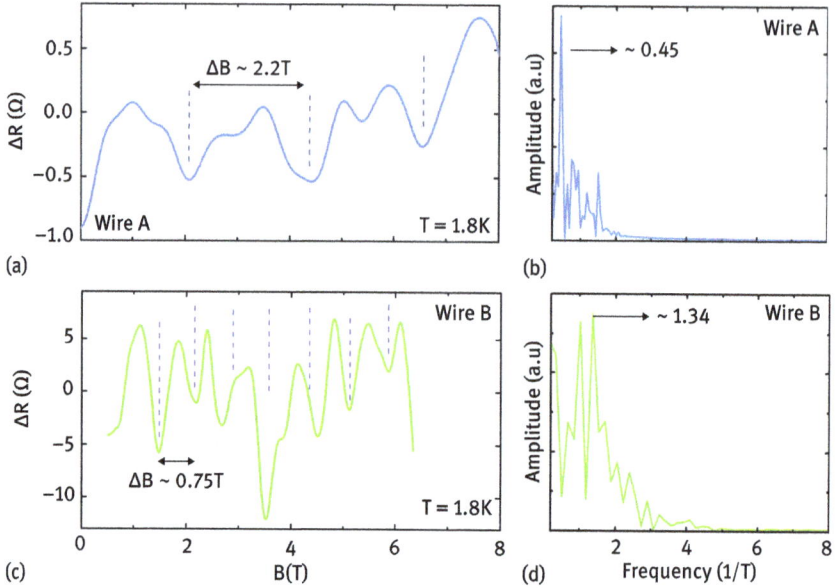

Figure 11.18: (a) Resistance oscillations ΔR of an Sb_2Te_3 nanowire (wire A) as a function of a magnetic field. The magnetic field was aligned along the nanowire axis. The cross section of the nanowires is shown in Figure 11.17b. The period of ΔB = 2.2 T fits to a single flux quantum penetrating the nanowire cross section. (b) Corresponding Fourier transform. (c) Magnetoresistance oscillations of wire B with larger cross section, i. e. 72 × 73 nm^2. The oscillation period is 0.75 T. (d) Corresponding Fourier transform. Figure provide by Y. Arango, Forschungszentrum Jülich.

occupied states varies periodically with the magnetic flux [190]. As already explained in connection with the semiconductor nanowires (cf. Section 7.7.3), this is expected to result in a flux-periodic oscillation of the nanowire resistance. As a matter of fact, the observed oscillations can only be explained if the carriers encircle a well-defined area. For a wire, which is conductive within the whole cross section, many closed loops of electron paths with different cross sectional areas would be possible. In this case oscillations with larger periods are expected, owing to areas smaller than the total cross section of the wire. Thus, no distinct frequency would be observed or even more the superposition of oscillations with different frequencies would average out the resistance modulations.

11.6.3 Landau quantization in two-dimensional topological surface states

In Section 2.8.1 we found that in a two-dimensional electron system with parabolic energy dispersion the application of a perpendicular magnetic field results in the so-called Landau quantization. Here, we will have a look at the situation when the energy dispersion is linear (cf. Figure 11.8). As for the system with parabolic dispersion, we have to take care that the magnetic field is properly included into the surface state Hamiltonian given by equation (11.15) by inserting the vector potential into the expression for the momentum

$$\hat{p}_x \to -i\hbar \frac{\partial}{\partial x}, \tag{11.25}$$

$$\hat{p}_y \to \hbar k_y + eA = \hbar k_y + eBx. \tag{11.26}$$

As done previously for the parabolic system, we separate the spatial part of the wave function

$$\Psi = e^{ik_y y} \begin{pmatrix} \chi_1(x) \\ \chi_2(x) \end{pmatrix}, \tag{11.27}$$

so that in the presence of a magnetic field the Hamiltonian given by equation (11.15) can be written as

$$\hat{H} = v_F \hbar \left[(k_y + eBx/\hbar)\sigma_x + i \frac{\partial}{\partial x} \sigma_y \right]. \tag{11.28}$$

By employing the relation we already used for the system with parabolic dispersion, i. e.

$$\hbar k_y + eBx = \frac{\hbar}{l_m^2}(x - x_0), \tag{11.29}$$

with l_m the magnetic length and the corresponding definition of $\hat{a}$ and $\hat{a}^\dagger$ given by equations (2.62) and (2.61), the Hamiltonian takes the following form

$$\hat{H} = v_F \hbar \left[\begin{pmatrix} 0 & -\partial_x \\ \partial_x & 0 \end{pmatrix} + \begin{pmatrix} 0 & k_y + eBx/\hbar \\ k_y + eBx/\hbar & 0 \end{pmatrix} \right]$$

$$= \frac{\sqrt{2}v_F \hbar}{l_m} \begin{pmatrix} 0 & \hat{a}^\dagger \\ \hat{a} & 0 \end{pmatrix}. \tag{11.30}$$

We now have to find the solution of the corresponding equation

$$\hat{H} \begin{pmatrix} \psi_1(x) \\ \psi_2(x) \end{pmatrix} = v_F \sqrt{2eB\hbar} \begin{pmatrix} 0 & \hat{a}^\dagger \\ \hat{a} & 0 \end{pmatrix} \begin{pmatrix} \psi_1(x) \\ \psi_2(x) \end{pmatrix} = E \begin{pmatrix} \psi_1(x) \\ \psi_2(x) \end{pmatrix}, \tag{11.31}$$

where the wave functions are spinors, with the components $\psi_1(x)$ and $\psi_2(x)$. For the sake of simplicity we omit to explicitly write the x-dependency of ψ_1 and ψ_2 from now on. Both components are coupled by

$$\hat{a}^\dagger \psi_2 = \tilde{E}\psi_1, \tag{11.32}$$

$$\hat{a}\psi_1 = \tilde{E}\psi_2, \tag{11.33}$$

where we used the normalized form $\tilde{E} = E/(v_F\sqrt{2eB\hbar})$. From that it follows directly

$$\psi_1 \sim |n\rangle \quad \leftrightarrow \quad \psi_2 \sim |n-1\rangle, \tag{11.34}$$

with the solutions given by equation (2.72). In order to obtain the energy eigenvalues, the Hamiltonian is applied once again

$$\hat{H}^2\begin{pmatrix}\psi_1\\\psi_2\end{pmatrix} = v_F^2 2eB\hbar\begin{pmatrix}\hat{a}^\dagger\hat{a} & 0\\0 & \hat{a}\hat{a}^\dagger\end{pmatrix}\begin{pmatrix}\psi_1\\\psi_2\end{pmatrix} = E^2\begin{pmatrix}\psi_1\\\psi_2\end{pmatrix}. \tag{11.35}$$

Using the normalized energy $\tilde{E}$, we find for the first component of the spinor

$$\hat{a}^\dagger a\psi_1 = \tilde{E}^2\psi_1. \tag{11.36}$$

Comparing to the harmonic oscillator we directly find

$$\hat{a}^\dagger \hat{a}\psi_1 = n\psi_1 = \tilde{E}^2\psi_1, \tag{11.37}$$

with $n = 1,2,3,\dots$. Thus, finally we obtain for the energy eigenvalues of a two-dimensional Dirac system

$$E_n = \pm v_F\sqrt{2eB\hbar n}. \tag{11.38}$$

In contrast to a system with a parabolic dispersion, here a state at zero energy exists. Furthermore, the states are no longer equally spaced in energy. The Landau level fan for the Dirac system is depicted in Figure 11.19.

11.6.4 Shubnikov–de Haas oscillations in Dirac systems

The Landau quantization in a two-dimensional Dirac system upon applying a perpendicular magnetic field results in magnetoresistance oscillations, i. e. Shubnikov–de Haas oscillations. However, in contrast to the parabolic system, as discussed in Section 2.8.2, here, the energy difference between adjacent Landau levels is no longer constant, i. e.

$$\Delta E = v_F\sqrt{2e\hbar B(n+1)} - v_F\sqrt{2e\hbar Bn}$$
$$= v_F\sqrt{2e\hbar B}(\sqrt{n+1} - \sqrt{n}). \tag{11.39}$$

In order to observed a maximum in the magnetoresistance R_{xx} the energy eigenvalues E_n have to match with the Fermi energy E_F

$$E_n \doteq E_F \Rightarrow E_F = v_F\sqrt{2e\hbar Bn}, \tag{11.40}$$

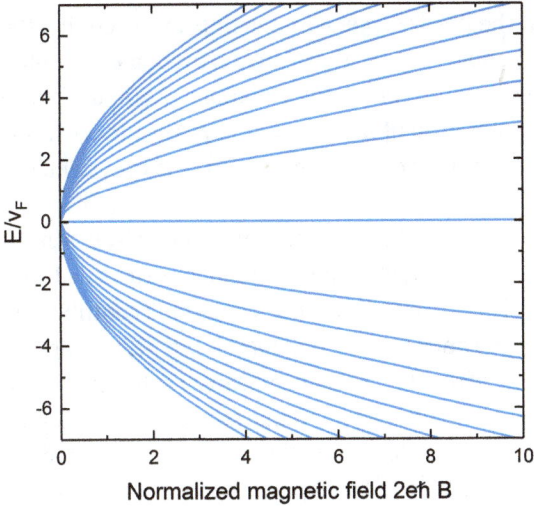

Figure 11.19: Landau level fan for a two-dimensional system with linear dispersion for levels up to $n = 12$.

which results for the level index n in

$$n = \frac{E_F^2}{v_F^2 2e\hbar} \frac{1}{B}. \tag{11.41}$$

If we compare this result with the situation of the parabolic case, cf. equation (2.80), we find that the peak positions in the resistance are shifted by a factor 1/2. The oscillation frequency when plotting vs. $1/B$ is given by

$$f_{1/B} = \frac{E_F^2}{v_F^2 2e\hbar}. \tag{11.42}$$

With the Fermi energy given by $E_F = v_F \hbar k_F$ and $k_F = \sqrt{4\pi n_{2D}/g_s}$ one finds

$$f_{1/B} = \frac{h}{2e} \frac{2}{g_s} n_{2D}. \tag{11.43}$$

For a Dirac system the spin degeneracy is usually lifted, i. e. $g_s = 1$. Once again the oscillations are periodic in $1/B$ and the oscillation frequency is proportional to the carrier concentration n_{2D}. In analogy to the parabolic system equation (11.41) corresponds to maxima in the magnetoresistance R_{xx}. Therefore, we can write for the Shubnikov–de Haas oscillations in the longitudinal magnetoresistance

$$\Delta R_{xx} \sim \cos\left(2\pi f_{1/B} \frac{1}{B}\right). \tag{11.44}$$

For the parabolic as well as for the linear dispersion it is thus expected that the magnetoresistance oscillations are periodic in $1/B$. Therefore, no fundamental difference is observed in this respect. However, by comparing equations (2.80) and (11.41) one finds that there is a phase shift of π present between both systems. Indeed, this phase shift can be utilized to distinguish in an experiment, which kind of system, i. e. parabolic or linear, is present.

In Figure 11.20 (a) Shubnikov–de Haas oscillations in the magnetoresistance measured on a Sb_2Te_3 Hall bar structure are shown [191]. The oscillations are revealed after subtracting a smooth background resistance, which is due to a large bulk contribution. From additional Hall measurements one finds that the transport is p-type, i. e. the current is carried by holes. From the slope of the Hall resistance a carrier concentration of $n_{Hall} = 5 \times 10^{13}$ cm^{-2} is extracted. One can identify five extrema starting at $B = 8.5$ T. By extracting the oscillation frequency of $f_{1/B} = 40.7$ T from the Fourier analysis one obtains a carrier concentration $n_{2D} = 9.85 \times 10^{11}$ cm^{-2} by employing equation (11.43). This is about two orders of magnitude lower the value extracted from Hall measurements. This discrepancy is found often in topological insulators like Bi_2Te_3 or Sb_2Te_3 where a bulk contribution dominates the transport. The measurements shown here indicate that at least two transport channels are present, i. e. a bulk channel with a relatively large carrier concentration in the order of 10^{13} cm^{-2} and a two-dimensional surface channel of surface channel with a carrier concentration in the order of 10^{12} cm^{-2}. In Figure 11.20 (b) the filling factor, i. e. the number of occupied Landau levels, is plotted as a function of inverse magnetic field. An integer filling factor corresponds to a minimum in the oscillations shown in Figure 11.20 (a). According to the theoretical considerations discussed above, for a Dirac system of surface carriers the filling factor should be shifted in the quantum limit ($1/B \to 0$) from 0 for a parabolic system to 1/2. Thus, in principle one should be able to identify the type of surface states by extrapolating the Landau level fan diagram to $1/B = 0$ T^{-1}, if the intersection with the filling factors is shifted from 0 to 1/2. For the case shown in Figure 11.20 (b) the extrapolation intersects at a filling factor of about 0.15. This is clearly different from the expected values of 0 for a two-dimensional system but also from 1/2 for ideal Dirac fermions. Possible reasons for the non-ideal behavior are deviations of the Dirac spectrum from a strict linear dispersion due to strong spin-orbit coupling and lack of inversion symmetry of the surface states, as well as Zeeman coupling of the carriers to the external magnetic field [192, 193].

11.7 Summary

- In three-dimensional topological insulators the bulk is insulating, while the surface is conducting. Typical materials are Bi_2Te_3, Bi_2Se_3, or Sb_2Te_3. The crystal is formed by quintuple layers which are bonded by van der Waals forces.

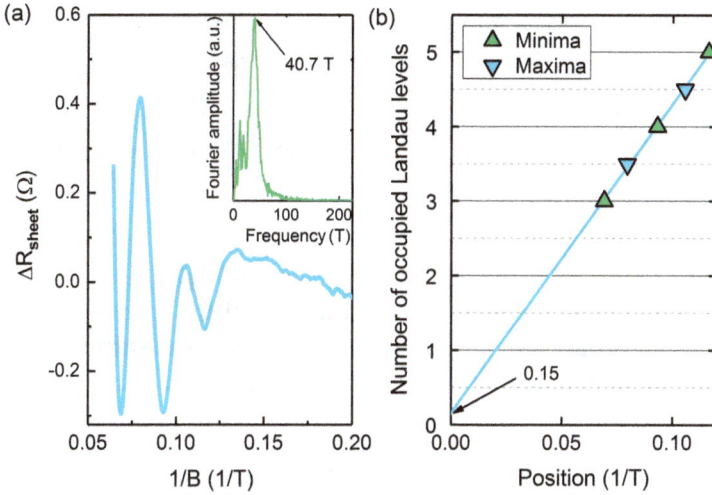

Figure 11.20: (a) Shubnikov–de Haas oscillations of a 13.6-nm-thick Sb_2Te_3 layer plotted as a function of inverse magnetic field. The inset shows the corresponding Fourier transform. (b) Number of occupied Landau levels, i. e. filling factor vs. inverse magnetic field (Figure provided by Christian Weyrich, Forschungszentrum Jülich).

- The surface states are topologically protected. They cannot be removed by potential shifts. The surface states are helical, i. e. the direction of propagation and the spin orientation are locked. The surface states form a Dirac cone in energy-momentum space.
- The band structure of topological insulators can be measured by angle-resolved photo-emission spectroscopy.
- In transport the existence of topologically protected surface states can be confirmed by quantum Hall effect measurements or by measuring magnetic flux-periodic conductance oscillations in nanoribbons.

Exercises

Problem 11.1. Confirm that the Dirac equation given by

$$v \begin{pmatrix} Mv & \hat{\sigma}\hat{p} \\ \hat{\sigma}\hat{p} & -Mv \end{pmatrix} \Psi = E\Psi,$$

with

$$\Psi = \begin{pmatrix} \chi \\ \psi \end{pmatrix}$$

a four-component spinor reduces for $M = 0$ to

$$\hat{\sigma}\hat{p}\chi = E\chi.$$

Problem 11.2. The Hamiltonian for the Jackiw–Rebbi model [194] can be expressed as

$$\hat{H}_{JR} = v\hat{p}_z\hat{\sigma}_z + M(z)\hat{\sigma}_x$$

with $\hat{p}_z$ being the momentum operator along the z-direction and $M(z)$ the z-dependent mass. A topological insulator is attached to a trivial insulator with the interface located at $z = 0$. For the topological insulator $(z < 0)$ we assume $M(z) < 0$, while for the trivial insulator $(z > 0)$ the mass is assumed to be larger than zero. Show that at the interface a state at zero energy $E = 0$ exists.

Problem 11.3. We assume a circular Sb_2Te_3 nanowire with a radius of $r_0 = 40\,nm$. The Dirac velocity v of the topologically protected surface states is assumed to be $3.8 \times 10^5\,m/s$. Calculate the energy gap between the states in eV at zero magnetic field. Which field is necessary to close the gap.

Problem 11.4. Calculate the eigenstates of a topological insulator nanowire with a circular cross section at $B = 0$. The radius of the wire is r_0. The Hamiltonian is given by equation (11.23). Make use of the fact that if $\chi = (\chi_1, e^{i\phi}\chi_2)^T$ is an eigenstate of

$$\begin{pmatrix} \alpha & e^{-i\phi}\beta \\ -e^{i\phi}\beta & -\alpha \end{pmatrix},$$

then $\tilde{\chi} = (\chi_1, \chi_2)^T$ is an eigenstate of

$$\begin{pmatrix} \alpha & \beta \\ -\beta & -\alpha \end{pmatrix}.$$

Problem 11.5. Calculate the density of states for a two-dimensional Dirac system with the energy dispersion $E_{\pm} = \pm v\hbar|k|$.

12 Quantum computation with electron spins

12.1 Introductory remarks

In the previous chapters we discussed the physical principles of different components of spin electronic devices, e. g. spin injection into a semiconductor or spin control by making use of the Rashba effect. However, although the performance of spintronic devices promise to be superior compared to their charge-based counterparts, the final application itself in an electronic circuit is often similar. In this chapter we will move one fundamental step further and discuss spin electronic devices in quantum information circuits [17]. Quantum information technology is based on a completely different way of data processing and communication. It relies on the coherent manipulation of quantum mechanical states. Instead of a bit with its values 0 and 1, its counterpart, the quantum bit, consists of a quantum mechanical two-level system. Owing to the special operation principle of quantum computers, quantum algorithms promise to have a much higher performance than the corresponding algorithms for digital computers. Various quantum mechanical systems are in principle suitable to realize a quantum processor, among them trapped ions [195] or superconducting circuits [196]. Here, we focus on semiconductor quantum dot based systems. In the most straightforward approach, a single electron is confined in a quantum dot and the Zeeman-split states are used as a two-level systems [19].

In order to give some basic insight in quantum computation, we first briefly introduce the different elements of a quantum processor, i. e. the quantum bit or quantum gates. In order to illustrate the operation of a quantum computer, we discuss a simple quantum algorithm. Finally, it is shown how the electron spin in a quantum dot can be employed to represent a quantum bit and how quantum operations can be implemented in these systems.

12.2 Basic elements of a quantum computer

The basic element of a quantum computer is the quantum bit, which is a quantum mechanical two-level system. Below, the quantum bit and its extension to larger systems, the quantum register, are introduced.

12.2.1 Quantum bit

The quantum bit, or qubit, can be viewed as an extension of the classical notion of a bit. The qubit is a quantum two-level system that in addition to the two eigenstates $|0\rangle$ and $|1\rangle$ can be configured in any superposition of these eigenstates

$$|Q\rangle = c_0|0\rangle + c_1|1\rangle. \tag{12.1}$$

https://doi.org/10.1515/9783110639001-012

In contrast to the classical bit, a qubit can be prepared in an infinitive number of appropriate quantum states by choosing different coefficients c_0 and c_1. The qubit state is normalized so that $|c_0|^2 + |c_1|^2 = 1$. To be more specific, the two eigenstates usually correspond to the ground and excited states of a two-level system. A possible realization is the spin-up ground state and spin-down excited state of an electron in a constant magnetic field along the z-direction

$$|\uparrow\rangle \leftrightarrow |0\rangle \leftrightarrow \begin{pmatrix} 1 \\ 0 \end{pmatrix}, \tag{12.2}$$

$$|\downarrow\rangle \leftrightarrow |1\rangle \leftrightarrow \begin{pmatrix} 0 \\ 1 \end{pmatrix}. \tag{12.3}$$

As illustrated in Figure 12.1 (a), by applying a magnetic field B_z along the z-direction, the energy of the two spin states are split by $\mu_B g B_z$. The states $|0\rangle$ and $|1\rangle$ serve as a basis to form any qubit state according to equation (12.1). The qubit state can best be visualized by means of the Bloch sphere shown in Figure 12.1(b). All possible spin states can be assigned to a point on the surface of the Bloch sphere, i. e. the upper and lower poles represent the $|0\rangle$ and $|1\rangle$ state, respectively. Furthermore, the spin state $(|0\rangle + |1\rangle)/\sqrt{2}$, which is an eigenstate of $\hat{\sigma}_x$ lies on the intersection of the Bloch sphere with the x-axis, while the state $(|0\rangle + i|1\rangle)/\sqrt{2}$ being an eigenstate of $\hat{\sigma}_y$ corresponds to the crossing with the y-axis.

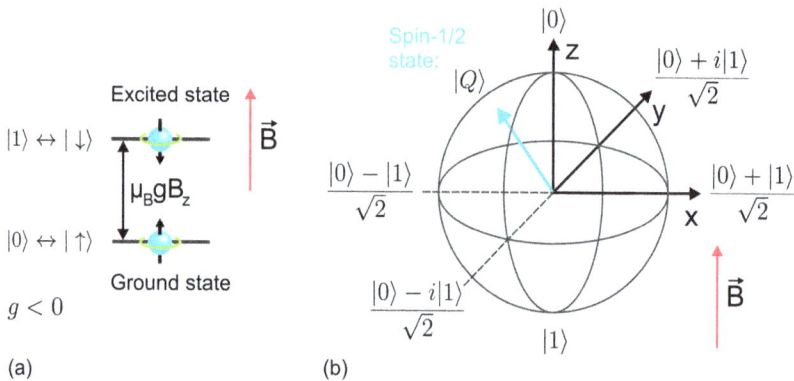

Figure 12.1: (a) Energy splitting of the states of an electron with spin 1/2 in a magnetic field along the z-axis. The ground $|0\rangle$ and excited $|1\rangle$ states correspond to the $|\uparrow\rangle$ and $|\downarrow\rangle$ states, respectively. The g-factor is assumed to be negative. (b) Bloch sphere for a spin 1/2 particle. The upper and lower poles correspond to the $|0\rangle$ and $|1\rangle$ states, respectively. Any qubit state $|Q\rangle$ can be represented by a point on the surface of the Bloch sphere.

Nevertheless, a qubit cannot be used to transmit more than one bit of information. This is due to the last axiom of quantum mechanics, where only eigenstates of the quantum system are finally detected by the measurement procedure. For a superposition state as given by equation (12.1) the final measurement will return 0 with probability $|c_0|^2$ and 1 with probability $|c_1|^2$. Thus, after the measurement the state of the qubit is either the eigenstate $|0\rangle$ or $|1\rangle$ but not in a superposition of both. Therefore, in order to obtain a meaningful result, one has to make sure that at the end of the calculation, the qubits are in an eigenstate.

A collection of qubits, which is usually called quantum register can be used to encode more complex information. As an example a two-qubit state can be composed of the following set of eigenvectors

$$|0\rangle \otimes |0\rangle = |00\rangle,$$
$$|0\rangle \otimes |1\rangle = |01\rangle,$$
$$|1\rangle \otimes |0\rangle = |10\rangle,$$
$$|1\rangle \otimes |1\rangle = |11\rangle. \tag{12.4}$$

Here, $\otimes$ denotes the direct product of the vectors. Explicitly, these eigenvectors can be written as

$$|00\rangle = \begin{pmatrix} 1 \\ 0 \\ 0 \\ 0 \end{pmatrix}, \quad |01\rangle = \begin{pmatrix} 0 \\ 1 \\ 0 \\ 0 \end{pmatrix},$$

$$|10\rangle = \begin{pmatrix} 0 \\ 0 \\ 1 \\ 0 \end{pmatrix}, \quad |11\rangle = \begin{pmatrix} 0 \\ 0 \\ 0 \\ 1 \end{pmatrix}. \tag{12.5}$$

By using this set of orthonormal basis vectors a general superposition of these state has the following form

$$|\psi\rangle = c_0|00\rangle + c_1|01\rangle + c_2|10\rangle + c_3|11\rangle. \tag{12.6}$$

12.2.2 Entangled states

A closer look on two or more qubit registers reveals, that states can be prepared, where the constituting qubits in a register are not independent. This situation is called entanglement. A typical two-qubit entangled state is given by

$$|\psi\rangle = \frac{1}{\sqrt{2}}(|1\rangle_1 \otimes |0\rangle_2 - |0\rangle_1 \otimes |1\rangle_2). \tag{12.7}$$

Here, $|0\rangle_1$ and $|1\rangle_1$ are orthogonal states in the space of particle 1 (qubit 1), while $|0\rangle_2$ and $|1\rangle_2$ are the corresponding states for particle 2 (qubit 2). An entangled state cannot be factorized into an expression like

$$|\psi\rangle = |m\rangle_1 \otimes |n\rangle_2. \tag{12.8}$$

A typical representative of a state which can be factorized is given by

$$|\psi\rangle = \frac{1}{\sqrt{2}}(|1\rangle_1 \otimes |0\rangle_2 - |1\rangle_1 \otimes |1\rangle_2), \tag{12.9}$$

since we can write

$$|\psi\rangle = \frac{1}{\sqrt{2}}|1\rangle_1 \otimes (|0\rangle_2 - |1\rangle_2). \tag{12.10}$$

What is the consequence an entangled state? Let us assume that we are able to measuring the state of the first qubit of the state given by equation (12.7). If it would be $|0\rangle_1$, it directly implies that the second qubit is in the state $|1\rangle_2$. This behavior is of special importance, if the two qubits are represented by two electrons with spin-1/2 propagating in two opposite directions. A measurement of the spin state of the first electron directly implies the spin state of the second electron, although the detectors of both electrons can be very far apart.

12.3 Basic quantum gates

So far, we only described how a quantum bit state in a quantum computer is represented. We also learned to group a number of states into a quantum register and to an entangled state. The next step is to perform gate operations on these qubits, similar to boolean operations in classical computers. An initial single qubit state or a qubit register is modified by these gates in a well-controlled fashion. This is done by an unitary operation U. In order to build a quantum computer, it is sufficient to have just a small set of elementary qubit gates, i. e. a single-qubit gate, which only modifies a single qubit, and a two-qubit gate, which operates on two qubits. Below we will explain both types of quantum gates.

12.3.1 Single-qubit gate

Let us start with a quantum gate operating on a single qubit, which is described by an unitary transformation U, as depicted in Figure 12.2 (a).

By this gate the qubit state $|a\rangle$ is transferred to

$$|a\rangle \rightarrow U|a\rangle. \tag{12.11}$$

The most general transformation on a single qubit is represented by a general rotation in the three-dimensional space on the surface of the Bloch sphere. For this rotation

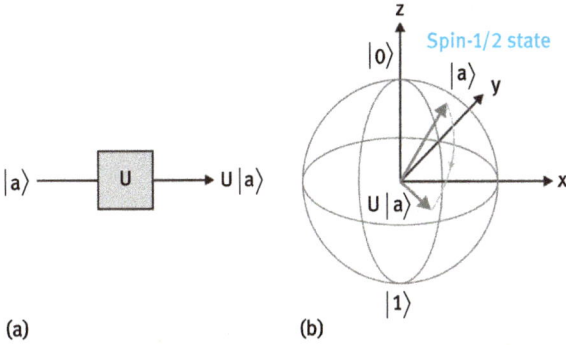

(a)   (b)

Figure 12.2: (a) Scheme of a single qubit operation. The unitary operation U transfers the state $|a\rangle$ into the final state $U|a\rangle$. (b) Single qubit operation of a spin-1/2 state on the Bloch sphere. By performing a single qubit operation on an initial state $|a\rangle$ it is moved on the Bloch sphere to $U|a\rangle$. The dashed line shows the net rotation path.

three angles β, θ, and α (Euler's angles) are sufficient. The first and third rotation by β and α are about the z-axis, while the second rotation by θ is about the x-axis:

$$U_{\mathrm{r}} = R_z(\alpha)R_y(\theta)R_z(\beta)$$
$$= \begin{pmatrix} e^{i\alpha/2} & 0 \\ 0 & e^{-i\alpha/2} \end{pmatrix} \begin{pmatrix} \cos\theta/2 & \sin\theta/2 \\ -\sin\theta/2 & \cos\theta/2 \end{pmatrix} \begin{pmatrix} e^{i\beta/2} & 0 \\ 0 & e^{-i\beta/2} \end{pmatrix}, \qquad (12.12)$$

where the unitary transformations $R_z(\alpha)$ and $R_z(\beta)$ describe a rotation about the z-axis and $R_y(\theta)$ a rotation about the y-axis. The angles α, θ, and β are real values. A typical rotation of an initial state $|a\rangle$ the state $U|a\rangle$ is illustrated in Figure 12.2 (b). The transformation can be further generalized by multiplying U_{r} by a phase factor δ

$$U_{\mathrm{p}}(\delta) = \begin{pmatrix} e^{i\delta} & 0 \\ 0 & e^{i\delta} \end{pmatrix}, \qquad (12.13)$$

so that we finally arrive at the general unitary matrix for a single-qubit operation

$$U_\theta = U_{\mathrm{p}}U_{\mathrm{r}} = \begin{pmatrix} e^{i(\delta+\alpha/2+\beta/2)}\cos\theta/2 & e^{i(\delta+\alpha/2-\beta/2)}\sin\theta/2 \\ -e^{i(\delta-\alpha/2+\beta/2)}\sin\theta/2 & e^{i(\delta-\alpha/2-\beta/2)}\cos\theta/2 \end{pmatrix}. \qquad (12.14)$$

A typical single-qubit operation is the π-rotation (π-ROT) about the y-axis,

$$U_\pi = \begin{pmatrix} 0 & 1 \\ -1 & 0 \end{pmatrix}, \qquad (12.15)$$

which is similar to the inverter in a boolean computer, since it transfers the state $|0\rangle$ to $-|1\rangle$ and the state $|1\rangle$ to $|0\rangle$.

12.3.2 Controlled-NOT gate

A typical two-qubit gate is the controlled-NOT (CNOT) gate. The circuit diagram is shown in Figure 12.3.

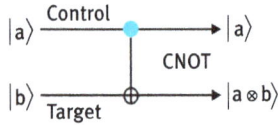

Figure 12.3: Quantum circuit diagram of a CNOT gate. The target qubit $|b\rangle$ is flipped if the control qubit $|a\rangle$ is set to $|1\rangle$.

Here, the so-called target qubit $|b\rangle$ is flipped if the other qubit, the control qubit, $|a\rangle$ is in the $|1\rangle$ state. As long as $|a\rangle$ is in the state $|0\rangle$ no action is taken on $|b\rangle$. The CNOT gate corresponds to the exclusive-OR gate (XOR) gate of a boolean computer. The corresponding truth table is given in Table 12.1.

Table 12.1: Truth table of a CNOT gate.

$|a, b\rangle$	$|a, a \oplus b\rangle$
$|00\rangle$	$|00\rangle$
$|01\rangle$	$|01\rangle$
$|10\rangle$	$|11\rangle$
$|11\rangle$	$|10\rangle$

The CNOT gate is an unitary transformation on states represented by a superposition of a set of four basis vectors define by equation (12.5). The corresponding unitary 4×4 matrix of the CNOT gate is given by

$$U_{\text{CNOT}} = \begin{pmatrix} 1 & 0 & 0 & 0 \\ 0 & 1 & 0 & 0 \\ 0 & 0 & 0 & 1 \\ 0 & 0 & 1 & 0 \end{pmatrix}. \tag{12.16}$$

The CNOT gate can be used to create an entangled state out of a formerly non-entangled state. For this purpose we set the first qubit $|a\rangle$ to a superposition state $(|0\rangle - |1\rangle)/\sqrt{2}$ and the second qubit to $|b\rangle = |1\rangle$. The input state is thus $(|01\rangle - |11\rangle)/\sqrt{2}$, which is non-entangled, since it can be factorized into $(|0\rangle - |1\rangle)/\sqrt{2} \otimes |1\rangle$. Applying a

CNOT gate we obtain

$$
\begin{pmatrix} 1 & 0 & 0 & 0 \\ 0 & 1 & 0 & 0 \\ 0 & 0 & 0 & 1 \\ 0 & 0 & 1 & 0 \end{pmatrix} \begin{pmatrix} 0 \\ 1/\sqrt{2} \\ 0 \\ -1/\sqrt{2} \end{pmatrix} = \begin{pmatrix} 0 \\ 1/\sqrt{2} \\ -1/\sqrt{2} \\ 0 \end{pmatrix}. \tag{12.17}
$$

Thus, the result is an entangled state given by $1/\sqrt{2}(|01\rangle - |10\rangle)$.

12.3.3 Realization of a quantum computer

The basic elements of a quantum computer are the quantum bits and the corresponding quantum gate operations. However, in order to realize a working quantum computer a number of other criteria have to be fulfilled:

– First, one needs to be able to initialize the states in a quantum computer, i. e. by setting all qubits to their ground state. This is necessary to have a well-defined starting condition before performing a number of qubit operations.
– The system has to be scalable, in order to have a useful quantum computer.
– Quantum operation must be possible by means of a set of basic quantum gates, e. g. a single qubit gate and a CNOT gate.
– It must be possible to read-out the result at the end of the quantum processing. For that the final state after the calculation has to be in an eigenstate, i. e. in the $|0\rangle$ or $|1\rangle$. This is necessary because only if the final state of a qubit is in an eigenstate one gets a unambiguous results. Just imagine that the state would be in a superposition like given by equation (12.1), with both c_0 and c_1 being non-zero. In that case one would obtain a result $|0\rangle$ with probability $|c_0|^2$ and $|1\rangle$ with a probability $|c_1|^2$.
– During the complete computation procedure the coherence of the states needs to be preserved.

The criteria listed above were first summarized by DiVincenzo [197] and are therefore called DiVincenzo criteria. For a physical representation of a quantum computer, it is necessary to check if it is in principle possible to fulfill all DiVincenzo criteria before starting to realize a system. As we have seen above, a spin-1/2 particle with its spin-up and spin-down states is a natural representation of a qubit state. There are several possibilities for implementation, i. e. electronic states of trapped ions [198], donor states in a semiconductor [199], or spin states in a semiconductor quantum dot [19]. Here, we will focus on the latter, where a single electron is bound in a quantum dot or a pair of electrons is confined in a double dot system.

12.4 Quantum algorithms

As a next step, we will briefly have a look on quantum algorithms. In Figure 12.4, a typical quantum computation sequence is depicted. Here, a system with a three-quantum-bit register is shown. First, the system is initialized to the ground state. In the course of time a number of quantum gate operations, single- and two-qubit operations, are performed. In the given example the single-qubit operations are the rotation by π (π-ROT), defined by equation (12.15), and the CNOT operation defined by equation (12.16). After all operations are finished the result is finally read-out. Note that the number of qubits is fixed, in contrast to the data lines in a classical computer. Usually the electron spins, representing the qubits remain on a confined position but the spin state itself is changed in the course of time by the quantum bit operations.

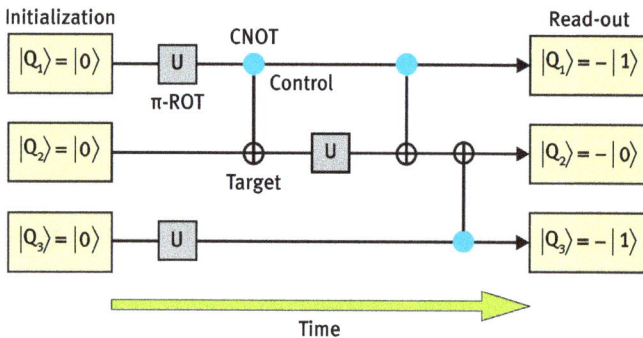

Figure 12.4: Data processing in a 3-qubit quantum computer: First, the qubits are initialized by setting them to the ground state $|0\rangle$. Subsequently, different quantum gates are applied. Here, unitary single-qubit π-rotations and two-qubit CNOT operations are performed. Finally, the qubits are read-out.

There are some quantum algorithms, which promise to solve computational problems substantially faster than the corresponding algorithms on a classical computer. One prominent example is the Shor algorithm [18], which extracts the prime factors of a large number. Using a classical computer, the computation time increases exponentially with the number of digits, whereas for the Shor algorithms the increase is only polynomially. Another example is the Grover search algorithm [200], which can speed up finding an element in a large data base. Below we will introduced the Deutsch–Josza algorithm, as a relatively simple example of a quantum algorithm [201, 202, 203].

12.4.1 Deutsch–Josza quantum algorithm

The Deutsch–Jozsa algorithm [201, 202, 203] can be employed to extract a fundamental property of a one-bit function very efficiently. Of course this statement sound rather abstract. However, the problem addressed here is actually a real world problem: Let us suppose we have to decide if a coin is fair i. e. obverse (eagle) on one side and reverse (number) on the other side, or fake, i. e. obverse or reverse on both sides (see Figure 12.5 (a)).

Obverse   Reverse

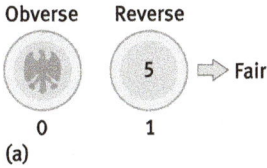

0   1
(a)

x	$f_{00}(x)$	$f_{01}(x)$	$f_{10}(x)$	$f_{11}(x)$
0	0	0	1	1
1	1	1	0	1
	Constant	Balanced	Balanced	Constant
	Fake	Fair	Fair	Fake
$f(0) \otimes f(1)$	0	1	1	0

(b)

Figure 12.5: (a) A fair coin with an eagle on the obverse and the number on the reverse. (b) Constant or balanced functions represented by fair or fake coins.

In order to gain information if both sides are identical, we need to have a look on both sides. It is not sufficient to get the information of only one side of the coin. Thus, in a classical algorithm at least two steps are required to decide if a coin is fair or fake, as illustrated in Figure 12.6 (a).

As we will see in detail below, by using a quantum algorithm the decision can be made after a single run, only (cf. Figure 12.6 (b)).

More formally the problem can be defined by considering a single-bit binary function $f(x)$. In principle only four different functions $f_{00}(x), f_{01}(x), f_{10}(x)$, and $f_{11}(x)$ can exist. The functions are listed in Figure 12.5 (b). There are two different types of functions. Either the function is constant, i. e. f_{00} and f_{11}, or a function is balanced, i. e. f_{01} and f_{10}. A constant function corresponds to a fake coin while a balanced function corresponds to a fair one. Let us suppose the function $f(x)$, i. e. the coin, is unknown, which we name test function (cf. Figure 12.6 (a)). Classically both inputs 0 and 1, representing each side of the coin, have to be applied sequentially, in order to decide if the function is constant or balanced. As illustrated in Figure 12.6 (a), the algorithm is processed as follows: After resetting, the input parameter is applied to the test func-

Classical algorithm

Two loops

(a)

Deutsch-Josza algorithm

Single loop

(b)

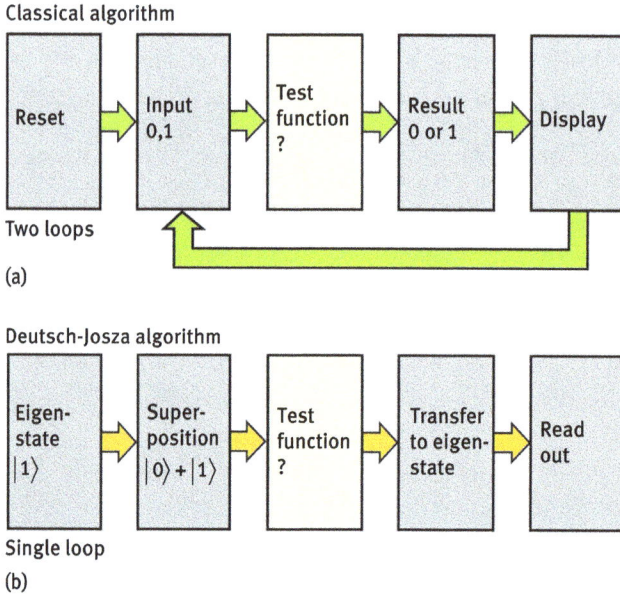

Figure 12.6: (a) Steps in a classical algorithm to decide if a function is balanced or constant, i. e. a coin is fair or fake, respectively. (b) Corresponding steps in a quantum algorithm.

tion $f(x)$. The result, either 0 or 1, is delivered and displayed. Two sequential runs are required, in order to make a decision if the coin is fair or fake.

By using a quantum algorithm the decision if a function is balanced or constant can be made within a single run. The first step is to define a quantum circuit, in order to determine $f(0)$ and $f(1)$. This is shown in Figure 12.6 (b).

The corresponding unitary transformation U_f can be expressed as

$$|x\rangle|y\rangle \xrightarrow{U_f} |x\rangle|y \oplus f(x)\rangle. \qquad (12.18)$$

The symbol $\oplus$ indicates addition modulo 2. The corresponding circuit element of U_f is shown in Figure 12.7 (a). For example, according to equation (12.18), if the $|y\rangle$ qubit is set to $|0\rangle$ the value of $f(x)$ appears at the output.

In order to decide if the function $f(x)$ is balanced or constant it is sufficient to calculate $f(0) \oplus f(1)$, as shown in Figure 12.5. If the function is constant $f(0) \oplus f(1)$ results in zero, while 1 is obtained for a balanced function. The problem connected to this is to determine $f(0) \oplus f(1)$ by only a single application of U_f [201, 202]. For a classical computer it is impossible to do, but a quantum computer can solve this problem by using the circuit shown in Figure 12.7 (b). Here, four Hadamard transformations are used in addition, in order to prepare a superposition of the input state. The Hadamard transformation is a single-qubit operation defined by the following transformation

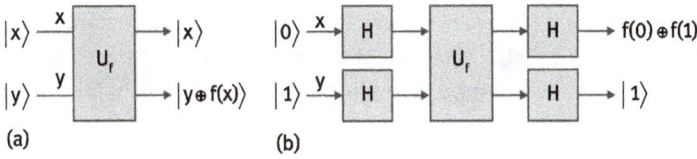

Figure 12.7: (a) Test function which determines $|x\rangle|y\oplus f(x)\rangle$ of a balanced or constant binary function. (b) A quantum circuit to determine $f(0)\oplus f(1)$ with a single application of U_f. H represents a single Hadamard transformation.

matrix

$$H = \frac{1}{\sqrt{2}}\begin{pmatrix} 1 & 1 \\ -1 & 1 \end{pmatrix}. \tag{12.19}$$

The preparation of a superposition state allows to perform the calculation in parallel and obtain the result in a single step. Let us begin by transforming the second qubit $|y\rangle$ which was initially set to the $|1\rangle$ state by a Hadamard transformation H_y and subsequently by U_f

$$|x\rangle|0\rangle \xrightarrow{H_y} |x\rangle\left(\frac{|0\rangle - |1\rangle}{\sqrt{2}}\right) \xrightarrow{U_f} |x\rangle\left(\frac{|0\oplus f(x)\rangle - |1\oplus f(x)\rangle}{\sqrt{2}}\right)$$

$$= \begin{cases} |x\rangle\frac{|0\rangle - |1\rangle}{\sqrt{2}} & \text{if } f(x) = 0 \\ |x\rangle\frac{|1\rangle - |0\rangle}{\sqrt{2}} & \text{if } f(x) = 1 \end{cases}$$

$$= (-1)^{f(x)}|x\rangle\frac{|0\rangle - |1\rangle}{\sqrt{2}}. \tag{12.20}$$

The value of $f(x)$ is now encoded into the overall phase of the result, with the qubit left unchanged otherwise. The calculation step becomes finally useful if the first qubit $|x\rangle$, which is initially set to state $|1\rangle$ (see Figure 12.7 (b)), is also transferred into a superposition $(|0\rangle + |1\rangle)/\sqrt{2}$ by a Hadamard transformation. In this case we obtain

$$\left(\frac{|0\rangle + |1\rangle}{\sqrt{2}}\right)\left(\frac{|0\rangle - |1\rangle}{\sqrt{2}}\right) \xrightarrow{U_f} \left(\frac{(-1)^{f(0)}|0\rangle + (-1)^{f(1)}|1\rangle}{\sqrt{2}}\right)\left(\frac{|0\rangle - |1\rangle}{\sqrt{2}}\right)$$

$$= (-1)^{f(0)}\left(\frac{|0\rangle + (-1)^{f(0)\oplus f(1)}|1\rangle}{\sqrt{2}}\right)$$

$$\otimes \left(\frac{|0\rangle - |1\rangle}{\sqrt{2}}\right). \tag{12.21}$$

By performing a second Hadamard transformation on the $|x\rangle$ qubit a back transformation from a superposition state into an eigenstate is obtained. In case of $f(x)$ being a balanced function the $|x\rangle$ qubit is in the state $(|0\rangle - |1\rangle)/\sqrt{2}$ and then converted by the Hadamard transformation to $|1\rangle$. The negative sign in the superposition state appears because for a balanced function $f(0) \oplus f(1) = 1$ gives $(-1)^1 = -1$. For a constant function the $|x\rangle$ qubit in the state $(|0\rangle + |1\rangle)/\sqrt{2}$ is transformed to $|1\rangle$. Thus, by measuring the state of the $|x\rangle$ qubit a decision can be made if the function $f(x)$ is balanced or constant.

12.5 Quantum dot spin qubits

After introducing the basic principles of quantum computation, we will now explain, how an electron in a semiconductor quantum dot can be employed as a quantum bit. First, the general concept, which was developed by Loss and DiVincenzo [19], is explained. Afterwards, the different physical realizations are introduced. Following the DiVincenzo criteria the different elements, e. g. initialization or quantum gate operation, for a successful implementation are discussed in detail.

12.5.1 General concept

The concept of using a quantum dot with a single electron as a quantum bit was first proposed by Loss and DiVincenzo [19]. A schematic illustration of the set-up is depicted in Figure 12.8.

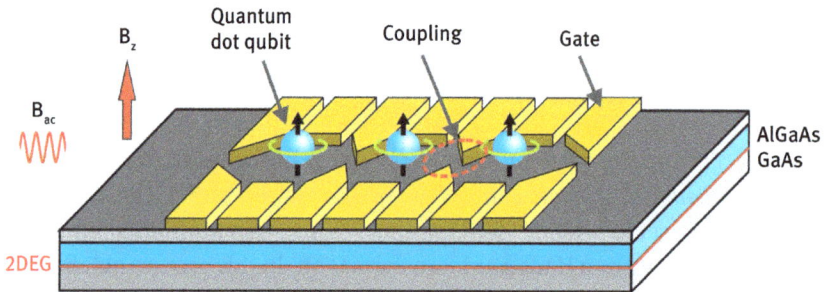

Figure 12.8: Concept of a quantum computer based on quantum dots in a semiconductor heterostructure. The quantum dots are defined by biasing the gate electrodes. The qubit is represented by the spin of a single electron in the quantum dot. The two-level state is formed by applying a fixed magnetic field B_z perpendicular to the sample surface. Single qubit gate operation is realized by irradiation an ac magnetic field B_{ac}, while a two-qubit gate operation is performed by coupling two adjacent quantum dots [19].

A series of quantum dots is formed by applying sufficiently large negative bias voltages to the gate fingers placed on a two-dimensional electron gas in an AlGaAs/GaAs heterostructure. Each quantum dot contains only a single electron. The qubit state is represented by the electron spin. Here, a two-level system is realized by splitting the energy of the spin-up and spin-down states by applying a constant magnetic field B_z perpendicularly to the sample surface. In Figure 12.9 the different situations of a single electron confined in a quantum dot are shown. The ground and excited states $|0\rangle$ and $|1\rangle$ are represented by an electron with spin-up and spin-down, respectively. The spin orientation of the ground state is inverse to the one assumed in Section 3.1.4, since GaAs has a negative g-factor of –0.44 for bulk material. In addition, a superposition state $c_0|0\rangle + c_1|1\rangle$ state can be formed, where the ground and excited states are occupied with a probability $|c_0|^2$ and $|c_1|^2$, respectively.

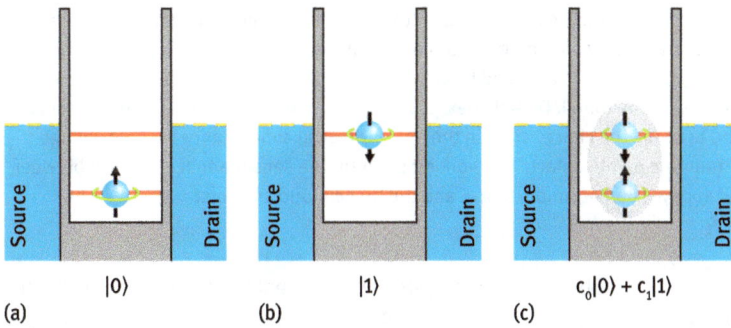

|0) |1) $c_0|0\rangle + c_1|1\rangle$
(a)   (b)   (c)

Figure 12.9: Quantum dot occupied with a single electron. The energy levels are split due to the Zeeman effect by applying a constant magnetic field. (a) Ground state $|0\rangle$ with an electron with spin-up. (b) Excited state $|1\rangle$ represented by a spin-down state. (c) Superposition state.

Single qubit gate operations are performed by irradiation with an ac magnetic field. As we will discuss below, this process corresponds to electron spin resonance (ESR) transitions. A two-qubit gate operation can be realized by coupling two adjacent quantum dots. Here, the barrier between neighboring quantum dots is lowered by reducing the negative gate voltage at the corresponding gate fingers. As a result, the electron states in both quantum dots get closer and couple.

12.5.2 Experimental realization of a quantum dot qubit

In most cases quantum dot qubits are realized by employing a two-dimensional electron gas in an AlGaAs/GaAs heterostructure. All operations are performed at low temperatures in a dilution refrigerator, with a typical temperature of the electronic reservoir of around 100 mK. The quantum dot is defined by means of a number of gate fingers. Figure 12.10 (a) shows a typical layout.

The first task is, to adjust the gate voltages on gates A, B, C, and P to appropriate voltages, so that only a single electron is trapped in the quantum dot. The quantum

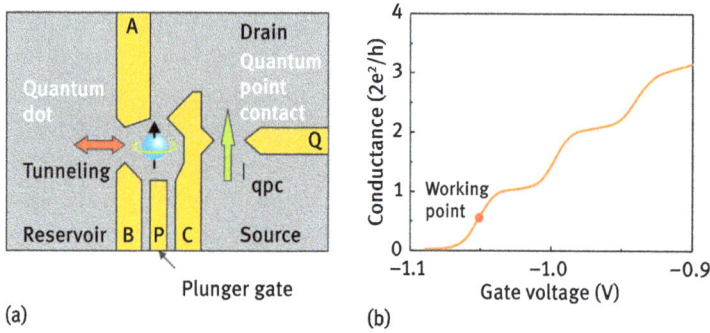

Figure 12.10: (a) Schematic illustration of a quantum dot defined by a number of gate fingers on a semiconductor heterostructure containing a two-dimensional electron gas. The tunneling from the reservoir into the quantum dot is adjusted by gates A and B. By means of the plunger gate P the quantum dot states can be tuned. Directly next to the quantum dot a quantum point contact is placed. This is used to probe the charge state in the dot. (b) Typical conductance vs. gate voltage characteristics of quantum point contact. The working point for the sensor application is in between pinch-off, i. e. total suppression of conductance, and the first conductance plateau.

state in the dot is fine-tuned by the plunger gate P. In contrast to the quantum dots discussed in Section 2.7, here the dot can only be filled from one side on the left. The tunneling probability for an electron tunneling from the reservoir into the quantum dot is adjusted by gates A and B. The reason for accessing the dot from only one side is, that the qubit state is not probed by a current flowing through the dot but rather by electrostatic means. This is achieved by employing a quantum point contact placed right next to the quantum dot, as illustrated in Figure 12.10 (a). As shown in Figure 12.10 (b) the conductance through the quantum point contact is quantized in units of $2e^2/h$. By monitoring the conductance of the quantum point contact, it can be found out if the quantum dot is empty or filled. This is due to the fact, that the charge of the electron in the dot slightly varies the potential profile of the quantum point contact opening. For the application as a charge sensor the voltage on gate Q, defining the opening width of the quantum point contact, is set to such a value that the working point is adjusted in between pinch-off and the first conductance plateau. A fixed voltage is applied between source and drain. Owing to the strongly non-linear characteristics, the conductance is most sensitive at a gate voltage in this range. The working principle of the quantum point contact sensor is, that the presence of an electron in the quantum dot makes the opening width of the quantum point contact slightly smaller due to the electrostatic interaction. Thus, the conductance of the quantum point contact decreases. If the electron is removed from the quantum dot, the conductance increases by a small amount due to the increased point contact opening width. The quantum point contact sensor as described above, works as a charge sensor. The spin state of the electron, which actually represents the qubit state in the dot is not probed. As we will see below, there are certain schemes, which allow to map the spin state to a charge state and by that get information of the spin state via the conductance of the quantum point contact.

12.5.3 Initialization

According to the DeVinzeno criteria, the first step to show is the initialization of the system, i. e. to transfer all qubits to a well-defined state before starting the quantum computation process. Usually, initialization means to bring all qubits into the ground state. In Figure 12.11 a possible initialization sequence is shown, which is based on spin-relaxation.

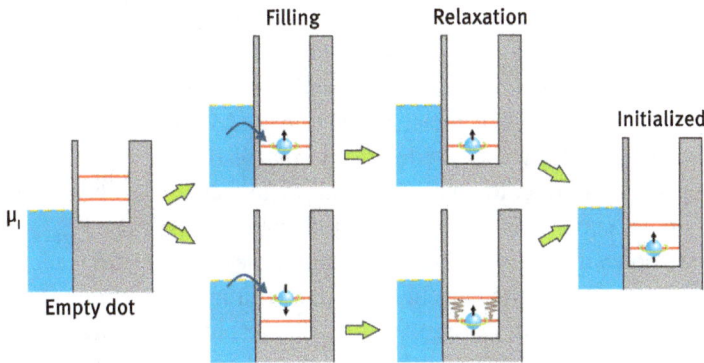

Figure 12.11: Qubit initialization by means of relaxation: First, the dot is emptied by biasing the plunger gate with a large negative voltage. Subsequently, the voltage at the plunger gate is adjusted, so that both Zeeman-split levels are below the electro-chemical potential μ_l of the reservoir on the left side. The quantum dot is either filled with a spin-up or spin-down electron. Since the spin-up state is the lowest state, the electron remains in that state. The spin-down state relaxes after a time larger than the relaxation time into the spin-up state. Thus, the qubit is finally initialized to the spin-up state.

The initialization sequence starts by emptying the quantum dot. Referring to the quantum dot shown in Figure 12.10 (a), this can be achieve by applying a sufficiently large negative voltage to the plunger gate, so that both levels are above the electro-chemical potential μ_l of the reservoir on the left side (cf. Figure 12.11). After a time longer than the tunneling time it is ensured that the quantum dot is empty. As a next step, the plunger gate is set to a voltage that both levels in the dot are below μ_l, so that an electron can tunnel from the reservoir on the left. The right barrier is that high that tunneling is prevented. The filling can be monitored by using a quantum point contact placed next to the quantum dot on the right side. However, although one can be sure that for the potential in the quantum dot only a single electron occupies the dot due to Coulomb blockade the spin state is not well-defined. This is because the spin-up as well as the spin-down state are below μ_1 and could have been filled. In order to make sure that only the spin-up ground state is occupied one has the wait a time longer then the spin relaxation time, so that the higher lying spin-down state relaxes into the spin-up ground state. If the spin-up state was initially occupied it will remain in that state during the waiting time. Thus, finally after a waiting time longer then the relaxation time the quantum dot qubit is in a well defined ground state.

12.5.4 Read-out

For reading-out the state of quantum dot qubit state after performing a quantum algorithm a quantum point contact placed next to the quantum dot is employed (cf. Figure 12.10 (a)).

   A general problem regarding the read-out is that the spin of the state cannot be measured directly. One has to transfer the spin information into a charge state. This can be achieve by means of the sequence depicted in Figure 12.12. What is show here is a full sequence to demonstrate the working principle, so that also some initialization is included. For the actual read-out after a computation cycle only the last two steps are performed. The whole sequence is controlled by applying a voltage pulse V_{pulse} to the plunger gate P, as shown in Figure 12.12 (a). The state of the quantum dot is monitored by measuring the change of the current ΔI_{QPC} through the adjacent quantum point contact (see Figure 12.12 (b)). In Figure 12.12 (c) the corresponding sequence for the quantum dot occupation is depicted. Similar to the initialization one starts with emptying the quantum dot completely by raising the two levels in the dot above the electro-chemical potential of the left reservoir. Subsequently, V_{pulse} is increased, thus lowering the potential of the quantum dot, in order to fill the dot by a single electron, either spin-up or spin-down. As can be seen in Figure 12.12 (b), changing the pulse voltage at gate P directly alters the current ΔI_{QPC} through the quantum dot owing to capacitive coupling. Even more, at the moment the dot is filled ΔI_{QPC} decreases, since the negative charge of the electron narrows the potential well of the quantum point contact. After a waiting time which depends on the tunneling rate through the potential barrier of the quantum dot, one can be confident that the dot is finally filled. However, the waiting time should not be that long that spin relaxation occurs. The actual read-out process is to find out, in which spin state the electron is. This is done by lowering V_{pulse}, so that the spin-up level lies above and the spin-down level below μ_1 of the left reservoir, respectively. Thus, the spin-down electron gets a chance to tunnel into the reservoir, while the spin-up electron remains in the dot. When the spin-down electron tunnels into the reservoir this can be seen directly as an increase of ΔI_{QPC}, because the point contact is opened slightly due to the missing negative charge in the dot. As a next step, the dot is refilled by an electron, but this time by a spin-down electron, since this state is the only one lying below μ_1. Consequently, ΔI_{QPC} jumps back

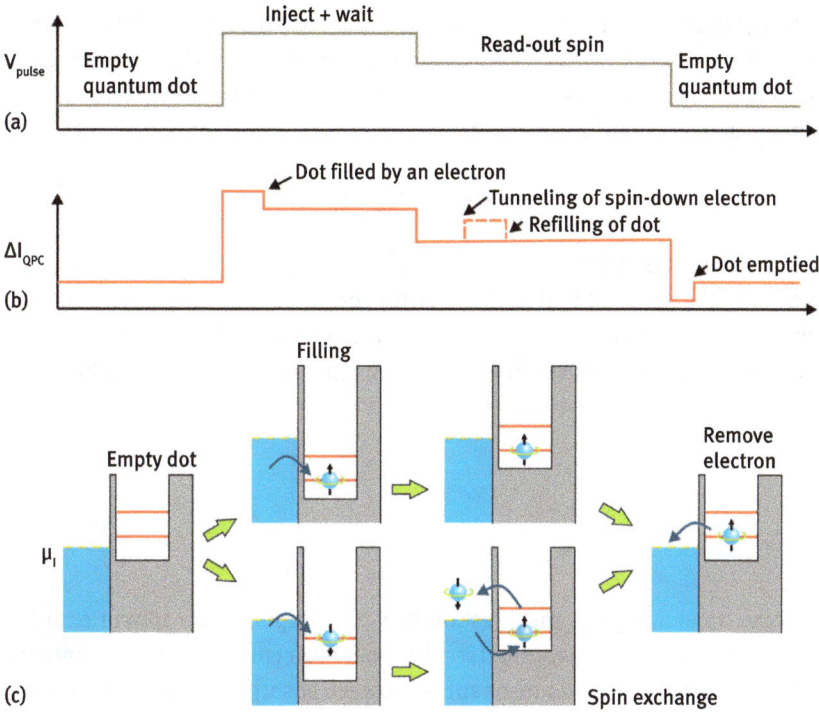

Figure 12.12: Read-out scheme of a quantum dot spin qubit: (a) Voltage pulse V_{pulse} sequence applied to the plunger gate. (b) Corresponding current flowing through the spin-gate point contact placed next the to quantum dot. (c) Sequence of level occupation in the quantum dot. First, the dot is emptied by raising the potential in the dot using the plunger gate. After lowering the potential of the dot, it is filled by a single electron. In the next step the potential is raised slightly so that only a spin-down electron in the upper level gets a chance to tunnel into the reservoir. Since the charge in the quantum dot changes during that process it can be detected by means of the quantum point contact. The quantum dot is subsequently refilled with a spin-up electron. To finish up the sequence the quantum dot is emptied (Figure after Elzerman et al. [204]).

to its initial value. The process described above gives information about the spin state of the electron in the dot. Here, the stepwise change of ΔI_{QPC} indicates that the electron in the dot was in a spin-down state. When the electron was in the spin-up state nothing happens to ΔI_{QPC}, since the electron remains in its state, no tunneling occurs. Thus, this process allows to gain information on the electron spin by monitoring the charge in the quantum dot. The whole sequence finishes by lowering V_{pulse} so that the dot is emptied. As can be seen in Figure 12.12 (b), the moment the electron is removed from the dot results in a small step in ΔI_{QPC}.

The sequence shown in Figure 12.12 can also be used to gain information on the spin relaxation time [204, 205]. This can be done by varying the waiting time t_{wait} after injecting an electron into the quantum dot. The longer the waiting time is, the more likely it is that an electron in the spin-down state relaxes into the spin-up state. Thus less events are found, where a spin exchange is measured by the quantum point contact during the read-out process. The probability $p^\downarrow$ of finding the quantum dot in the spin-down state can be described by

$$p^\downarrow = \alpha + C \exp\left(-t_{\text{wait}}/T_1\right), \tag{12.22}$$

with T_1 the spin relaxation time, with α and C being fitting parameter. The parameter α corresponds to the probability that the detector current exceeds the threshold even though the electron was actually in the $|\uparrow\rangle$ state, for example due to thermally activated tunneling or electrical noise. Spin relaxation times in the order of a millisecond are found experimentally [204]. The relaxation time decreases with increasing external magnetic field.

12.5.5 Electron spin resonance

As a next step, we will address the question how to manipulate the spin orientation of an electron confined in a quantum dot. This operation employs electron spin resonance (ESR), which is based on resonant microwave absorption. The phenomenon itself is known since long and is used e. g. to study chemical reactions. Here, we will first introduce the basic principle of operation and later on will discuss how this mechanism can be implemented in a quantum dot structure.

Electron spin resonance in a quantum dot is based on a spin-split two-level system, as shown in Figure 12.13.

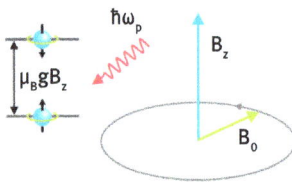

Figure 12.13: Electron spin resonance: By applying a constant magnetic field B_z along the z-axis the states for spin up and down are split by $\hbar\omega_p = g\mu_B B_z$ due to the Zeeman effect. In addition, a field B_0 rotating with the precession frequency ω_p is applied.

This can be achieved by applying a constant magnetic field along the z-axis. The energy difference for both spin orientations due to the Zeeman energy splitting is then given by

$$\hbar\omega_p = g\mu_B B_z, \tag{12.23}$$

with ω_p the precession frequency, g the gyromagnetic factor, and μ_B the Bohr magneton defined by equation (3.14). As shown in Figure 12.13, by applying an additional magnetic field rotating in the plane perpendicular to the z-axis with a rotation frequency matching to ω_p, coherent transitions between the two Zeeman-split levels are induced.

The switching of the electron spin can be explained by solving the appropriate Schrödinger equation for an electron spin in a magnetic field. We can express the wave function of the electron spin by a spinor

$$\phi(t) = \begin{pmatrix} c_0(t) \\ c_1(t) \end{pmatrix}, \tag{12.24}$$

where $|c_0|^2$ and $|c_1|^2$ are the probabilities of the electron spin to be aligned up or down along the z-axis. As done before, we assume a constant magnetic field B_z along the z-direction. The energy difference for both spin orientations due to the Zeeman energy splitting is given by equation (12.23). In order to switch the spin orientation a small rotating field within the xy-plane needs to be applied in addition

$$B_x = B_0 \cos \omega_p t, \tag{12.25}$$
$$B_y = B_0 \sin \omega_p t, \tag{12.26}$$

with ω_p the frequency of the rotating field, i. e. the precession frequency as defined above.

In Section 3.1.4, the Hamiltonian, equation (3.25), for the Zeeman effect was introduced. The corresponding time-dependent Schrödinger equation for an electron spin in an external magnetic field is given by

$$i\hbar \frac{\partial}{\partial t} \begin{pmatrix} c_0(t) \\ c_1(t) \end{pmatrix} = \beta \begin{pmatrix} B_z & B_x - iB_y \\ B_x + iB_y & -B_z \end{pmatrix} \begin{pmatrix} c_0(t) \\ c_1(t) \end{pmatrix}, \tag{12.27}$$

with $\beta = sg\mu_B$ and $s = 1/2$ the spin quantum number. We begin by writing the Schrödinger equation explicitly as

$$(\hbar\omega_p/2)c_0 + \beta B_0 \exp(-i\omega_p t)c_1 = i\hbar \frac{\partial c_0}{\partial t}, \tag{12.28}$$
$$\beta B_0 \exp(i\omega_p t)c_0 - (\hbar\omega_p/2)c_1 = i\hbar \frac{\partial c_1}{\partial t}. \tag{12.29}$$

The coefficients $c_0(t)$ and $c_1(t)$ can be subdivided in two components

$$c_0(t) = d_0(t)\exp(-i\omega_p t/2) \quad \text{and} \quad c_1(t) = d_1(t)\exp(i\omega_p t/2). \tag{12.30}$$

By inserting this ansatz in equations (12.28) and (12.29) we arrive at

$$\beta B_0 d_1 = i\hbar \frac{\partial d_0}{\partial t}, \tag{12.31}$$

$$\beta B_0 d_0 = i\hbar \frac{\partial d_1}{\partial t}. \tag{12.32}$$

Differentiating the first equation with respect to t and inserting the result in the second equation we obtain

$$\frac{\partial^2 d_0}{\partial t^2} + \frac{\beta^2 B_0^2}{\hbar^2} d_0 = 0. \tag{12.33}$$

The solution of this equation is given by

$$d_0(t) = a \cos (\Omega t + \varphi), \tag{12.34}$$

where a is the amplitude and φ is a phase factor. The frequency Ω is given by

$$\Omega = \beta B_0 / \hbar. \tag{12.35}$$

Inserting this expression into equation (12.32) results in

$$d_1(t) = -ia \sin (\Omega t + \varphi). \tag{12.36}$$

Normalization of the wave function requires $a = 1$. Without loosing generality the phase factor φ can be set to zero. Inserting the above expressions for $d_0(t)$ and $d_1(t)$ into equation (12.30) results in the spinor given by

$$\phi(t) = \begin{pmatrix} \cos (\Omega t) \exp (-i\omega_p t/2) \\ -i \sin (\Omega t) \exp (i\omega_p t/2) \end{pmatrix}. \tag{12.37}$$

From equation (12.37) the expectation value of the electron spin in z-direction can be determined

$$\begin{aligned}
\langle s_z \rangle &= (\hbar/2)\langle \phi(t)|\hat{\sigma}_z|\phi(t)\rangle \\
&= \frac{\hbar}{2}[\cos^2(\Omega t) - \sin^2(\Omega t)] \\
&= \frac{\hbar}{2} \cos (2\Omega t).
\end{aligned} \tag{12.38}$$

Thus, the expectation value of the electron spin in z-direction oscillates between $\pm\hbar/2$ with the frequency 2Ω if a rotating field of strength B_0 is applied. At the beginning at $t = 0$ when the oscillating magnetic field is switched on, the spin is aligned upwards $\langle s_z \rangle = \hbar/2$. After a time $t = \pi/4\Omega = \pi\hbar/4\beta B_0$ the expectation value $\langle s_z \rangle$ vanishes. The electron spin is aligned horizontally in the xy-plane. A pulse of this duration is called $\pi/2$ pulse, since the spin is rotated by $\pi/2$. As can be inferred from equation (12.38) a pulse duration of $t = \pi/2\Omega = \pi\hbar/2\beta B_0$ leads to a switching of the spin opposite to the initial orientation (π-pulse).

Similarly to equation (12.38), the expectation value of the electron spin along the x- and y-directions can be determined

$$\langle s_x \rangle = (\hbar/2)\langle \phi(t)|\hat{\sigma}_x|\phi(t)\rangle = \frac{\hbar}{2}\sin(2\Omega t)\sin(\omega_p t), \qquad (12.39)$$

$$\langle s_y \rangle = (\hbar/2)\langle \phi(t)|\hat{\sigma}_y|\phi(t)\rangle = -\frac{\hbar}{2}\sin(2\Omega t)\cos(\omega_p t). \qquad (12.40)$$

For these orientations the expectation values are also affected by the precession of the spin about the z-axis. At $t = 0$ the expectation values $\langle s_x \rangle$ and $\langle s_y \rangle$ vanish, if we assume an initial upwards alignment of $\langle s_z \rangle = \hbar/2$. In Figure 12.14, it is shown how the spin evolves in the course of time.

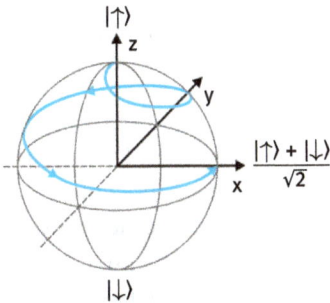

Figure 12.14: Change of spin orientation during application of a rotating magnetic field. The spin was initially in the $|\uparrow\rangle$ state, corresponding to the spin-up orientation, and evolves into the $(|\uparrow\rangle + |\downarrow\rangle)/\sqrt{2}$ state.

12.5.6 Spin-control in a double dot system

After discussing the general mechanisms of spin control by applying an alternating magnetic field, we will explain now, how this scheme can be implemented in a quantum dot system.

The experiments are performed in a double quantum dot, which is shown in Figure 12.15 (a). The double quantum dot is filled from the left side, while the electrons leave in the right side so that a net current flows. This current I_{dot} is controlled by charging effects, as discussed in Section 2.7. However, since here we are dealing with a double dot a spin-related effect, the so-called Pauli blockade, has to be taken into account in addition. In Figure 12.16, a schematic of the double quantum dot system is shown. Each dot is separately controlled by gate voltages V_L and V_R, respectively (see also Figure 12.15 (a)). Current flow is enabled by applying a small bias voltage V_{sd} between the left and right reservoir. Let us assume a situation, where the right quantum dot is occupied by a spin-up electron. The left dot is either occupied by a spin-down

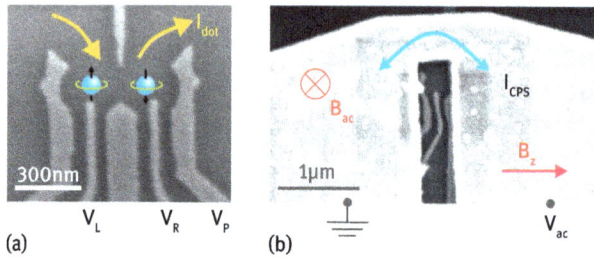

Figure 12.15: (a) Scanning electron beam micrograph of a double quantum dot structure. The two-/dimensional electron gas in an AlGaAs/GaAs heterostructure is located 90 nm below the surface. The arrows indicate the current flow. The gate voltage pulses V_p are applied on the rightmost gate. The quantum dot states are tuned by V_L and V_R applied to the left and right plunger gates, respectively. (b) Double quantum dot structures with a 400 nm thick gold stripline on top of the quantum dot gates. The stripline is separated from the gate electrodes by a 100 nm thick dielectric layer. I_{CPS} is the alternating current through the stripline. A quantizing magnetic field B_z is applied parallel to the surface [206] (Images provide by L. M. K. Vandersypen, Delft University of Technology).

or spin-up electron, as shown in Figure 12.16 (a) and (b), respectively. As long as each dot is occupied by a single electron and the tunnel barrier between these dots is sufficiently high, the mutual spin orientation does not matter. Let us ask the question, under which condition the electron in the left dot can be transferred to the right one, so that afterwards two electrons are residing in the right dot. Here, we have to obey the rule for electrons as fermions that the total wave function has to be antisymmetric under particle exchange. In the first case, with both spins aligned anti-parallel a spin singlet state $|S(0,2)\rangle$ is formed in the right quantum dot, with $(0,2)$ meaning that the left dot is empty and two electrons occupy the right dot. The spin state is antisymmetric therefore the orbital wave function needs to be symmetric. In principle it can be written as the product of two single particle ground states $\psi_0(x_1)\psi_0(x_2)$, i. e. belonging to the ground state quantum dot level E_0. Thus, the total energy for the two electrons would be $2E_0$ under the condition that interactions are neglected. If interactions between the two electrons are included in a single electron picture, this implies that the state belonging to the singlet state $|S(0,2)\rangle$, which is occupied by the second electron is energetically slightly higher than the state of the first electron. This situation is illustrated in the scheme shown in Figure 12.16 (a). Here, we assumed that the initially occupied state in the left quantum dot is energetically higher than the vacant state in the right dot, thus the electron transfer from left to right is allowed. In case that the electro-chemical potential of the right reservoir is sufficiently low, the second electron can tunnel into the right reservoir, so that a net current I_{dot} flows.

Let us now consider the case, where the two electron spins are parallel, as depicted in Figure 12.16 (b). Here, a triplet state $|T(1,1)\rangle$ forms after electron transfer into the right quantum dot, which is symmetric under particle exchange. Consequently, the corresponding orbital wave function for two electrons needs to be antisymmet-

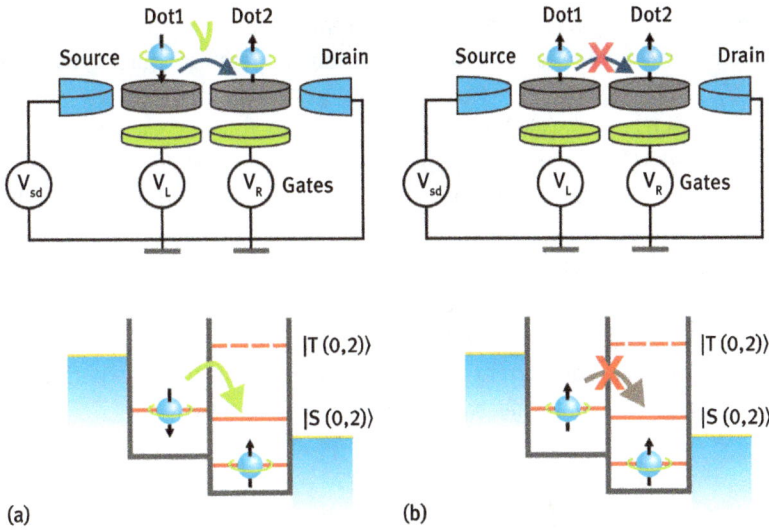

Figure 12.16: Schematics of an electron transfer in a double dot system. Current flow is enabled by applying a source drain voltage V_{sd}, while the quantum dot potential can be tuned by applying appropriate gate voltages V_L and V_R. (a) The electrons in the left and right quantum dots have opposite spin orientations. As shown in the schematics below, electron transfer is allowed, since an energetically lower lying state belonging to the singlet configuration $|S(0,2)\rangle$ can be occupied in the right quantum dot. The state which can be occupied by the second electron belonging to the triplet configuration $|T(0,2)\rangle$ is energetically inaccessible. (b) The electron spins in the left and right dot are aligned in the same direction. Electron transfer is forbidden, since the energy level corresponding triplet state $|T(0,2)\rangle$ in the right quantum dot is too high. The singlet state cannot be formed because of the parallel spin configuration.

ric. To be non-zero, this requires a second state with a higher energy E_1 so that the orbital state can be expressed as $[\psi_0(x_1)\psi_1(x_2) - \psi_0(x_2)\psi_1(x_1)]/\sqrt{2}$. Neglecting interactions between the electrons, the corresponding energy is $E_0 + E_1$, being by the amount $\Delta E = E_1 - E_0$ larger than the energy of the non-interacting singlet state. Generally, one finds that the triplet state is energetically considerably higher than the singlet state, so that it cannot be occupied by transferring a spin-up electron from the right side. As illustrated in the schematics shown in Figure 12.16 (b), the electron transfer is prohibited, since the single state is not available due to the parallel spin configuration. This phenomenon is called Pauli blockade.

The Pauli-blockade introduced above will now be employed to demonstrate coherent spin control. For this experiment, the double quantum dot shown in Figure 12.15 (a) is used. In order to apply an ac magnetic field, the whole double dot structure is covered by a gold coplanar stripline (CPS), as depicted in Figure 12.15 (b). The stripline is insulated by a dielectric from the underlying gate fingers. An alternating current I_{CPS} driven through that conductor generates an oscillating magnetic field B_{ac}, which is oriented perpendicularly to the constant in-plane magnetic field B_z. The potentials of the quantum dots are adjusted by applying appropriate gate voltages V_L and V_R. In addition, the outermost gate is used to change the gate potential quickly by gate voltages pulse V_p.

A typical electron transfer cycle is depicted in Figure 12.17. Here, we assume that the right dot is already filled with a spin-up electron and that an electron with the same spin orientation is transferred from the left reservoir into the left dot. Since Pauli-blockade applies, a further transfer into the right dot is prohibited. Only after applying an oscillating magnetic field so that the spin orientation of the electron in the left dot is inversed, eventually an electron transfer can occur. We assumed that only the left dot is in resonance with the oscillating field. Finally, the electron tunnels into the right reservoir and contributes to the current through the double dot. In the experiment, this sequence is repeated so that a measurable net current I_{dot} flows.

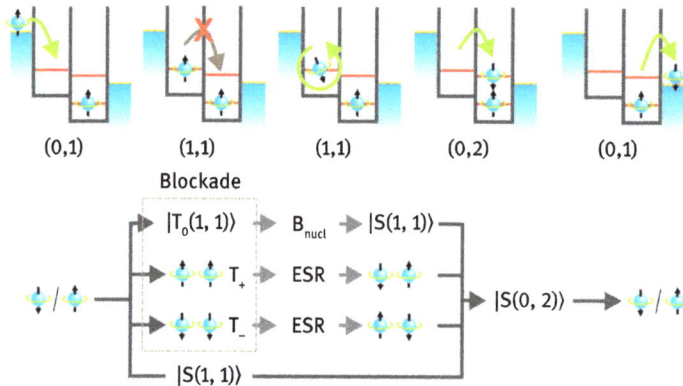

Figure 12.17: Transport cycle in the spin blockade regime: At first the left dot is initially empty, i. e. $(0, 1)$. Subsequently, it is filled by an electron from the left reservoir. We assume that the added electron has the same spin orientation as the electron in the right dot, so that Pauli blockade occurs. By applying an oscillating field B_{ac} the spin is rotated in the left dot. Eventually Pauli blockade is lifted so that an electron transfer into the right dot is allowed, corresponding to a $(0, 2)$ occupation. Finally, the electron in the upper level tunnels into the right reservoir: $(0, 1)$. The lower scheme shows all possible electron transfer paths. If the system is in the triplet $|T_0\rangle$ state the presence of a residual nuclear field can lead to a decay into the $|S(1, 1)\rangle$ state, so that Pauli blockade is lifted. Alternatively, the double dot can either be in the $|T_+\rangle$ or $|T_-\rangle$ triplet state, corresponding to the magnetic quantum numbers $m = +1$ or -1, respectively. These states are transferred by an ESR pulse into a $|\downarrow\uparrow\rangle$ or a $|\uparrow\downarrow\rangle$, which is subsequently coupled to a $|S(1, 1)\rangle$ state by the nuclear field. As a last option, the system can be in an $|S(1, 1)\rangle$ state right after electron transfer, so that a direct transition into $|S(0, 2)\rangle$ is possible. (Figure adapted from Koppens et al. [206]).

In order to understand the electron transfer in more detail we have to consider the influence of the nuclear magnetic field. This is due to the fact, that the Ga and As cores possess a finite nuclear spin. Each electron is coupled to about 10^6 GaAs nuclear spins via the hyperfine interaction, as schematically illustrated in Figure 12.18. The random

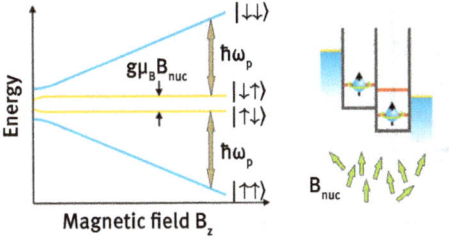

Figure 12.18: Energy splitting of a double dot as a function of an external magnetic field B_z under the influence of a nuclear magnetic field B_{nuc}. The sketch on the right side illustrates the nuclear magnetic field affecting the spin states in the quantum dots.

nuclear field B_{nuc} leads to an energy splitting of the electron state by $g\mu_B B_{nuc}$. More important, the nuclear field is inhomogeneous, so that there is a field gradient ΔB_{nuc} between the left and right dot. As shown in Figure 12.18, in case that the z-component $B_{nuc,z}$ is assumed to be stronger on the left side. Instead of $|S(1,1)\rangle$ and $|T_0(1,1)\rangle$ the eigenstates are rather given by $|\uparrow\downarrow\rangle$ and $|\downarrow\uparrow\rangle$, with the previous being the lower level. This leads to a coupling of the $|S(1,1)\rangle$ with the $|T_0(1,1)\rangle$ state, since the lowest level $|\uparrow\downarrow\rangle$ can be expressed as superposition state

$$|\uparrow\downarrow\rangle = \frac{1}{\sqrt{2}}(|S(1,1)\rangle + |T_0(1,1)\rangle). \tag{12.41}$$

The spin-parallel states $|T_-(1,1)\rangle$ and $|T_+(1,1)\rangle$ with magnetic quantum numbers $m = -1$ and $+1$, respectively, are basically not affected by B_{nuc}, since the Zeeman energy splitting $\hbar\omega_p$ is much larger.

After introducing the effect of the nuclear magnetic field we can now return to the electron transfer through the double dot structure. All four possible transport cycle are depicted in the lower scheme in Figure 12.17. As a first option, the double dot can be in a Pauli blocked $|T_0(1,1)\rangle$ triplet state, corresponding to the magnetic quantum number $m = 0$. Here, we have to take into account that the nuclear field B_{nuc} is present, which couples the $|T_0(1,1)\rangle$ state to the $|S(1,1)\rangle$ state via the $|\uparrow\downarrow\rangle$ and $|\downarrow\uparrow\rangle$ states so that an electron transfer into the right dot is possible. Alternatively, the double quantum dot can be in the blocked $|T_+(1,1)\rangle$ or $|T_-(1,1)\rangle$ state. By an ESR pulse for a certain period these states can be transferred into the $|\downarrow\uparrow\rangle$ or $|\uparrow\downarrow\rangle$ state by rotation the left spin. Once again, due to the presence of B_{nuc} they can decay into the $|S(1,1)\rangle$ and subsequently into the $|S(0,2)\rangle$ state. As a last possibility, the double dot can be right away in the $|S(1,1)\rangle$ state, which directly transfers into the $|S(0,2)\rangle$ state.

In order to coherently manipulate the spin orientation, the sequence described above is slightly modified by employing Coulomb blockade in addition. As shown in Figure 12.19, the system is initialized by filling the left dot with an electron. As outlined above, electrons will be transmitted through the double dot as long as the system is in a $|S(1,1)\rangle$ or a $|T_0(1,1)\rangle$ state. This electron transfer is stopped abruptly when the

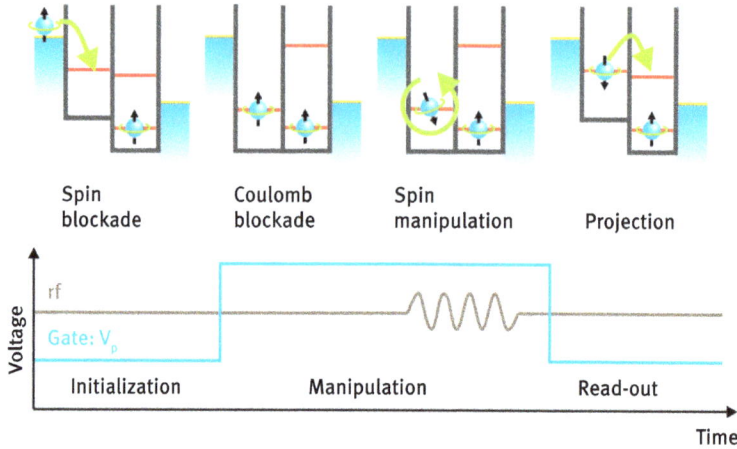

Figure 12.19: Sequence for coherent manipulation of the electron spin. During the initialization stage the double-dot is tuned in the spin blockade regime. During initialization, the double dot is filled with two parallel spins, i. e. either $|\uparrow\uparrow\rangle$ or $|\downarrow\downarrow\rangle$. Subsequently, the system is put into the Coulomb blockade regime, by raising the gate voltage V_p. The orientation of electron spin in the left dot is manipulated by a rf burst. Finally, the Coulomb blockade is lifted by lowering V_p, so that the spin state can be read out. (Figure adapted from Koppens et al. [206]).

system is either in the state $|\uparrow\uparrow\rangle$ or $|\downarrow\downarrow\rangle$, i. e. until Pauli blockade is reached. In order to stabilize the system, V_p is raised, so that the double dot system is in the Coulomb blockade regime. Within that period an radio frequency (rf) burst is applied via the stripling, generating an oscillating field B_{ac}. This oscillating field coherently rotates the spin orientation in the left dot, where the rotation angle depends on the rf burst time and the intensity of the field, as discussed in Section 12.5.5. Subsequently, the gate voltage V_p is lowered so that Coulomb blockade is lifted. Only if the spins are anti-parallel an electron transfer to the right lead occurs.

The corresponding measurements are shown in Figure 12.20. The dot current oscillates as a function of the rf burst time. At time zero, the current is in a minimum, since the system remains in the Pauli blockade regime. When the first maximum is reached, the spin in the left dot is rotated by π and an anti-parallel configuration is achieved, so that the electron can leave the dot to the right reservoir and increase the current by that. For longer burst times, next a 2π rotation is achieved, resulting once again in a Pauli blockade. If the excitation power of the rf burst is increased, the oscillation frequency increases, as explained in Section 12.5.5.

In the example given above an oscillating magnetic field was directly employed for spin manipulation. Regarding possible up-scaling of the systems this approach might cause problems. The reason is that the setup with the stripline is bulky. Furthermore, the alternating current used to generate the oscillating magnetic field is relatively large. As an alternative, one can employ spin-orbit coupling to generate an ac magnetic field. This scheme is called electron-dipole spin resonance [207]. Indeed this

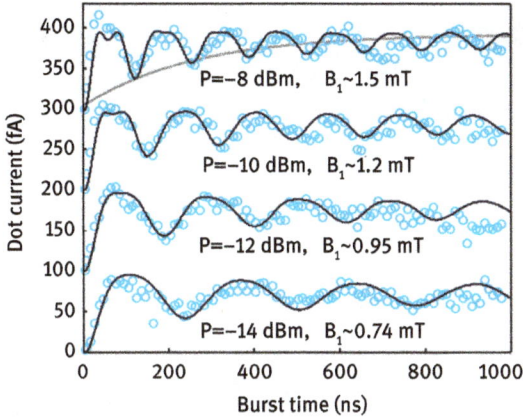

Figure 12.20: Oscillating dot current giving information on the spin state at the end of the radio frequency burst as a function of RF burst length. The curves are offset for clarity. The frequency of B_{ac} is set at the spin resonance frequency of 200 MHz, while an external field B_z of 41 mT was applied. The period of the oscillation increases for decreasing RF power. P is the estimated power and B_1 the corresponding amplitude of the ac field. Each measurement point is averaged over 15 s. The gray line corresponds to an exponentially damped envelope [206] (Figure provide by L. M. K. Vandersypen, Delft University of Technology).

mechanism was successfully employed in an AlGaAs/GaAs quantum dot [208] and in a quantum dot in an InAs nanowire [209]. Here, on one of the confining gates a high-frequency voltage was applied. The corresponding oscillating electric field the electron in the quantum dot experiences results in an effective magnetic field via spin-orbit coupling, i. e. Rashba and Dresselhaus contributions. This effective magnetic field is finally responsible for changing the spin state of the electron, similar to the classical electron spin resonance discussed above.

12.5.7 Singlet-triplet qubit

We have seen that by applying an ac magnetic or electric field, it is possible to manipulate a quantum dot qubit constituted by a single confined electron. However, the gate operation tends to be slow and is thus prone to decoherence. Furthermore, an ac field has to be applied for the single qubit operation, which is technically intricate. In this section we will introduce a different type of qubit, the so-called singlet-triplet (S-T_0) qubit, which is based on two electrons confined in a double quantum dot. For its implementation the double dot structure shown in Figure 12.15 (a) can be taken. The dot potential can be tuned, e. g. by applying an appropriate voltages V_L and V_R to the left and right plunger gates. We will first start with a simple situation, where we neglect tunnel coupling between the two quantum dots. Two cases can be distinguished, which are depicted in Figure 12.21 (a). First, a single electron is confined in

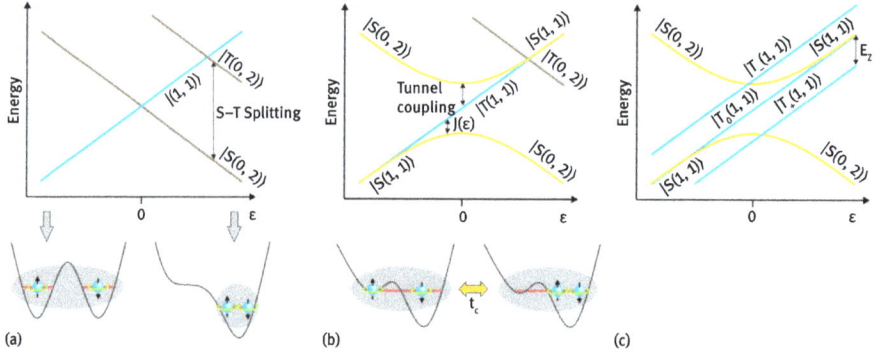

Figure 12.21: Energy and potential diagrams for the singlet-triplet qubit in a double quantum dot. (a) Situation where the tunnel coupling between both dots is neglected. The $|(1,1)\rangle$ state corresponds to a case that in each dot a single electron is confined, while for the $|(0,2)\rangle$ two electrons are on the right dot. This charge configuration is energetically split into the singlet state $|S(0,2)\rangle$ and the triplet state $|T(0,2)\rangle$. (b) Situation, where tunnel coupling between the quantum dots is included. An avoided crossing occurs between the two singlet states $|S(1,1)\rangle$ and $|S(0,2)\rangle$. The two states are illustrated at the bottom. The triplet state $|T(1,1)\rangle$ is not affected. (c) Energy diagram after applying an external magnetic field. The triplet state $T(0,1)$ is split into the $|T_-(1,1)\rangle$, $|T_0(1,1)\rangle$, and $|T_+(1,1)\rangle$ separated by the Zeeman energy E_z.

each quantum dot, corresponding to a $|(1,1)\rangle$ state. As long as the two electrons are sufficiently well separated, the spin state, i.e. singlet or triplet, does not matter. The $|(1,1)\rangle$ state is achieved if both quantum dots are symmetrically biased. However, it is also possible to raise the potential in the left dot and lower the potential in the right dot, so that eventually, a state with both electrons residing in the right dot is in favor $|(0,2)\rangle$. Here, the spin state matters. As outlined in Section 12.5.6, the spin-singlet state $|S(0,2)\rangle = (|\uparrow\downarrow\rangle - |\downarrow\uparrow\rangle)/\sqrt{2}$ has a lower energy than the spin-triplet state. As shown in Figure 12.21(a), the singlet $|S(0,2)\rangle$ and the triplet $|T(0,2)\rangle$ state are separated by the S-T splitting. For the remaining discussion the $|T(0,2)\rangle$ state is neglected, since it is energetically not accessible.

So, far we discussed the two extreme cases that the electrons are either located symmetrically on both dots or that they are confined in a single dot. By tuning the gate voltages V_L and V_R a gradual transition between both cases can be achieved. We define the detuning ε as the energy difference between the $|(1,1)\rangle$ and the $|(0,2)\rangle$ state

$$\varepsilon = E_{|(1,1)\rangle} - E_{|(0,2)\rangle}. \tag{12.42}$$

The detuning is directly related to the difference of the gate voltages $V_R - V_L$. For $\varepsilon > 0$ the ground state configuration is $|(0,2)\rangle$, while for $\varepsilon < 0$ the state $|(1,1)\rangle$ is the energetically lowest state. At $\varepsilon = 0$ both states are degenerated.

As a next step tunneling between the $S|(1,1)\rangle$ and $S|(0,2)\rangle$ will be allowed, as depicted in Figure 12.21 (b). This can be achieved by lowering the barrier between the two quantum dots or by lifting the potential of the left dot, so that the electron from the left can tunnel more easily into the right dot. This process can be described by means of a finite tunneling probability t_c between both states, so that the corresponding Hamiltonian can be written as

$$\hat{H} = \begin{pmatrix} \varepsilon/2 & t_c \\ t_c & -\varepsilon/2 \end{pmatrix}, \tag{12.43}$$

which result in the eigenvalues

$$E_\pm = \pm\sqrt{t_c^2 + \frac{\varepsilon^2}{4}}. \tag{12.44}$$

Thus, at $\varepsilon = 0$ the two superpositions of the singlet states are separated by t_c. For gate operations, we are interested in the energy dependent exchange splitting $J(\varepsilon)$ between the lower singlet superposition state, labeled $|S\rangle$, and the triplet state $|T(1,1)\rangle$ (cf. Figure 12.21 (b)).

Finally, the degeneracy of the triplet state is lifted by applying an external magnetic field, as shown in Figure 12.21 (c). This splits the triplet states into $|T_-(1,1)\rangle$, $|T_0(1,1)\rangle$, and $|T_+(1,1)\rangle$ according to their magnetic quantum number $m = -1$, 0, and +1. For the qubit operation the lower superposition $|S\rangle$ state and the $|T_0(1,1)\rangle$ state are taken as the two-level system. The corresponding Bloch sphere is depicted in Figure 12.22. The basis states $|S\rangle$ and $|T_0\rangle = |T_0(1,1)\rangle$ are placed on the north and south pole, respectively. The states $|\uparrow \downarrow\rangle$ and $|\downarrow \uparrow\rangle$ are formed by a superposition of $|S\rangle$ and $|T_0\rangle$.

$$|S\rangle = (|\uparrow\downarrow\rangle - |\downarrow\uparrow\rangle)/\sqrt{2}$$

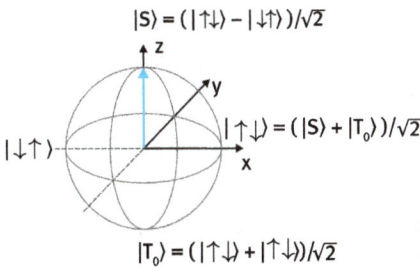

$$|\uparrow\downarrow\rangle = (|S\rangle + |T_0\rangle)/\sqrt{2}$$

$$|\downarrow\uparrow\rangle$$

$$|T_0\rangle = (|\uparrow\downarrow\rangle + |\uparrow\downarrow\rangle)/\sqrt{2}$$

Figure 12.22: Bloch sphere for a S-T_0 qubit: The basis states $|S\rangle$ and $|T_0\rangle$ are located on the north and south pole, respectively. The $|\uparrow\downarrow\rangle$ and $|\downarrow\uparrow\rangle$ states oriented along the x-axis are formed by a superposition of the $|S\rangle$ and $|T_0\rangle$ states.

12.5.8 Control of the S-T_0 qubit

After defining the S-T_0 qubit, we are now in the position to discuss the corresponding single qubit operation. For that, we have to consider another ingredient, i. e. the influence of the nuclear magnetic field via the hyperfine interaction, as illustrated in the sketch shown in Figure 12.23.

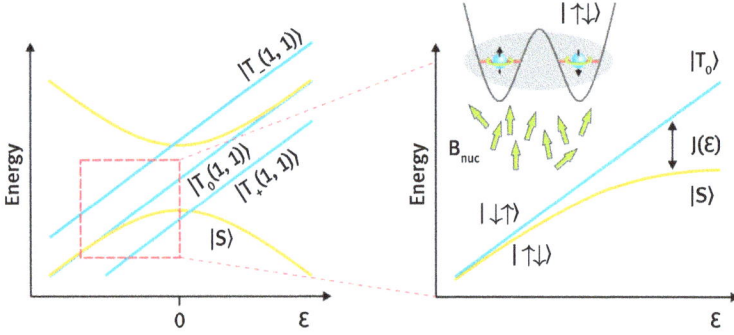

Figure 12.23: Detail of the energy diagram for $\varepsilon < 0$. Due to the fluctuations of field B_{nuc} induced by the nuclear spins at small values of ε the eigenstates are $|\uparrow\downarrow\rangle$ and $|\uparrow\downarrow\rangle$.

The random nuclear field B_{nuc} changes only slowly ($> 10\,\mu\text{s}$) compared to typical pulse sequences during computation and results in an energy splitting of the electron state of $g\mu_B B_{\text{nuc}}$ [210]. In case of large negative detuning, so that $|J(\varepsilon)| < |g\mu_B B_{\text{nuc}}|$ the nuclear field mixes the $|S\rangle$ and $|T_0\rangle$ states. Moreover, since the nuclear field fluctuates the two separated electrons experience a difference in the local magnetic field ΔB_{nuc}. As a consequence, for small $J(\varepsilon)$ the eigenstates will be $|\uparrow\downarrow\rangle$ and $|\downarrow\uparrow\rangle$, where we assumed the first state to be the lower one (cf. Figure 12.23). Summing up all contributions we end up with the following Hamiltonian in the $|S\rangle$ and $|T_0\rangle$ basis

$$\hat{H} = \begin{pmatrix} -J(\varepsilon)/2 & g\mu_B \Delta B_{\text{nuc}}/2 \\ g\mu_B \Delta B_{\text{nuc}}/2 & J(\varepsilon)/2 \end{pmatrix} = \frac{g\mu_B \Delta B_{\text{nuc}}}{2} \hat{\sigma}_x - \frac{J(\varepsilon)}{2} \hat{\sigma}_z. \tag{12.45}$$

For $|\uparrow\downarrow\rangle = |S\rangle + |T_0\rangle$ being the ground state at $J(\varepsilon) = 0$ we have to assume $g\mu_B \Delta B_{\text{nuc}} < 0$.

For a typical single qubit operation one can initialize the system by means of a large positive detuning $\varepsilon > 0$. As shown in Figure 12.24, at point (1) the $|S\rangle = S(0,2)$ state is occupied, where both electrons reside as a singlet state in the right dot.

Subsequently, the system is transfered to point (2), by slowly changing the detuning ε to negative values. As mentioned above, owing to the nuclear field, here the stable ground state is the $|\uparrow\downarrow\rangle$ state. Thus the $|S\rangle$ state is adiabatically transferred to this state. This situation is depicted in Figure 12.25 (a).

As a next step the exchange splitting $J(\varepsilon)$ is switched on by changing ε abruptly to a less negative value (cf. Figure 12.24 at (3)) for a fixed period of time. As a consequence, the state will precess around the z-axis, as shown in Figure 12.25 (b). We assumed that $J(\varepsilon) \gg g\mu_B B_{\text{nucl}}$, so that a precession about the x-axis can be neglected. The accumulated precession angle about the z-axis is determined by the time $J(\varepsilon)$ is switched on.

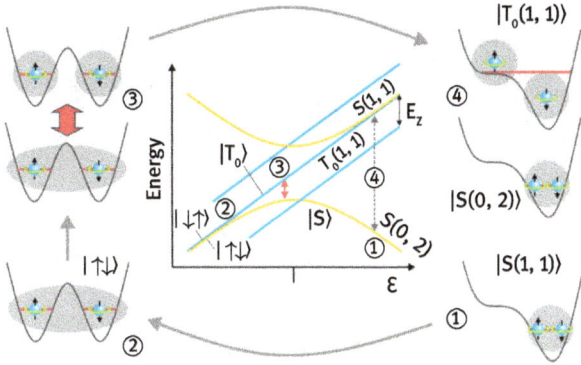

Figure 12.24: Control sequence in a S-T_0 quantum bit: (1) At large detuning $\varepsilon > 0$ the dot is transferred into the $|S\rangle \approx |S(0,2)\rangle$ state. (2) Changing the detuning to negative gate voltages so that the initial state is transferred to the $|\uparrow\downarrow\rangle$ state. (3) Transfer to finite exchange coupling $J(\varepsilon)$ and coherent oscillation between $|S\rangle$ and $|T_0\rangle$ state for a specific time interval. (4) Read-out of the qubit state by transferring to large positive detuning $\varepsilon > 0$.

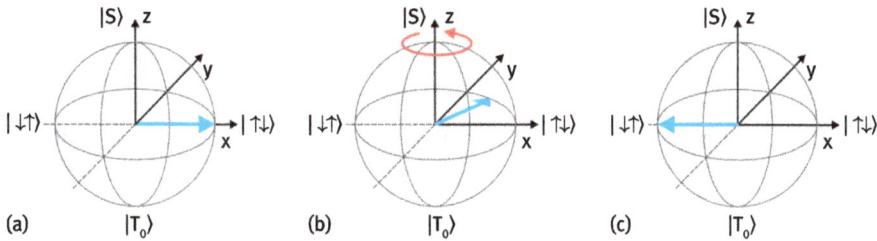

Figure 12.25: Single qubit operation of a S-T_0 qubit on a Bloch sphere. (a) Initial state $|\uparrow\downarrow\rangle$. (b) Rotation in the xy-plane as long as $J(\varepsilon)$ is switched on. (c) Final position after $J(\varepsilon)$ is set back to zero. In total the state was rotated by π.

For the sequence shown in Figure 12.25 (c), we have assumed that the state precessed by π so that it ends up in $|\downarrow\uparrow\rangle$. For read-out the double dot is set to large positive detuning, as shown in Figure 12.24 at (4). For the specific case discussed here returning with ε to large positive value would transfer the system into the $|T_0(1,1)\rangle$ state. The transfer has to be adiabatic, i. e. the state $|\downarrow\uparrow\rangle$ is gradually transferred to $|T_0(1,1)\rangle$. A final state $|\uparrow\downarrow\rangle$, would have been returned adiabatically to $|S\rangle$. Generally, depending on the period of time the exchange interaction was turned on, the final state is either the $|S\rangle = |S(0,2)\rangle$ or the $|T_0\rangle$ state or in a superposition state of both. In the latter case $|S\rangle = |S(0,2)\rangle$ and $|T_0\rangle$ would be measured by a certain probability. Which of these states is actually occupied is probed by a quantum point contact placed next to the right dot, see Section 12.5.2. By measuring its conductance it can be distinguished if two or one electrons are confined in the right dot, corresponding to the $|S\rangle$ or $|T_0\rangle$ state, respectively.

12.6 Summary

- A quantum bit is a quantum mechanical two-level system, consisting of a ground and an excited state. Superposition states are possible.
- A quantum computer can be realized based on an elementary set of quantum gates, i. e. a single- and a two-qubit gate, respectively.
- A quantum dot containing a single electron or a double dot with two electrons can be used to represent a spin qubit. In the latter case, the qubit is defined by a singlet and triplet states, i. e. S-T_0.
- The state of the qubit in a quantum dot can be read out by means of an adjacent quantum point contact.
- Single-qubit operation in a quantum dot can be realized by electron spin resonance using an external rf magnetic field or by electron-dipole spin resonance based on spin-orbit coupling.
- In S-T_0 single qubit operation is realized by exchange coupling.
- In quantum dots based on III-V semiconductors the nuclear magnetic field results in spin dephasing.

Exercises

Problem 12.1. Show that the controlled-not (CNOT) two-qubit gate can be used to disentangle an entangled state.

Problem 12.2. Show that the controlled-rotation (CROT) two-qubit gate defined by

$$U_{\text{CROT}} = \begin{pmatrix} 1 & 0 & 0 & 0 \\ 0 & 1 & 0 & 0 \\ 0 & 0 & 1 & 0 \\ 0 & 0 & 0 & -1 \end{pmatrix},$$

can be transferred into the controlled-NOT (CNOT) gate by applying $\pi/2$-operation on the first qubit before and a $-\pi/2$-operation after the CROT operation.

Problem 12.3. Solve the Schrödinger equation for a spin-1/2 particle, if only a fixed magnetic field along the z-direction is applied. Calculate the spin expectation value along the x- and y-direction as a function of time, assuming an initial state

$$\phi(0) = \frac{1}{\sqrt{2}} \begin{pmatrix} 1 \\ 1 \end{pmatrix}$$

aligned along the x-direction.

Problem 12.4. Let us assume a S-T_0 qubit systems, with an exchange coupling $J(\varepsilon) = 0$. How does the $|S\rangle$ state evolve in time?

13 Majorana fermions

13.1 Overview

In this chapter we will address a subject which is intimately connected to spin-orbit coupling. By combining a material containing spin-orbit coupling with a superconductor, so-called Majorana fermions can be created. This unique type of particles goes back to Ettore Majorana, who found a special kind of the Dirac equation, which gives pure real solutions [211]. This has drastic implications such that Majorana fermions are their own antiparticles. From that it directly follows that these particles should have zero energy, no charge, and no spin. It was anticipated that Majorana fermions might be present in solid state materials. As already mentioned above, the ingredients are a material with spin-orbit coupling which is combined with a superconductor. For the spin-orbit materials we basically have two choices, i. e. a semiconductor with Rashba spin-orbit coupling (cf. Chapter 7) or a topological material, such as two- or three-dimensional topological insulators (cf. Chapters 10 and 11).

The qubit realizations we discussed in the previous chapter are quite sensitive to decoherence, where the quantum state is destroyed by coupling to the environment. This limits the maximum operation time of a quantum computer. A possible solution is to apply error correction schemes. However, the price to pay is that for representing a logical qubit for performing a quantum algorithm, many physical qubits are required for a proper operation. This makes the hardware of a quantum processor much more complex. A way-out of that dilemma might be to use topologically protected states, consisting of pairs of the above mentioned Majorana fermions, which are located at a large distance from each other. Since the two states are stored at two well-separated sites in local space, it makes them more robust against distortions from outside.

In this chapter we first outline how Majorana fermions can be created as quasi-particles in condensed matter systems. Subsequently, we will give a concrete recipe how to realize Majorana states by combining a spin-orbit material with a superconductor. Furthermore, we will discuss what is happening when these Majorana states are exchanged, i. e. are braided. Finally, we will show how these Majoranas can be employed to define a so-called topological qubit. For further reading the review articles of Alicea [212], Leijnse and Flensberg [213], Aguado [214], and Lutchyn et al. [215] are recommended.

13.2 Majorana's version of the Dirac equation

Let us begin with the Dirac Hamiltonian we already introduced in Section 11.6.1

$$\hat{H} = c \begin{pmatrix} mc & \vec{\sigma}\vec{p} \\ \vec{\sigma}\vec{p} & -mc \end{pmatrix}. \tag{13.1}$$

https://doi.org/10.1515/9783110639001-013

Here, m is the particle mass, $\vec{p} = -i\hbar\nabla$ is the momentum operator, and $\vec{\sigma}$ is the vector with Pauli matrices. We inserted the velocity of light c instead of the Fermi velocity which was previously used in the Dirac Hamiltonian for treating topological band structures. The Dirac equation

$$i\hbar\frac{\partial}{\partial t}\Psi = \hat{H}\Psi, \tag{13.2}$$

is a wave equation for relativistic particles in contrast to the Schrödinger equation which deals with nonrelativistic particles. We explicitly included the time-dependence by using the quantum mechanical energy operator $i\hbar(\partial/\partial t)$. In order to rewrite the Dirac equation in a different, more compact form we define the following 4×4 matrices

$$\alpha_j = \sigma_x \otimes \sigma_j = \begin{pmatrix} 0 & \sigma_j \\ \sigma_j & 0 \end{pmatrix}, \quad \beta = \begin{pmatrix} 1 & 0 \\ 0 & -1 \end{pmatrix}, \tag{13.3}$$

with σ_j being Pauli matrices. These matrices were actually found by Dirac to express the square root of the relativistic energy-momentum relation $E = \sqrt{p^2c^2 + m^2c^4}$ by an ansatz linear in momentum. The matrices α_j and β obey the following relations

$$\alpha_j^2 = 1, \tag{13.4}$$
$$\beta^2 = 1, \tag{13.5}$$
$$\alpha_j\beta - \beta\alpha_j = 0, \tag{13.6}$$
$$\alpha_j\alpha_k - \alpha_k\alpha_j = 0, \tag{13.7}$$

with $j, k = 1, 2, 3$ corresponding to the three spatial directions. Using these matrices we can write the Dirac equation in the following form

$$i\hbar\frac{\partial}{\partial t}\Psi = (c\vec{\alpha}\vec{p} + \beta mc^2)\Psi, \quad \text{with } \Psi = \begin{pmatrix} \psi_1 \\ \psi_2 \\ \psi_3 \\ \psi_4 \end{pmatrix}. \tag{13.8}$$

Since the Dirac equation originates from the square root of the Laplacian operator, it also implies that positive and negative energies are allowed. Generally, there are four independent solutions of that system, two with $E > 0$ and two with $E < 0$. Negative-energy solutions belong to antiparticles. Defining the so-called gamma matrices, given by

$$\gamma^\mu = (\beta, \beta\vec{\alpha}), \quad \mu = 1, 2, 3, 4, \tag{13.9}$$

we can write the Dirac equation in an even more compact form

$$(i\gamma^\mu\partial_\mu - m)\Psi = 0, \quad \text{with } \hbar = c = 1. \tag{13.10}$$

The γ matrices generate the Clifford algebra, i. e. the anticommutator relation $\{\gamma^{\mu}, \gamma^{\nu}\} = \gamma^{\mu}\gamma^{\nu} + \gamma^{\nu}\gamma^{\mu} = 2\eta_{\mu\nu}$ holds, with the metric tensor given by $\eta_{\mu\nu} = \mathrm{diag}(1, -1, -1, -1)$. The gamma matrices are explicitly written by

$$\gamma^0 = \begin{pmatrix} 1 & 0 & 0 & 0 \\ 0 & 1 & 0 & 0 \\ 0 & 0 & -1 & 0 \\ 0 & 0 & 0 & -1 \end{pmatrix}, \quad \gamma^1 = \begin{pmatrix} 0 & 0 & 0 & 1 \\ 0 & 0 & 1 & 0 \\ 0 & -1 & 0 & 0 \\ -1 & 0 & 0 & 0 \end{pmatrix}, \tag{13.11}$$

$$\gamma^2 = \begin{pmatrix} 0 & 0 & 0 & -i \\ 0 & 0 & i & 0 \\ 0 & i & 0 & 0 \\ -i & 0 & 0 & 0 \end{pmatrix}, \quad \gamma^3 = \begin{pmatrix} 0 & 0 & 1 & 0 \\ 0 & 0 & 0 & -1 \\ -1 & 0 & 0 & 0 \\ 0 & 1 & 0 & 0 \end{pmatrix}. \tag{13.12}$$

The matrices γ^{μ} contain both real and imaginary numbers, which means that the solutions of the Dirac equations are necessarily also complex. Majorana found a different set of gamma matrices:

$$\tilde{\gamma}^0 = \begin{pmatrix} 0 & 0 & 0 & -i \\ 0 & 0 & -i & 0 \\ 0 & i & 0 & 0 \\ i & 0 & 0 & 0 \end{pmatrix}, \quad \tilde{\gamma}^1 = \begin{pmatrix} 0 & 0 & i & 0 \\ 0 & 0 & 0 & i \\ i & 0 & 0 & 0 \\ 0 & i & 0 & 0 \end{pmatrix}, \tag{13.13}$$

$$\tilde{\gamma}^2 = \begin{pmatrix} i & 0 & 0 & 0 \\ 0 & i & 0 & 0 \\ 0 & 0 & -i & 0 \\ 0 & 0 & 0 & -i \end{pmatrix}, \quad \tilde{\gamma}^3 = \begin{pmatrix} 0 & 0 & 0 & -i \\ 0 & 0 & i & 0 \\ 0 & i & 0 & 0 \\ -i & 0 & 0 & 0 \end{pmatrix}. \tag{13.14}$$

resulting in a modified Dirac equation, also called the Majorana equation

$$(i\tilde{\gamma}^{\mu}\partial_{\mu} - m)\tilde{\Psi} = 0, \quad \text{with } \hbar = c = 1. \tag{13.15}$$

Since the matrices $i\tilde{\gamma}^{\mu}$ are all real, the solutions are also real. Thus, we can write

$$\Psi = \Psi^*. \tag{13.16}$$

This directly implies that the particle described by Ψ is its own antiparticle. Since usually the antiparticle has the opposite charge, e. g. electron and positron, these so-called Majorana fermions, or in short Majoranas, must be neutral. In addition, the spin is also required to be zero.

13.3 Majorana modes in solid-state systems

We are interested to transfer the concept of Majorana particles to condensed matter systems. From semiconductors, we know that the carrier transport can be maintained by negatively charged electrons or positively charged holes for n- and p-type doping, respectively. In a sense, a hole is the antiparticle of an electron. However, this directly implies that considering pure electrons and holes is not sufficient to create Majorana fermions in solid state systems, since they have opposite charge, i. e. they are not neutral as required. A way out of that dilemma is to form so-called quasiparticles, which are composed by electron and hole contributions. The idea behind that is to compensate the negative charge of the electron component by the positive charge of the hole contribution, so that in total a zero charge state is created, as requested for a Majorana particle. In order to implement that idea we have to include superconductivity in our system. Here, it is known that excitations are described by quasiparticles, which are neither pure electrons nor holes but coherent superpositions of both. The Bogoliubov–de Gennes equation, which is explained in detail below, is the right tool to deal with these quasiparticles.

However, including superconductivity comes with the price that we have to consider many-particle states, since the superconducting state is maintained by pairs of electrons, so-called Cooper pairs. Therefore, we will first pick-up once again the concept of creation and annihilation operators, which was introduced in connection with the harmonic oscillator (cf. Section 2.8.1). The creation and annihilation operators will later also be used to describe Majorana states and the basic operations of topological quantum computation.

13.3.1 Many-particle states

In this section we briefly sketch how many-particle states can be handled on the basis of creation and annihilation operators. In order to do so we have to introduce field operators, which describe electrons as an excitation of a many-body field. Since for now it is sufficient to catch the basic ideas, we refrain from providing a rigorous derivation. More detailed information can be found, e. g. in Ref. [40].

A many-particle state can be constructed by making use of single-particle states, e. g. the normalized solutions $\varphi_i(\vec{r})$ of the single-particle Schrödinger equation

$$\left[\frac{\hbar^2}{2m}\Delta + V(\vec{r})\right]\varphi_i(\vec{r}) = \varepsilon_i\varphi_i(\vec{r}), \tag{13.17}$$

with $V(\vec{r})$ the potential the particles are exposed to. Employing the single particle wave functions $\varphi_i(\vec{r})$ the corresponding many-particle state $|\phi\rangle$ can be written as

$$|\phi\rangle = |\ldots, n_i, \ldots\rangle, \tag{13.18}$$

with the state occupation numbers n_i being either 0 or 1, showing if the corresponding state $\varphi_i(\vec{r})$ is occupied or not. The fact that we are dealing with fermions requires that each state is either empty or occupied with only a single electron. In order to build up the many particle state out of a vacuum state $|0\rangle$, we can apply the creation operators $\hat{c}_i^\dagger$ for a fermion occupying the states $\varphi_i(\vec{r})$. The operator $\hat{c}_i^\dagger$ is defined in the same sense as the creation operator for excitations in a harmonic oscillator, as introduced in Section 2.8.1. Thus, we can create the many-particle state by leaving out or applying the creation operators $\hat{c}_i^\dagger$:

$$|\phi\rangle = \prod_i (\hat{c}_i^\dagger)^{n_i} |0\rangle, \quad n_i = 0, 1, \tag{13.19}$$

i.e. for $n_i = 0$ no particle is created while for $n_i = 1$ the operator $\hat{c}_i^\dagger$ creates one. Due to the fact that we are dealing with fermions, we have to ensure that under exchange of two particles a sign change takes place. This implies the following anticommutator relations [40]

$$\{\hat{c}_i^\dagger, \hat{c}_j\} = \hat{c}_i^\dagger \hat{c}_j + \hat{c}_j \hat{c}_i^\dagger = \delta_{ij}, \tag{13.20}$$

$$\{\hat{c}_i, \hat{c}_j\} = \hat{c}_i \hat{c}_j + \hat{c}_j \hat{c}_i = 0, \tag{13.21}$$

$$\{\hat{c}_i^\dagger, \hat{c}_j^\dagger\} = \hat{c}_i^\dagger \hat{c}_j^\dagger + \hat{c}_j^\dagger \hat{c}_i^\dagger = 0, \tag{13.22}$$

with $\hat{c}_i$ the annihilation operator of the state $\varphi_i(\vec{r})$.

For many-particle systems the expectation value of physical observables can be calculated by making use of the field operators defined as

$$\hat{\psi}(\vec{r}) = \sum_i \hat{c}_i \varphi_i(\vec{r}), \tag{13.23}$$

$$\hat{\psi}^\dagger(\vec{r}) = \sum_i \hat{c}_i^\dagger \varphi_i^*(\vec{r}). \tag{13.24}$$

As an example of how to employ the field operators $\hat{\psi}(\vec{r})$ and $\hat{\psi}^\dagger(\vec{r})$, the expectation value of the energy

$$E = \langle \phi | \hat{\mathcal{H}} | \phi \rangle \tag{13.25}$$

can be calculated by using the Hamilton operator $\hat{H}$ of the many-particle system defined as

$$\hat{\mathcal{H}} = \int d\vec{r} \hat{\psi}^\dagger(\vec{r}) \left[\frac{\hbar^2}{2m} \Delta + V(\vec{r}) \right] \hat{\psi}(\vec{r}). \tag{13.26}$$

By explicitly inserting the field operators given by equations (13.23) and (13.24) one obtains for the Hamilton operator

$$\hat{\mathcal{H}} = \sum_{ij} \int d\vec{r} \hat{c}_i^\dagger \varphi_i^*(\vec{r}) \left[\frac{\hbar^2}{2m} \Delta + V(\vec{r}) \right] \hat{c}_j \varphi_j(\vec{r})$$

$$= \sum_{ij} \int d\vec{r} \hat{c}_i^\dagger \varphi_i^*(\vec{r}) \varepsilon_j \hat{c}_j \varphi_j(\vec{r})$$

$$= \sum_{ij} \hat{c}_i^\dagger \hat{c}_j \varepsilon_j \delta_{ij} = \sum_i \epsilon_i \hat{c}_i^\dagger \hat{c}_i$$

$$= \sum_i \epsilon_i \hat{n}_i. \tag{13.27}$$

Here, $\hat{n}_j = \hat{c}_i^\dagger \hat{c}_i$ is the number operator, i. e. in case that the state i is occupied by an electron or empty it equals one or zero, respectively. For the sake of simplicity, if we assume that the state i is occupied, i. e. $|\phi\rangle = \hat{c}_i^\dagger |0\rangle$, we obtain for the corresponding energy expectation value

$$\langle \phi | \hat{\mathcal{H}} | \phi \rangle = \left\langle 0 \left| \hat{c}_j \sum_i \epsilon_i \hat{c}_j^\dagger \hat{c}_j \hat{c}_i^\dagger \right| 0 \right\rangle$$

$$= \left\langle 0 \left| \hat{c}_j \sum_i \epsilon_i \hat{c}_j^\dagger (\delta_{ij} - \hat{c}_i^\dagger \hat{c}_j) \right| 0 \right\rangle$$

$$= \langle 0 | \hat{c}_i \epsilon_i \hat{c}_i^\dagger | 0 \rangle = \langle \phi | \epsilon_i | \phi \rangle = \epsilon_i. \tag{13.28}$$

In a solid state system at zero temperature, the states in an energy band are occupied up to the Fermi level, i. e. starting from the vacuum state $|0\rangle$ the many-particle state is build up by creating electron excitations using $\hat{c}_j^\dagger$ with energies up to E_F. A hole in a fully occupied Fermi sea can be created by applying an annihilation operator $\hat{c}_i$, which removes an electron at i and leaves an empty state.

13.3.2 Bogoliubov–de Gennes equation

In a normal so-called s-type superconductor the attractive interaction between a pair of electrons mediated by phonon scattering leads to an energy lowering of the system. According to the Bardeen–Cooper–Schrieffer (BCS) theory a pair of electrons with opposite wave vectors $(\vec{k}, -\vec{k})$ is scattered into the final unoccupied state $(\vec{k}', -\vec{k}')$ [216]. The electrons of these pairs have opposite spin, i. e. there is singlet pairing. The formation of these Cooper pairs is accompanied by the appearance of an energy gap Δ_0 at the Fermi energy. The probability that a pair state $(\vec{k}, -\vec{k})$ is occupied is given by

$$v_{\vec{k}}^2 = \frac{1}{2} \left(1 - \frac{\varepsilon_{\vec{k}}}{E_{\vec{k}}} \right), \tag{13.29}$$

with the kinetic energy $\varepsilon_{\vec{k}}$ referred to chemical potential μ given by

$$\varepsilon_{\vec{k}} = \frac{\hbar^2 k^2}{2m} - \frac{\hbar^2 k_F^2}{2m} = \frac{\hbar^2 k^2}{2m} - \mu, \tag{13.30}$$

and $E_{\vec{k}}$ defined by

$$E_{\vec{k}} = \sqrt{\varepsilon_{\vec{k}}^2 + \Delta_0^2}. \tag{13.31}$$

The probability $v_{\vec{k}}^2$ to find a Cooper pair at $\vec{k}$ is shown in Figure 13.1. It can be seen that it is smeared out even at $T = 0$ in contrast to the Fermi distribution of free electrons which is abrupt at zero temperature. The reason for the smearing is, that it pays off to invest kinetic energy in order to create unoccupied states. The latter can be used to allow the phonon scattering which then finally leads to an effective lowering of the total energy due to the Cooper pairing.

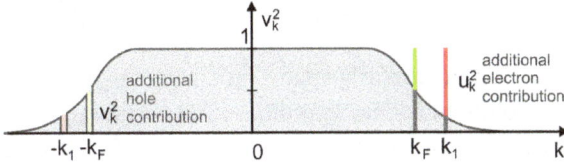

Figure 13.1: Probability $v_{\vec{k}}^2$ that a pair state $(\vec{k}, -\vec{k})$ is occupied. Two single-particle excitations are shown, with an electron placed at k_1 and k_F, respectively. Since these states do not form a Cooper pair, the states at $-k_1$ and $-k_F$ must be empty.

Let us have a look on what happens if we add a single unpaired electron to a superconducting system. In Figure 13.1 we added an electron at k_1 above the Fermi wave vector k_F to the system. Since this electron should not be paired with another one, i. e. it shall not form a Cooper pair, the state at $-k_1$ necessarily has to be empty. The whole situation can be described by effectively adding an electron contribution $u_{\vec{k}}^2 = 1 - v_{\vec{k}}^2$ to the system, with reference to the Cooper pair distribution function, while at $-k_1$ a hole contribution of $v_{\vec{k}}^2$ is formed. Thus, adding an unpaired electron to the systems ends up with a state having a mixture of electron and hole character. Since for the case shown in Figure 13.1 the additional electron contribution dominates, we name that case electron-like quasiparticle excitation. If the k-vector of the unpaired single electron is smaller than k_F, the state has a larger hole contribution at the opposite k-vector site, i. e. we have a hole-like quasiparticle. The electron-hole quasiparticles are called Bogoliubov quasiparticles. An interesting situation occurs if we place an additional electron at k_F, as shown in Figure 13.1. In that case the electron and hole contribution is balanced, which implies that the charge of that quasiparticle is zero.

The single particle excitations described above are captured by the Bogoliubov–de Gennes equation which is an extension of the Schrödinger equation coupling electron and hole states by the superconducting pair potential $\Delta(\vec{r})$. The Bogoliubov–de Gennes equation is defined as [214]

$$\begin{pmatrix} \hat{H}_0(\vec{r}) & \Delta(\vec{r}) \\ \Delta^*(\vec{r}) & -\sigma_y \hat{H}_0^*(\vec{r})\sigma_y \end{pmatrix} \Phi(\vec{r}) = E\Phi(\vec{r}), \tag{13.32}$$

where $\hat{H}_0$ contains the kinetic and single-electron potentials of spin-up and spin-down electrons. The term $-\sigma_y \hat{H}_0^*(\vec{r})\sigma_y$ is the time-reversed of $\hat{H}_0^*$, corresponding to holes being the time-reversed of electrons. As previously mentioned, a hole is formed by removing an electron from the filled Fermi sea. In the Bogoliubov–de Gennes equation, as given above, the spin is included explicitly. The solutions can be written as

$$
\Phi(\vec{r}) = \begin{pmatrix} u_\uparrow(\vec{r}) \\ u_\downarrow(\vec{r}) \\ v_\uparrow(\vec{r}) \\ v_\downarrow(\vec{r}) \end{pmatrix},
\tag{13.33}
$$

with $u_\uparrow(\vec{r})$ and $u_\downarrow(\vec{r})$ the amplitudes of the electron contributions, whereas $v_\uparrow(\vec{r})$ and $v_\downarrow(\vec{r})$ are the ones of the holes.

In case of a homogeneous superconductor the spatial component of $u_\uparrow(\vec{r})$, $u_\downarrow(\vec{r})$, $v_\uparrow(\vec{r})$, and $v_\downarrow(\vec{r})$, e. g. a plane wave solution, can be split off, so that we end up with the factors $u_\uparrow, u_\downarrow, v_\uparrow$, and $v_\downarrow$. In order to create or annihilate Bogoliubov quasiparticle states we make use of creation and annihilation operators of electrons, i. e. $\hat{c}_\sigma^\dagger$ and $\hat{c}_\sigma$ with σ the spin. As an example, the annihilation operator of a Bogoliubov quasiparticle can be expressed by

$$
\hat{b} = u_\uparrow \hat{c}_\uparrow^\dagger + v_\downarrow \hat{c}_\downarrow,
\tag{13.34}
$$

with the creation operator $\hat{c}^\dagger$ for the electron contribution and the annihilation $\hat{c}_\downarrow$ operator creating a hole by removing an electron from the Fermi sea. As mentioned above, depending on the ratio between $u_\uparrow$ and $v_\downarrow$ the quasiparticle has either more hole or more electron character. In order to get to a Majorana fermion we have to do some modifications. Since it is required that the particle be its own antiparticle, we need $|u|$ to be equal to $|v|$, to maintain charge neutrality. Furthermore, we need to have the same spin for both contributions

$$
\hat{\gamma} = u \hat{c}_\sigma^\dagger + u^* \hat{c}_\sigma.
\tag{13.35}
$$

The argument about the spin can be understood in the sense that if a system only has one spin orientation, the spin is irrelevant and can effectively be left out. Since the electron and hole contribution is the same $|u| = |u^*| = 1/2$, the net charge is zero. The specific form of $\hat{\gamma}$ directly implies that the operator is Hermitian

$$
\hat{\gamma} = \hat{\gamma}^\dagger,
\tag{13.36}
$$

i. e. the particle is its own antiparticle (cf. equation (13.16)). A comparison of the operators expressed by equations (13.34) and (13.35) reveals, that in an ordinary s-type superconductor a Majorana fermion cannot be created, since the quasiparticle operator $\hat{b}$ contains contributions with opposite spins. One would need to have a spinless superconductor, which basically means a superconductor with electrons having only one kind of spins so that the spin orientation doesn't matter at the end. This type of superconductor has a triplet pairing symmetry instead of singlet pairing. In one-dimensional systems this would be a so-called p-wave pairing, while in two dimensions it would be a $p_x \pm p_y$ pairing symmetry. So far, there is no experimental evidence that materials with triplet pairing exist. However, there are ways to get an effective triplet pairing, i. e. by having a semiconductor in proximity to an s-type superconductor and taking care that the spins are aligned by applying an external magnetic field and making use of spin-orbit coupling. As explained in detail in Section 13.5, under these conditions the proximity effect induces a superconducting state within the semiconductor with the required pairing symmetry.

13.4 The Kitaev chain

The are a number of ways to realize Majorana fermions in solid-state systems. Here, we will focus on a very simple intuitive model, which was introduced by Kitaev [217]. We are dealing with a one-dimensional chain of electrons represented by $\hat{c}_j$, as illustrated in Figure 13.2 (a). We assume that the spins are effectively aligned in one direction only, which implies a p-wave superconducting pairing. Since the spins are fixed to a single direction, we neglect it further on and consider a spinless system. The tight-binding Hamiltonian of the chain including the chemical potential, the kinetic energy, and superconducting pairing can be described by

$$H = -\mu \sum_{j=1}^{N}\left(\hat{c}_j^\dagger \hat{c}_j - \frac{1}{2}\right) + \sum_{j=1}^{N-1}[-t(\hat{c}_j^\dagger \hat{c}_{j+1} + \hat{c}_{j+1}\hat{c}_j^\dagger) + \Delta(\hat{c}_j\hat{c}_{j+1} + \hat{c}_{j+1}^\dagger\hat{c}_j^\dagger)]. \tag{13.37}$$

Here, μ is the chemical potential, $\hat{c}_i^\dagger$ and $\hat{c}_i$ are the electron creation and annihilation operators for site j, respectively, and $n_j = \hat{c}_j^\dagger \hat{c}_j$ is the corresponding number operator which expresses that a state j is occupied. The superconducting gap is given by Δ, while t expresses the hopping to the next neighbor, e. g. annihilation $\hat{c}_{j+1}$ on site $j+1$ and creation $\hat{c}_j^\dagger$ on site j.

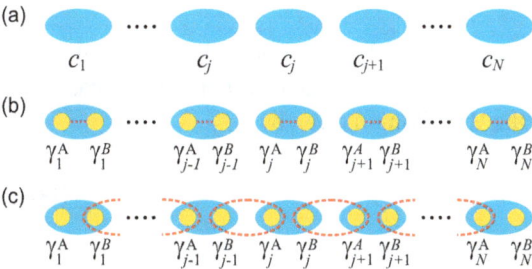

Figure 13.2: (a) Schematic illustration of a chain of fermions $\hat{c}_j$. (b) Kitaev chain with the fermion operator on site j split into two Majorana operators $\hat{\gamma}_j^A$ and $\hat{\gamma}_j^B$. (c) Situation where Majoranas are combined on neighboring sites $\hat{\gamma}_{j+1}^A$ and $\hat{\gamma}_j^B$.

The Hamiltonian given above can be rewritten in terms of Majorana fermions. In the spirit of the Majorana operator given by equation (13.35) we can define the operators

$$\hat{\gamma}_j^A = \hat{c}_j^\dagger + \hat{c}_j, \tag{13.38}$$

$$\hat{\gamma}_j^B = i(\hat{c}_j^\dagger - \hat{c}_j). \tag{13.39}$$

These obviously Hermitian operators are obtained by splitting the operator of a fermion into its real and imaginary parts

$$\hat{c}_j = \frac{1}{2}(\hat{\gamma}_j^A + i\hat{\gamma}_j^B), \tag{13.40}$$

$$\hat{c}_j^\dagger = \frac{1}{2}(\hat{\gamma}_j^A - i\hat{\gamma}_j^B). \tag{13.41}$$

The splitting of the fermion operators into two Majorana operators is illustrated in Figure 13.2 (b). From the anticommutator relations for fermions given by equations (13.20)–(13.22) the following anticommutator relation for Majorana operators can be derived

$$\{\hat{\gamma}_i^A, \hat{\gamma}_j^B\} = 2\delta_{ij}\delta_{AB}. \tag{13.42}$$

Using the definitions of the Majorana operators one can write the Hamiltonian of the Kitaev chain as

$$H = -\frac{i\mu}{2} \sum_{j=1}^N \hat{\gamma}_j^B \hat{\gamma}_j^A + \frac{i}{2} \sum_{j=1}^{N-1} [w_+ \hat{\gamma}_j^B \hat{\gamma}_{j+1}^A + w_- \hat{\gamma}_{j+1}^B \hat{\gamma}_j^A], \tag{13.43}$$

where $w_\pm = \Delta \pm t$ are the hopping amplitudes between Majorana fermions on neighboring sites. For the special case of no hopping and no pairing, i. e. $t = \Delta = 0$, we get the trivial case

$$H = -\frac{i\mu}{2} \sum_{j=1}^N \hat{\gamma}_j^B \hat{\gamma}_j^A, \tag{13.44}$$

which is just the case shown in Figure 13.2 (b) with Majorana operators paired on the same site j to form a fermion at j. In contrast, if we go to the other limiting case with $\mu = 0$ and setting $t = \Delta$ we arrive at the Hamiltonian given by

$$H = -it \sum_{j=1}^{N-1} \hat{\gamma}_j^B \hat{\gamma}_{j+1}^A. \tag{13.45}$$

Here, Majorana operators of neighboring sites are coupled, whereas there is no coupling between Majorana operators of the same site. Furthermore, the Majorana operators at the end of the chain $\hat{\gamma}_1^A$ and $\hat{\gamma}_N^B$ are missing in the summation. The corresponding situation is illustrated in Figure 13.2 (c). One can combine Majorana operators of neighboring sites to form a new set of fermion operators by

$$\hat{\tilde{c}}_j = \frac{1}{2}(\hat{\gamma}_j^B + i\hat{\gamma}_{j+1}^A), \tag{13.46}$$

$$\hat{c}_j^\dagger = \frac{1}{2}(\hat{\gamma}_j^B - i\hat{\gamma}_{j+1}^A).$$ (13.47)

Using these operators one ends up with the Hamiltonian

$$H = 2t \sum_{j=1}^{N-1} \left(\hat{c}_j^\dagger \hat{c}_j + \frac{1}{2} \right).$$ (13.48)

The new fermion operators diagonalize the problem. They describe Bogoliubov quasi-particles with energy t. Interestingly, the new diagonalized Hamiltonian contains only $N-1$ quasiparticle operators, whereas the original chain itself contains N sites. Thus, as illustrated in Figure 13.2 (c), on each terminal of the chain a single Majorana operator is not considered. These two remaining Majorana operators can be described by a single fermion operator expressed by

$$\hat{c}_M = \frac{1}{2}(\hat{\gamma}_N^B + i\hat{\gamma}_1^A).$$ (13.49)

This state is highly delocalized, since $\hat{\gamma}_1^A$ and $\hat{\gamma}_N^B$ are from opposite ends of the chain. Since they are also not included in equation (13.48) the state requires zero energy. This is actually one of the crucial points. Usually, in a superconducting condensate with Cooper pairs, it costs energy to add an additional single quasiparticle to the system. In contrast, in the present case, it does not matter energetically if that quasiparticle is present or not. Thus, we have a twofold degenerate ground state with an even or odd total number of electrons in the superconductor.

13.5 Majorana zero modes in semiconductor nanowires

There are several proposals on how to realize Majorana fermions in condensed-matter systems [212, 214]. Here, we will consider a system where a semiconductor nanowire is combined with a superconducting electrode [218, 219, 220]. The first challenge is to obtain an effective p-wave pairing in a semiconductor. We will see that this can be achieved by having a semiconductor system with strong spin-orbit coupling. Subsequently, we will show how a pair of Majorana states are formed in a semiconductor nanowire partly covered by a superconducting electrode. From now one we will also name the Majorana fermions Majorana zero modes due to the fact that we are dealing with collective excitations in a solid state material.

13.5.1 p-wave pairing in a proximitized semiconductor nanowire

We consider a one-dimensional system, i. e. a semiconductor nanowire in proximity to a superconductor. In zinc-blende III-V semiconductors spin-orbit coupling is present in addition to the kinetic term. Thus, the Hamiltonian can be written as [218, 219]

$$\hat{\mathcal{H}}_0 = \int dx \hat{\psi}^\dagger \left[-\frac{\hbar^2 \partial_x^2}{2m^*} - \mu - i\alpha_R \sigma_y \partial_x \right] \hat{\psi}, \tag{13.50}$$

with m^* the effective mass, μ the chemical potential, α_R the Rashba spin-orbit coupling strength, and $\hat{\psi}$ and $\hat{\psi}^\dagger$ the field operators.

As we know from Chapter 7, the Rashba term leads to a spin splitting, as illustrated in Figure 13.3 (a). The spin is always oriented perpendicularly to the momentum. As a next step, an external magnetic field B_x is applied, represented by the Zeeman Hamiltonian

$$\hat{\mathcal{H}}_Z = E_Z \int dx \hat{\psi}^\dagger \sigma_x \hat{\psi} = \frac{g\mu_B}{2} B_x \int dx \hat{\psi}^\dagger \sigma_x \hat{\psi}, \tag{13.51}$$

with E_Z the Zeeman energy. Next, the complete Hamiltonian will be diagonalized. In the k-space it can be expressed by a superposition of the annihilation states $\hat{\psi}_\pm$ for the upper and lower bands

$$\hat{\psi}(k) = \varphi_-(k)\hat{\psi}_-(k) + \varphi_+(k)\hat{\psi}_+(k), \tag{13.52}$$

with $\varphi_\pm(k)$ the corresponding normalized wave functions

$$\varphi_\pm(k) = \frac{1}{\sqrt{2}} \begin{pmatrix} \pm \eta_k \\ 1 \end{pmatrix}, \tag{13.53}$$

and

$$\eta_k = \frac{i\alpha_R k + E_Z}{\sqrt{E_Z^2 + \alpha_R^2 k^2}}. \tag{13.54}$$

The index $+/-$ corresponds to the helicity of the two subbands, cf. Section 7.7.2. Employing $\hat{\psi}_\pm(k)$ in the Hamiltonian it becomes

$$\mathcal{H}_0 + \mathcal{H}_Z = \int \frac{dk}{2\pi} [E_+(k)\hat{\psi}_+^\dagger(k)\hat{\psi}_+(k) + E_-(k)\hat{\psi}_-^\dagger(k)\hat{\psi}_-(k)], \tag{13.55}$$

with the eigenenergies given by

$$E_\pm = \frac{k^2}{2m} - \mu \pm \sqrt{E_Z^2 + \alpha_R^2 k^2}. \tag{13.56}$$

The additional contribution of the Zeeman term results in the emergence of a gap around zero momentum, as can be seen in Figure 13.3 (b). A crucial point for the subsequent discussion is that the chemical potential μ is adjusted to be located within the so-called helical gap at $k = 0$, so that only the lower helical band $E_-(k)$ is occupied.

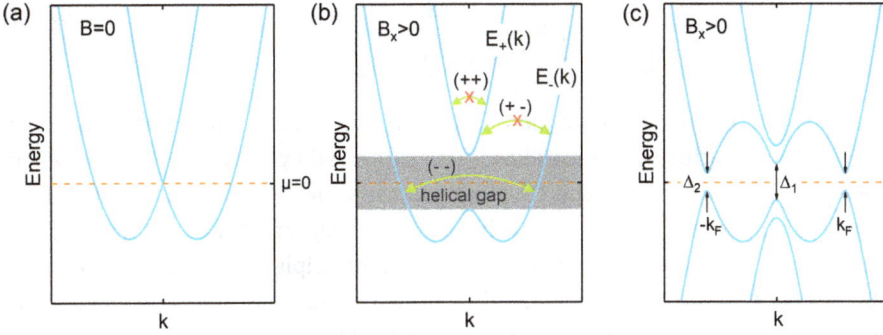

Figure 13.3: (a) Schematic energy-momentum dispersion of a wire structure with Rashba spin-orbit coupling included. The chemical potential μ is set to zero. (b) Corresponding dispersion at finite magnetic field ($B_x > 0$) including the Zeeman effect in addition. $E_+(k)$ and $E_-(k)$ are the dispersions for the upper and lower energy eigenvalues, respectively. The arrows indicate the possible couplings between electrons in the bands for the superconducting coupling, i. e. ($--$) stands for pairing of electrons originating both from the $E_-(k)$ subbands. In (c) the wire is exposed to a magnetic field and is in proximity to an s-type superconductor. A superconducting gap is formed around $\mu = 0$. Two characteristic gaps can be identified, i. e. a gap Δ_1 at $k = 0$ and a gap Δ_2 around the Fermi wave number $\pm k_F$.

When the semiconductor nanowire gets into contact with an s-type superconductor, a pairing term is generated in the semiconductor due to the proximity effect, which is expressed by the contribution

$$\hat{\mathcal{H}}_{sc} = \int dx [\Delta \hat{\psi}_\uparrow^\dagger \hat{\psi}_\downarrow^\dagger + \text{H.c.}], \tag{13.57}$$

with H. c. the hermitian conjugate. Proximity effect means that the superconduction correlation is transferred across the interface into the semiconductor. In the Hamiltonian given above we expressed the inducted pairing by the creation of two electronic states with opposite spins, i. e. s-type pairing. However, we can also describe the system in the helical basis. Expressing $\hat{\mathcal{H}}_{sc}$ in momentum space in terms of $\hat{\psi}_\pm(k)$ by using the wave functions $\varphi_\pm(k)$ gives

$$\hat{\mathcal{H}}_{sc} = \int \frac{dk}{2\pi} \frac{\Delta_{++}^p}{2} \{\hat{\psi}_+^\dagger(k)\hat{\psi}_+^\dagger(-k) + \text{H.c.}\} + \frac{\Delta_{--}^p}{2} \{\hat{\psi}_-^\dagger(k)\hat{\psi}_-^\dagger(-k) + \text{H.c.}\}$$
$$+ \Delta_{+-}^s \{\hat{\psi}_+^\dagger(k)\hat{\psi}_-^\dagger(-k) + \hat{\psi}_-^\dagger(-k)\hat{\psi}_+^\dagger(k)\}, \tag{13.58}$$

with the effective gaps

$$\Delta_{++}^p = \frac{-i\alpha_R k \Delta}{\sqrt{E_Z^2 + \alpha_R^2 k^2}}, \tag{13.59}$$

$$\Delta_{--}^p = \frac{i\alpha_R k \Delta}{\sqrt{E_Z^2 + \alpha_R^2 k^2}}, \tag{13.60}$$

$$\Delta^s_{+-} = \frac{E_Z\Delta}{\sqrt{E_Z^2 + \alpha_R^2 k^2}}. \tag{13.61}$$

One finds that the projection of the BCS-pairing onto helical bands results in an effective p-wave pairing by intraband coupling expressed by Δ^p_{++} and Δ^p_{--} and an effective s-wave pairing by interband coupling expressed by Δ^s_{+-}. A characteristics of the p-wave pairing is that Δ^p_{++} and Δ^p_{--} are linear in k taking care that the total wave function of the pair is antisymmetric to obey the Pauli principle. If the chemical potential μ is located within the helical gap, only pairing between electrons of a single band with the same helicity occurs, i. e. only electrons of the lower helical band $E_-(k)$ are pairing. The upper helical band $E_+(k)$ is empty, since it is located above μ. Therefore, no intraband pairing of electrons within $E_+(k)$ or interband pairing between electrons from $E_-(k)$ and $E_+(k)$ take place, as illustrated in Figure 13.3 (b). As a consequence, the Hamiltonian given by equation (13.58) reduces to

$$\hat{\mathcal{H}}_{sc} = \int \frac{dk}{2\pi} \frac{\Delta^p_{--}}{2} \{\hat{\psi}^\dagger_-(k)\hat{\psi}^\dagger_-(-k) + \text{H.c.}\}. \tag{13.62}$$

Finally, we reached a scenario which is very similar to the Kitaev chain [217]. We effectively get a p-wave pairing, however, the pairing does not take place between electrons with the same spin but rather with electrons having the same helicity.

13.5.2 Majorana zero modes

By adding the Hamiltonian describing the superconductive coupling within a single helical band given by equation (13.62) to the total Hamiltonian one gets

$$\hat{\mathcal{H}} = \hat{\mathcal{H}}_0 + \hat{\mathcal{H}}_Z + \hat{\mathcal{H}}_{sc}. \tag{13.63}$$

By solving the corresponding Bogoliubov–de Gennes equation one finds for the energy eigenvalues [214]

$$E^2_\pm(k) = \left(\frac{\hbar^2 k^2}{2m^*} - \mu\right)^2 + \alpha_R k^2 + E_Z^2 + \Delta^2$$

$$\pm 2\sqrt{E_Z^2\Delta^2 + \left(\frac{\hbar^2 k^2}{2m^*} - \mu\right)^2 (E_Z^2 + \alpha_R^2 k^2)}. \tag{13.64}$$

In Figure 13.3 (c) an exemplary energy-momentum dispersion is depicted for the case that $E_Z > \Delta$ with the chemical potential μ set to zero. Owing to the superconducting coupling the dispersion comprises an energy gap around zero energy. As we will discuss in detail below, the gaps Δ_1 at $k = 0$ and Δ_2 at $\pm k_F$ are affected by the ratio between Δ and E_Z but also by the chemical potential μ. By setting k equal to zero in equation (13.64) we obtain the following relation for Δ_1

$$\Delta_1 = 2|E_Z - \sqrt{\Delta^2 + \mu^2}|. \tag{13.65}$$

We are particularly interested in the case that the gap Δ_1 located at $k = 0$ closes, i.e. $\Delta_1 = 0$, since this situation can be put into relation with a Majorana state. According to equation (13.65) three parameters can be varied to bring Δ_1 to zero. By setting Δ_1 in equation (13.65) to zero one obtains the following relation for the characteristic critical chemical potential

$$\mu_c = \sqrt{E_Z^2 - \Delta^2}. \tag{13.66}$$

In Figure 13.4 (a) the critical chemical potential is plotted as a function of E_Z/Δ. Inside the area with the boundary μ_c the system is in a topological superconductive state, whereas outside it, the system is in a trival state. As we will see in detail below, similar to topological insulators we discussed before, the subbands are crossing each other around $k = 0$ when the boundary of μ_c is passed from the trival to the topological range.

In order to elucidate the emergence of a Majorana zero mode, we refer to the structure sketched in Figure 13.4 (b). A semiconductor nanowire is partly covered by a superconductor in order to induced a superconducting state in the nanowire by means of the proximity effect. An external magnetic field B_x is applied along the x-axis. For the sake of simplicity we keep μ equal to zero. We ask ourself under which condition a Majorana state is formed and how does the energy-momentum dispersion looks like at different ratios between the Zeeman energy E_Z and the induced superconducting gap Δ in the semiconductor. By moving from the green dot to the red dot in Figure 13.4 (a), first a trivial state is formed, since $E_Z < \Delta$. The corresponding dispersion is depicted in Figure 13.4 (c). Owing to the proximity effect a gap Δ_1 is opened at $k = 0$. Furthermore, at $k = \pm k_F$ an energy gap Δ_2 is present. As can be inferred from Figure 13.4 (d), for $E_Z = \Delta$, the gap at $k = 0$ is closed, while Δ_2 remains open. Here, at the boundary between the trivial and the topological state a Majorana zero mode is formed. Upon further increasing E_Z with respect to Δ, the gap Δ_1 at $k = 0$ reopens. During the crossing of μ_c and the accompanied gap closing and reopening, the bands in the vicinity of $k = 0$ cross each other and a topological state is formed.

A viable way to obtain Majorana zero modes would be to modulate the external magnetic field, i.e. to change E_Z. By that way, trival and topological sections in the nanowire are formed with a Majorana mode at their boundary. However, in practice the external magnetic field is usually kept constant. If we refer to the sketch in Figure 13.4 (b), Majorana zero modes are formed at the terminals of the nanowire. This can be explained in the following way. In the nanowire covered by the superconducting electrode a topological state is formed at sufficiently large Zeeman energies E_Z. The

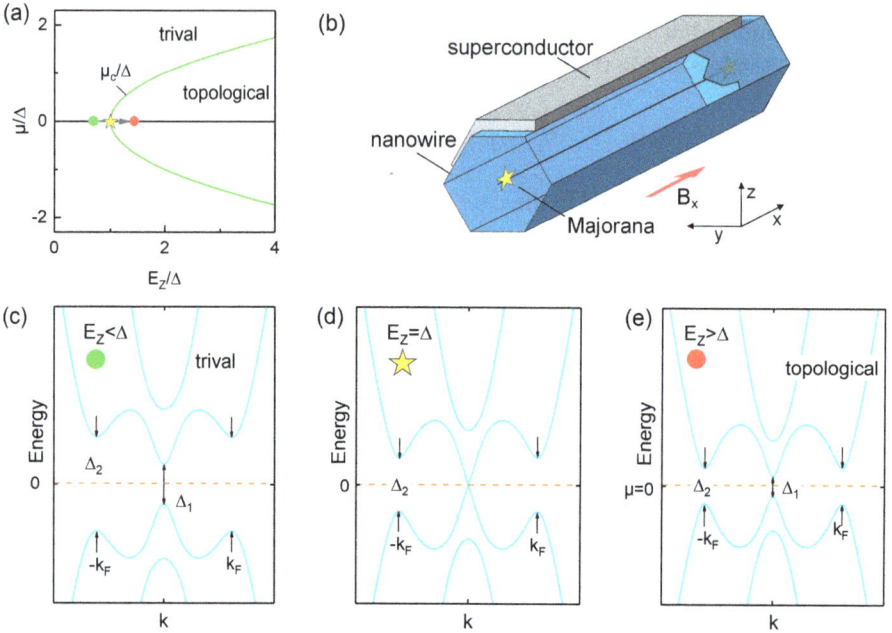

Figure 13.4: (a) Normalized critical chemical potential μ_c/Δ as a function of E_z/Δ. For μ in the inside area the system is in a topological state, while outside it, the state is trival. The trace from the red to the green dot illustrates the shift of E_z/Δ. At the crossing with μ_c a Majorana zero mode is formed (yellow asterix). (b) Schematic illustration of a nanowire partly covered by a superconducting electrode. At the terminal of the wire a Majorana zero mode (yellow asterix) is formed. (c)–(e) Subband dispersion for $E_z < \Delta$, $E_z = \Delta$, and $E_z > \Delta$, respectively, at $\mu = 0$. For the case $E_z = \Delta$ the gap at $k = 0$ is closed, while the gap Δ_2 remains open.

terminal of the wire are in contact to vacuum, which is a trival, nontopological state by nature. As we discussed above, the transition from a topological state to a trival state is accompanied by a closing of the gap and the formation of a Majorana state at the terminals of the nanowire. Since the Majorana zero modes have a spatial extension, one has to make sure that the nanowire and its superconductor coverage is sufficiently long. Otherwise, the two Majorana zero modes overlap with the consequence that the degeneracy is lifted.

13.5.3 Experimental realization in nanowires

The first experiments indicating the existence of Majorana zero modes was reported by Mourik and Zuo et al. [221]. A schematics of the sample layout is shown in Figure 13.5 (a). A semiconductor nanowire is partly covered by a superconducting electrode on one side, while on the other side the nanowire is contacted by a normal metal. As semiconductor material InSb or InAs are suitable, because of the strong Rashba

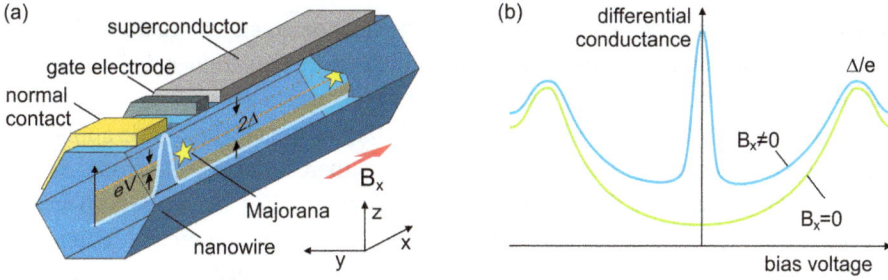

Figure 13.5: (a) Schematic illustration of a nanowire-superconductor hybrid structure used to detect Majorana zero modes. Between the normal and superconducting contact a narrow gate electrode is place. A negative gate bias creates a tunnel barrier. The states in the superconducting region are detected by tunnel spectroscopy. Here, a bias voltage is applied between the normal and superconducting region and the tunnel current is measured. (b) Qualitative traces of the differential conductance at zero magnetic field and at a finite field B_x along the x direction. The field B_x needs to be larger than the critical field in order to create a topological superconducting state underneath the superconductor electrode. The presence of a Majorana zero mode leads to a conductance peak at zero bias.

spin-orbit coupling and the large g-factor [95]. The latter allows to obtain large Zeeman energies at moderate magnetic fields so that the superconductivity is not suppressed. In between the normal contact and the superconducting electrode one finds a gate, which allows to control the coupling between both sides by raising a tunnel barrier. The existence of the Majorana zero mode is indicated by performing tunneling spectroscopy experiments. Here, the energy eV of the left electrode is varied by applying a bias voltage (cf. Figure 13.5 (a)). In case that states are available in the semiconductor section covered by the superconductor a tunneling current flows. Usually, in the nanowire area which is covered by the superconductor, a gap 2Δ is induced by the proximity effect. As a consequence, no states or only a few are available in that energy range. The former case, called hard gap, occurs if the coupling between the superconductor and semiconductor is ideal, while in the latter case, called soft gap, the coupling is nonideal [222].

As we know from the previous discussion, at zero magnetic field the semiconductor section underneath the superconducting electrode is in a trivial superconducting state, i. e. no Majorana zero modes are expected. In Figure 13.5 (b) a schematics of the measurement outcome is shown. In these measurements a bias voltage is applied between the normal and superconducting sections, which are separated by a tunnel barrier. The barrier is induced by applying a sufficiently large negative gate voltage. The differential conductance usually shows features at $\pm\Delta/e$, indicating the presence of superconductivity. In between, the differential conductance is reduced. For the case illustrated here, the conductance is still finite, i. e. the induced gap is soft. Owing to the smaller density of states within the gap region, the tunnel current and thus the differential conductance is smaller in that energy window. In case of a hard gap, the differential conductance would even drop to zero since no states are accessible for tunneling [222]. However, since at zero magnetic field we only induce a trivial super-conducting state, no feature originating from the presence of a Majorana zero mode is expected. The situation changes, if a sufficiently large magnetic field B_x is applied so that a topological state is induced underneath the superconducting electrode. As discussed in the previous section and illustrated in Figure 13.5 (a), Majorana zero modes appear in the semiconductor nanowire below the edges of the superconducting electrode. These modes are located at zero energy. As a consequence, a peak is observed in the differential conductance at zero bias, which is due to tunneling via a Majorana zero mode. The appearance of the peak can be interpreted as an indication that a Majorana state is present, although it is not a definite proof. Since Majorana fermions come in pairs, a possible follow-up experiment would be to simultaneously perform tunneling current measurements at both ends of the nanowire.

13.6 Topological quantum computing

Here we discuss how a topological qubit can be defined based on Majorana zero modes. Furthermore, it will be shown how a single-qubit operation can be realized.

13.6.1 Fermion parity and degenerate ground state

In Section 13.4 we learned that we have a degeneracy for the superconducting condensate, i. e. for no or one single extra electron on top of all Cooper pairs forming the condensate. This extra electron is, e. g., responsible that a pair of Majoranas is formed at the terminals of a nanowire which is in a topological superconductive state. For the remainder we will only have a look on a particular electron which is added to the electrons of the condensate and forms a pair of Majorana fermions or in short a pair of Majoranas. Thus, the superconducting condensate is put into background. Later on we will generalize this concept by allowing systems with more than one pair of Majoranas. In order to simplify the description of the system, we will use a slightly different notation similar to equation (13.49) to define fermion operators in terms of two Majoranas located at the terminals of a topological superconductor chain

$$\hat{c} = \frac{1}{2}(\hat{\gamma}_1 + i\hat{\gamma}_2), \tag{13.67}$$

$$\hat{c}^\dagger = \frac{1}{2}(\hat{\gamma}_1 - i\hat{\gamma}_2). \tag{13.68}$$

It must be stressed that the fermion operators $\hat{c}$ and $\hat{c}^\dagger$, do not represent a fermion in the normal sense, i. e. with Majoranas next to each other. They rather describe a fermion which is constituted by two Majoranas far apart at the terminals of a chain.

The distinction between absence and presence of a fermion forming a pair of Majoranas can be made by applying the number operator

$$\hat{n} = \hat{c}^\dagger \hat{c}, \tag{13.69}$$

which gives $n = 0$ and 1, respectively. The degenerate ground state can be expressed by the number states

$$|0\rangle = |0\rangle,$$
$$|1\rangle = \hat{c}^\dagger |0\rangle, \tag{13.70}$$

for having an extra electron or not. The state $|1\rangle$ was obtained by applying the creation operator to the ground state $\hat{c}^\dagger |0\rangle$. Instead of dealing directly with the Majorana fermions it is often more practical to use the two basis states $|0\rangle$ and $|1\rangle$, i. e. for representing a topological qubit or to perform qubit operations. In an ideal system, where the pair of Majoranas is well separated, no distinction with respect to energy can be made between $|0\rangle$ and $|1\rangle$. However, if the two Majoranas approach each other and there is an overlap between both, the degeneracy is lifted. This property is used in practical implementations to measure if the state $|0\rangle$ or $|1\rangle$ was realized. The parity operator given by

$$\hat{P} = 1 - 2\hat{n} = 1 - 2\hat{c}^\dagger \hat{c} = -i\hat{\gamma}_1\hat{\gamma}_2 \tag{13.71}$$

tells if an extra electron is present or not. In order to get the last term the anticommutator relations for Majorana fermions were applied. The parity operators has the eigenvalues ± 1 for even and odd fermion parity, respectively. The parity also holds if we take the superconducting condensate into account, since with electrons coming in Cooper pairs the parity is always even. Thus, adding an extra electron switches the parity of the whole system to odd.

We can generalize the system by having in total N fermions described by the many-particle state $|\phi\rangle = |n_1, n_2, \ldots, n_N\rangle$, each n_i being zero or one. This ground state has a 2^N-fold degeneracy. The degenerate ground state will be the basis of topological quantum computation. As we will see in the following section, the quantum computational operations are performed by moving around Majorana fermions, e. g. exchanging them. In contrast to the spin qubits we don't have a quantum mechanical two-level system with a ground and an excited state, since here all states are at zero energy.

13.6.2 Braiding

As a next step, we will discuss, what happens when Majorana fermions are exchanged. This is best explained if we extend the system to two dimensions, i. e. we refer to a two-dimensional p-wave superconductor, a so-called $p_x \pm p_y$-wave superconductor, as illustrated in Figure 13.6 (a). The concept was first introduced by Ivanov [223]. The Majoranas are located in vortices in the superconducting plane. In a vortex the superconductor is penetrated by a magnetic flux. Inside the vortices the superconducting gap vanishes. Encircling the vortex with a Cooper pair adds up to a phase change of 2π. To simplify the system, we shrink the continuous phase change to a branch cut with an abrupt phase change of 2π for the Cooper pairs when crossing that cut. If we exchange two vortices containing the Majoranas, one of them has to cross the branch cut of the other, whereas the other one can be moved without crossing a branch cut. Since we are dealing with fermions rather than with Cooper pairs, the phase change is only π. This can be understood as follows. A phase change of ϕ for Cooper pairs translates in a phase change of $\phi/2$ for single fermion operators, thus $\hat{c} \to e^{i\phi/2}\hat{c}$ and $\hat{c}^\dagger \to e^{-i\phi/2}\hat{c}^\dagger$ which yields equation (13.38) for $\phi = 2\pi$: $\hat{\gamma} = (\hat{c} + \hat{c}^\dagger) \to e^{i\pi}(\hat{c} + \hat{c}^\dagger)$. Thus, we arrive at a transformation rule, where one of the Majoranas acquires a phase change of π while the other does not

$$\hat{\gamma}_1 \to -\hat{\gamma}_2, \tag{13.72}$$

$$\hat{\gamma}_2 \to +\hat{\gamma}_1. \tag{13.73}$$

The exchange process of Majorans is called braiding. The braiding operation is illustrated in Figure 13.6 (b). The unitary operator representing the braiding operation reads

$$\hat{B}_{12} = e^{-\frac{\pi}{4}\hat{\gamma}_1\hat{\gamma}_2} = e^{i\frac{\pi}{4}\hat{P}} = \frac{1}{\sqrt{2}}(1 + i\hat{P}) = \frac{1}{\sqrt{2}}(1 + \hat{\gamma}_1\hat{\gamma}_2), \tag{13.74}$$

with $\hat{P}$ the parity operator defined by equation (13.71). The transformation rule given above can then be written as

$$\hat{\gamma}_i \to \hat{B}_{12}\hat{\gamma}_i\hat{B}_{12}^\dagger, \tag{13.75}$$

with $i = 1, 2$. We can also apply the braiding operator to the number states

$$\hat{B}_{12}|0\rangle = e^{i\frac{\pi}{4}\hat{P}}|0\rangle = e^{i\frac{\pi}{4}(1-2\hat{n})}|0\rangle = e^{+i\frac{\pi}{4}}|0\rangle = \frac{1}{\sqrt{2}}(1 + i)|0\rangle, \tag{13.76}$$

$$\hat{B}_{12}|1\rangle = e^{i\frac{\pi}{4}\hat{P}}|1\rangle = e^{i\frac{\pi}{4}(1-2\hat{n})}|1\rangle = e^{-i\frac{\pi}{4}}|1\rangle = \frac{1}{\sqrt{2}}(1 - i)|1\rangle. \tag{13.77}$$

Thus, braiding only adds a phase factor to the state. The parity of the state is not changed. Alternatively, one can also use the Pauli matrices to define the braiding operator as a rotation operator acting on $|0\rangle = (1, 0)^T$ and $|1\rangle = (0, 1)^T$

$$\hat{B}_{12} = e^{i\frac{\pi}{4}\sigma_z} = \cos\left(\frac{\pi}{4}\right)I_0 + i\sin\left(\frac{\pi}{4}\right)\sigma_z = \frac{1}{\sqrt{2}}(I_0 + i\sigma_z), \tag{13.78}$$

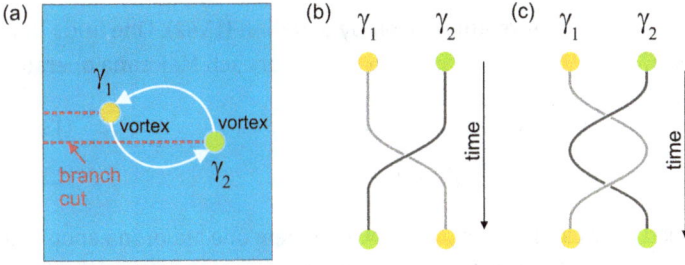

Figure 13.6: (a) Exchange of two Majoranas $\hat{\gamma}_1$ and $\hat{\gamma}_2$. The red lines represent the abrupt phase change, branch cut. During the exchange process $\hat{\gamma}_1$ crosses the branch cut and acquires a phase factor −1. (b) Illustration of the braiding process of two Majoranas $\hat{\gamma}_1$ and $\hat{\gamma}_2$. (c) Operation of two subsequent braidings.

where σ_z is the Pauli matrix in the basis $|0\rangle$ and $|1\rangle$ and I_0 is the identity matrix.[1] Here, for the series expansion of $\exp[i(\pi/4)\sigma_z]$ we used the property $\sigma_z^2 = I_0$ of the Pauli matrix. Thus, comparing with equation (13.74) we see that the Pauli matrix represents a product of Majorana operators

$$i\hat{\gamma}_1\hat{\gamma}_2 = \sigma_z. \tag{13.79}$$

For later use we can express the braiding operator applied to the number states in terms of fermion creation and annihilation operators. According to equations (13.67) and (13.68) we can write the braiding operator given by equation (13.74) as

$$\hat{B}_{12} = \frac{1}{\sqrt{2}}(1 + \hat{\gamma}_1\hat{\gamma}_2)$$

$$= \frac{1}{\sqrt{2}}[1 + i(\hat{c} + \hat{c}^\dagger)(\hat{c}^\dagger - \hat{c})]$$

$$= \frac{1}{\sqrt{2}}[1 + i(\hat{c}\hat{c}^\dagger - \hat{c}^\dagger\hat{c})]. \tag{13.80}$$

Here, we made use of the anticommutator relations for fermions given by equations (13.21) and (13.22). Applying this form of the operator to $|0\rangle$ and $|1\rangle$ immediately gives the results of equations (13.76) and (13.77).

Two successive exchanges, i. e. braidings, are described by $\hat{B}_{12}^2$. The corresponding operation is illustrated in Figure 13.6 (c). The two step operation can be reduced to

$$\hat{B}_{12}^2 = \frac{1}{2}(1 + \hat{\gamma}_1\hat{\gamma}_2)(1 + \hat{\gamma}_1\hat{\gamma}_2)$$

$$= \frac{1}{2}(1 + 2\hat{\gamma}_1\hat{\gamma}_2 + \hat{\gamma}_1\hat{\gamma}_2\hat{\gamma}_1\hat{\gamma}_2)$$

$$= \hat{\gamma}_1\hat{\gamma}_2, \tag{13.81}$$

---

1 There is a negative sign in Refs. [213] and [214]. The definition of $\hat{B}_{12}$ fits to the original paper of Ivannov [223].

where we used the anticommutator relation given by equation (13.42). One finds that applying the operator $\hat{B}_{12}^2$ just results in phase factor of -1 for each Majorana operator

$$\hat{\gamma}_1 \rightarrow (\hat{\gamma}_1\hat{\gamma}_2)\hat{\gamma}_1(\hat{\gamma}_1\hat{\gamma}_2)^\dagger = -\hat{\gamma}_1, \tag{13.82}$$

$$\hat{\gamma}_2 \rightarrow (\hat{\gamma}_1\hat{\gamma}_2)\hat{\gamma}_2(\hat{\gamma}_1\hat{\gamma}_2)^\dagger = -\hat{\gamma}_2. \tag{13.83}$$

The operation described here can be seen as a process where one Majorana encircles the other one. Effectively each Majorana is crossing the branch cut of the other Majorana once, thus acquiring a phase factor of -1.

13.6.3 Braiding with two pairs of Majoranas

So far the braiding of a pair of Majoranas only resulted in a phase factor for one of the Majoranas. In order to gain a substantial change of a state, which is relevant for topological quantum computing, we have to extend our systems to two pairs of Majoranas. Thus, we consider a state with two fermions, i. e. $|\phi\rangle = |n_2, n_1\rangle$, so that the ground state is 4-fold degenerate. In total we have to deal with four Majoranas defined by

$$\hat{c}_1 = \frac{1}{2}(\hat{\gamma}_1 + i\hat{\gamma}_2), \quad \hat{c}_1^\dagger = \frac{1}{2}(\hat{\gamma}_1 - i\hat{\gamma}_2), \tag{13.84}$$

$$\hat{c}_2 = \frac{1}{2}(\hat{\gamma}_3 + i\hat{\gamma}_4), \quad \hat{c}_2^\dagger = \frac{1}{2}(\hat{\gamma}_3 - i\hat{\gamma}_4). \tag{13.85}$$

In a straightforward manner we can define the braiding operators $\hat{B}_{12}$ and $\hat{B}_{34}$ acting on $\hat{\gamma}_1, \hat{\gamma}_2$ and on $\hat{\gamma}_3, \hat{\gamma}_4$, respectively. The four fermion basis states are now given by

$$|00\rangle = |0\rangle,$$
$$|01\rangle = c_1^\dagger|0\rangle,$$
$$|10\rangle = c_2^\dagger|0\rangle,$$
$$|11\rangle = c_1^\dagger c_2^\dagger|0\rangle. \tag{13.86}$$

In terms of the creation and annihilation operators we can express the braiding operator according to equation (13.80)

$$\hat{B}_{12} = \frac{1}{\sqrt{2}}[1 + i(\hat{c}_1\hat{c}_1^\dagger - \hat{c}_1^\dagger\hat{c}_1)], \tag{13.87}$$

$$\hat{B}_{34} = \frac{1}{\sqrt{2}}[1 + i(\hat{c}_2\hat{c}_2^\dagger - \hat{c}_2^\dagger\hat{c}_2)]. \tag{13.88}$$

As an example, we can have a look on the effect on the $|00\rangle$ state

$$\hat{B}_{12}|00\rangle = \frac{1}{\sqrt{2}}(1 + i)|00\rangle, \tag{13.89}$$

$$\hat{B}_{34}|00\rangle = \frac{1}{\sqrt{2}}(1+i)|00\rangle. \tag{13.90}$$

In Figure 13.7 the subsequent operation of $\hat{B}_{12}$ and $\hat{B}_{34}$ and the corresponding operation in reverse order is illustrated. The order of these operations does not matter. The more interesting case is, if two Majoranas of different pairs are braided, e. g. $\hat{\gamma}_2$ and $\hat{\gamma}_3$, as illustrated in Figure 13.8. The corresponding braiding operator reads

$$\hat{B}_{23} = e^{-\frac{\pi}{4}\hat{\gamma}_2\hat{\gamma}_3} = \frac{1}{\sqrt{2}}(1+\hat{\gamma}_2\hat{\gamma}_3). \tag{13.91}$$

With the definitions of the fermion operators $\hat{c}_1$ and $\hat{c}_2$, we can also write

$$\begin{aligned}
\hat{B}_{23} &= \frac{1}{\sqrt{2}}[1+i(\hat{c}_1^\dagger - \hat{c}_1)(\hat{c}_2^\dagger + \hat{c}_2)] \\
&= \frac{1}{\sqrt{2}}[1+i(\hat{c}_1^\dagger\hat{c}_2^\dagger - \hat{c}_1^\dagger\hat{c}_2 + \hat{c}_1\hat{c}_2^\dagger - \hat{c}_1\hat{c}_2)].
\end{aligned} \tag{13.92}$$

For example, if we apply this operator to the states $|00\rangle$ and $|11\rangle$ we get

$$\hat{B}_{23}|00\rangle = \frac{1}{\sqrt{2}}(|00\rangle + i|11\rangle), \tag{13.93}$$

$$\hat{B}_{23}|11\rangle = \frac{i}{\sqrt{2}}(|00\rangle - i|11\rangle). \tag{13.94}$$

For the latter we made use of the anticommutator relation for fermions given by equation (13.20)

$$\begin{aligned}
-i\hat{c}_1\hat{c}_2|11\rangle &= -i\hat{c}_1\hat{c}_2\hat{c}_1^\dagger\hat{c}_2^\dagger|00\rangle = i\hat{c}_1\hat{c}_1^\dagger\hat{c}_2\hat{c}_2^\dagger|00\rangle \\
&= i(1-\hat{c}_1^\dagger\hat{c}_1)(1-\hat{c}_2^\dagger\hat{c}_2)|00\rangle \\
&= i(1-\hat{n}_1)(1-\hat{n}_2)|00\rangle = i|00\rangle.
\end{aligned} \tag{13.95}$$

Thus, by applying $\hat{B}_{23}$, we get a superposition of the states $|00\rangle$ and $|11\rangle$. Interestingly, the total parity of the system is maintained after the operation, i. e. no odd parity states $|01\rangle$ or $|10\rangle$ are produced, or no electron is lost. However, one might wonder, where the two electrons go to when after applying $\hat{B}_{23}$ to $|11\rangle$ a superposition state containing $|00\rangle$ is gained. In fact, this is possible by depositing the two electrons in the superconducting condensate.

13.6.4 Non-Abelian anyons

The exchange statistics of Majorana quasiparticles is rather different from other particles, like bosons or fermions. By applying an exchange operation for bosons or

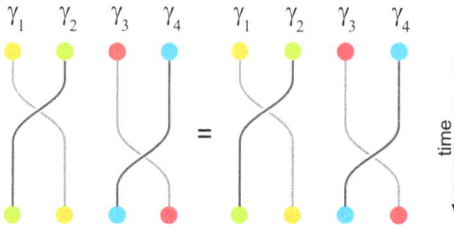

Figure 13.7: Independent braiding operations $\hat{B}_{12}$ and $\hat{B}_{34}$. Performing the operations in reversed order does not change the outcome.

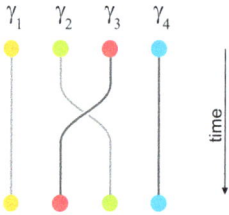

Figure 13.8: Braiding operation $\hat{B}_{23}$ between $\hat{\gamma}_2$ and $\hat{\gamma}_3$.

fermions, the corresponding wavefunction is multiplied by a factor +1 or −1, respectively, i. e.

$$|\psi_1\psi_2\rangle = \pm|\psi_2\psi_1\rangle. \tag{13.96}$$

Particles with different exchange operations are called anyons. The name originates from *any-* and *-ons*. They only exists in two-dimensional systems. For Abelian anyons an exchange of particles comes with a phase factor $e^{i\theta}$

$$|\psi_1\psi_2\rangle = e^{i\theta}|\psi_2\psi_1\rangle. \tag{13.97}$$

For our Majorana quasi-particles the exchange is even more intrigue. They belong to the so-called non-Abelian anyons. Instead of just multiplying a phase factor an exchange can result in a fundamentally different quantum state. The name non-Abelian comes from the fact, that subsequent exchanges do not generally commute. This means that it makes a difference in which order the exchange operations are performed. If we have two pairs of Majorana fermions and perform a braiding operation on each pair, these operations commute

$$[\hat{B}_{12}, \hat{B}_{34}] = 0. \tag{13.98}$$

However, when two exchanges involve some of the same Majorana fermions, the braiding operators do not commute

$$[\hat{B}_{i-1,i}, \hat{B}_{i,i+1}] = \hat{\gamma}_{i-1}\hat{\gamma}_{i+1}. \tag{13.99}$$

It implies that it makes a difference in which order the exchange was performed. For example, for the first three Majoranas one gets

$$[\hat{B}_{12}, \hat{B}_{23}] = \hat{\gamma}_1\hat{\gamma}_3, \tag{13.100}$$

thus, giving different results for the braiding operations shown in Figure 13.9. This explicitly shows the non-Abelian statistics.

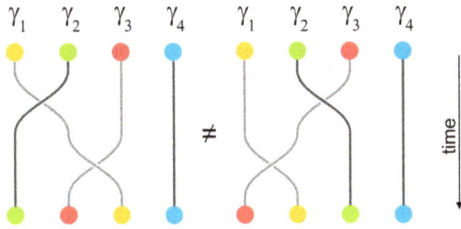

Figure 13.9: Sequence of two subsequent braidings with a difference in braiding orders of $\hat{B}_{12}$ and $\hat{B}_{23}$.

13.6.5 Majorana qubits

In order to use Majorana fermions as qubits we need a system, which allows to change the state by braiding operations, similar to a spin qubit, where the spin orientation can be changed by a single qubit operation. However, as we recognized for a braiding operation on a pair of Majoranas, cf. equations (13.76) and (13.77), the state is unchanged apart from a phase factor. In order to fix that problem we have to extend the system to four Majoranas. As illustrated by equation (13.93), by performing a braiding operation between $\hat{\gamma}_2$ and $\hat{\gamma}_3$, one can transfer the state $|00\rangle$ to a superposition of $|00\rangle$ and $|11\rangle$. By that operation the parity is conserved. Thus, the two state system of the Majorana qubit can be defined by the even parity subspace

$$|\bar{0}\rangle \equiv |00\rangle, \tag{13.101}$$
$$|\bar{1}\rangle \equiv |11\rangle. \tag{13.102}$$

Equivalently, one could also use the odd parity subspace $|01\rangle$ and $|10\rangle$.

The qubit operations can be mapped to different braidings. Let us begin with the braiding operation B_{12} defined by equation (13.74). According to equation (13.78) we can express the braiding by a rotation by the fixed angle $\pi/2$ in the subspace $\{|00\rangle = (1,0)^T, |11\rangle = (0,1)^T\}$

$$\hat{B}_{12} = e^{i\frac{\pi}{4}\sigma_z}. \tag{13.103}$$

The effect of the braiding operation is illustrated on the Bloch sphere shown in Figure 13.10. Let us assume an initial state along the x-axis $(|00\rangle + |11\rangle)/\sqrt{2}$. Applying the

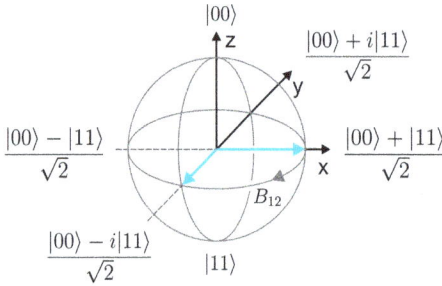

Figure 13.10: Bloch sphere for topological qubits with even basis states $|00\rangle$ and $|11\rangle$. The blue arrows shows the evolution of the state $(|00\rangle + |11\rangle)/\sqrt{2}$ after applying $\hat{B}_{12}$.

braiding operator B_{12} gives

$$
\begin{aligned}
\hat{B}_{12}\frac{1}{\sqrt{2}}(|00\rangle + |11\rangle) &= \frac{1}{\sqrt{2}}\begin{pmatrix} e^{i\pi/4} & 0 \\ 0 & e^{-i\pi/4} \end{pmatrix}\begin{pmatrix} 1 \\ 1 \end{pmatrix} \\
&= \frac{1}{\sqrt{2}}e^{i\pi/4}\begin{pmatrix} 1 & 0 \\ 0 & e^{-i\pi/2} \end{pmatrix}\begin{pmatrix} 1 \\ 1 \end{pmatrix} \\
&= \frac{1}{2}(1+i)\begin{pmatrix} 1 \\ -i \end{pmatrix} \\
&= \frac{1+i}{2}(|00\rangle - i|11\rangle).
\end{aligned}
\tag{13.104}
$$

Thus, the state on the Bloch sphere located at the intersection with the x-axis is rotated by $\pi/2$ to a state along the y-axis.

Similar to the Hadamard transformation introduced in Section 12.4.1 the braiding operator $\hat{B}_{23}$ can be employed to transfer the initial state $|00\rangle$ to a superposition state, according to equation (13.93). As outlined in the previous chapter, bringing the initial state into superposition is often the first step in a quantum algorithm. The braiding operation $\hat{B}_{23}$ is equivalent to a rotation by $\pi/2$ about the x-axis, i. e. similar to equation (13.79) we can write $i\hat{y}_2\hat{y}_3 = \sigma_x$. Obviously, the single qubit operations described here are not universal, since the rotations are only performed for discrete rotations by $\pi/2$. However, universal quantum computation can be achieved by nontopological operations [224] or by coupling Majorana qubits to other qubit systems [225, 226].

13.7 Summary

- Majorana fermions are charge- and spinless particles which are their own antiparticles. These unique particles are deduced from a special form of the Dirac equation found by Ettore Majorana, which gives pure real solutions.
- In solid state systems Majorana states can be realized by combining a spin-orbit material with a superconductor under additional application of a magnetic field.

- Majorana fermions belong to the group of non-Abelian anyons. Here, the exchange of particles is not only accompanied by multiplying a phase factor but results in a fundamentally different quantum state. The exchange of two Majorana fermions is called braiding.
- Braiding of Majoranas can be utilized to realize qubit operations.

Exercises

Problem 13.1. Derive the anticommutator relation for Majorana fermions, given by

$$\{\hat{\gamma}_i^A, \hat{\gamma}_j^B\} = 2\delta_{ij}\delta_{AB},$$

from the anticommutator relations of fermions.

Problem 13.2. According to equation (13.71) the parity operator $\hat{P}$ is defined as

$$\hat{P} = 1 - 2\hat{n} = 1 - 2\hat{c}^\dagger\hat{c} = -i\hat{\gamma}_1\hat{\gamma}_2.$$

Show explicitly, using the definitions for Majorana fermions and the corresponding anticommutator relations, that $-i\hat{\gamma}_1\hat{\gamma}_2$ follows from $1 - 2\hat{c}^\dagger\hat{c}$.

Problem 13.3. Show that by applying the braiding operators $\hat{B}_{12}$ and $\hat{B}_{23}$ to the state $|00\rangle$ in reversed order a different final state is obtained.

Solutions

Chapter 2. Low-dimensional semiconductor structures

Problem 2.1
The Schrödinger equation of our system is given by

$$\left(-\frac{\hbar^2}{2m^*}\frac{\partial^2}{\partial x^2} + V(x)\right)\psi(x) = E\psi(x).$$

With the ansatz

$$\psi(x) = A\sin(kx) + B\cos(kx),$$

one finds after inserting into the Schrödinger equation

$$-\frac{\hbar^2}{2m^*}(-k^2)\psi(x) = E\psi(x).$$

This results in

$$E = \frac{\hbar^2 k^2}{2m^*}.$$

Because of the infinitively high barriers one finds

$$\psi(x) = 0, \quad \text{for } x < 0,\ x > a.$$

From the first boundary condition $\psi(0) = 0$ we find that the solution is restricted to $\psi(x) = A\sin(kx)$. Regarding the second boundary condition $\psi(a) = 0$ one finds $A\sin(ka) = 0$, which implies $ka = n\pi$, with $n = 1, 2, 3, \ldots$ Inserting this into the above expression of the energy gives

$$E_n = \frac{\hbar^2 \pi^2}{2m^* a^2}n^2.$$

The energy eigenvalues for the first three states are given by $E_1 = 60.7\,\text{meV}$, $E_2 = 242.8\,\text{meV}$, and $E_3 = 546.3\,\text{meV}$.

Problem 2.2
For periodic boundary conditions the number of states in a two-dimensional disk in k-space is given by

$$Z(k) = \pi k^2/(2\pi/L)^2,$$

with L the length of the box. With $k^2 = 2m^*E/\hbar^2$ one finds

$$Z(E) = L^2\frac{m^*E}{2\pi}\hbar^2.$$

https://doi.org/10.1515/9783110639001-014

The density of states per area L^2 is given by

$$D(E) = \frac{1}{L^2} \frac{dZ(E)}{dE} = \frac{m^*}{2\pi\hbar^2}.$$

For a spin degenerate system we have to multiply the spin-degeneracy factor g_s, i.e. $g_s = 2$ for electrons in the conduction band of a semiconductor. The expression obtained here corresponds to the formula given in Table 2.1.

Problem 2.3

The mobility can be calculated from equation (2.36): $\sigma = \rho^{-1} = en\mu_e$ thus $\mu_e = 1/(\rho en)$. Inserting the values given above results in $\mu_e = 0.437\,\mathrm{m^2/Vs}$. The elastic scattering time can be obtained from $\tau_e = \mu_e m_0/e$, which gives $\tau_e = 2.5\,\mathrm{ps}$.

Problem 2.4

The absolute value Fermi wave vector is given by

$$k_F = \frac{1}{\hbar} \sqrt{2m^* E_F}.$$

In order to overcome the barrier at the point contact opening the wave vector perpendicular to the point contact has to be at least

$$k_x = \frac{1}{\hbar} \sqrt{2m^* V_0}.$$

The angle of acceptance can be calculated from

$$\cos\theta = \frac{k_x}{k_F} = \frac{\frac{1}{\hbar}\sqrt{2m^* V_0}}{\frac{1}{\hbar}\sqrt{2m^* E_F}},$$

which results in

$$\theta = \arccos\sqrt{V_0/E_F}.$$

Problem 2.5

We begin with applying the operator $\hat{a}$ to the Schrödinger equation of the harmonic oscillator

$$\hat{a}\hat{H}|n\rangle = \hbar\omega_c \hat{a}\left(\hat{a}^\dagger \hat{a} + \frac{1}{2}\right)|n\rangle = E_n \hat{a}|n\rangle,$$

$$= \hbar\omega_c\left(\hat{a}\hat{a}^\dagger \hat{a} + \frac{1}{2}\hat{a}\right)|n\rangle = E_n \hat{a}|n\rangle.$$

Making use of the commutator relation $[\hat{a}, \hat{a}^\dagger] = \hat{a}\hat{a}^\dagger - \hat{a}^\dagger\hat{a} = 1$ gives

$$\hat{a}\hat{H}|n\rangle = \hbar\omega_c\left[(\hat{a}^\dagger\hat{a} + 1)\hat{a} + \frac{1}{2}\hat{a}\right]|n\rangle = E_n\hat{a}|n\rangle,$$

$$= \hbar\omega_c\left(\hat{a}^\dagger\hat{a} + 1 + \frac{1}{2}\right)\hat{a}|n\rangle = E_n\hat{a}|n\rangle,$$

which results in

$$\hbar\omega_c\left(\hat{a}^\dagger\hat{a} + \frac{1}{2}\right)\hat{a}|n\rangle = (E_n - \hbar\omega_c)\hat{a}|n\rangle.$$

Thus, the energy of the state $\hat{a}|n\rangle$ is lower by $\hbar\omega_c$ compared to the energy E_n of the state $|n\rangle$.

Chapter 3. Magnetism in solids

Problem 3.1

The Pauli spin matrix is given by

$$\sigma_y = \begin{pmatrix} 0 & -i \\ i & 0 \end{pmatrix},$$

which results in the eigenwert equation

$$\begin{pmatrix} -\lambda & -i \\ i & -\lambda \end{pmatrix}\begin{pmatrix} c_0 \\ c_1 \end{pmatrix} = 0.$$

From that we obtain from the second equation

$$c_0 = -i\lambda c_1,$$

and after inserting into the first equation

$$(\lambda^2 - 1)c_1 = 0.$$

Thus the eigenvalues are ± 1. Furthermore, by assuming $c_0 = 1$ the normalized eigenvectors are finally

$$|\psi_\pm\rangle = \frac{1}{\sqrt{2}}\begin{pmatrix} 1 \\ \pm i \end{pmatrix}.$$

Problem 3.2
We need to calculate the product

$$\hat{H}_Z|\psi_+\rangle = s\mu_B g_0 |B| \begin{pmatrix} \cos\theta & \sin\theta e^{-i\phi} \\ \sin\theta e^{i\phi} & -\cos\theta \end{pmatrix} \begin{pmatrix} \cos(\theta/2) \\ e^{i\phi}\sin(\theta/2) \end{pmatrix}$$

$$= s\mu_B g_0 |B| \begin{pmatrix} \cos(\theta/2) \\ e^{i\phi}\sin(\theta/2) \end{pmatrix},$$

with $s = 1/2$ being the spin quantum number. For the first component of the resulting spinor we obtain

$$\cos\theta\cos(\theta/2) + \sin\theta\sin(\theta/2) = [\cos^2(\theta/2) - \sin^2(\theta/2)]\cos(\theta/2)$$
$$+ 2\sin(\theta/2)\cos(\theta/2)\sin(\theta/2)$$
$$= \cos(\theta/2)[\cos^2(\theta/2) + \sin^2(\theta/2)]$$
$$= \cos(\theta/2),$$

where we made use of

$$\cos\theta = \cos^2(\theta/2) - \sin^2(\theta/2) \quad \text{and} \quad \sin\theta = 2\sin(\theta/2)\cos(\theta/2).$$

The second component of the spinor can be calculated in an analog fashion. Similarly, one finds

$$\hat{H}_Z|\psi_-\rangle = -s\mu_B g_0 |B|\,|\psi_-\rangle.$$

Problem 3.3
The current I for an electron with charge $-e$ can be written as

$$I = -\frac{e}{t} = -\frac{e\omega}{2\pi},$$

with t being the period of the orbital motion and ω the orbital frequency. With $F = \pi r^2$ and the definition for the angular momentum $L = m_0 \omega r^2$ we arrive at

$$\mu = -\frac{e}{2m_0}L.$$

Problem 3.4
According to Table 2.1, the Fermi wave number is given by

$$k_F = \sqrt{\frac{4\pi n_{2D}}{g_s}},$$

with g_s being the spin degeneracy factor and n_{2D} the carrier concentration. In case of a spin degenerate system $g_s = 2$ one finds

$$k_F^{\uparrow\downarrow} = \sqrt{2\pi n_{2D}},$$

while for the spin-polarized system $g_s = 1$ one gets

$$k_F^{\uparrow} = \sqrt{4\pi n_{2D}}.$$

Thus, one obtains

$$k_F^{\uparrow} = \sqrt{2}k_F^{\uparrow\downarrow}.$$

For free electrons with

$$E_{\text{kin}} = \frac{\hbar^2 k_F^2}{2m_0},$$

the kinetic energy at the Fermi energy differs by a factor of two.

Chapter 4. Diluted magnetic semiconductors

Problem 4.1

The expression for the energy splitting is given by

$$\Delta E_{sd} = -N_0 \alpha x \langle S_z \rangle,$$

with $\langle S_z \rangle$ being the thermal averaged spin of the localized moments of spin-1/2 particles ($S = 1/2$). Assuming that the magnetization in the material is dominated by localized spins the magnetization can be expressed as

$$M = g\mu_B S \tanh\left(\frac{g\mu_B S B_0}{k_B T}\right),$$

which gives for the thermal averaged spin

$$\langle S_z \rangle = S \tanh\left(\frac{g\mu_B S B_0}{k_B T}\right).$$

The final expression is given by

$$\Delta E_{sd} = -N_0 \alpha x S \tanh\left(\frac{g\mu_B S B_0}{k_B T}\right).$$

Inserting the values above gives for $B_0 = 1\,\text{T}$

$$\Delta E_{sd} = 0.025\,\text{eV} \tanh\left(\frac{0.67 \cdot 1}{5}\right) = 3.3\,\text{meV}$$

and for $B_0 = 10\,\text{T}$ a value of $\Delta E_{sd} = 22\,\text{meV}$. For a field of $10\,\text{T}$ the splitting is close to a saturation value.

Problem 4.2

The magnetization is given by

$$M = M_0 \tanh\left(\frac{g\mu_B S B_0}{k_B T}\right).$$

We are first interested in the magnetic field B_0 at which M/M_0 is 0.90,

$$\frac{g\mu_B S B_0}{k_B T} = \tanh^{-1}(0.9) = 1.47.$$

Inserting the values given above results in a field of $B_0 = 10.97$ T. For $M/M_0 = 0.99$ we obtain a field of 19.7 T.

Problem 4.3

For Mn we assume a spin of $S = 3/2$. For the density of states at the Fermi energy $D(E_F)$ we use the following formula:

$$D(E_F) = \frac{1}{2\pi} \frac{m^*}{\hbar^3} \sqrt{2m^* E_F}$$

Inserting all values given above results in

$$T_C = 9.4 \text{ K}.$$

One should take care that eV is converted into Joule during the calculation.

Chapter 5. Magnetic electrodes

Problem 5.1

The expression for the domain wall thickness is given by

$$d_{dw} = \pi \sqrt{A/K_u}.$$

Inserting the values above results in 811 and 397 nm for Fe and permalloy, respectively.

Problem 5.2

From equation (5.13) we get the expression

$$A\left(\frac{d\phi}{dx}\right)^2 = K_u \sin^2 \phi.$$

This can be inserted into the integral for the energy per unit area

$$\gamma_B = \int_{-\infty}^{\infty} \left[K_u \sin^2 \phi + A\left(\frac{d\phi}{dx}\right)^2 \right] dx,$$

which results in

$$\gamma_B = \int_{-\infty}^{\infty} 2K_u \sin^2 \phi \, dx.$$

Using equation (5.15) given by

$$dx = \pm\sqrt{\frac{A}{K_u}} \frac{d\phi}{\sin \phi},$$

we can write the integral as

$$\gamma_B = \int_0^{\pi} 2\sqrt{AK_u} \sin \phi \, d\phi.$$

The integration gives

$$\gamma_B = 4\sqrt{AK_u}.$$

Problem 5.3
The expression for the Hall resistance $R_H = V_H/I$ is given by

$$R_H = \frac{1}{en_{2D}} B.$$

Inserting the values given above gives a Hall voltage of $V_H = 3.9 \, \text{mV}$.

Chapter 6. Spin injection

Problem 6.1
With the time evolution in the Heisenberg picture

$$i\hbar \frac{d}{dt} \hat{S}_i = [\hat{S}_i, \hat{H}_Z]$$

and the Zeeman Hamiltonian we obtain

$$i\hbar \frac{d}{dt} \hat{S}_i = \frac{g\mu_B}{\hbar} \sum_j B_j [\hat{S}_i, \hat{S}_j]$$

$$= \frac{g\mu_B}{\hbar} \sum_j B_j i\hbar \epsilon_{ijk} \hat{S}_k,$$

where we used the commutator relation for spins

$$[\hat{S}_i, \hat{S}_j] = \epsilon_{ijk} \hat{S}_k.$$

Finally, the sum given above can be expressed as a vector product so that we get

$$\frac{d}{dt}\hat{S} = -\frac{g\mu_B}{\hbar}(\hat{S} \times \vec{B}).$$

The relation is also valid for the expectation values

$$\frac{d}{dt}\langle\vec{S}\rangle = -\frac{g\mu_B}{\hbar}(\langle\vec{S}\rangle \times \vec{B}).$$

Problem 6.2

The current spin polarization is defined as

$$\alpha = \frac{j^\uparrow - j^\downarrow}{j^\uparrow + j^\downarrow}.$$

The resistances of the ferromagnet for spin-up and spin-down are given by

$$R^\uparrow_{FM} = \frac{2R_{FM}}{1+\beta}, \quad R^\downarrow_{FM} = \frac{2R_{FM}}{1-\beta},$$

with β being the spin polarization in the ferromagnet and R_{FM} the total resistance of the ferromagnet. The resistance of the semiconductor is R_{SC}. The current densities are proportional to

$$j^\uparrow \propto (R_{SC} + R^\uparrow_{FM})^{-1}$$

and

$$j^\downarrow \propto (R_{SC} + R^\downarrow_{FM})^{-1}.$$

With α written as

$$\alpha = \frac{1 - \frac{j^\downarrow}{j^\uparrow}}{1 + \frac{j^\uparrow}{j^\uparrow}},$$

and inserting the expressions for the resistances we get

$$\alpha = \frac{1 - \frac{R_{SC}+R^\uparrow_{FM}}{R_{SC}+R^\downarrow_{FM}}}{1 + \frac{R_{SC}+R^\uparrow_{FM}}{R_{SC}+R^\downarrow_{FM}}}.$$

Inserting the expressions for $R^\uparrow_{FM}$ and $R^\downarrow_{FM}$ gives

$$\alpha = \frac{2R_{FM}\frac{1}{1-\beta} - 2R_{FM}\frac{1}{1+\beta}}{2R_{SC} + 2R_{FM}\frac{1}{1+\beta} + 2R_{FM}\frac{1}{1-\beta}}.$$

This can be simplified to

$$\alpha = R_{FM} \frac{\frac{1+\beta-1+\beta}{1-\beta^2}}{R_{SC} + R_{FM}\frac{1+\beta+1-\beta}{1-\beta^2}}$$

and

$$\alpha = R_{FM} \frac{2\beta}{R_{SC}(1-\beta^2) + 2R_{FM}}.$$

From this we finally get the desired expression

$$\alpha = \beta \frac{R_{FM}}{R_{SC}} \frac{2}{2\frac{R_{FM}}{R_{SC}} + (1-\beta^2)}.$$

Chapter 7. Spin transistor

Problem 7.1

The magnetic field in the own frame of reference of an electron propagating with velocity $\vec{v}$ in an electric field $\vec{\mathcal{E}}$ is given by the Lorentz transformation

$$\vec{B} = -\frac{1}{c^2}\vec{v} \times \vec{\mathcal{E}},$$

with c being the velocity of light. For an electron with charge $-e$ ($e > 0$) the electric field is given by $\vec{\mathcal{E}} = (1/e)\vec{\nabla}V$, with V the potential. The magnetic moment of an electron is given by

$$\vec{\mu} = -\frac{e}{2m_0}\vec{S},$$

with $\vec{S}$ being the electron spin. We omitted the g-factor. Hamiltonian for the magnetic energy is given by

$$\hat{H} = -\vec{\mu}\vec{B}.$$

Inserting the expression given above results in

$$\hat{H}_{so} = +\frac{e}{2m_0}\vec{S}\left(-\frac{1}{c^2}\right)(\vec{v} \times \vec{\mathcal{E}}).$$

With $\vec{S} = (\hbar/2)\vec{\sigma}$ we arrive at

$$\hat{H}_{so} = -\frac{\hbar}{4m_0^2c^2}\vec{\sigma}(\hat{\vec{p}} \times \vec{\nabla}V),$$

where $\vec{p}$ is the momentum and $\vec{\sigma}$ the Pauli spin matrices.

Problem 7.2

We define the spinor $|\psi\rangle$ as

$$|\psi\rangle = \begin{pmatrix} a \\ b \end{pmatrix},$$

so that we can write

$$|\hat{\mathcal{T}}\psi\rangle = \begin{pmatrix} 0 & 1 \\ -1 & 0 \end{pmatrix} \hat{\mathcal{K}} \begin{pmatrix} a \\ b \end{pmatrix} = \begin{pmatrix} b^* \\ -a^* \end{pmatrix}.$$

Furthermore, for the state in the dual vector space we get

$$\langle \hat{\mathcal{T}}\psi| = (a^*, b^*)\hat{\mathcal{T}}^\dagger = (a^*, b^*)\begin{pmatrix} 0 & -1 \\ 1 & 0 \end{pmatrix}\hat{\mathcal{K}} = (b^*, -a^*)\hat{\mathcal{K}} = (b, -a),$$

with $\hat{\mathcal{T}}^\dagger = \hat{\mathcal{T}}^{*^T}$. The inner product of these states is then given by

$$\langle \hat{\mathcal{T}}\psi|\hat{\mathcal{T}}\psi\rangle = (b, -a)\begin{pmatrix} b^* \\ -a^* \end{pmatrix} = bb^* + aa^* = \langle \psi|\psi\rangle.$$

Problem 7.3

The time-reversal operator does not change the position

$$\hat{\mathcal{T}}\hat{x}\hat{\mathcal{T}}^{-1} = \hat{x},$$

with $\hat{\mathcal{T}}\hat{\mathcal{T}}^{-1} = 1$. Similar expressions hold for $\hat{y}$ and $\hat{z}$. Since the momentum operator is proportional to the velocity, the sign of the momentum operator is reversed:

$$\hat{\mathcal{T}}\hat{p}_x\hat{\mathcal{T}}^{-1} = -\hat{p}_x.$$

Applying $\hat{\mathcal{T}}$ to the commutator $[\hat{x}, \hat{p}_x]$ results in

$$\begin{aligned} \hat{\mathcal{T}}[\hat{x}, \hat{p}_x]\hat{\mathcal{T}}^{-1} &= \hat{\mathcal{T}}(\hat{x}\hat{p}_x - \hat{p}_x\hat{x})\hat{\mathcal{T}}^{-1} \\ &= \hat{\mathcal{T}}\hat{x}\hat{\mathcal{T}}^{-1}\hat{\mathcal{T}}\hat{p}_x\hat{\mathcal{T}}^{-1} - \hat{\mathcal{T}}\hat{p}_x\hat{\mathcal{T}}^{-1}\hat{\mathcal{T}}\hat{x}\hat{\mathcal{T}}^{-1} \\ &= -\hat{x}\hat{p}_x + \hat{p}_x\hat{x} \\ &= -[\hat{x}, \hat{p}_x]. \end{aligned}$$

Because of the uncertainty relation, we know $[\hat{x}, \hat{p}_x] = i\hbar$, so that one finds

$$\hat{\mathcal{T}}i\hbar\hat{\mathcal{T}}^{-1} = -i\hbar.$$

Thus

$$\hat{\mathcal{T}}i\hat{\mathcal{T}}^{-1} = -i.$$

This result implies that the time-reversal operator $\hat{\mathcal{T}}$ must be proportional to the complex conjugate operator.

Problem 7.4

We can write

$$e^{-i\pi\sigma_y/2} = \sum_{n=0}^{\infty} \frac{1}{n!} \left(\frac{-i\pi\sigma_y}{2} \right)^n = -i\sigma_y,$$

where we used

$$\sigma_y^2 = 1 \quad \text{and} \quad \exp(ix) = \cos x + i \sin x,$$

with

$$\cos(\pi/2) = 0 \quad \text{and} \quad \sin(\pi/2) = 1.$$

Problem 7.5

We start from the Hamiltonian and insert the expressions for $\tilde{x}_0$, x_0, and ω^2:

$$\hat{H} = -\frac{\hbar^2}{2m^*}\frac{\partial^2}{\partial x^2} + \frac{m^*}{2}\omega_c^2(x - \tilde{x}_0)^2 + \frac{\omega_0^2}{\omega^2}\frac{\hbar^2 k_y^2}{2m^*}$$

$$= -\frac{\hbar^2}{2m^*}\frac{\partial^2}{\partial x^2} + \frac{m^*}{2}\omega^2\left(x^2 - 2\frac{\omega_c^2}{\omega^2}xx_0 + \frac{\omega_c^4}{\omega^4}x_0^2 + \frac{\omega_0^2\omega_c^2}{\omega^4}x_0^2\right)$$

$$= -\frac{\hbar^2}{2m^*}\frac{\partial^2}{\partial x^2} + \frac{m^*}{2}(\omega_c^2 x^2 + \omega_0^2 x^2 - 2\omega_c^2 xx_0 + \omega_c^2 x_0^2)$$

$$= -\frac{\hbar^2}{2m^*}\frac{\partial^2}{\partial x^2} + \frac{m^*}{2}\omega_c^2(x - x_0)^2 + \frac{m^*}{2}\omega_0^2 x^2.$$

Problem 7.6

Generally, the density of states of a two-dimensional system in k-space is given by

$$D(k) = \frac{dZ(k)}{dk} = \frac{1}{2\pi}k,$$

with $Z(k)$ the number of states per unit area. From that expression the density of states $D(E) = dZ(E)/dE$ can be determined using

$$\frac{dZ(k)}{dk} = \frac{dZ(E)}{dE}\frac{dE}{dk}$$

which gives

$$D(E) = \frac{1}{2\pi}\frac{k}{dE(k)/dk}$$

In the presence of Rashba effect the energy momentum dispersion is given by

$$E = \frac{\hbar^2 k^2}{2m^*} \pm \alpha_R k.$$

The dispersion is depicted in Fig. 7.4. We have to consider two branches for $k \geq 0$. The first branch with larger energy is given by

$$E = \frac{\hbar^2 k^2}{2m^*} + \alpha_R k.$$

For this case we get for $k \geq 0$

$$k(E) = -\frac{\alpha_R m^*}{\hbar^2} + \sqrt{\left(\frac{\alpha_R m^*}{\hbar^2}\right)^2 + \frac{2m^* E}{\hbar^2}}.$$

Inserting this expression and the derivative

$$\frac{dE}{dk} = \frac{\hbar^2 k}{m^*} + \alpha_R$$

into the above expression for the density of states one finally obtains for the density of states

$$D_+(E) = \frac{1}{2\pi} \frac{m^*}{\hbar^2} \left(1 - \frac{1}{\sqrt{1 + \frac{2E\hbar^2}{\alpha_R m^*}}}\right).$$

This branch of the dispersion is restricted to $E \geq 0$. For the other one the energy dispersion is given by

$$E = \frac{\hbar^2 k^2}{2m^*} - \alpha_R k,$$

with

$$k(E) = +\frac{\alpha_R m^*}{\hbar^2} \pm \sqrt{\left(\frac{\alpha_R m^*}{\hbar^2}\right)^2 + \frac{2m^* E}{\hbar^2}}.$$

The energy can also be negative in this case. The minimum energy is given by

$$E_{min} = -\frac{\alpha_R^2 m^*}{2\hbar^2}.$$

In complete analogy we obtain

$$D_+(E) = \frac{1}{2\pi} \frac{m^*}{\hbar^2} \left|1 \pm \frac{1}{\sqrt{1 + \frac{2E\hbar^2}{\alpha_R m^*}}}\right|.$$

Here, we made sure that the density of states is always positive. The part $D_-(E)$ covers the range from $0 \leq k < (m^* \alpha_R)/\hbar^2$. It is only valied for $E_{min} \leq E \leq 0$. The part $D_+(E)$ covers the range $k \geq (m^* \alpha_R)/\hbar^2$ with the full energy range starting at E_{min}. Both contributions of the density of states have to be added to obtain the complete density of states of this branch of the dispersion.

Chapter 8. Spin interference

Problem 8.1

For both matrices we obtain identical eigenvalues:

$$E_\pm = \alpha \pm \beta.$$

Inserting the eigenvalue in the eigenvalue equation for the first case gives

$$0 = \pm\beta\chi_1 + e^{-i\phi}\beta(e^{i\phi}\chi_2) = \pm\beta\chi_1 + \beta\chi_2,$$

where the right side corresponds to the relation of the matrix without the phase factors. The same holds for the second case. Here, the phase factors cancel each other.

Problem 8.2

The spin matrices are given by

$$\sigma_r = \cos\phi\,\sigma_x + \sin\phi\,\sigma_y = \begin{pmatrix} 0 & e^{-i\phi} \\ e^{i\phi} & 0 \end{pmatrix},$$

$$\sigma_\phi = -\sin\phi\,\sigma_x + \cos\phi\,\sigma_y = \begin{pmatrix} 0 & -ie^{-i\phi} \\ ie^{i\phi} & 0 \end{pmatrix},$$

and

$$\sigma_z = \begin{pmatrix} 1 & 0 \\ 0 & -1 \end{pmatrix}.$$

For the expectation value along $\vec{e}_r$ we obtain

$$\langle \psi_+^{ccw}|\hat{\sigma}_r|\psi_+^{ccw}\rangle = e^{-il\phi}\left(\sin\frac{\xi}{2}, e^{-i\phi}\cos\frac{\xi}{2}\right)\begin{pmatrix} 0 & e^{-i\phi} \\ e^{i\phi} & 0 \end{pmatrix}e^{il\phi}\begin{pmatrix} \sin\frac{\xi}{2} \\ e^{i\phi}\cos\frac{\xi}{2} \end{pmatrix}$$

$$= \left(\sin\frac{\xi}{2}, e^{-i\phi}\cos\frac{\xi}{2}\right)\begin{pmatrix} \cos\frac{\xi}{2} \\ e^{i\phi}\sin\frac{\xi}{2} \end{pmatrix}$$

$$= 2\sin\frac{\xi}{2}\cos\frac{\xi}{2}$$

$$= \sin\xi.$$

For the expectation value along $\vec{e}_\phi$ one finds

$$\langle \psi_+^{ccw}|\hat{\sigma}_\phi|\psi_+^{ccw}\rangle = e^{-il\phi}\left(\sin\frac{\xi}{2}, e^{-i\phi}\cos\frac{\xi}{2}\right)\begin{pmatrix} 0 & -ie^{-i\phi} \\ ie^{i\phi} & 0 \end{pmatrix}e^{il\phi}\begin{pmatrix} \sin\frac{\xi}{2} \\ e^{i\phi}\cos\frac{\xi}{2} \end{pmatrix}$$

$$= \left(\sin\frac{\xi}{2}, e^{-i\phi}\cos\frac{\xi}{2}\right)\begin{pmatrix} -i\cos\frac{\xi}{2} \\ ie^{i\phi}\sin\frac{\xi}{2} \end{pmatrix}$$

$$= 0.$$

For the expectation value along $\vec{e}_z$ one gets

$$\langle \psi_+^{ccw}|\hat{\sigma}_z|\psi_+^{ccw}\rangle = e^{-il\phi}\left(\sin\frac{\xi}{2}, e^{-i\phi}\cos\frac{\xi}{2}\right)\begin{pmatrix} 1 & 0 \\ 0 & -1 \end{pmatrix} e^{il\phi}\begin{pmatrix} \sin\frac{\xi}{2} \\ e^{i\phi}\cos\frac{\xi}{2} \end{pmatrix}$$

$$= \sin^2\frac{\xi}{2} - \cos^2\frac{\xi}{2}$$

$$= -\cos\xi.$$

Problem 8.3

The products of rotation matrices are given by

$$\hat{R}_z(\alpha)\hat{R}_y(\theta) = \begin{pmatrix} e^{i\alpha/2} & 0 \\ 0 & e^{-i\alpha/2} \end{pmatrix}\begin{pmatrix} \cos\theta/2 & \sin\theta/2 \\ -\sin\theta/2 & \cos\theta/2 \end{pmatrix}$$

$$= \begin{pmatrix} e^{i\alpha/2}\cos\theta/2 & e^{i\alpha/2}\sin\theta/2 \\ -e^{-i\alpha/2}\sin\theta/2 & e^{-i\alpha/2}\cos\theta/2 \end{pmatrix}$$

and

$$\hat{R}_y(\theta)\hat{R}_z(\alpha) = \begin{pmatrix} \cos\theta/2 & \sin\theta/2 \\ -\sin\theta/2 & \cos\theta/2 \end{pmatrix}\begin{pmatrix} e^{i\alpha/2} & 0 \\ 0 & e^{-i\alpha/2} \end{pmatrix}$$

$$= \begin{pmatrix} e^{i\alpha/2}\cos\theta/2 & e^{-i\alpha/2}\sin\theta/2 \\ -e^{i\alpha/2}\sin\theta/2 & e^{-i\alpha/2}\cos\theta/2 \end{pmatrix}.$$

Thus for $\alpha \neq 0$ or $\theta \neq 0$ the products are usually different. Therefore, for the commutator we get

$$[\hat{R}_y(\theta), \hat{R}_z(\alpha)] = \hat{R}_y(\theta)\hat{R}_z(\alpha) - \hat{R}_z(\alpha)\hat{R}_y(\theta) \neq 0.$$

Problem 8.4

We assume a ring-shaped conductor formed in a semiconductor heterostructure. In the presence of Rasbha effect, the effective magnetic field B_{eff} will point outwards of the ring within the plane of the two-dimensional electron gas. The magnetic field is perpendicular to the direction of motion. We assume the adiabatic limit. Calculate the Berry phase the electron acquires along a full round in counter-clockwise direction.

The effective magnetic field can be written as

$$\vec{B}_{eff} = B_0\begin{pmatrix} \cos\phi \\ \sin\phi \\ 0 \end{pmatrix},$$

with ϕ being the azimuthal angle. We assume that the spin orientation during the propagation follows the effective magnetic field. The corresponding eigenstates are

$$|n_+\rangle = \frac{1}{\sqrt{2}}\begin{pmatrix} 1 \\ e^{i\phi} \end{pmatrix}$$

and

$$|n_-\rangle = \frac{1}{\sqrt{2}} \begin{pmatrix} -1 \\ e^{i\phi} \end{pmatrix} ;.$$ (1)

In order to obtain the Berry phase, first, we have to determine the Berry connection $\vec{A}_\pm$. For that we need to calculate $\vec{\nabla}|n_\pm\rangle$ according to equation (8.92). Here we assume a fixed polar angle of $\theta = \pi/2$ and a fixed magnitude of the magnetic field, so that we get

$$\vec{\nabla}|n_\pm\rangle \rightarrow \frac{1}{B_0} \frac{\partial}{\partial\phi} \vec{e}_\phi |n_\pm\rangle.$$

This gives for both states the same result:

$$\frac{1}{B_0} \frac{\partial}{\partial\phi} \vec{e}_\phi |n_+\rangle = \frac{1}{\sqrt{2}B_0} \begin{pmatrix} 0 \\ ie^{i\phi} \end{pmatrix} \vec{e}_\phi$$

and

$$\frac{1}{B_0} \frac{\partial}{\partial\phi} \vec{e}_\phi |n_-\rangle = \frac{1}{\sqrt{2}B_0} \begin{pmatrix} 0 \\ ie^{i\phi} \end{pmatrix} \vec{e}_\phi.$$

By calculating the scalar product with $\langle n_\pm |$ the Berry connections $\vec{A}_\pm$

$$\vec{A}_+ = i\langle n_+|\vec{\nabla}|n_+\rangle = -\frac{1}{2B_0} \vec{e}_\phi$$

and

$$\vec{A}_- = i\langle n_-|\vec{\nabla}|n_-\rangle = -\frac{1}{2B_0} \vec{e}_\phi$$

are obtained. Integration along curve C over ϕ from 0 to 2π, with radius B_0 gives

$$\gamma_\pm = \oint_C \vec{A}_\pm B_0 d\phi \vec{e}_\phi$$

$$= \oint_C i\langle n_\pm|\vec{\nabla}|n_\pm\rangle B_0 d\phi \vec{e}_\phi$$

$$= -\pi$$

for the Berry phase.

Chapter 9. Spin Hall effect

Problem 9.1

With the magnetic field along the z-direction we obtain the following set of equations:

$$\frac{d}{dt}\langle S_x\rangle = -\omega_L\langle S_y\rangle,$$

$$\frac{d}{dt}\langle S_y\rangle = \omega_L\langle S_x\rangle,$$

$$\frac{d}{dt}\langle S_z\rangle = 0.$$

Combining the first equations results in

$$\frac{d^2\langle S_x\rangle}{dt^2} = -\omega_L^2\langle S_x\rangle,$$

which give the general solution

$$\langle S_x\rangle = \langle S_x\rangle_0\cos(\omega_L t) + \langle S_y\rangle_0\sin(\omega_L t),$$

with $\langle S_x\rangle_0$ and $\langle S_y\rangle_0$ being the initial values at $t = 0$. Differentiation with respect to t gives the corresponding solution for $\langle S_y\rangle$:

$$\langle S_y\rangle = \langle S_y\rangle_0\cos(\omega_L t) - \langle S_x\rangle_0\sin(\omega_L t).$$

The z-component of the spin is not changed in the course of time, and thus

$$\langle S_z\rangle = \langle S_z\rangle_0.$$

Problem 9.2
The integral can be solved by partial integration:

$$\int_0^\infty e^{-t/T_2^*}\cos(\omega_L t)\,dt = -T_2^* e^{-t/T_2^*}\cos(\omega_L t)\Big|_0^\infty - T_2^*\omega_L\int_0^\infty e^{-t/T_2^*}\sin(\omega_L t)\,dt,$$

which simplifies to

$$\int_0^\infty e^{-t/T_2^*}\cos(\omega_L t)\,dt = -T_2^*\omega_L\int_0^\infty e^{-t/T_2^*}\sin(\omega_L t)\,dt + T_2^*.$$

This is followed by another partial integration:

$$-T_2^*\omega_L\int_0^\infty e^{-t/T_2^*}\sin(\omega_L t)\,dt + T_2^* = (T_2^*)^2\omega_L e^{-t/T_2^*}\sin(\omega_L t)\Big|_0^\infty$$

$$- (T_2^*)^2\omega_L^2\int_0^\infty e^{-t/T_2^*}\cos(\omega_L t)\,dt + T_2^*,$$

which can be rearranged to

$$\int_0^{\infty} e^{-t/T_2^*} \cos(\omega_L t)\, dt = \frac{T_2^*}{1 + (T_2^* \omega_L)^2}.$$

With the definition of $B_{1/2}$ we get

$$T_2^* \omega_L = \frac{B}{B_{1/2}}$$

and thus the desired result.

Problem 9.3
In contrast to a non-spin-polarized current, here an imbalance of the spin-dependent scattering occurs. Thus, more electrons with one spin orientation are deflected to one side, compared to electrons with the opposite spin orientation to the other side. This gives an imbalance of the electrochemical potential. As a consequence, a voltage drop is measured. The effect is called inverse spin Hall effect.

Chapter 10. Quantum spin Hall effect

Problem 10.1
The energy eigenvalues E are obtained by calculating the determinant

$$\det\left(\hat{h}(\vec{k}) - E I_{2\times2}\right) = 0,$$

which results in

$$E_\pm = \epsilon(\vec{k}) \pm \sqrt{A^2(\sin^2 k_x a + \sin^2 k_y a) + M^2(\vec{k})}.$$

If $M \to 0$, which is the case for a quantum well thickness approaching the critical thickness ($d = d_c$), the dispersion can be approximated by the linear dispersion for massless Dirac fermions. Thus, $\sin(k_{x,y} a)$ is approximated by $k_{x,y} a$.

Problem 10.2
The two-terminal resistance between contacts 1 and 4 is given by

$$R_{14,14} = \frac{V_{14}}{I_{14}} = \frac{(\mu_1 - \mu_4)/e}{I_{14}}.$$

As given in Section 10.6, the current can be expressed as

$$I_{14} = \frac{e}{h}(2\mu_1 - 2\mu_2).$$

From equations (10.31) and (10.32) one obtains

$$\mu_2 = (\mu_4 + 2\mu_1)/3,$$

which has to be inserted into the expression for I_{14}. The 2-terminal resistance is then given by

$$R_{14,14} = \frac{V_{14}}{I_{14}} = \frac{3}{2}\frac{h}{e^2}. \qquad (2)$$

This value differs from the value of $h/2e^2$, which is expected for the 2-terminal resistance in the quantum Hall effect regime. The difference is due to the fact that in the quantum spin Hall effect the carriers in the edge channels on each side propagate in opposite directions, while in the quantum Hall regime the carriers move in the same direction.

Problem 10.3

Since the solution given above corresponds to zero energy states for the Hamiltionian for the x-direction equation (10.10) one has to deal with equation (10.11):

$$\hat{H}_1 = Dk_y^2 I_{4\times 4} + \begin{pmatrix} -Bk_y^2 & iAk_y & 0 & 0 \\ -iAk_y & Bk_y^2 & 0 & 0 \\ 0 & 0 & -Bk_y^2 & -iAk_y \\ 0 & 0 & +iAk_y & Bk_y^2 \end{pmatrix}.$$

We can neglect the first term and obtain for the upper 2×2 matrix:

$$\int dx \frac{1}{2} \xi^*(x) \xi(x)(1,i)\begin{pmatrix} -Bk_y^2 & iAk_y \\ -iAk_y & Bk_y^2 \end{pmatrix}\begin{pmatrix} 1 \\ -i \end{pmatrix} = Ak_y.$$

We assumed that $\xi(x)$ was properly normalized. Similarly, one gets for the lower 2×2 matrix the result $-Ak_y$. Since the contributions of the upper matrix corresponding to spin-up and contribution of the lower matrix corresponding to spin-down have the opposite sign, we can combine both by using the σ_z Pauli spin matrix

$$\hat{H}_{edge} = Ak_y\hat{\sigma}_z.$$

Problem 10.4

For a rectangular well with infinitively high barriers the energy eigenvalues for electrons and holes are given by

$$E_{n,e} = \frac{\hbar^2}{2m_e^*}\frac{\pi^2}{d^2}n^2 \quad \text{and} \quad E_{n,h} = \frac{\hbar^2}{2m_h^*}\frac{\pi^2}{d^2}n^2,$$

respectively. For the lowest levels this results in

$$E_{1,e} = 0.94\,\text{eVnm}^2\frac{1}{d^2} \quad \text{and} \quad E_{1,h} = 0.47\,\text{eVnm}^2\frac{1}{d^2}$$

for the electron and hole system, respectively. The widths d is given in nanometers. The energy difference between the electron and hole levels can be expressed by

$$\Delta E = E_g + (E_{1,e} + E_{1,h}).$$

For the critical width $\Delta E = 0$, which results in a value of $d_c = 1.9\,\text{nm}$.

Chapter 11. Topological insulators

Problem 11.1
Applying the Hamiltonian to the two components of Ψ results in

$$\hat{\vec{\sigma}}\hat{\vec{p}}\chi = E\psi$$

and

$$\hat{\vec{\sigma}}\hat{\vec{p}}\,\psi = E\chi.$$

Combining both gives

$$(\hat{\vec{\sigma}}\hat{\vec{p}})(\hat{\vec{\sigma}}\hat{\vec{p}})\chi = E^2\chi,$$
$$[(\hat{\vec{\sigma}}\hat{\vec{p}})^2 - E^2]\chi = 0.$$

This can be written as

$$(\hat{\vec{\sigma}}\hat{\vec{p}} + E)(\hat{\vec{\sigma}}\hat{\vec{p}} - E)\chi = 0.$$

It is sufficient if

$$(\hat{\vec{\sigma}}\hat{\vec{p}} - E)\chi = 0.$$

Problem 11.2
For $E = 0$ the corresponding Dirac equation is given by

$$\hat{H}_{JR}\psi_0 = 0,$$

which results in

$$\left[-iv\hat{\sigma}_z\frac{\partial}{\partial z} + M(z)\hat{\sigma}_x\right]\psi_0 = 0,$$

and finally in

$$\frac{\partial}{\partial z}\psi_0 = \frac{M(z)}{v}\hat{\sigma}_y\psi_0,$$

where we multiplied the Schrödinger equation by σ_z. The solution of this differential equation can be obtained by integration

$$\psi_0 \propto \exp\left[-\int_0^z dz' M(z')/v\right]\phi_-,$$

with ϕ_- being the eigenvector of the Pauli spin matrix σ_y:

$$\phi_- = \frac{1}{\sqrt{2}}\begin{pmatrix}1\\-i\end{pmatrix}.$$

The sign in front of the integral makes sure that the wave function vanishes for $|z| \to \infty$.

Problem 11.3
According to equation (11.24) the energy of the lowest state with positive energy at $k = 0$ is given by

$$E_{00} = \frac{\hbar v}{2r_0},$$

which results in 3.13 meV. The gap is twice that value thus 6.26 meV. The energy gap is closed at a flux of $\Phi_0/2 = h/2e$. With the cross sectional area given by πr_0^2 this results in a field of 0.41 T.

Problem 11.4
Inserting the eigenvalues given by equation (11.24) into eigenvalue equation gives for $k \neq 0$

$$\frac{\hbar v}{r_0}\left[\left(l+\frac{1}{2}\right)\mp\sqrt{(kr_0)^2+\left(l+\frac{1}{2}\right)^2}\right]\chi_1 + i\hbar k v\chi_2 = 0.$$

This can be simplified to

$$\left[\left(l+\frac{1}{2}\right)\mp\sqrt{(kr_0)^2+\left(l+\frac{1}{2}\right)^2}\right]\chi_1 + ikr_0\chi_2 = 0.$$

Assuming $\chi_1 = 1$, neglecting normalization first, results in

$$\chi_2 = -i\frac{1}{kr_0}\left[\left(l+\frac{1}{2}\right)\mp\sqrt{(kr_0)^2+\left(l+\frac{1}{2}\right)^2}\right].$$

The normalization factor is given by $y = 1/\sqrt{\chi_1^2 + \chi_2^2} = 1/\sqrt{1 + \chi_2^2}$. The final state is given by

$$\chi = y \begin{pmatrix} 1 \\ e^{i\phi}\chi_2 \end{pmatrix},$$

with χ_2 and y given above.

Problem 11.5

The density of states in k-space is given by

$$\frac{dN(k)}{dk} = g_s \frac{1}{(2\pi)^2} 2\pi k = g_s \frac{k}{2\pi}.$$

Here, dN corresponds to the number of states in a ring in k space with circumference $2\pi k$ of width dk and g_s is the spin degeneracy factor being 1 for topological insulators. The density of states with respect to energy is given by

$$D(E) = \frac{dN}{dE} = \frac{dN}{dk}\frac{dk}{dE}.$$

With $dE/dk = v\hbar$ we finally arrive at

$$D(E) = \frac{k}{2\pi}\frac{1}{v\hbar} = \frac{1}{2\pi}\frac{|E|}{(v\hbar)^2}.$$

Thus, the density of states is linear in E.

Chapter 12. Quantum dot spin qubits

Problem 12.1

Let us assume as an entangled state

$$|\psi\rangle = \frac{1}{\sqrt{2}}(|00\rangle + |11\rangle),$$

which corresponds to the vector

$$|\psi\rangle = \frac{1}{\sqrt{2}} \begin{pmatrix} 1 \\ 0 \\ 0 \\ 1 \end{pmatrix}.$$

Applying this vector to the matrix for the CNOT operation

$$U_{\text{CNOT}} = \begin{pmatrix} 1 & 0 & 0 & 0 \\ 0 & 1 & 0 & 0 \\ 0 & 0 & 0 & 1 \\ 0 & 0 & 1 & 0 \end{pmatrix}$$

results in the state

$$|\psi\rangle = \frac{1}{\sqrt{2}}(|00\rangle + |10\rangle),$$

which is a nonentangled state, since it can be written as

$$|\psi\rangle = \frac{1}{\sqrt{2}}(|0\rangle + |1\rangle) \otimes |0\rangle.$$

Problem 12.2
The matrices of the $\pi/2$-rotation and $-\pi/2$-rotation are

$$U_{\pi/2,1} = \begin{pmatrix} 1 & 1 & 0 & 0 \\ -1 & 1 & 0 & 0 \\ 0 & 0 & 1 & 1 \\ 0 & 0 & -1 & 1 \end{pmatrix} \quad \text{and} \quad U_{-\pi/2,1} = \begin{pmatrix} 1 & -1 & 0 & 0 \\ 1 & 1 & 0 & 0 \\ 0 & 0 & 1 & -1 \\ 0 & 0 & 1 & 1 \end{pmatrix},$$

respectively. Performing the required operation sequence results in

$$U_{-\pi/2,1} U_{\mathrm{CROT}} U_{\pi/2,1} = \begin{pmatrix} 1 & -1 & 0 & 0 \\ 1 & 1 & 0 & 0 \\ 0 & 0 & 1 & -1 \\ 0 & 0 & 1 & 1 \end{pmatrix} \begin{pmatrix} 1 & 0 & 0 & 0 \\ 0 & 1 & 0 & 0 \\ 0 & 0 & 1 & 0 \\ 0 & 0 & 0 & -1 \end{pmatrix} \begin{pmatrix} 1 & 1 & 0 & 0 \\ -1 & 1 & 0 & 0 \\ 0 & 0 & 1 & 1 \\ 0 & 0 & -1 & 1 \end{pmatrix}$$

$$= \begin{pmatrix} 1 & 0 & 0 & 0 \\ 0 & 1 & 0 & 0 \\ 0 & 0 & 0 & 1 \\ 0 & 0 & 1 & 0 \end{pmatrix} = U_{\mathrm{CNOT}}.$$

Problem 12.3
The Schrödinger equation for a spin-1/2 particle in a magnetic field along the z-axis is given by

$$i\hbar \frac{\partial}{\partial t} \begin{pmatrix} c_0(t) \\ c_1(t) \end{pmatrix} = sg\mu_B \begin{pmatrix} B_z & 0 \\ 0 & -B_z \end{pmatrix} \begin{pmatrix} c_0(t) \\ c_1(t) \end{pmatrix},$$

with s the spin quantum number. The two components are decoupled. The solutions are exponential functions. With the initial value given by

$$\phi(0) = \begin{pmatrix} c_0(0) \\ c_1(0) \end{pmatrix},$$

the solution is given by

$$\phi(t) = \begin{pmatrix} c_0(0)\exp(-i\omega_p t/2) \\ c_1(0)\exp(i\omega_p t/2) \end{pmatrix}.$$

Here, $\omega_p = g\mu_B B_z/\hbar$ is the precession frequency. With the initial value given above the solution for any time t is given by

$$\phi(t) = \frac{1}{\sqrt{2}}\begin{pmatrix} \exp\left(-i\omega_p t/2\right) \\ \exp\left(i\omega_p t/2\right) \end{pmatrix}.$$

The expectation values of the spin along the x- and y-directions are given by

$$\langle s_x \rangle = \frac{\hbar}{2}\langle\phi(t)|\hat{\sigma}_x|\phi(t)\rangle = \frac{\hbar}{2}\cos\left(\omega_p t\right)$$

and

$$\langle s_y \rangle = \frac{\hbar}{2}\langle\phi(t)|\hat{\sigma}_y|\phi(t)\rangle = \frac{\hbar}{2}\sin\left(\omega_p t\right),$$

respectively. Thus, the spin is precessing in the xy-plane, with the precession frequency ω_p.

Problem 12.4

From equation (12.45) and assuming $J(\varepsilon)$ the corresponding Hamiltonian is given by

$$\hat{H} = \begin{pmatrix} 0 & g\mu_B\Delta B_{nuc}/2 \\ g\mu_B\Delta B_{nuc}/2 & 0 \end{pmatrix} = \frac{g\mu_B\Delta B_{nuc}}{2}\hat{\sigma}_x.$$

The Schrödinger equation can then be written as

$$i\hbar\frac{\partial}{\partial t}\begin{pmatrix} c_0(t) \\ c_1(t) \end{pmatrix} = \begin{pmatrix} 0 & g\mu_B\Delta B_{nuc}/2 \\ g\mu_B\Delta B_{nuc}/2 & 0 \end{pmatrix}\begin{pmatrix} c_0(t) \\ c_1(t) \end{pmatrix},$$

which results in

$$i\hbar\frac{\partial}{\partial t}c_0(t) = \frac{g\mu_B\Delta B_{nuc}}{2}c_1(t)$$

and

$$i\hbar\frac{\partial}{\partial t}c_1(t) = \frac{g\mu_B\Delta B_{nuc}}{2}c_0(t).$$

Substituting the derivative of the first equation into the second one gives

$$-\hbar^2\frac{\partial^2}{\partial t^2}c_0(t) = \frac{g\mu_B\Delta B_{nuc}}{2}^2 c_0(t).$$

Assuming for the initial state $(1,0)^T$ implies

$$c_0(t) = \cos\left(\Omega t\right),$$

with the frequency given by

$$\Omega = \frac{g\mu_B \Delta B_{\text{nuc}}}{2\hbar}.$$

For the second component we obtain

$$c_1(t) = -i \sin(\Omega t).$$

After a $\pi/2$-rotation the state developes into

$$\frac{1}{\sqrt{2}} \begin{pmatrix} 1 \\ -i \end{pmatrix},$$

which is an eigenstate of $\hat{\sigma}_y$. Thus the spin is aligned along the y-axis.

Chapter 13. Majorana fermions

Problem 13.1

With the definion for the Majorana operators given by $\hat{\gamma}_i^A = \hat{c}_i + \hat{c}_i^\dagger$ and $\hat{\gamma}_j^A = \hat{c}_j + \hat{c}_j^\dagger$ we can write

$$
\begin{aligned}
\{\hat{\gamma}_i^A, \hat{\gamma}_j^A\} &= (\hat{c}_i + \hat{c}_i^\dagger)(\hat{c}_j + \hat{c}_j^\dagger) + (\hat{c}_j + \hat{c}_j^\dagger)(\hat{c}_i + \hat{c}_i^\dagger) \\
&= \hat{c}_i \hat{c}_j + \hat{c}_i \hat{c}_j^\dagger + \hat{c}_i^\dagger \hat{c}_j + \hat{c}_i^\dagger \hat{c}_j^\dagger + \hat{c}_j \hat{c}_i + \hat{c}_j \hat{c}_i^\dagger + \hat{c}_j^\dagger \hat{c}_i + \hat{c}_j^\dagger \hat{c}_i^\dagger \\
&= (\hat{c}_i \hat{c}_j + \hat{c}_j \hat{c}_i) + (\hat{c}_i \hat{c}_j^\dagger + \hat{c}_j \hat{c}_i^\dagger) + (\hat{c}_i^\dagger \hat{c}_j + \hat{c}_j^\dagger \hat{c}_i) + (\hat{c}_i^\dagger \hat{c}_j^\dagger + \hat{c}_j^\dagger \hat{c}_i^\dagger) \\
&= 0 + \delta_{ij} + \delta_{ij} + 0 = 2\delta_{ij}.
\end{aligned}
$$

Here, we used the anticommutator relation for fermions given by equations (13.20)–(13.22). In a similar way we obtain

$$
\begin{aligned}
\{\hat{\gamma}_i^B, \hat{\gamma}_j^B\} &= (-1)[(\hat{c}_i - \hat{c}_i^\dagger)(\hat{c}_j - \hat{c}_j^\dagger) + (\hat{c}_j - \hat{c}_j^\dagger)(\hat{c}_i - \hat{c}_i^\dagger)] \\
&= 2\delta_{ij}.
\end{aligned}
$$

For the third case we find

$$
\begin{aligned}
\{\hat{\gamma}_i^A, \hat{\gamma}_j^B\} &= (\hat{c}_i + \hat{c}_i^\dagger)(\hat{c}_j - \hat{c}_j^\dagger) + (\hat{c}_j - \hat{c}_j^\dagger)(\hat{c}_i + \hat{c}_i^\dagger) \\
&= (\hat{c}_i \hat{c}_j^\dagger + \hat{c}_j^\dagger \hat{c}_i) - (\hat{c}_i \hat{c}_j + \hat{c}_j \hat{c}_i) + (\hat{c}_i^\dagger \hat{c}_j^\dagger + \hat{c}_j^\dagger \hat{c}_i^\dagger) - (\hat{c}_i^\dagger \hat{c}_j + \hat{c}_j \hat{c}_i^\dagger) \\
&= \delta_{ij} + 0 + 0 - \delta_{ij} = 0.
\end{aligned}
$$

Thus in total we obtain

$$\{\hat{\gamma}_i^A, \hat{\gamma}_j^B\} = 2\delta_{ij}\delta_{AB}. \tag{3}$$

Problem 13.2

By using the definitions of the Majorana operators given by equations (13.67) and (13.68) we can write

$$\hat{P} = 1 - 2\hat{c}^\dagger\hat{c} = 1 - \frac{1}{2}(\hat{\gamma}_1 - i\hat{\gamma}_2)(\hat{\gamma}_1 + i\hat{\gamma}_2)$$

$$= 1 - \frac{1}{2}(\hat{\gamma}_1\hat{\gamma}_1 + i\hat{\gamma}_1\hat{\gamma}_2 - i\hat{\gamma}_2\hat{\gamma}_1 + \hat{\gamma}_1\hat{\gamma}_1).$$

By making use of the anticommutator relations we can replace $\hat{\gamma}_1\hat{\gamma}_1$ and $\hat{\gamma}_2\hat{\gamma}_2$ by 1 and $\hat{\gamma}_2\hat{\gamma}_1$ by $-\hat{\gamma}_1\hat{\gamma}_2$ which results in

$$1 - 2\hat{c}^\dagger\hat{c} = 1 - \frac{1}{2}(1 + i\hat{\gamma}_1\hat{\gamma}_2 - i\hat{\gamma}_2\hat{\gamma}_1 + 1)$$

$$= -\frac{i}{2}(\hat{\gamma}_1\hat{\gamma}_2 - \hat{\gamma}_2\hat{\gamma}_1)$$

$$= -\frac{i}{2}(\hat{\gamma}_1\hat{\gamma}_2 + \hat{\gamma}_1\hat{\gamma}_2)$$

$$= -i\hat{\gamma}_1\hat{\gamma}_2.$$

Problem 13.3

By applying first $\hat{B}_{12}$ and then $\hat{B}_{23}$ to the state $|00\rangle$ we obtain

$$\hat{B}_{23}\hat{B}_{12}|00\rangle = \hat{B}_{23}\frac{1}{\sqrt{2}}(1 + i)|00\rangle$$

$$= \frac{1}{2}(1 + i)(|00\rangle + i|11\rangle),$$

which is different from the result when applying the operators in the opposite order

$$\hat{B}_{12}\hat{B}_{23}|00\rangle = \hat{B}_{12}\frac{1}{\sqrt{2}}(|00\rangle + i|11\rangle)$$

$$= \frac{1}{2}[(1 + i)|00\rangle + i(1 - i)|11\rangle]$$

$$= \frac{1}{2}[(1 + i)|00\rangle + (1 + i)|11\rangle]$$

$$= \frac{1}{2}(1 + i)(|00\rangle + |11\rangle).$$

Bibliography

[1]   G. E. Moore. Progress in digital integrated electronics. In *IEEE International Electron Devices Meeting, Technical Digest*, pages 11–13, 1975.

[2]   Semiconductor Industry Association. The international technology roadmap for semiconductors. Technical report, Semiconductor Industry Association, 2013.

[3]   M. N. Baibich, J. M. Broto, A. Fert, F. Nguyen Van Dau, F. Petroff, P. Etienne, G. Creuzet, A. Friederich, and J. Chazelas. Giant magnetoresistance of (001)Fe/(001) Cr magnetic superlattices. *Phys. Rev. Lett.*, 61:2472–2475, 1988.

[4]   G. Binasch, P. Grünberg, F. Saurenbach, and W. Zinn. Enhanced magnetoresistance in layered magnetic structures with antiferromagnetic interlayer exchange. *Phys. Rev. B*, 39:4828–4830, 1989.

[5]   K. Bernstein, R. K. Cavin, W. Porod, A. Seabaugh, and J. Welser. Device and architecture outlook for beyond CMOS switches. *Proc. IEEE*, 98(12):2169–2184, 2010.

[6]   S. Datta and B. Das. Electronic analog of the electro-optic modulator. *Appl. Phys. Lett.*, 56(7):665–667, 1990.

[7]   M. E. Flatté and G. Vignale. Unipolar spin diodes and transistors. *Appl. Phys. Lett.*, 78(9):1273–1275, 2001.

[8]   K. C. Hall, W. H. Lau, K. Gündoğdu, M. E. Flatté, and T. F. Boggess. Nonmagnetic semiconductor spin transistor. *Appl. Phys. Lett.*, 83(14):2937–2939, 2003.

[9]   J. C. Egues, G. Burkard, and D. Loss. Datta-Das transistor with enhanced spin control. *Appl. Phys. Lett.*, 82(16):2658–2660, 2003.

[10]   J. Schliemann, J. C. Egues, and D. Loss. Nonballistic spin-field-effect transistor. *Phys. Rev. Lett.*, 90:146801, 2003.

[11]   B. Behin-Aein, D. Datta, S. Salahuddin, and S. Datta. Proposal for an all-spin logic device with built-in memory. *Nat. Nanotechnol.*, 5:266, 2010.

[12]   H. C. Koo, J. H. Kwon, J. Eom, J. Chang, S. H. Han, and M. Johnson. Control of spin precession in a spin-injected field effect transistor. *Science*, 325(5947):1515–1518, 2009.

[13]   Yu. A. Bychkov and E. I. Rashba. Oscillatory effects and the magnetic susceptibility of carriers in inversion layers. *J. Phys. C, Solid State Phys.*, 17(33):6039–6045, 1984.

[14]   M. König, H. Buhmann, L. W. Molenkamp, T. Hughes, C.-X. Liu, X.-L. Qi, and S.-C. Zhang. The quantum spin Hall effect: theory and experiment. *J. Phys. Soc. Jpn.*, 77(3):031007, 2008.

[15]   X.-L. Qi and S.-C. Zhang. Topological insulators and superconductors. *Rev. Mod. Phys.*, 83:1057–1110, 2011.

[16]   Y. Ando. Topological insulator materials. *J. Phys. Soc. Jpn.*, 82(10):102001, 2013.

[17]   M. A. Nielsen and I. L. Chuang. Quantum Computation and Quantum Information. Cambridge University Press, 2010.

[18]   P. W. Shor. Algorithms for quantum computation: discrete logarithms and factoring. In S. Goldwasser, editor, *Proc. 35th Annual Symposium on the Foundations of Computer Science*, pages 124–134. IEEE Computer Society Press, Los Alamitos, CA, 1994.

[19]   D. Loss and D. P. DiVincenzo. Quantum computation with quantum dots. *Phys. Rev. A*, 57:120, 1998.

[20]   R. Hanson, L. P. Kouwenhoven, J. R. Petta, S. Tarucha, and L. M. K. Vandersypen. Spins in few-electron quantum dots. *Rev. Mod. Phys.*, 79:1217–1265, 2007.

[21]   P. Yu and M. Cardona. *Fundamentals of Semiconductors*. Springer, Berlin Heidelberg, 2010.

[22]   T. Heinzel. *Mesoscopic Electronics in Solid State Nanostructures*. Wiley-VCH, 2006.

[23]   T. Ihn. *Semiconductor Nanostructures*. Oxford University Press, 2009.

[24]   H. Ibach and H. Lüth. *Solid-State Physics*. Springer, Berlin Heidelberg, 2009.

[25]   S. Datta. *Electronic Transport in Mesoscopic Systems*. Cambridge University Press, Cambridge, 1995.

https://doi.org/10.1515/9783110639001-015

[26] H. Lüth. *Surfaces and Interfaces of Solid Materials*. Springer, Berlin, 1996.

[27] C. T. Foxon, J. J. Harris, D. Hilton, J. Hewitt, and C. Roberts. Optimisation of (Al, Ga)As/GaAs two-dimensional electron gas structures for low carrier densities and ultrahigh mobilities at low temperatures. *Semicond. Sci. Technol.*, 4:582, 1989.

[28] L. Pfeiffer, K. W. West, H. L. Störmer, and K. W. Baldwin. Electron mobilities exceeding 10^7 cm^2/Vs in modulation-doped GaAs. *Appl. Phys. Lett.*, 55:1888–1890, 1989.

[29] H. Hardtdegen, R. Meyer, H. Løken-Larsen, J. Appenzeller, Th. Schäpers, and H. Lüth. Extremely high mobilities in modulation doped InGaAs/InP heterostructures grown by LP MOVPE. *J. Cryst. Growth*, 116:521–523, 1992.

[30] H. Hardtdegen, R. Meyer, M. Hollfelder, Th. Schäpers, J. Appenzeller, H. Løken-Larsen, Th. Klocke, Ch. Dieker, B. Lengeler, H. Lüth, and W. Jäger. Optimization of modulation doped InGaAs/InP heterostructures towards extremely high mobilities. *J. Appl. Phys.*, 73:4489–4493, 1993.

[31] S. P. Beaumont, P. G. Bower, T. Tamamara, and C. D. W. Wilkinson. Sub-20-nm-wide metal lines by electron-beam exposure of thin poly(methyl methacrylate) films and liftoff. *Appl. Phys. Lett.*, 38:436–439, 1981.

[32] C. Thelander, P. Agarwal, S. Brongersma, J. Eymery, L. F. Feiner, A. Forchel, M. Scheffler, W. Riess, B. J. Ohlsson, U. Gösele, and L. Samuelson. Nanowire-based one-dimensional electronics. *Mater. Today*, 9:28–35, 2006.

[33] H. Lüth. *Solid Surfaces, Interfaces and Thin Films*. Springer, Berlin, Heidelberg, New York, 2010.

[34] K. Tomioka, P. Mohan, J. Noborisaka, S. Hara, J. Motohisa, and T. Fukui. Growth of highly uniform InAs nanowire arrays by selective-area MOVPE. *J. Cryst. Growth*, 298:644, 2007.

[35] M. Akabori, K. Sladek, H. Hardtdegen, Th. Schäpers, and D. Grützmacher. Influence of growth temperature on the selective area MOVPE of InAs nanowires on GaAs (111) B using N_2 carrier gas. *J. Cryst. Growth*, 311(15):3813–3816, 2009.

[36] B. J. van Wees, H. van Houten, C. W. J. Beenakker, J. G. Willamson, L. P. Kouwenhoven, D. van der Marel, and C. T. Foxon. Quantized conductance in point contacts in a two/dimensional electron gas. *Phys. Rev. Lett.*, 60:848–850, 1988.

[37] D. A. Wharam, T. J. Thornton, R. Newbury, M. Pepper, H. Ahmed, J. E. F. Frost, D. G. Hasko, D. C. Peacock, D. A. Ritchie, and G. A. C. Jones. One-dimensional transport and the quantisation of the ballistic resistance. *J. Phys. C*, 21:209–214, 1988.

[38] R. Landauer. Spatial variations of currents and fields due to localized scatterers in metallic conduction. *IBM J. Res. Dev.*, 1:223–231, 1957.

[39] M. Büttiker. Absence of backscattering in the quantum Hall effect in multiprobe conductors. *Phys. Rev. B*, 38:9375–9389, 1988.

[40] H. Lüth. *Quantum Physics in the Nanoworld*. Springer, 2015.

[41] M. Büttiker, Y. Imry, R. Landauer, and S. Pinhas. Generalized many-channel conductance formula with application to small rings. *Phys. Rev. B*, 31:6207–6215, 1985.

[42] K. von Klitzing, G. Dorda, and M. Pepper. New method for high-accuracy determination of the fine-structure constant based on quantized Hall resistance. *Phys. Rev. Lett.*, 45(6):494–497, 1980.

[43] M. A. Ruderman and C. Kittel. Indirect exchange coupling of nuclear magnetic moments by conduction electrons. *Phys. Rev.*, 96:99–102, 1954.

[44] T. Kasuya. A theory of metallic ferro- and antiferromagnetism on Zener's model. *Prog. Theor. Phys.*, 16:45, 1956.

[45] K. Yosida. Magnetic properties of Cu-Mn alloys. *Phys. Rev.*, 106:893–898, 1957.

[46] J. F. Janak. Uniform susceptibilities of metallic elements. *Phys. Rev. B*, 16:255–262, 1977.

[47] V. L. Moruzzi, J. F. Janak, and A. R. Williams. *Calculated Electronic Properties of Metals*. Pergamon Press, NY, 1978.

[48]   J. K. Furdyna. Diluted magnetic semiconductors. *J. Appl. Phys.*, 64(4):R29–R64, 1988.

[49]   R. L. Aggarwal, S. N. Jasperson, P. Becla, and J. K. Furdyna. Optical determination of the antiferromagnetic exchange constant between nearest-neighbor Mn^{2+} ions in $Zn_{0.95}Mn_{0.05}Te$. *Phys. Rev. B*, 34:5894–5896, 1986.

[50]   H. Munekata, H. Ohno, S. von Molnar, A. Segmüller, L. L. Chang, and L. Esaki. Diluted magnetic III-V semiconductors. *Phys. Rev. Lett.*, 63:1849–1852, 1989.

[51]   H. Ohno. Making nonmagnetic semiconductors ferromagnetic. *Science*, 281:951–956, 1998.

[52]   H. Ohno, A. Shen, F. Matsukura, A. Oiwa, A. Endo, S. Katsumoto, and Y. I. Ga. MnAs: a new diluted magnetic semiconductor based on GaAs. *Appl. Phys. Lett.*, 69(3):363–365, 1996.

[53]   T. Dietl, H. Ohno, F. Matsukura, J. Cibert, and D. Ferrand. Zener model description of ferromagnetism in zinc-blende magnetic semiconductors. *Science*, 287(5455):1019–1022, 2000.

[54]   H. Ohno. *Semiconductor Spintronics and Quantum Computation, Chapter Ferromagnetic I-V Semiconductors and Their Heterostructures*, page 1. Springer, 2002.

[55]   T. Dietl and H. Ohno. Dilute ferromagnetic semiconductors: physics and spintronic structures. *Rev. Mod. Phys.*, 86:187–251, 2014.

[56]   T. Jungwirth, K. Y. Wang, J. Mašek, K. W. Edmonds, J. König, J. Sinova, M. Polini, N. A. Goncharuk, A. H. MacDonald, M. Sawicki, A. W. Rushforth, R. P. Campion, L. X. Zhao, C. T. Foxon, and B. L. Gallagher. Prospects for high temperature ferromagnetism in (Ga, Mn)As semiconductors. *Phys. Rev. B*, 72:165–204, 2005.

[57]   K. Sato, L. Bergqvist, J. Kudrnovský, P. H. Dederichs, O. Eriksson, I. Turek, B. Sanyal, G. Bouzerar, H. Katayama-Yoshida, V. A. Dinh, T. Fukushima, H. Kizaki, and R. Zeller. First-principles theory of dilute magnetic semiconductors. *Rev. Mod. Phys.*, 82:1633–1690, 2010.

[58]   Y. Ohno, D. K. Young, B. Beschoten, F. Matsukara, H. Ohno, and D. D. Awschalom. Electrical spin injection in a ferromagnetic semiconductor heterostructure. *Nature*, 402:790–792, 1999.

[59]   D. Chiba, F. Matsukura, and H. Ohno. Electric-field control of ferromagnetism in (Ga, Mn). *Appl. Phys. Lett.*, 89(16):162505, 2006.

[60]   H. Ohno, F. Matsukura, T. Omiya, and N. Akiba. Spin-dependent tunneling and properties of ferromagnetic (Ga, Mn)As (invited). *J. Appl. Phys.*, 85(8):4277–4282, 1999.

[61]   H. Ohno, D. Chiba, F. Matsukura, T. Omiya, E. Abe, T. Dietl, Y. Ohno, and K. Ohtani. Electric-feld control of ferromagnetism. *Nature*, 408:944, 2000.

[62]   A. Yamaguchi, T. Ono, S. Nasu, K. Miyake, K. Mibu, and T. Shinjo. Real-space observation of current-driven domain wall motion in submicron magnetic wires. *Phys. Rev. Lett.*, 92:077205, 2004.

[63]   D. A. Allwood, G. Xiong, C. C. Faulkner, D. Atkinson, D. Petit, and R. P. Cowburn. Magnetic domain-wall logic. *Science*, 309(5741):1688–1692, 2005.

[64]   S. S. P. Parkin, M. Hayashi, and L. Thomas. Magnetic domain-wall racetrack memory. *Science*, 320:190, 2008.

[65]   J. Nitta, Th. Schäpers, H. B. Heersche, T. Koga, Y. Sato, and H. Takayanagi. Investigation of ferromagnetic microstructures by local Hall effect and magnetic force microscopy. *Jpn. J. Appl. Phys.*, 41:2497–2500, 2002.

[66]   M. Johnson, B. R. Bennett, M. J. Yang, M. M. Miller, and B. V. Shanabrook. Hybrid Hall effect device. *Appl. Phys. Lett.*, 71(7):974–976, 1997.

[67]   F. G. Monzon, M. Johnson, and M. L. Roukes. Strong Hall voltage modulation in hybrid ferromagnet/semiconductor microstructures. *Appl. Phys. Lett.*, 71(21):3087–3089, 1997.

[68]   S. Heedt, C. Morgan, K. Weis, D. E. Bürgler, R. Calarco, H. Hardtdegen, D. Grützmacher, and Th. Schäpers. Electrical spin injection into InN semiconductor nanowires. *Nano Lett.*, 12(9):4437–4443, 2012.

[69] M. J. Donahue and D. G. Porter. Oommf user's guide, version 1.0, interagency report NISTIR 6376. Technical report, National Institute of Standards and Technology, Gaithersburg, MD (Sept 1999), 1999.

[70] M. Hayashi, L. Thomas, R. Moriya, C. Rettner, and S. S. P. Parkin. Current-controlled magnetic domain-wall nanowire shift register. *Science*, 320:209, 2008.

[71] G. Schmidt and L. W. Molenkamp. Spin injection into semiconductors, physics and experiments. *Semicond. Sci. Technol.*, 17(4):310, 2002.

[72] G. Schmidt, D. Ferrand, L. W. Molenkamp, A. T. Filip, and B. J. van Wees. Fundamental obstacle for electrical spin injection from a ferromagnetic metal into a diffusive semiconductor. *Phys. Rev. B*, 62:4790–4793, 2000.

[73] C. Weisbuch and B. Vinter. *Quantum Semiconductor Structures: Fundamentals and Applications*. Academic, Boston, 1991.

[74] R. Fiederling, M. Keim, G. Reuscher, W. Ossau, G. Schmidt, A. Waag, and L. W. Molenkamp. Injection and detection of a spin-polarized current by a light emitting diode. *Nature*, 402:787–790, 1999.

[75] E. I. Rashba. Theory of electrical spin injection: tunnel contacts as a solution of the conductivity mismatch problem. *Phys. Rev. B*, 62:R16267–R16270, 2000.

[76] H. B. Heersche, Th. Schäpers, J. Nitta, and H. Takayanagi. Enhancement of spin injection from ferromagnetic metal into a two-dimensional electron gas using a tunnel barrier. *Phys. Rev. B*, 64(16):161307, 2001.

[77] A. Fert and H. Jaffrès. Conditions for efficient spin injection from a ferromagnetic metal into a semiconductor. *Phys. Rev. B*, 64(18):184420, 2001.

[78] G. E. Blonder, M. Tinkham, and T. M. Klapwijk. Transition from metallic to tunneling regimes in superconducting microconstrictions: excess current, charge imbalance, and supercurrent conversion. *Phys. Rev. B*, 25:4515–4532, 1982.

[79] T. Valet and A. Fert. Theory of the perpendicular magnetoresistance in magnetic multilayers. *Phys. Rev. B*, 48:7099–7113, 1993.

[80] T. Manago and H. Akinaga. Spin-polarized light-emitting diode using metal/insulator/semiconductor structures. *Appl. Phys. Lett.*, 81:694–696, 2002.

[81] A. T. Hanbicki, B. T. Jonker, G. Itskos, G. Kioseoglou, and A. Petrou. Efficient electrical spin injection from a magnetic metal/tunnel barrier contact into a semiconductor. *Appl. Phys. Lett.*, 80(7):1240–1242, 2002.

[82] A. T. Hanbicki, O. M. J. van 't Erve, R. Magno, G. Kioseoglou, C. H. Li, B. T. Jonker, G. Itskos, R. Mallory, M. Yasar, and A. Petrou. Analysis of the transport process providing spin injection through an Fe/AlGaAs Schottky barrier. *Appl. Phys. Lett.*, 82(23):4092–4094, 2003.

[83] F. G. Monzon and M. L. Roukes. Spin injection and the local Hall effect in InAs quantum wells. *J. Magn. Magn. Mater.*, 198–199:632–635, 1999.

[84] N. Tombros, S. J. van der Molen, and B. J. van Wees. Separating spin and charge transport in single-wall carbon nanotubes. *Phys. Rev. B*, 73:233–403, 2006.

[85] M. Johnson and R. H. Silsbee. Interfacial charge-spin coupling: injection and detection of spin magnetization in metals. *Phys. Rev. Lett.*, 55:1790–1793, 1985.

[86] M. E. Flatté, J. M. Byers, and W. H. Lau. In *Spin Dynamics in Semiconductors, Semiconductor Spintronics and Quantum Computation*, pages 107–145. Springer, 2002.

[87] J. M. Kikkawa and D. D. Awschalom. Resonant spin amplification in n-type GaAs. *Phys. Rev. Lett.*, 80:4313–4316, 1998.

[88] K. Schmalbuch, S. Göbbels, Ph. Schäfers, Ch. Rodenbücher, P. Schlammes, Th. Schäpers, M. Lepsa, G. Güntherodt, and B. Beschoten. Two-dimensional optical control of electron spin orientation by linearly polarized light in InGaAs. *Phys. Rev. Lett.*, 105:246603, 2010.

[89] M. Heidkamp. Spin-coherence and -dephasing of donor and free conduction band electrons across the metal-insulator transistion in Si:GaAs. *PhD. thesis, RWTH Aachen University*, 2004.

[90] P. W. Anderson. Absence of diffusion in certain random lattices. *Phys. Rev.*, 109:1492, 1958.

[91] N. F. Mott. *Metal-Insulator Transitions*. Taylor & Francis, 1997.

[92] M. Oestreich, S. Hallstein, A. P. Heberle, K. Eberl, E. Bauser, and W. W. Rühle. Temperature and density dependence of the electron Landé g factor in semiconductors. *Phys. Rev. B*, 53:7911–7916, 1996.

[93] S. Bandyopadhyay and M. Cahay. Alternate spintronic analog of the electro-optic modulator. *Appl. Phys. Lett.*, 85:1814, 2004.

[94] J. Nitta, T. Akazaki, H. Takayanagi, and T. Enoki. Gate control of spin-orbit interaction in an inverted $In_{0.53}Ga_{0.47}As/In_{0.52}Al_{0.48}As$ heterostructure. *Phys. Rev. Lett.*, 78:1335–1338, 1997.

[95] R. Winkler. *Spin Orbit Coupling Effects in Two-Dimensional Electron and Hole Systems*. Springer, Berlin, Heidelberg, New York, 2003.

[96] R. Winkler. Spin orientation and spin precession in inversion-asymmetric quasi-two-dimensional electron systems. *Phys. Rev. B*, 69(4):45317, 2004.

[97] E. O. Kane. Energy band theory. In T. S. Moss, editor, *Handbook on Semiconductors*, pages 193–217. North-Holland, Amsterdam, 1982.

[98] P.-O. Löwdin. A note on the quantum-mechanical perturbation theory. *J. Chem. Phys.*, 19(11):1396–1401, 1951.

[99] R. Lassnig. $k \cdot p$ theory, effective-mass approach, and spin splitting for two-dimensional electrons in GaAs-GaAlAs heterostructures. *Phys. Rev. B*, 31(12):8076–8086, 1985.

[100] Th. Schäpers, G. Engels, J. Lange, Th. Klocke, M. Hollfelder, and H. Lüth. Effect of the heterointerface on the spin splitting in modulation doped $In_xGa_{1-x}As/InP$ quantum wells for $B \rightarrow 0$. *J. Appl. Phys.*, 83(8):4324–4333, 1998.

[101] Landolt-Börnstein. *Group III Condensed Matter, volume III/41*. Springer, 2002.

[102] R. Winkler. Rashba spin splitting and Ehrenfest's theorem. *Physica E*, 22:450–454, 2004.

[103] G. Engels, J. Lange, Th. Schäpers, and H. Lüth. Experimental and theoretical approach to spin splitting in modulation-doped $In_xGa_{1-x}As/InP$ quantum wells for $B \rightarrow 0$. *Phys. Rev. B*, 55:R1958–R1961, 1997.

[104] D. Grundler. Large Rashba splitting in InAs quantum wells due to electron wave function penetration into the barrier layers. *Phys. Rev. Lett.*, 84:6074–6077, 2000.

[105] G. Dresselhaus. Spin-orbit coupling effects in zinc-blende structures. *Phys. Rev.*, 100:580, 1955.

[106] B. A. Bernevig, T. L. Hughes, and S. C. Zhang. Quantum spin Hall effect and topological phase transition in HgTe quantum wells. *Science*, 314:1757–1761, 2006.

[107] J. D. Koralek, C. P. Weber, J. Orenstein, B. A. Bernevig, S.-C. Zhang, S. Mack, and D. D. Awschalom. Emergence of the persistent spin helix in semiconductor quantum wells. *Nature*, 458:610–613, 2009.

[108] M. P. Walser, C. Reichl, W. Wegscheider, and G. Salis. Direct mapping of the formation of a persistent spin helix. *Nat. Phys.*, 8:757–762, 2012.

[109] A. Sasaki, S. Nonaka, Y. Kunihashi, M. Kohda, T. Bauernfeind, T. Dollinger, K. Richter, and J. Nitta. Direct determination of spin-orbit interaction coefficients and realization of the persistent spin helix symmetry. *Nat. Nanotechnol.*, 9:703, 2014.

[110] M. Governale and U. Zülicke. Spin accumulation in quantum wires with strong Rashba spin-orbit coupling. *Phys. Rev. B*, 66(7):073311, 2002.

[111] J. Knobbe and Th. Schäpers. Magnetosubbands of semiconductor quantum wires with Rashba spin-orbit coupling. *Phys. Rev. B*, 71(3):35311, 2005.

[112] P. Středa and P. Šeba. Antisymmetric spin filtering in one-dimensional electron systems with uniform spin-orbit coupling. *Phys. Rev. Lett.*, 90:256601, 2003.

[113] Y. V. Pershin, J. A. Nesteroff, and V. Privman. Effect of spin-orbit interaction and in-plane magnetic field on the conductance of a quasi-one-dimensional system. *Phys. Rev. B*, 69:121306, 2004.

[114] S. Heedt, N. Traverso Ziani, F. Crépin, W. Prost, St. Trellenkamp, J. Schubert, D. Grützmacher, B. Trauzettel, and Th. Schäpers. Signatures of interaction-induced helical gaps in nanowire quantum point contacts. *Nat. Phys.*, 13:563, 2017.

[115] T. Richter, Ch. Blömers, H. Lüth, R. Calarco, M. Indlekofer, M. Marso, and Th. Schäpers. Flux quantization effects in InN nanowires. *Nano Lett.*, 8:2834–2838, 2008.

[116] Ö. Gül, N. Demarina, C. Blömers, T. Rieger, H. Lüth, M. I. Lepsa, D. Grützmacher, and Th. Schäpers. Flux periodic magnetoconductance oscillations in GaAs/InAs core/shell nanowires. *Phys. Rev. B*, 89:045417, 2014.

[117] A. Bringer and Th. Schäpers. Spin precession and modulation in ballistic cylindrical nanowires due to the Rashba effect. *Phys. Rev. B*, 83(11):115305, 2011.

[118] Y. Aharonov and D. Bohm. Significance of electromangnetic potentials in the quantum theory. *Phys. Rev.*, 115:485–491, 1959.

[119] R. G. Chambers. Shift of an electron interference pattern by enclosed magnetic flux. *Phys. Rev. Lett.*, 5(1):3–5, 1960.

[120] A. Tonomura, T. Matsuda, R. Suzuki, A. Fukuhara, N. Osakabe, H. Umezaki, J. Endo, K. Shinagawa, Y. Sugita, and H. Fujiwara. Observation of Aharonov-Bohm effect by electron holography. *Phys. Rev. Lett.*, 48(21):1443–1446, 1982.

[121] R. A. Webb, S. Washburn, C. P. Umbach, and R. B. Laibowitz. Observation of h/e Aharonov-Bohm oscillations in normal-metal rings. *Phys. Rev. Lett.*, 54(25):2696–2699, 1985.

[122] B. Krafft, A. Förster, A. van der Hart, and Th. Schäpers. Control of Aharonov–Bohm oscillations in an AlGaAs/GaAs ring by asymmetric and symmetric gate biasing. *Physica E*, 9(4):635–641, 2001.

[123] J. Appenzeller, Th. Schäpers, H. Hardtdegen, B. Lengeler, and H. Lüth. Aharonov-Bohm effect in quasi-one-dimensional $In_{0.77}Ga_{0.23}As/InP$ rings. *Phys. Rev. B*, 51:4336–4342, 1995.

[124] B. L. Al'tshuler, A. G. Aronov, and B. Z. Spivak. Aharonov-Bohm effect in disordered conductors. *Pis'ma Zh. Eksp. Teor. Fiz.*, 33:101–103, 1981. *JETP Lett.*, 33:94, 1981.

[125] D. Yu. Sharvin and Yu. V. Sharvin. Quantisation of the magnetic flow in a normal metal cylindrical film. *Pis'ma Zh. Eksp. Teor. Fiz.*, 34:285–288, 1981. *JETP Lett.*, 34:272, 1981.

[126] G. J. Dolan, J. C. Licini, and D. J. Bishop. Quantum interfernce effects in lithium ring arrays. *Phys. Rev. Lett.*, 56(14):1493–1496, 1986.

[127] S. Chakravarty and A. Schmid. Weak localization: the quasiclassical theory of electrons in a random potential. *Phys. Rep.*, 140:193–236, 1986.

[128] C. W. J. Beenakker and H. van Houten. Semiconductor heterostructures and nanostructures (see also: http://de.arxiv.org/abs/cond-mat/0412664v1). In H. Ehrenreich and D. Turnbull, editors, *Solid State Physics*, volume 44, page 1. Academic, New York, 1991.

[129] C. W. J. Beenakker and H. van Houten. Boundary scattering and weak localization of electrons in a magnetic field. *Phys. Rev. B*, 38(5):3232–3240, 1988.

[130] B. L. Altshuler, D. Khmel'nitzkii, A. I. Larkin, and P. A. Lee. Magnetoresistance and Hall effect in a disordered two-dimensional electron gas. *Phys. Rev. B*, 22:5142–5153, 1980.

[131] S. Hikami, A. I. Larkin, and Y. Nagaoka. Spin-orbit interaction and magnetoresistance in the two dimensional random system. *Prog. Theor. Phys.*, 63(2):707–710, 1980.

[132] R. J. Elliott. Theory of the effect of spin-orbit coupling on magnetic resonance in some semiconductors. *Phys. Rev.*, 96:266–279, 1954.

[133] G. Bergmann. Weak anti-localization-an experimental proof for the destructive interference of rotated spin 1/2. *Solid State Commun.*, 42:815–817, 1982.

[134] W. Knap, C. Skierbiszewski, A. Zduniak, E. Litwin-Staszewska, D. Bertho, F. Kobbi, J. L. Robert, G. E. Pikus, F. G. Pikus, S. V. Iordanskii, V. Mosser, K. Zekentes, and Yu. B. Lyanda-Geller. Weak antilocalization and spin precession in quantum wells. *Phys. Rev. B*, 53(7):3912–3924, 1996.

[135] M. I. D'yakonov and V. I. Perel'. Spin relaxation of conduction electrons in noncentrosymmetric semiconductors. *Fiz. Tverd. Tela*, 13:3581, 1971. *Sov. Phys. Solid State*, 13:3023, 1971.

[136] Y. Yafet. g factors and spin-lattice relaxation of conduction electrons. In *Solid State Physics*, volume 14, pages 1–98. Academic Press, 1963.

[137] J. N. Chazalviel. Spin relaxation of conduction electrons in n-type indium antimonide at low temperature. *Phys. Rev. B*, 11:1555–1562, 1975.

[138] G. L. Bir, A. G. Aronov, and G. E. Pikus. Spin relaxation of electrons due to scattering by holes. *Sov. Phys. JETP*, 42:705, 1976.

[139] N. Thillosen, S. Cabañas, N. Kaluza, V. A. Guzenko, H. Hardtdegen, and Th. Schäpers. Weak antilocalization in gate-controlled $Al_xGa_{1-x}N$/GaN two-dimensional electron gases. *Phys. Rev. B*, 73(24):241311, 2006.

[140] S. V. Iordanskii, Yu. B. Lyanda-Geller, and G. E. Pikus. Weak localization in quantum wells with spin-orbit interaction. *JETP Lett.*, 60(3):206–211, 1994.

[141] V. A. Guzenko, T. Schäpers, and H. Hardtdegen. Weak antilocalization in high mobility $Ga_xIn_{1-x}As$/InP two-dimensional electron gases with strong spin-orbit coupling. *Phys. Rev. B*, 76(16):165301, 2007.

[142] L. E. Golub. Weak antilocalization in high-mobility two-dimensional systems. *Phys. Rev. B*, 71:235310, 2005.

[143] Ç. Kurdak, A. M. Chang, A. Chin, and T. Y. Chang. Quantum interference effects and spin-orbit interaction in quasi-one-dimensional wires and rings. *Phys. Rev. B*, 46:6846–6856, 1992.

[144] Th. Schäpers, V. A. Guzenko, M. G. Pala, U. Zülicke, M. Governale, J. Knobbe, and H. Hardtdegen. Suppression of weak antilocalization in $Ga_xIn_{1-x}As$/InP narrow quantum wires. *Phys. Rev. B*, 74(8):081301, 2006.

[145] V. A. Guzenko, J. Knobbe, H. Hardtdegen, Th. Schäpers, and A. Bringer. Rashba effect in parallel InGaAs/InP quantum wires. *Appl. Phys. Lett.*, 88:032102, 2006.

[146] I. L. Aleiner and V. I. Fal'ko. Spin-orbit coupling effects on quantum transport in lateral semiconductor dots. *Phys. Rev. Lett.*, 87(25):256801, 2001.

[147] J. B. Miller, D. M. Zumbühl, C. M. Marcus, Y. B. Lyanda-Geller, D. Goldhaber-Gordon, K. Campman, and A. C. Gossard. Gate-controlled spin-orbit quantum interference effects in lateral transport. *Phys. Rev. Lett.*, 90:076807, 2003.

[148] S. Kettemann. Dimensional control of antilocalization and spin relaxation in quantum wires. *Phys. Rev. Lett.*, 98(17):176808, 2007.

[149] M. V. Berry. Quantal phase factors accompanying adiabatic changes. *Proc. R. Soc. Lond. Ser. A*, 392:45, 1984.

[150] J. Nitta, F. E. Meijer, and H. Takayanagi. Spin-interference device. *Appl. Phys. Lett.*, 75:695–697, 1999.

[151] F. E. Meijer, A. F. Morpurgo, and T. M. Klapwijk. One-dimensional ring in the presence of Rashba spin-orbit interaction: derivation of the correct Hamiltonian. *Phys. Rev. B*, 66:033107, 2002.

[152] D. Frustaglia and K. Richter. Spin interference effects in ring conductors subject to Rashba coupling. *Phys. Rev. B*, 69(23):235310, 2004.

[153] Y. Aharonov and J. Anandan. Phase change during a cyclic quantum evolution. *Phys. Rev. Lett.*, 58:1593–1596, 1987.

[154] T. Koga, Y. Sekine, and J. Nitta. Experimental realization of a ballistic spin interferometer based on the Rashba effect using a nanolithographically defined square loop array. *Phys. Rev. B*, 74(4):041302, 2006.

[155] M. I. D'yakonov and V. I. Perel'. Possibility of orienting electron spins with current. *JETP Lett.*, 13:46, 1971.

[156] J. E. Hirsch. Spin Hall effect. *Phys. Rev. Lett.*, 83:1834–1837, 1999.

[157] H.-A. Engel, B. I. Halperin, and E. I. Rashba. Theory of spin Hall conductivity in n-doped GaAs. *Phys. Rev. Lett.*, 95:166605, 2005.

[158] J. Sinova, D. Culcer, Q. Niu, N. A. Sinitsyn, T. Jungwirth, and A. H. MacDonald. Universal intrinsic spin Hall effect. *Phys. Rev. Lett.*, 92:126603, 2004.

[159] J. Smit. The spontaneous Hall effect in ferromagnetics-II. *Physica*, 24:39, 1958.

[160] N. F. Mott and H. S. W. Massey. *The Theory of Atomic Collisions*. Oxford University Press, London, 1965.

[161] L. Berger. Side-jump mechanism for the Hall effect of ferromagnets. *Phys. Rev. B*, 2:4559–4566, 1970.

[162] G. Vignale. Ten years of spin Hall effect. *J. Supercond. Nov. Magn.*, 23(1):3–10, 2010.

[163] E. M. Hankiewicz and G. Vignale. Coulomb corrections to the extrinsic spin-Hall effect of a two-dimensional electron gas. *Phys. Rev. B*, 73:115339, 2006.

[164] Y. K. Kato, R. C. Myers, A. C. Gossard, and D. D. Awschalom. Observation of the spin Hall effect in semiconductors. *Science*, 306:1910–1913, 2004.

[165] J. Wunderlich, B. Kaestner, J. Sinova, and T. Jungwirth. Experimental observation of the spin-Hall effect in a two-dimensional spin-orbit coupled semiconductor system. *Phys. Rev. Lett.*, 94:047204, 2005.

[166] B. Yan and S.-C. Zhang. Topological materials. *Rep. Prog. Phys.*, 75(9):096501, 2012.

[167] M. König, S. Wiedmann, C. Brüne, A. Roth, H. Buhmann, L. W. Molenkamp, X.-L. Qi, and S.-C. Zhang. Quantum spin Hall insulator state in HgTe quantum wells. *Science*, 318(5851):766–770, 2007.

[168] C. Brüne, A. Roth, H. Buhmann, E. M. Hankiewicz, L. W. Molenkamp, J. Maciejko, X.-L. Qi, and S.-C. Zhang. Spin polarization of the quantum spin Hall edge states. *Nat. Phys.*, 8:485, 2012.

[169] H. Zhang, C.-X. Liu, X.-L. Qi, X. Dai, Z. Fang, and S.-C. Zhang. Topological insulators in Bi_2Se_3, Bi_2Te_3 and Sb_2Te_3 with a single Dirac cone on the surface. *Nat. Phys.*, 5:438–442, 2009.

[170] J. Krumrain, G. Mussler, S. Borisova, T. Stoica, L. Plucinski, C. M. Schneider, and D. Grützmacher. MBE growth optimization of topological insulator Bi_2Te_3 films. *J. Cryst. Growth*, 324(1):115–118, 2011.

[171] C.-X. Liu, X.-L. Qi, H. Zhang, X. Dai, Z. Fang, and S.-C. Zhang. Model Hamiltonian for topological insulators. *Phys. Rev. B*, 82:045122, 2010.

[172] A. Herdt, L. Plucinski, G. Bihlmayer, G. Mussler, S. Döring, J. Krumrain, D. Grützmacher, S. Blügel, and C. M. Schneider. Spin-polarization limit in Bi_2Te_3 Dirac cone studied by angle- and spin-resolved photoemission experiments and ab initio calculations. *Phys. Rev. B*, 87:035127, 2013.

[173] I. A. Nechaev, R. C. Hatch, M. Bianchi, D. Guan, C. Friedrich, I. Aguilera, J. L. Mi, B. B. Iversen, S. Blügel, Ph. Hofmann, and E. V. Chulkov. Evidence for a direct band gap in the topological insulator Bi_2Se_3 from theory and experiment. *Phys. Rev. B*, 87:121111, 2013.

[174] M. Michiardi, I. Aguilera, M. Bianchi, V. E. de Carvalho, L. O. Ladeira, N. G. Teixeira, E. A. Soares, C. Friedrich, S. Blügel, and P. Hofmann. Bulk band structure of Bi_2Te_3. *Phys. Rev. B*, 90:075105, 2014.

[175] L. Fu and C. L. Kane. Topological insulators with inversion symmetry. *Phys. Rev. B*, 76:045302, 2007.

[176] L. Plucinski, G. Mussler, J. Krumrain, A. Herdt, S. Suga, D. Grützmacher, and C. M. Schneider. Robust surface electronic properties of topological insulators: Bi_2Te_3 films grown by molecular beam epitaxy. *Appl. Phys. Lett.*, 98(22):222503, 2011.

[177] J. Kampmeier, S. Borisova, L. Plucinski, M. Luysberg, G. Mussler, and D. Grützmacher. Suppressing twin domains in molecular beam epitaxy grown Bi_2Te_3 topological insulator thin films. *Cryst. Growth Des.*, 15(1):390–394, 2015.

[178] Y. L. Chen, J. G. Analytis, J. H. Chu, Z. K. Liu, S. K. Mo, X. L. Qi, H. J. Zhang, D. H. Lu, X. Dai, Z. Fang, S. C. Zhang, I. R. Fisher, Z. Hussain, and Z. X. Shen. Experimental realization of a three-dimensional topological insulator, Bi_2Te_3. *Science*, 325:178, 2009.

[179] D. Hsieh, Y. Xia, D. Qian, L. Wray, J. H. Dil, F. Meier, J. Osterwalder, L. Patthey, J. G. Checkelsky, N. P. Ong, A. V. Fedorov, H. Lin, A. Bansil, D. Grauer, Y. S. Hor, R. J. Cava, and M. Z. Hasan. A tunable topological insulator in the spin helical Dirac transport regime. *Nature*, 460:1101, 2009.

[180] J. Zhang, C.-Z. Chang, Z. Zhang, J. Wen, X. Feng, K. Li, M. Liu, K. He, L. Wang, X. Chen, Q.-K. Xue, X. Ma, and Y. Wang. Band structure engineering in $(Bi_{1-x}Sb_x)_2Te_3$ ternary topological insulators. *Nat. Commun.*, 2:574, 2011.

[181] D. Kong, Y. Chen, J. J. Cha, Q. Zhang, J. G. Analytis, K. Lai, Z. Liu, S. S. Hong, K. J. Koski, S.-K. Mo, Z. Hussain, I. R. Fisher, Z.-X. Shen, and Y. Cui. Ambipolar field effect in the ternary topological insulator $(Bi_xSb_{1-x})_2Te_3$ by composition tuning. *Nat. Nanotechnol.*, 6:705–709, 2011.

[182] J. G. Analytis, R. D. McDonald, S. C. Riggs, J.-H. Chu, G. S. Boebinger, and I. R. Fisher. Two/dimensional surface state in the quantum limit of a topological insulator. *Nat. Phys.*, 6:960–964, 2010.

[183] D.-X. Qu, Y. S. Hor, J. Xiong, R. J. Cava, and N. P. Ong. Quantum oscillations and Hall anomaly of surface states in the topological insulator Bi_2Te_3. *Science*, 329:821–824, 2010.

[184] Z. Ren, A. A. Taskin, S. Sasaki, K. Segawa, and Y. Ando. Large bulk resistivity and surface quantum oscillations in the topological insulator Bi_2Te_2Se. *Phys. Rev. B*, 82:241306, 2010.

[185] Y. Xu, I. Miotkowski, C. Liu, J. Tian, H. Nam, N. Alidoust, J. Hu, C.-K. Shih, M. Z. Hasan, and Y. P. Chen. Observation of topological surface state quantum Hall effect in an intrinsic three-dimensional topological insulator. *Nat. Phys.*, 10:956–963, 2014.

[186] H. Peng, K. Lai, D. Kong, S. Meister, Y. Chen, X.-L. Qi, S.-C. Zhang, Z.-X. Shen, and Y. Cui. Aharonov-Bohm interference in topological insulator nanoribbons. *Nat. Mater.*, 9:225–229, 2010.

[187] F. Xiu, L. H. Wang, L. Cheng, L.-T. Chang, M. Lang, G. Huang, X. Kou, Y. Zhou, X. Jiang, Z. Chen, J. Zou, A. Shailos, and K. L. Wang. Manipulating surface states in topological insulator nanoribbons. *Nat. Nanotechnol.*, 6:216–221, 2011.

[188] P. M. Ostrovsky, I. V. Gornyi, and A. D. Mirlin. Interaction-induced criticality in Z_2 topological insulators. *Phys. Rev. Lett.*, 105:036803, 2010.

[189] G. Rosenberg, H.-M. Guo, and M. Franz. Wormhole effect in a strong topological insulator. *Phys. Rev. B*, 82:041104, 2010.

[190] J. H. Bardarson, P. W. Brouwer, and J. E. Moore. Aharonov-Bohm oscillations in disordered topological insulator nanowires. *Phys. Rev. Lett.*, 105:156803, 2010.

[191] C. Weyrich, T. Merzenich, J. Kampmeier, I. E. Batov, G. Mussler, J. Schubert, D. Grüzmacher, and Th. Schäpers. Magnetoresistance oscillations in MBE-grown Sb_2Te_3 thin films. *Appl. Phys. Lett.*, 110:092104, 2017.

[192] A. A. Taskin, Z. Ren, S. Sasaki, K. Segawa, and Y. Ando. Observation of Dirac holes and electrons in a topological insulator. *Phys. Rev. Lett.*, 107:016801, 2011.

[193] G. P. Mikitik and Yu. V. Sharlai. Berry phase and the phase of the Shubnikov-de Haas oscillations in three-dimensional topological insulators. *Phys. Rev. B*, 85:033301, 2012.

[194] R. Jackiw and C. Rebbi. Solitons with fermion number $\frac{1}{2}$. *Phys. Rev. D*, 13:3398–3409, 1976.

[195] H. Häffner, C. F. Roos, and R. Blatt. Quantum computing with trapped ions. *Phys. Rep.*, 469(4):155–203, 2008.

[196] J. Clarke and F. K. Wilhelm. Superconducting quantum bits. *Nature*, 453:1031, 2008.

[197] D. P. DiVincenzo. The physical implementation of quantum computation. *Fortschr. Phys.*, 48(9–11):771–783, 2000.

[198] C. Monroe, D. M. Meekhof, B. E. King, W. M. Itano, and D. J. Wineland. Demonstration of a fundamental quantum logic gate. *Phys. Rev. Lett.*, 75(25):4714–4717, 1995.

[199] R. Vrijen, E. Yablonovitch, K. Wang, H. W. Jiang, A. Balandin, V. Roychowdhury, T. Mor, and D. DiVincenzo. Electron-spin-resonance transistors for quantum computing in silicon-germanium heterostructures. *Phys. Rev. A*, 62:012306, 2000.

[200] L. K. Grover. Quantum mechanics helps in searching for a needle in a haystack. *Phys. Rev. Lett.*, 79(2):325–328, 1997.

[201] D. Deutsch. Quantum theory, the Church-Turing principle and the universal quantum computer. *Proc. R. Soc. Lond. A*, 400:97, 1985.

[202] D. Deutsch. Quantum computational networks. *Proc. R. Soc. Lond. A*, 425:73, 1989.

[203] D. Deutsch and R. Jozsa. Rapid solution of problems by quantum computation. *Proc. R. Soc. Lond. Ser. A*, 439(1907):553–558, 1992.

[204] J. M. Elzerman, R. Hanson, W. van Beveren, B. Witkamp, L. M. K. Vandersypen, and L. P. Kouwenhoven. Single-shot read-out of an individual electron spin in a quantum dot. *Nature*, 430:431, 2004.

[205] R. Hanson, L. H. Willems van Beveren, I. T. Vink, J. M. Elzerman, W. J. M. Naber, F. H. L. Koppens, L. P. Kouwenhoven, and L. M. K. Vandersypen. Single-shot readout of electron spin states in a quantum dot using spin-dependent tunnel rates. *Phys. Rev. Lett.*, 94(19):196802, 2005.

[206] F. H. L. Koppens, C. Buizert, K. J. Tielrooij, I. T. Vink, K. C. Nowack, T. Meunier, L. P. Kouwenhoven, and L. M. K. Vandersypen. Driven coherent oscillations of a single electron spin in a quantum dot. *Nature*, 442:766, 2006.

[207] V. N. Golovach, M. Borhani, and D. Loss. Electric-dipole-induced spin resonance in quantum dots. *Phys. Rev. B*, 74:165319, 2006.

[208] K. C. Nowack, F. H. L. Koppens, Yu. V. Nazarov, and L. M. K. Vandersypen. Coherent control of a single electron spin with electric fields. *Science*, 318(5855):1430–1433, 2007.

[209] S. Nadj-Perge, S. M. Frolov, E. P. A. M. Bakkers, and L. P. Kouwenhoven. Spin-orbit qubit in a semiconductor nanowire. *Nature*, 468:1084–1087, 2010.

[210] J. R. Petta, A. C. Johnson, J. M. Taylor, E. A. Laird, A. Yacoby, M. D. Lukin, C. M. Marcus, M. P. Hanson, and A. C. Gossard. Coherent manipulation of coupled electron spins in semiconductor quantum dots. *Science*, 309:2180, 2005.

[211] E. Majorana. Teoria simmetrica dell'elettrone e del positrone. *Nuovo Cimento*, 14:171, 1937.

[212] J. Alicea. New directions in the pursuit of Majorana fermions in solid state systems. *Rep. Prog. Phys.*, 75:076501, 2012.

[213] M. Leijnse and K. Flensberg. Introduction to topological superconductivity and Majorana fermions. *Semicond. Sci. Technol.*, 27:124003, 2012.

[214] R. Aguado. Majorana quasiparticles in condensed matter. *Riv. Nuova Cimento*, 40:523–593, 2017.

[215] R. M. Lutchyn, E. P. A. M. Bakkers, L. P. Kouwenhoven, P. Krogstrup, C. M. Marcus, and Y. Oreg. Majorana zero modes in superconductor–semiconductor heterostructures. *Nat. Rev. Mater.*, 3:52–68, 2018.

[216] J. Bardeen, L. N. Cooper, and J. R. Schrieffer. Theory of superconductivity. *Phys. Rev.*, 108:1175–1204, 1957.

[217] A. Y. Kitaev. Unpaired Majorana fermions in quantum wires. *Phys. Usp.*, 44(10S):131, 2001.

[218] Y. Oreg, G. Refael, and F. von Oppen. Helical liquids and Majorana bound states in quantum wires. *Phys. Rev. Lett.*, 105:177002, 2010.

[219] R. M. Lutchyn, J. D. Sau, and S. Das Sarma. Majorana fermions and a topological phase transition in semiconductor-superconductor heterostructures. *Phys. Rev. Lett.*, 105:077001, 2010.

[220] J. Alicea. Majorana fermions in a tunable semiconductor device. *Phys. Rev. B*, 81:125318, 2010.

[221] V. Mourik, K. Zuo, S. M. Frolov, S. R. Plissard, E. P. A. M. Bakkers, and L. P. Kouwenhoven. Signatures of Majorana fermions in hybrid superconductor-semiconductor nanowire devices. *Science*, 336(6084):1003–1007, 2012.

[222] W. Chang, S. M. Albrecht, T. S. Jespersen, F. Kuemmeth, P. Krogstrup, J. Nygård, and C. M. Marcus. Hard gap in epitaxial semiconductor-superconductor nanowires. *Nat. Nanotechnol.*, 10:232, 2015.

[223] D. A. Ivanov. Non-Abelian statistics of half-quantum vortices in *p*-wave superconductors. *Phys. Rev. Lett.*, 86:268–271, 2001.

[224] A. Yu. Kitaev. Fault-tolerant quantum computation by anyons. *Ann. Phys.*, 303:2–30, 2003.

[225] F. Hassler, A. R. Akhmerov, C.-Y. Hou, and C. W. J. Beenakker. Anyonic interferometry without anyons: how a flux qubit can read out a topological qubit. *New J. Phys.*, 12(12):125002, 2010.

[226] J. D. Sau, S. Tewari, and S. Das Sarma. Universal quantum computation in a semiconductor quantum wire network. *Phys. Rev. A*, 82:052322, 2010.

Index

www.ingramcontent.com/pod-product-compliance
Lightning Source LLC
Chambersburg PA
CBHW080647220326

41598CB00033B/5133